$\frac{3}{4}$

FLORA

OF THE

SANTA CRUZ

MOUNTAINS

OF

CALIFORNIA

FLORA

of the

SANTA CRUZ
MOUNTAINS
of
CALIFORNIA

A Manual of the Vascular Plants

John Hunter Thomas

STANFORD UNIVERSITY PRESS
Stanford, California

Stanford University Press
Stanford, California
© 1961 by the Board of Trustees of the
Leland Stanford Junior University
Printed in the United States of America

Cloth ISBN 0-8047-0017-6
Paper ISBN 0-8047-1862-8

Original printing 1961

Last figure below indicates year of this printing:

00   99   98   97   96   95   94   93   92   91

PREFACE

Although there have been resident botanists in central California for over one hundred years, none of them has previously prepared a flora of the Santa Cruz Mountains. This is somewhat surprising, since the Santa Cruz Mountains constitute a well-defined area, form a distinct geographical and phytogeographical unit, and are heavily populated. That central California is still a pleasant place to live is attested to by the rapid expansion of human population. This increase in population has caused many irreversible effects on the vegetation, and the rate of disturbance is increasing at an accelerated rate. Much of the vegetation cover has been altered or contaminated by alien plants. Many areas bear little resemblance to what they were prior to the establishment of the Roman Catholic Missions. Permanent conservation measures, beyond the reach of local politicians, are as desperately needed locally as they are throughout the world. Perhaps conservation measures are needed more in central California than in some other areas. Certainly it is desirable to preserve the beauty of mountains, forests, beaches, chaparral, and grasslands close to a major population center. Many people are more apt to conserve the things they know about than to conserve the things that are foreign to them. This flora will, I hope, acquaint at least a few more people with the plants around them, and perhaps thus serve as a stimulus, however slight, toward more permanent protection of our environment.

This flora is intended for both professional botanists and amateurs. A person not trained in botany may be bewildered by the length and seeming complexity of some of the keys. Unfortunately there are no short cuts in preparing keys to nearly 1800 taxa. I have attempted to make the keys as usable as possible. They are artificial and do not necessarily reflect evolutionary relationships. The keys were compiled with specimens from the Santa Cruz Mountains. Plants from other areas, although perhaps belonging to families represented locally, cannot necessarily be determined to the level of family or genus. In many cases, it is essential to have the entire plant, with both flowers and fruits, in order to determine it. The line drawings, many of which are of the more common plants, are from the *Illustrated Flora of the Pacific States*. I hope they will be useful. Unless otherwise indicated, all the line drawings are reduced by a factor of two.

All the native plants of the Santa Cruz Mountains have been included, whether or not they are now extinct locally. Their general habitat and distribution within the Santa Cruz Mountains is indicated. The range of species beyond the Santa Cruz Mountains is not listed, but it can be found in such regional or sectional floras as *A California Flora, A Manual of the Flowering Plants of California*, and the *Illustrated Flora of the Pacific States*.

It is often very difficult to decide whether or not to include a particular non-native species. With such species as *Avena fatua* and *Brassica campestris* there is no problem. Both are thoroughly established and exceedingly abundant. *Ornithopus roseus*, a native of Europe, is known from only one small area in Santa

Cruz County, but since it has maintained itself for at least twenty years it must therefore be included. *Danthonia pilosa*, a native of Australia, has been collected only once in the Santa Cruz Mountains, but it is known from several localities to the north and can reasonably be expected at additional localities locally in the future, and is thus included. A number of species such as *Berberis darwinii*, *Glaucium flavum*, *Sedum dendroideum*, *Acaena sanguisorbae*, *Lens culinaris*, and *Viburnum tinus* have been observed growing spontaneously only once or a few times. They are included because they are garden plants or economically important plants and can be expected again occasionally. One can predict that in the next few years quite a large number of additional plants will be found as occasional escapes from cultivation or as accidental or deliberate introductions.

Synonyms are included only for convenience in referring to the same taxon in other regional and sectional floras. The type localities of many proposed taxa occur within the Santa Cruz Mountains. I have made no attempt to list or include these taxa unless they are still considered to be valid. The arrangement of families in general follows the sequence proposed by Engler and Prantl. The specimens on which this flora is based are for the most part to be found in the following herbaria: the Dudley Herbarium of Stanford University, the herbarium of the California Academy of Sciences, the Jepson Herbarium and the herbarium of the University of California at Berkeley, and my personal herbarium. No new names or combinations are presented in this book.

During the preparation of this flora, I have received help, advice, and encouragement from many people. I would particularly like to express my thanks to Professor Ira L. Wiggins, under whose guidance this work was originally carried out; to Professor Richard W. Holm, who helped me in numerous ways; and to Mrs. Roxana S. Ferris, who answered many questions. Mr. John Thomas Howell of the California Academy of Sciences has given me much valuable assistance. Miss Vesta F. Hesse of Boulder Creek made her notes on the flora of Santa Cruz County available to me, accompanied me on several field trips, sent me numerous specimens, and kindly read a preliminary copy of my manuscript. Mr. Wallace R. Ernst gave advice on the Papaveraceae, Mr. Dennison H. Morey aided with the Valerianaceae, and Dr. Rimo Bacigalupi of the Jepson Herbarium kindly determined some of my collections of *Castilleja*. I am indebted to Dr. Earl E. Brabb for contributing the section on geology and for suggestions concerning the section on geography. The curators of the herbaria I utilized have been most helpful. I wish especially to express my appreciation to Mrs. Barbara W. Law, who typed two versions of my mansucript, prepared the indices to the common and scientific names, and read through the proofs with me. Numerous persons, the Supervisors of Santa Cruz County, and the members of the State Park Commission, have given me permission to collect on the land they own or administer.

August 1960

CONTENTS

Preface, v

Part I. Introduction and Keys
Introduction, 3
Keys to Divisions, Classes, Subclasses, and Families, 37

Part II. Annotated Catalogue of Vascular Plants
Division Lepidophyta, 53
Division Calamophyta, 54
Division Pterophyta, 55

Part III. References and Glossary
General References, 383
Glossary of Technical Terms, 385

Part IV. Indexes
Index of Place Names, 395
Index of Common Plant Names, 403
Index of Scientific Plant Names, 419

ILLUSTRATIONS,
MAPS, AND CHARTS

Map of the Santa Cruz Mountains area, 4

Stratigraphic column of rocks, 6

Fog-filled coastal valleys, 11

Coastal sand dunes near Watsonville, 13

Watsonville Slough, 14

Searsville Lake, 15

Redwoods and Douglas Firs, 17

Interdigitation of plant communities, 19

Coastal grasslands near Pescadero, 20

Foothill woodland on Jasper Ridge, 21

Streambank vegetation near La Honda, 22

A. D. E. Elmer, A. A. Heller, and L. Abrams, 33

V. Rattan, W. R. Dudley, and C. L. Anderson, 34

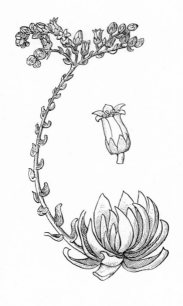

PART I

Introduction and Keys

INTRODUCTION

Description of the Santa Cruz Mountains

The Santa Cruz Mountains are part of the Coast Range of California. They are separated from the North Coast Ranges by the Golden Gate, from the Diablo Range to the east by San Franciso Bay and the Santa Clara Valley, from the Santa Lucia Mountains and the Salinas Valley to the south by the Pajaro River, and the Pacific Ocean forms the western boundary. The Diablo Range, which in central California includes Mount Diablo, the Oakland-Berkeley Hills, and the Mount Hamilton Range, is part of the Inner Coast Range, while the North Coast Ranges as represented in Marin County, the Santa Cruz Mountains, and the Santa Lucia Mountains are part of the Outer Coast Range. The eastern terrestrial boundary of the area included in this flora extends from Alviso to the center of San Jose, thence southward through the center of the Santa Clara Valley to Gilroy, and from Gilroy south to the Pajaro River. The area thus outlined is about 1386 square miles. The descriptive term, the Santa Cruz Mountains, is usually used in this flora to include the entire area as delimited above.

The distance from the Golden Gate to the Pajaro River is about 74 miles. The minimum distance across the mountains, at the latitude of Daly City, is 5.5 miles; the maximum width is 29 miles. Several counties lie partly or entirely within the Santa Cruz Mountains. San Mateo and Santa Cruz counties lie completely within the Santa Cruz Mountains, as does San Francisco County with the exception of the Farallon Islands and some of the islands in San Francisco Bay. About one-third of Santa Clara County lies within this area.

The main axis of the Santa Cruz Mountains tends in a general northwest–southeast direction. At the north, the hills of San Francisco and San Bruno Mountain are separated from the main body of the Santa Cruz Mountains by Merced Valley, which extends obliquely across the San Francisco Peninsula from San Francisco Bay to the Pacific Ocean. From just south of San Francisco to the southern end of Castle Rock Ridge, the crests form an almost straight line and are parallel to the San Andreas Fault. A gap, 1800 feet in elevation and about 6 miles south of Los Gatos, separates the northern part of the Santa Cruz Mountains from the southern portion. The northern part is sometimes called the Sierra Morena and the southern part the Sierra Azule. The southern part forms a ridge which is about 25 miles long, and contains the highest point in the Santa Cruz Mountains, Loma Prieta. Loma Prieta is a distinctive, flat-topped peak 3806 feet high and is located about 15 miles south of San Jose. Among the higher peaks are the following: Mount Umunhum, 3442 feet; Castle Rock, 3214 feet; Black Mountain on Monte Bello Ridge, 2810 feet; Ben Lomond Mountain, 2600 feet; Eagle Rock, 2488 feet; Kings Mountain, 2315 feet; Mount Madonna, 1897 feet; and Montara Mountain, 1750 feet. Ben Lomond Mountain in Santa Cruz County extends from near Big Basin to Santa Cruz and is almost parallel to the main axis of the Santa Cruz Mountains.

N

SAN FRANCISCO BAY COUNTIES

GOLDEN GATE

SAN FRANCISCO

Brisbane

San Pedro Point

Montara Mountain

Moss Beach

Pillar Point

Pilarcitos

Half Moon Bay

Kings Mountain

Searsville Lake

Portola

San Gregorio

Pescadero

Pigeon Point

Año Nuevo Point

Davenport

SAN FRANCISCO BAY

San Andreas Lake

Crystal Springs Lakes

San Mateo

Belmont

Redwood City

Woodside

Palo Alto
Stanford

Los Altos

La Honda

San Gregorio Creek

Pescadero Creek

Butano Creek

Big Basin

Eagle Rock

Waddell Creek

Mill Creek

Boulder Creek

Ben Lomond

Scott Creek

Swanton

Felton

Bonny Doon

Santa Cruz

Laguna Creek

Tunitas Creek

PACIFIC OCEAN

San Lorenzo River

Bear Creek

Zayante Creek

Camp Evers

Soquel

Black Mountain

Permanente Creek

Stevens Creek

Saratoga Summit

Santa Clara

San Jose

Campbell

Saratoga

Los Gatos

Campbell Creek

Los Gatos Creek

Guadalupe Creek

Coyote Creek

Alviso

Sunnyvale

Edenvale

Alamitos Creek

Guadalupe

Almaden

Wrights

Loma Prieta

Glenwood

Hester Creek

Soquel Creek

Aptos Creek

Aptos

Corralitos Creek

Corralitos

Mount Madonna

Uvas Creek

Little Arthur Creek

Coyote

Madrone

Morgan Hill

Gilroy

Llagas Creek

Carnadero Creek

Robroy

Watsonville

Soda Lake

Pajaro River

MARIN

SAN FRANCISCO

CONTRA COSTA

San Francisco Bay

ALAMEDA

SAN MATEO

SANTA CLARA

SANTA CRUZ

SAN BENITO

MONTEREY

5   0   5   10 miles

MAP OF THE SANTA CRUZ MOUNTAINS AREA

Rivers and streams are for the most part more or less at right angles to the main axis of the Santa Cruz Mountains. The most prominent is the San Lorenzo River, which has its headwaters near Saratoga Summit and its mouth at Santa Cruz. The San Lorenzo River drains the region between Ben Lomond Mountain and the main crests to the east. Fault-trellis drainage patterns are well developed in San Mateo County. The rivers and creeks on the western slopes are perennial in most years, while those on the eastern slopes are usually dry during the summer and early fall. The eastern slopes are steeper than the western. On the eastern side is a series of low rolling foothills. Gently sloping valleys occur along the major streams near their mouths on the western side of the Santa Cruz Mountains. In southern Santa Cruz County, the Pajaro River has formed an extensive valley. Streams on the eastern side of the Santa Cruz Mountains have formed alluvial fans that extend toward the Bay and the center of the Santa Clara Valley. Most of the Santa Clara Valley and the watercourses that empty into it drain into San Francisco Bay via the Guadalupe River. The southern end of the Santa Clara Valley drains into the Pajaro River via Llagas Creek.

There are many small lakes in the Santa Cruz Mountains. San Andreas and the Crystal Springs lakes lie along the San Andreas Fault and are now part of the San Francisco water supply. The storage capacity of the Crystal Springs Lakes has been increased by a dam across San Mateo Creek. Pilarcitos Lake is also part of the San Francisco water supply system and is also partly artificial. Searsville Lake and Felt Lake on Stanford University lands are artificial and supply irrigation water to the University. Several artificial reservoirs have been built on the eastern slopes of the Santa Cruz Mountains. The largest are on Stevens Creek, Los Gatos Creek, Guadalupe Creek, and Alamitos Creek. In southern Santa Cruz County there are several shallow lakes, some of which have been drained for agricultural land. Soda Lake in the southeast corner of Santa Cruz County has no outlet channel and is alkaline. In San Francisco, drifting sands have formed a number of dune lakes, of which Lake Merced is the largest. In addition, there are a number of smaller lakes and ponds in various parts of the Santa Cruz Mountains, some natural and some artificial. Additional information concerning the geography and geology of the Santa Cruz Mountains will be found in the *Geologic Guidebook of San Francisco Bay Counties* edited by O. P. Jenkins.

Geology of the Santa Cruz Mountains

EARL E. BRABB
School of Mineral Sciences, Stanford University

The rocks of the Santa Cruz Mountains were described by naturalists as early as 1816. They have been chipped, examined, and mapped intermittently since that time, but many of their secrets still lie buried beneath the cover of redwood, chaparral, and alluvium. That they show a nearly continuous geologic record from Late Cretaceous time to the present is certain, but the beds are so intensely fractured and compressed, and their outcrops so few, that only the most refined geologic techniques are successful in mapping them. These methods

Era	System	Series	Stage	Age (millions of years)	Rock unit	Member or lithology	Max. thickness (feet)	Areas where rock units are well exposed
CENOZOIC	Quaternary	Pleist. & Rec.				gravel, sand and clay		
					Colma sand	sand	50	Colma and the San Francisco area
					Santa Clara formation	gravel, sand and clay	500	Santa Clara Valley, Alpine Rd., Matadero Ck. San Andreas Fault zone near Woodside
				1	Merced ss.	sandstone	5,000	Southwest San Francisco
	Tertiary	Pliocene			Purisima formation	"Tunitas" ss.	400	Tunitas Ck., San Gregorio Ck.
						"Lobitos" mds.	450	Lobitos Ck., Tunitas Ck.
						"San Gregorio" ss.	350	San Gregorio Ck., Lobitos Ck.
						"Pomponio" mds.	2,300	Pomponio Ck., Half Moon Bay area
						"Tahana" sls.	2,150	Tahana Gulch, La Honda, Portola Park
		Miocene	Delmontian	10	Monterey group	ss.	3,000	West flank of Ben Lomond Mountain and "The Chalks." Sandstone (Santa Margarita) forms conspicuous white hills near Scott Valley
			Mohnian			siliceous ss.	2,000	Not exposed. Reported in an oil test well northwest of Davenport
			Luisian			mudstone ss.	3,500	San Lorenzo River and Route 9 near Brookdale, Half Moon Bay Road
			Relizian	30		ss.	2,000	Woodhams Creek, Peters Creek, Oil Creek, and Love Creek
		Oligocene	Saucesian		Sandholdt siltstone	siltstone	2,800	Hester Ck., Slate Ck., Peters Ck., Russian Ridge, Kings Mt. school area
						"Mindego" volcanics	6,000	Mindego Hill, Langley Hill, La Honda, Slate Ck., Skyline Blvd., Peters Ck.
			Zemorrian		Vaqueros sandstone	sandstone	4,500	Bear Ck., Kings Ck., Castle Rock Ridge, Waterman Gap, Table Mt., Saratoga Gap
			Refugian	40	San Lorenzo formation	"Rices" mudstone	1,700	Kings Ck., Bear Ck., San Lorenzo River, Big Basin, La Honda Ck., Soquel Ck., Newell Ck.
						"Twobar" shale	800	Kings Ck., San Lorenzo River, Bear Ck., Butano Ridge, La Honda Ck., Soquel Ck., Newell Ck.
		Eocene	Narizian		Butano sandstone	sh. sandstone cgl. cgl.	9,000	Butano Ridge, Pine Mountain, Kings Ck., Bear Ck., San Lorenzo River, Newell Ck., Route 17 at Mount Charlie Rd., La Honda Ck. at Skylonda, Durham and Star Hill Roads
			Ulatisian		missing			(A shale possibly of Ulatisian age crops out one mile south of Los Trancos Woods)
			Penutian	60				
		Paleocene	Bulitian					
			Ynezian		"Locatelli" fm.	siltstone and ss.	500	Scott Ck., Smith Grade Rd., Jamison Rd.
			Cheneyan	70	missing			
MESOZOIC	Cretaceous	Upper	Maestrichtian		Pigeon Point formation	sandstone, shale and conglomerate	8,000	Along the beach between Año Nuevo Point and Pescadero Point
			Senonian					
			Turonian		missing			
			Cenomanian	90	granitic intrusive rocks			Ben Lomond Mountain, Montara Mountain, Scott Valley, Sugarloaf Mountain
		Lower	Albian					
			Aptian		Franciscan formation	sandstone, shale, radiolarian chert, limestone, greenstone, and schist	more than 9,000	San Francisco area, Monte Bello Ridge, San Bruno Mountain, Sweeney Ridge, Sawyer Ridge, Buri Buri Ridge, Cahill Ridge, Belmont Hill, Black Mountain
			Neocomian	130				
	Jurassic			175	missing			
	Triassic			200				
PALEOZOIC (?)					Sur series	quartzite, schist gneiss, and marble	not known	Ben Lomond Mountain and Glenwood Basin

COMPOSITE STRATIGRAPHIC COLUMN OF ROCKS IN
THE SANTA CRUZ MOUNTAINS

Notes: (1) ⌇⌇⌇ = diastrophism or mountain building. (2) Rock units in quotation marks have not been formally proposed. (3) sh = shale; mds = mudstone; sls = siltstone; ss = sandstone; cgl = conglomerate; fm = formation.

are so painstaking that several months or even a year or two are required to map a small area. The reward for careful work is more than academic, for the region has mineral wealth and economic potential. More than $600,000,000 has been realized from mineral deposits in the Santa Cruz Mountains. Cement, lime, crushed stone, sand, and gravel are the main products, and such minerals as quicksilver, clay, bituminous sand, and oil are important locally. The location of these mineral deposits and geologic maps of the counties may be obtained from the California Division of Mines, Ferry Building, San Francisco. Undeveloped resources of the Santa Cruz Mountains include land—for homesites and parks—water, and petroleum. The location and evaluation of these undeveloped resources are problems that currently concern several geologists. Discussions of the economic potential of a few areas can be found in theses by Stanford University students and in reports by the California Division of Mines.

The geologic record of the Santa Cruz Mountains is summarized in the chart. This generalized column is pieced together from several localities, and is not representative of any one. Nevertheless, it shows the many rock units, termed formations, and the similarities in formations of different age. Correspondingly, rock units are easily misidentified unless they contain fossils that tell us their age.

The distribution of formations is of interest to the soil scientist, construction engineer, petroleum geologist, cement manufacturer, hydrologist, and farmer, to mention a few people who exploit or investigate the earth. For example, slopes underlain by the San Lorenzo formation and other shale units tend to creep and slide in wet weather, whereas those underlain by sandstone are relatively stable. Therefore, houses built on the San Lorenzo formation may require more expensive foundations than those built on the Vaqueros sandstone. The Butano and Vaqueros sandstones resemble each other in outcrop, but the former is generally tight at depth, whereas the latter is a fresh water aquifer. Therefore, water wells drilled into the Vaqueros sandstone are more likely to be successful than those drilled into the Butano sandstone.

The hiker looks at the distribution of the formations from a different perspective. Terrain underlain by shale of the Monterey group is characterized by steep, narrow canyons and high ridges with steep slopes, whereas Quaternary deposits cover relatively flat land. The massive sandstones of the Butano and Vaqueros formations produce a rugged topography with a few cliffs, while on the contrary regions underlain by the Purisima formation are characterized by well-rounded hills and evenly sloping valleys.

Rock distribution is best shown on a geologic map, but it can also be summarized in written form. Montara and Ben Lomond mountains are made up of granitic rocks and—with respect to the latter—marble, quartzite, and schist of the Sur Series. Butano, Castle Rock, and Cahill ridges consist mainly of Butano and Vaqueros sandstone, whereas Monte Bello, Sawyer, Buri ridges, and San Bruno Mountain are formed from sandstones of the Franciscan group. Oligocene lava flows comprise Langley and Mindego hills. The Santa Clara Valley and the flanks of the Santa Cruz Mountains are covered with Quaternary gravel, sand, and clay, some of which contain the remains of elephants, camels, ground sloths, and other creatures no longer native to North America.

Structure, including folds and faults, and differential erosion control the distribution of formations. In general, upwarps or anticlines expose older rocks that resist erosion and form hills and ridges. The major upwarps are the Sierra Morena, Ben Lomond, Montara, Butano, La Honda, Haskin Hill, Johansen, Mindego Creek, Oil Creek, and Camp Campbell anticlines. Downwarps or synclines preserve younger rocks that generally erode easily and form valleys. The principal downwarps are the Big Basin, San Gregorio, Pescadero, Slate Creek, Oil Creek, Riverside, and San Lorenzo synclines. Faulting may produce cliffs and valleys directly by differential movement, or indirectly by differential erosion of the displaced formations. San Francisco Bay and the Santa Clara Valley, for example, outline a trough that is probably a downdropped fault block. The steep, northeast flank of Ben Lomond Mountain, on the other hand, is due to the more rapid erosion of the Monterey group, where it has been faulted against resistant quartz diorite. The faults themselves are zones of weakness that are rapidly attacked by water and other agents of erosion. The trace of the San Andreas, San Gregorio, and Pilarcitos faults, for instance, is marked by linear valleys, sag ponds, and lakes. San Andreas Lake, Crystal Springs Reservoir, and Pilarcitos Lake are good examples of lakes along fault zones.

The San Andreas Fault deserves special mention. It is as famous as Mrs. O'Leary's cow, and for a similar reason; both started fires that nearly destroyed major cities of the United States. The fault slices through the Santa Cruz Mountains from Mussel Rock near San Francisco to Mount Madonna near Gilroy; it has been mapped southeastward to Mexico and northwestward to Point Arena, a distance of 600 miles. Some geologists believe that the region west of the San Andreas Fault, including most of the Santa Cruz Mountains, has moved northwestward several hundred miles with respect to the eastern side. They postulate that this movement began in pre-Cretaceous time, and that it continued intermittently during the late Mesozoic and Cenozoic eras. Other geologists, including the writer, believe that the lateral displacement—in the Santa Cruz Mountains, at least—is only a recent feature, and that pre-Pliocene movement, if any, was an up-and-down slippage. Irrespective of the antiquity and type of movement, the San Andreas Fault is one of the most imposing features of the California landscape.

The geologic history of the Santa Cruz Mountains is interpreted from: (1) the nature and distribution of formations and fossils; (2) the relation of rock units to one another; and (3) folds and faults. This history is briefly summarized in the following paragraphs.

The Paleozoic (?) marble, quartzite, schist, and gneiss of the Sur series (see chart) on Ben Lomond Mountain are metamorphosed sedimentary rocks. The sediments comprising these rocks were laid down more than two hundred million years ago. They may have been metamorphosed at the end of the Paleozoic Era, or perhaps by heat and pressure accompanying the intrusion of granitic rocks during the Cretaceous Period.

Rocks of early Mesozoic age have not been found in the Santa Cruz Mountains, but they may lie beneath the cover of the younger strata. Or, the area may have been above sea level and undergoing erosion in Triassic and early Jurassic time. During the Late Jurassic and Early Cretaceous, "dirty" sandstone (graywacke), shale, radiolarian chert, limestone, pyroclastics, and lava flows of the Franciscan group were laid down in a large geosyncline or trough that extended north–south along the western border of North America. The intrusion of granitic rocks in mid-Cretaceous time marked the end of Franciscan-like deposition and the beginning—in the Santa Cruz Mountains, at least—of the deposition of "clean" sandstone and shale. The Late Cretaceous and Tertiary sandstones generally lack the clay that is abundant in the matrix of Franciscan sandstone, and have larger amounts of potash and plagioclase feldspar. The Upper Cretaceous Epoch also marks the beginning of richly fossiliferous deposits that enable the geologist to make reasonable inferences about the depth and environment of the ocean floor and the character of the bordering land masses. Most of the Tertiary marine rocks, for example, were probably laid down three to six thousand feet below sea level, in water temperatures colder than 40° F., and in ocean water of normal salinity. Leaves and other plants fragments in the marine strata of the Santa Cruz Mountains and in non-marine rocks elsewhere in California show that the climate was wet and tropical or subtropical during the Eocene Epoch, and that the climate was considerably colder and drier in late Tertiary time. The change in climate had such a pronounced effect on the plants and animals that it merits further consideration.

The warm and wet Eocene climate extended as far north as the southeastern tip of Alaska. Fig, palm, magnolia, cinnamon, and avocado trees flourished in the northern and west-central United States, far north of their present habitat. Corals and crocodiles were among the tropical and subtropical animals that lived in the ocean covering parts of California and Oregon, or on the land bordering this sea. By Late Oligocene time, however, the climate was cooler. The subtropical forests of the Pacific Northwest were replaced by redwoods, tan oaks, maples, and other elements of a warm, temperate flora. Some of the less hardy tropical plants growing on the forested slopes east and west of the sea covering the Santa Cruz Mountains probably died out. In late Miocene or Pliocene time, the redwood–tan oak forest probably extended as far south as central California and the Bay Area. As a result of the cooler climate and diminished rainfall, semiarid communities like chaparral, oak savanna, and grassland were introduced in the district. During the Quaternary Period or "Ice Age," forests at the latitude of the Santa Cruz Mountains "migrated" south and north, corresponding to the advance and retreat of the continental ice sheet. Today, they are either re-established floras or vestiges of communities that formerly had a wide distribution.

The Tertiary climate is only a small part of the geologic history of the Santa Cruz Mountains. During episodes of mountain building (see chart) the rocks were uplifted, folded, faulted, and eroded. The diastrophism at the end of the Eocene also changed the environment of deposition from open ocean conditions to an embayment partly protected by a land mass to the west. As a result of the mid-Pleistocene "revolution," all the folds in the area were further compressed, and the rocks intensely faulted. The frequent earthquakes in the region show that mountain building forces are still at work.

Climate

Much of California has a Mediterranean climate. A Mediterranean climate is characterized by at least three times as much rainfall in the wettest winter month as in the driest summer month and by average temperatures for the coolest month ranging between 32° and 64° F. The Santa Cruz Mountains have what may be called a cool summer Mediterranean climate, characterized by low average summer temperatures.

The annual average temperature for nine localities in the Santa Cruz Mountains ranges from 55.3° to 59.2° F. The range of temperature is from about 25° to about 102° F. with the extremes occurring rarely. The warmest months are July, August, and September; the coldest are December, January, and February. Table 1 shows the monthly and annual average temperatures at seven localities in the Santa Cruz Mountains. The localities are arranged from north to south.

TABLE 1

MONTHLY AND YEARLY AVERAGE TEMPERATURES AT SEVEN LOCALITIES IN THE
SANTA CRUZ MOUNTAINS IN DEGREES FAHRENHEIT

	Jan.	Feb.	Mar.	Apr.	May	June	July	Aug.	Sept.	Oct.	Nov.	Dec.	Yearly
San Francisco ...	50.1	53.0	54.9	55.7	57.1	59.1	58.9	59.3	61.6	61.0	57.2	56.1	56.7
Half Moon Bay ..	49.7	–	–	52.7	55.0	57.1	58.2	58.5	51.7	56.4	54.1	51.6	–
Redwood City ...	47.8	50.8	54.0	57.1	61.1	65.1	67.8	67.2	66.2	61.2	54.3	49.6	58.5
San Jose	49.3	52.8	55.2	57.6	61.4	65.3	67.7	67.2	66.6	62.3	56.0	50.5	59.3
Los Gatos	47.4	50.3	53.1	56.5	60.1	65.1	68.2	67.6	66.1	61.1	54.5	48.5	58.2
Ben Lomond	45.8	47.7	50.8	54.6	50.2	62.0	64.9	64.6	64.3	59.6	52.4	48.1	56.1
Santa Cruz	49.9	51.7	53.7	56.4	58.5	61.8	63.1	63.4	62.8	59.6	54.9	51.1	57.2

Coastal areas have relatively uniform temperatures. In San Francisco, for instance, January is the coldest month of the year and has an average temperature of 50.1° F. September is the hottest month of the year in San Francisco and the average temperature is 61.6° F. The range of average monthly temperatures is 11.5° F. At Los Gatos, on the eastern slope of the Santa Cruz Mountains, the coldest month is also January with an average temperature of 47.4° F. The hottest month is July with an average temperature of 68.2° F. The range for Los Gatos is 20.8° F. or nearly twice that of San Francisco.

Table 2 shows the average monthly and yearly precipitation at seven localities.

TABLE 2

MONTHLY AND YEARLY AVERAGE RAINFALL AT SEVEN LOCALITIES IN THE
SANTA CRUZ MOUNTAINS IN INCHES

	Jan.	Feb.	Mar.	Apr.	May	June	July	Aug.	Sept.	Oct.	Nov.	Dec.	Yearly
San Francisco ...	4.03	3.91	2.78	1.49	0.59	0.15	0.01	0.00	0.13	1.07	2.27	4.07	20.51
Half Moon Bay ..	4.27	4.38	3.34	1.87	0.68	0.20	0.02	0.01	0.11	1.48	2.66	4.34	23.46
Redwood City ...	4.08	3.76	2.83	1.21	0.45	0.12	0.02	0.03	0.06	0.95	1.74	3.70	18.95
San Jose	2.37	2.53	1.89	1.05	0.43	0.10	0.00	0.02	0.07	0.67	1.17	2.39	12.69
Los Gatos	6.61	5.78	4.75	1.88	0.82	0.10	0.00	0.03	0.40	1.38	2.76	5.60	30.11
Ben Lomond	10.94	11.90	9.70	3.42	1.46	0.24	0.03	0.02	0.13	2.88	5.79	9.93	56.34
Santa Cruz	5.77	5.20	4.14	1.86	0.97	0.22	0.02	0.03	0.47	1.43	2.74	5.40	28.25

The rainy season begins in October and continues through April and sometimes into the first part of May. June, July, August, and September are usually rainless.

Fog-filled valleys on the western side of the Santa Cruz Mountains; Monterey Bay and the Santa Lucia Mountains in the distance.

The distribution of rain throughout the Santa Cruz Mountains is not uniform. Rainfall varies from about 15 inches at Palo Alto, to 18 inches at Searsville Lake (6 miles southwest of Palo Alto), to over 60 inches at Big Basin on the western side of the main crest, to about 20–30 inches along the coast.

During the summer months coastal fogs are common and extend inland to the crests of the Santa Cruz Mountains and often pour through the passes onto the eastern slopes. In summer months the fog often hangs over the Santa Cruz Mountains, usually as far south as Kings Mountain, like a gigantic billowy blanket with fingers or ribbons extending eastward through the lower gaps. Occasionally updrafts on the eastern side of the mountains cause the formation of towering spires and peaks at the edge of the cloud bank.

The limit of well-developed redwood forests coincides with the limit of the summer coastal fogs. Local topographical conditions prevent the fog from regularly reaching some of the higher ridges, as for example Black Mountain. Black Mountain does not have a well-developed redwood forest except along streams. Kings Mountain, which in many respects is similar to Black Mountain, does have a well-developed redwood forest.

George T. Oberlander found that the fog drip on Cahill Ridge in northern San Mateo County from July 20 to August 28, 1951, amounted to 58.8 inches under *Lithocarpus densiflorus*, 17.1 inches under *Pseudotsuga menziesii*, and 1.8 inches under *Sequoia sempervirens*. The differences were explained on the basis of the degree of protection from the prevailing winds by the surrounding terrain and other vegetation.

Snow occasionally falls on the higher crests of the Santa Cruz Mountains, but usually melts within a day or less.

Plant Communities of the Santa Cruz Mountains

The Santa Cruz Mountains lie within two life zones, the Upper Sonoran and the Transition. In general, the western slopes of the Santa Cruz Mountains are in the Transition zone while the eastern slopes and the Pajaro River Valley are in the Upper Sonoran zone. In southern San Mateo County the Transition zone extends onto the eastern slope of the mountains. To the west of the crests are pockets of Upper Sonoran zone on dry ridges and on some south-facing slopes.

Hall and Grinnell have compiled lists of indicator species for the various life zones in California. Of their list of Upper Sonoran life zone plant-indicator species, 45 per cent occur in the Santa Cruz Mountains. The corresponding figure for the Transition zone is 48 per cent. Considering the size of California these figures indicate that the two life zones can be recognized in the Santa Cruz Mountains. Nevertheless, to attempt to assign the 1799 taxa of vascular plants in the Santa Cruz Mountains to one or another of these two life zones would be to ignore obvious characteristic assemblages of plants. On the other hand, it would be of little value to classify the vegetation according to some of the classical ecological categories of formation, association, faciation, and lociation. This would be not only inappropriate but almost impossible, since the vegetation changes markedly and rapidly from the edge of a stream in the redwood forest to the dry open grassland on the top of a ridge. The transition from streams to crest is marked, furthermore, by an overlapping and intergrading of plant species.

Characteristic assemblages of plant species or communities do exist. As considered here communities are the sum of their component taxonomic entities. The community is not an organism, a quasi-organism, or a super-organism, but rather an assemblage of plants with similar or complementary ranges of tolerances to various environmental factors.

Munz and Keck have recently proposed a system of classification for the vegetation of California. Some of the plant communities as characterized by them are recognizable in the Santa Cruz Mountains. In general, the communities they call chaparral, grassland, and foothill woodland are in the Upper Sonoran life zone. The Transition life zone consists of the coastal strand, coastal scrub, redwood forest, mixed evergreen forest, and some grassland. The salt and fresh water marshes may occur in either life zone. Their distribution is determined primarily by proximity to and availability of water rather than by temperature.

Some parts of the Santa Cruz Mountains cannot now be assigned to one community or another owing to the serious and continuing disturbance of the vegetation by man and his domesticated animals. Agricultural areas are located on coastal headlands, in stream and river valleys near the coast, and in the Santa Clara and Pajaro River valleys. The latter two valleys are among the best agricultural regions in California. Unfortunately much of the farm land in these valleys is being subdivided for housing and industrial sites. In some parts of the Santa Cruz Mountains some of the more gentle slopes have been cleared for orchards and vineyards. Many orchards and vineyards have been abandoned, and, depending upon the location, terrain, surrounding vegetation, and such factors as erosion, have reverted partly or in some cases entirely to the original vegetation. Many disturbed areas in towns, cities, along railroad tracks, road embankments, dumps, etc., now support a flora composed almost entirely of alien plants.

Coastal sand dunes near Watsonville with *Ammophila arenaria, Mesembryanthemum chilense, M. edule, Oenothera cheiranthifolia,* and *Artemisia pycnocephala* as the common sand-binding plants.

COASTAL STRAND

The coastal strand consists of the vegetation of sandy beaches and coastal sand dunes. In general, vascular plants on the coastal strand of the Santa Cruz Mountains are not subject to inundation at high tides. The coastal strand is more or less continuous from the Golden Gate to the mouth of the Pajaro River, except for short stretches where the coast line is formed of sheer cliffs and bluffs. Extensive sand dunes once occurred in western San Francisco. Small areas of sand dunes are still to be found, however, in San Francisco and in adjacent San Mateo County. Other dunes occur from Pillar Point to the mouth of Tunitas Creek, in the vicinity of San Gregorio and Pescadero creeks, at Año Nuevo Point, in the vicinity of Waddell Creek, near Davenport, at Santa Cruz, and along most of the coast line from Robroy to the mouth of the Pajaro River. Most of the dunes are now relatively stable and shift little with the wind.

Among the more characteristic coastal strand plants are those in the following list:

Bromus maritimus
Poa douglasii
Elymus mollis
Ammophila arenaria
Chorizanthe robusta
Eriogonum latifolium
Polygonum paronychia
Atriplex leucophylla
A. californica
Abronia latifolia
A. umbellata
Tetragonia tetragonioides
Mesembryanthemum chilense

Cakile edentula ssp.
californica
C. maritima
Fragaria chiloensis
Lupinus chamissonis
Lathyrus littoralis
Croton californicus
Oenothera contorta
var. *strigulosa*
O. micrantha var.
micrantha
O. cheiranthifolia

Convolvulus soldanella
Phacelia ramosissima
Amsinckia spectabilis
Castilleja latifolia
var. *latifolia*
Haplopappus ericoides
ssp. *ericoides*
Franseria chamissonis
Eriophyllum
staechadifolium
Achillea millefolium
var. *arenicola*
Artemisia pycnocephala

Watsonville Slough near the mouth of the Pajaro River; coastal dunes in the foreground with marsh and cultivated fields in the distance.

SALT MARSH

The salt marshes are most common along the margin of San Francisco Bay, in the vicinity of Santa Cruz, and along Watsonville Slough in southern Santa Cruz County. The marshes of San Francisco Bay are rapidly disappearing as more and more land is being filled. They will disappear almost entirely should the current proposals for conversion of large parts of San Francisco Bay into fresh water lakes become realities. The local salt marshes need extensive study and collecting before it is too late. Several salt-marsh species are known locally from only a few specimens or in a few cases from only one.

The species in the following list are among the more commonly encountered ones in the salt marshes and along their borders:

Triglochin striata
T. maritima
T. concinna
Distichlis spicata
  var. *stolonifera*
Spartina foliosa
Monerma cylindrica
Polygonum patulum
Salicornia pacifica
S. depressa
Beta vulgaris
Atriplex patula var. *hastata*

Echinopsilon hyssopifolium
Suaeda californica
Chenopodium macrospermum
  var. *farinosum*
Tetragonia expansa
Potentilla egedii var.
  grandis
Frankenia grandifolia
Limonium californicum
Cuscuta salina var. *major*
  (parasitic on *Salicornia*
  spp.)

Cressa truxillensis
Cordylanthus maritimus
Plantago juncoides var.
  juncoides
Grindelia humilis
G. latifolia
Aster exilis
Jaumea carnosa
Lasthenia glabrata
Cotula coronopifolia
Senecio hydrophilus

Three species of vascular plants occur locally in a marine habitat. They are: *Phyllospadix scouleri*, *P. torreyi*, and *Zostera marina*.

Searsville Lake and a portion of its freshwater marsh in the center, Jasper Ridge, Stanford University, and San Francisco Bay in the distance.

FRESH WATER MARSH

Numerous small fresh water marshes occur along the edges of lakes, ponds, and streams. Rather extensive fresh water marshes are found along the streams near the coast, as for example at Pescadero Creek, Waddell Creek, Scott Creek, and in the vicinity of Santa Cruz. The lower portions of some marshes are brackish and support a flora composed of both fresh and salt water marsh species. Fresh water marshes have in many cases suffered the same fate as the salt marshes, and mosquito control measures have eliminated many species from some. Roadside drainage ditches and low depressions in which water stands during the winter and early spring often present habitats similar to fresh water marshes, and some fresh water marsh species are found in them. The following is a list of the more common fresh water marsh species:

Marsilea vestita
Typha latifolia
T. angustifolia
T. domingensis
Sparganium eurycarpum
Potamogeton spp.
Lilaeq scillioides
Alisma plantago-aquatica
Echinodorus berteroi
Sagittaria latifolia
Polypogon monspeliensis
Carex spp.
Cyperus eragrostis
Eleocharis macrostachya
Scirpus koilolepis
S. americanus

S. olneyi
S. californicus
S. robustus
Lemna spp.
Juncus effusus
Rumex crispus
R. conglomeratus
Polygonum coccineum
P. natans
P. pesicaria
Arenaria paludicola
Ranunculus lobbii
R. aquatilis
R. bloomeri
Rorippa spp.
Lupinus polyphyllus
var. *grandifolius*

Hypericum anagalloides
Lythrum hyssopifolia
Jussiaea repens var.
  peploides
Epilobium franciscanum
E. adenocaulon var.
  occidentale
Boisduvalia densiflora
Hydrocotyle verticillata
Berula erecta
Allocarya chorisiana
Phyla nodiflora var. *rosea*
Bidens laevis
Helenium puberulum
Lasthenia glaberrima

Allied to the fresh water marsh vegetation is that of lakes and ponds. A number of ponds and lakes have fluctuating water levels. In some this is due to evaporation as the season progresses, while in others the fluctuation is due to water use or in some cases deliberate draining. Among the species in fresh water lakes and ponds are the following: *Elodea canadensis, Ceratophyllum demersum, Ranunculus aquatilis, Nymphaea polysepala, Myriophyllum exalbescens,* and *Hippuris vulgaris.*

COASTAL SCRUB

The coastal scrub is a low vegetation cover, usually less than six feet tall, occurring on coastal bluffs, coastal hills, and often on the tops of wind-swept summits. The growth is often dense and difficult to penetrate. The coastal scrub extends locally from San Francisco southward to the vicinity of Santa Cruz. South of Santa Cruz there are only a very few areas of this vegetation type. Throughout much of the year, the coastal scrub is subject to fog and strong winds.

The more conspicuous coastal scrub species are listed below:

Polypodium scouleri
Zygadenus fremontii
  var. *minor*
Urtica californica
Eriogonum latifoliúm
Dudleya farinosa
Fragaria chiloensis
Horkelia californica
Potentilla glandulosa
Rubus ursinus
Lupinus variicolor
L. arboreus
L. chamissonis

Rhus diversiloba
Ligusticum apiifolium
Heracleum maximum
Lomatium caruifolium
Armeria maritima var.
  californica
Phacelia malvaefolia
Scrophularia californica
Diplacus aurantiacus
Orthocarpus purpurascens
  var. *latifolius*
Castilleja latifolia var.
  wightii

Lonicera involucrata
Baccharis pilularis var.
  consanguinea
Erigeron glaucus
Anaphalis margaritacea
Layia platyglossa ssp.
  platyglossa
Eriophyllum
  staechadifolium
Achillea millefolium var.
  californica
Artemisia californica

REDWOOD FOREST

The redwood community extends more or less continuously from the vicinity of Pilarcitos Lake southward to Mount Madonna, mainly west of the Santa Cruz Mountains crest. Heavy stands of redwood do occur, however, on the eastern slopes in areas moistened by summer fogs. Individual trees sometimes occur along streams far removed from the general body of the redwood forest. A good example is the redwood tree along San Francisquito Creek near the railroad tracks between Menlo Park and Palo Alto.

In deeper canyons and ravines, *Sequoia sempervirens* forms nearly pure stands. In somewhat drier areas, it is associated with *Pseudotsuga menziesii*. The floor of the redwood forest supports relatively few species. The number of species of shrubs in the understory is likewise few.

Much of the redwood forest in the Santa Cruz Mountains has been cut. The first sawmill was located near the present town of Felton and dates from 1842. Cutting has continued in various parts of the mountains until the present. A few virgin stands still exist, notably along Butano Creek and in Big Basin.

Redwoods and Douglas firs near La Honda (photograph by W. R. Dudley, about 1900).

The redwood forest is characterized by the following taxa of vascular plants:

Equisetum telmateia var.
  braunii
Woodwardia fimbriata
Polystichum munitum
Adiantum pedatum var.
  aleuticum
Torreya californica
Pseudotsuga menziesii
Sequoia sempervirens
Bromus vulgaris
Melica geyeri
Festuca occidentalis
Trisetum canescens
Calamagrostis rubescens
Hierochloe occidentalis
Trillium ovatum

Scoliopus bigelovii
Disporum smithii
D. hookeri
Myrica californica
Lithocarpus densiflora
Asarum caudatum
Montia parvifolia
Anemone quinquefolia
  var. grayi
Dicentra formosa
Dentaria californica
Boykinia elata
Heuchera micrantha
Tiarella unifoliata
Whipplea modesta
Oxalis oregana

Euonymus occidentalis
Viola sempervirens
V. glabella
Aralia californica
Hemitomes congestum
Rhododendron occidentale
Gaultheria shallon
Vaccinium ovatum
V. parvifolium
Trientalis latifolia
Myosotis latifolia
Adenocaulon bicolor
Madia madioides
Petasites frigidus var.
  palmatus

MIXED EVERGREEN FOREST

The mixed evergreen forest usually occurs adjacent to redwood forests, but on drier sites and generally in more inland localities. The mixed evergreen forest occupies a considerable amount of the forested region of the Santa Cruz Mountains. This community could be subdivided into a number of subordinate ones. Among these would be the oak-madrone, oak-buckeye, tanbark oak, madrone, Douglas fir–tanbark, etc. These, however, replace one another, often within a very short distance.

The number of species occurring in the mixed evergreen forest is larger than in the redwood forest, mainly owing to the greater diversity of habitat. The more common and conspicuous species are enumerated in the following list:

Polypodium californicum
Pityrogramma triangularis
Dryopteris arguta
Pteridium aquilinum
  var. pubescens
Pseudotsuga menziesii
Elymus glaucus
Zygadenus fremontii var.
  fremontii
Chlorogalum
  pomeridianum
Fritillaria lanceolata
Calochortus albus
Corallorhiza maculata
Alnus oregona
Corylus californica
Lithocarpus densiflora
Quercus agrifolia
Q. wislizenii
Q. chrysolepis
Aquilegia formosa var.
  truncata

Delphinium nudicaule
Ranunculus hebecarpus
Umbellularia californica
Lithophragma affinis
Rubus spectabilis var.
  franciscanus
Rosa spithamea
R. gymnocarpa
Psoralea physodes
Acer macrophyllum
Aesculus californica
Rhus diversiloba
Rhamnus crocea ssp.
  crocea
Ceanothus sorediatus
C. thyrsiflorus
Viola ocellata
Dirca occidentalis
Clarkia purpurea ssp.
  quadrivulnera
C. rubicunda ssp.
  rubicunda

Osmorhiza chilensis
Arbutus menziesii
Pholistoma auritum
Nemophila heterophylla
Phacelia nemoralis
Cynoglossum grande
Scutellaria tuberosa
Stachys bullata
Solanum umbelliferum
Collinsia heterophylla
Pedicularis densiflora
Galium nuttallii
G. triflorum
G. aparine
Sambucus mexicana
Symphoricarpos albus
  var. laevigatus
S. mollis
Asyneuma prenanthoides
Hieracium albiflorum
Wyethia glabra

Interdigitation of grassland, chaparral, oak-woodland, and coniferous forest on the western slope of the Santa Cruz Mountains near Saratoga Summit.

CHAPARRAL

The largest areas of chaparral lie on the eastern slopes of the Santa Cruz Mountains. On the western slopes chaparral occupies some of the higher ridges and some of the steep, south-facing slopes. The chaparral in the Santa Cruz Mountains is by no means uniform. On the eastern slopes of Mount Umunhum it is composed of a nearly pure stand of *Arctostaphylos glauca*. On parts of Jasper Ridge *Adenostoma fasciculatum* forms nearly pure stands. Near Eagle Rock the chaparral consists of *Arctostaphylos glutinosa, Dendromecon rigida, Pinus attenuata*, and *Adenostoma fasciculatum*.

The following list gives some of the more common chaparral species:

Pityrogramma triangularis	*Adenostoma fasciculatum*	*A. crustacea*
Pellaea andromedaefolia	*Prunus ilicifolia*	*Vaccinium ovatum*
P. mucronata	*Photinia arbutifolia*	*Navarretia heterodoxa*
Pinus attenuata	*Pickeringia montana*	*N. mellita*
Zygadenus fremontii var.	*Lotus scoparius*	*Phacelia rattanii*
fremontii	*Rhus diversiloba*	*Emmenanthe penduliflora*
Chlorogalum	*Rhamnus californica*	*Eriodictyon californicum*
pomeridianum	*R. crocea*	*Cryptantha micromeres*
Castanopsis chrysophylla	*Ceanothus sorediatus*	*C. muricata* var. *jonesii*
var. *minor*	*C. papillosus*	*Salvia mellifera*
Quercus agrifolia	*C. cuneatus*	*Lepechinia calycina*
Q. wislizenii	*Helianthemum scoparium*	*Antirrhinum multiflorum*
Q. dumosa	*Zauschneria californica*	*Diplacus aurantiacus*
Q. durata	*Garrya elliptica*	*Castilleja foliolosa*
Calandrinia breweri	*G. fremontii*	*Baccharis pilularis* var.
Clematis lasiantha	*Arctostaphylos glauca*	*consanguinea*
Dendromecon rigida	*A. andersonii*	*Haplopappus arborescens*
Holodiscus discolor	*A. canescens*	*Gnaphalium californicum*
Cercocarpus betuloides	*A. glandulosa*	*Eriophyllum confertiflorum*

Coastal grasslands in the vicinity of Pescadero; most of the land has been cultivated or overgrazed and is rapidly being eroded.

GRASSLAND

Grasslands occupy large areas mainly on the eastern side of the Santa Cruz Mountains. Much grassland has given way to urban development or has been converted to agricultural lands. Almost without exception, grasslands have been heavily grazed by cattle and in some cases by sheep. The floristic composition of grasslands consequently has been altered greatly. In many areas, especially those that have been heavily overgrazed, few native species are to be found.

Areas of grassland occupy the tops of some ridges and are often adjacent to but sharply delimited from chaparral. The factors responsible for the sharp line of demarcation between grassland and chaparral need study. Scattered throughout the mixed evergreen forests are small areas of grassland. Most of the grassland species also occur in the foothill woodland. Serpentine areas are covered either by grassland or by chaparral. The species occurring on serpentine are discussed in a subsequent section.

The more common native and introduced grassland species are included in the following list:

Briza minor	*F. myuros*	*Agrostis diegoensis*
Bromus carinatus	*F. dertonensis*	*Juncus bufonius*
B. mollis	*F. rubra*	*Calochortus luteus*
B. rigidus	*Hordeum leporinum*	*Allium lacunosum*
Poa unilateralis	*H. hystrix*	*Brodiaea pulchella*
P. scabrella	*Avena fatua*	*B. laxa*
Melica californica	*Aira caryophyllea*	*B. terrestris*
Festuca megalura	*Stipa pulchra*	*Sisyrinchium bellum*

Open foothill woodland on Jasper Ridge, the Biological Reserve of Stanford University, in February.

Delphinium hesperium
Ranunculus californicus
Capsella bursa-pastoris
Lepidium nitidum
Lupinus succulentus
L. nanus
L. bicolor
Lotus purshianus
L. micranthus
L. subpinnatus
Trifolium amplectens
T. barbigerum
T. grayi
T. microdon
T. microcephalum
T. variegatum
T. dichotomum
Astragalus gambellianus
Geranium molle
G. dissectum
Eriodium obtusiplicatum
Erodium botrys
E. cicutarium
Sidalcea diploscypha

S. malvaeflora
Viola adunca
V. pedunculata
Clarkia unguiculata
C. rubicunda ssp.
  rubicunda
C. purpurea
Oenothera ovata
Sanicula arctopoides
Lomatium utriculatum
Androsace acuta
Microcala quadrangularis
Convolulus subacaulis
Phlox gracilis
Linanthus androsaceus
Gilia tricolor
G. clivorum
Nemophila menziesii
N. pedunculata
Amsinckia intermedia
Plagiobothrys nothofulvus
Orthocarpus
  lithospermoides

O. purpurascens var.
  purpurascens
O. densiflorus
O. attenuatus
O. pusillus
Plantago erecta
Microseris douglasii
Agoseris heterophylla
A. apargioides ssp.
  apargioides
Chaetopappa bellidiflora
C. exilis
C. alsinoides
Filago gallica
F. californica
Micropus californicus
Wyethia helenioides
W. angustifolia
Achyrachaena mollis
Lagophylla ramosissima
Hemizonia corymbosa
H. luzulaefolia
Baeria chrysostoma
Soliva sessilis

Streambank vegetation near La Honda (photograph by W. R. Dudley, about 1900).

FOOTHILL WOODLAND

Foothill woodland occurs as a discontinuous band from the Crystal Springs region southward to Mount Madonna. Most of the foothill woodland occurs to the east of the crests of the Santa Cruz Mountains. It may be relatively dense or rather open and form oak-savannas or oak-parklands. Transitions to grassland and mixed evergreen forest are common. Most grassland species occur in the woodland and will not be listed again. The more common foothill woodland species are included in the following list:

Pinus sabiniana	Q. lobata	Aesculus californica
Stipa lepida	Q. chrysolepis	Pholistoma auritum
Quercus kelloggii	Phoradendron villosum	Nemophila heterophylla
Q. agrifolia	Montia perfoliata	Pedicularis densiflora
Q. wislizenii	Umbellularia californica	Galium nuttallii
Q. douglasii	Photinia arbutifolia	Wyethia glabra
	Rhus diversiloba	

STREAMBANK VEGETATION

Small streams usually do not have floras conspicuously differing from those on the adjacent slopes. Along the larger streams, however, and especially along the lower portions of their courses a number of species are found which are not common elsewhere. Among the more common are:

Populus trichocarpa	Platanus racemosa	Cornus glabrata
Salix ssp.	Acer negundo var.	C. californica
Alnus oregona	californica	Baccharis viminea
A. rhombifolia	Fraxinus latifolia	

PONDEROSA PINE FOREST

The ponderosa pine forest in Santa Cruz County is so different from any of the other locally occurring communities that it should be recognized as distinct. In both general localities in which it occurs, the Ben Lomond Sand Hills and at Bonny Doon, the soil is very sandy. In addition to *Pinus ponderosa*, *P. attenuata*, *Quercus agrifolia*, and various chaparral species are common.

Discussion of the Flora

The vascular flora of the Santa Cruz Mountains consists of 167 families, 650 genera, and 1799 species, subspecies, varieties, forms, and hybrids. Table 3 is a tabulation of the flora by divisions, classes, and subclasses.

TABLE 3

TABULATION OF THE VASCULAR PLANTS

Division Class Subclass	Family	Genus	Species, Subspecies Varieties, Forms, and Hybrids		
			Total	Native	Introduced
Lepidophyta	2	2	3	3	0
Calamophyta	1	1	5	5	0
Pterophyta					
Filicinae	4	16	27	27	0
Gymnospermae	4	5	10	9	1
Angiospermae					
Monocotyledoneae	24	132	396	270	126
Dicotyledoneae	132	494	1358	932	426
Total	167	650	1799	1246	553

The ten largest families contain 44 per cent of the genera and 53 per cent of the specific and infraspecific taxa. Table 4 is a list of these families.

Thirty-six genera contain 10 or more entities each. These genera are listed in Table 5, together with the number of taxa below the rank of genus in each.

TABLE 4

LIST OF THE TEN LARGEST FAMILIES OF VASCULAR PLANTS IN THE SANTA CRUZ MOUNTAINS

Family	Number of Genera	Number of Taxa
Compositae	91	256
Gramineae	65	197
Fabaceae	19	126
Scrophulariaceae	22	77
Cruciferae	29	64
Cyperaceae	6	60
Umbelliferae	28	49
Onagraceae	8	45
Polygonaceae	7	45
Caryophyllaceae	12	38

TABLE 5

THE THIRTY-SIX LARGEST GENERA

Genus	Number of Taxa Below the Rank of Genus
Carex	32
Trifolium	31
Lupinus	26
Bromus, Juncus	19
Festuca	18
Polygonum, Lotus, Vicia	15
Solanum, Mimulus	14
Agrostis, Clarkia, Arctostaphylos, Hemizonia, Senecio, Cirsium	13
Quercus, Rumex, Chenopodium, Ceanothus, Orthocarpus	12
Scirpus, Atriplex, Ranunculus, Epilobium, Oenothera, Linanthus, Phacelia, Galium	11
Poa, Elymus, Brodiaea, Lathyrus, Euphorbia, Veronica	10

The vascular flora of the Santa Cruz Mountains is composed of several major floristic elements. These elements may be divided into a series of distributional patterns. Campbell and Wiggins have recognized 16 distinct distribution patterns among the plants occurring in California. In a smaller area, the number of such patterns is less. I consider that there are five main distribution patterns represented by the plants in the Santa Cruz Mountains. The following lists these patterns together with the percentage of species in each. The figures given by Campbell and Wiggins for California are shown for comparison:

Distribution Pattern	Santa Cruz Mountains %	California %
California	33	36
Pacific Coast	16	17
North America	13.5	32
Intercontinental	6.5	7
Introduced	31	8

The two sets of percentages agree closely in three of the distribution patterns. In the Santa Cruz Mountains there are fewer plants with widespread distributions in North America than in California as a whole. This difference can be explained on the basis of the lack of such regions as deserts, high mountains, and high arid plateaus in the Santa Cruz Mountains.

Plants with a Pacific Coast distribution are those with ranges from Alaska southward to southern California and northern Baja California, mainly through British Columbia, Washington, and Oregon, but often extending eastward to Idaho and western Montana. Typical taxa of the Pacific Coast distribution pattern are *Polypodium glycyrrhiza, Carex phyllomanica, Lysichiton americanum, Tellima grandiflora, Viola glabella, Vaccinium parvifolium,* and *Galium trifidum* var. *subtriflorum.* Included within the Pacific Coast distribution pattern are a number of other distinctive distribution patterns. One of the most common includes those plants that occur from southern British Columbia south to central or southern California. The following represent examples of taxa with a British Columbia–California distribution: *Polypodium scouleri, Juncus occidentalis, Rosa gymnocarpa, Oxalis oregana, Viola sempervirens,* and *Arbutus menziesii.* Plants with a Pacific Coast distributional pattern represent mainly a northern floristic element although a few represent a southern floristic element.

Plants with a California distributional pattern are endemic to California or range northward into the southwestern corner of Oregon or extend southward into northern Baja California. This distribution pattern excludes plants of the high Sierra Nevada, the northeastern corner of California, and in general the deserts. Examples of species with this distributional pattern are: *Lithocarpus densiflora, Delphinium nudicaule, Rosa californica, Linanthus ciliatus, Allocarya bracteata, Microseris douglasii* ssp. *douglasii,* and *Baccharis douglasii.*

The California distributional pattern includes within it a number of characteristic patterns. One of the more important of these is called the Coast Range endemic pattern. Thirty-seven per cent of the plants within the California distributional pattern are restricted to the Coast Ranges from Humboldt County south through San Luis Obispo County. Of these, 27 taxa are endemic within the Santa Cruz Mountains and will be discussed later. Representative plants with a Coast Range distributional pattern are: *Carex harfordii, Horkelia californica, Lotus junceus, Navarretia heterodoxa, Stachys chamissonis, Antirrhinum vexillo-calyculatum,* and *Wyethia glabra.*

The distribution pattern that I have called the North American is capable of repeated subdivision. In general, however, the taxa are widespread in western North America, extend across the United States, or extend southward and eastward to Arizona, New Mexico, Texas, and Mexico. Many plants in this group are of the southern floristic element. A sampling of species with a North American distribution pattern is as follows: *Scirpus koilolepis, Corallorhiza maculata, Polygonum coccineum, Sida leprosa* var. *hederacea, Clarkia rhomboidea, Lycopus americanus,* and *Filago californica.*

Plants with a widespread distribution pattern are cosmopolitan; they occur widely in the northern hemisphere, or are common both to North and South America. Many of these are aquatics or plants of very moist areas. Typical examples are *Luzula multiflora, Polygonum natans, Tillaea aquatica, Lythrum hyssopifolia, Berula erecta,* and *Limosella aquatica.*

Of the total number of vascular plants in the Santa Cruz Mountains, 553, or 31 per cent, are introduced. The majority of the alien plants are of Eurasian or North African origin. The others are from different parts of Africa, notably South Africa, South America, Australia, other parts of North America, and even a few from other parts of California. Many of the introduced species became established at the time of the founding of the Spanish Missions. Some of the introductions were accidental, some were deliberate. The fact that over one-quarter of the kinds of plants in the flora today are introduced shows the profound way in which the vegetation has been altered by man and his activities in the course of only about three centuries. Some introductions have been sporadic, persisting for only short periods. Others have become widespread, displacing some of the original plant species. A number of species have been introduced recently; for example, *Trifolium angustifolium, T. subterraneum, Ornithopus pinnatus, O. roseus, Oxalis pes-caprae,* and *Crepis vesicaria* ssp. *taraxacifolia.* Some introductions, as for example *Tribulus terrestris, Lavatera arborea, L. cretica, Buddleia davidii,* and *Kickxia spuria,* seem to be spreading rapidly today.

Every new garden plant is a potential addition to the flora. Every load of hay, sand, gravel, or fill from other areas is a possible source of new weeds. Keeping track of new introductions and noting their subsequent dispersal is a fascinating study in itself and an important means of learning more about the factors influencing plant migration.

Twenty-seven species and infraspecific taxa are considered to be endemic in the Santa Cruz Mountains. The endemics constitute 1.5 per cent of the total number of vascular plants and are listed below:

Pinus × *attenuradiata*	*A. glutinosa*
Cupressus abramsiana	*A. silvicola*
Silene verecunda ssp. *verecunda*	*A. crustacea* var. *rosei*
Erysimum teretifolium	*Allocarya chorisiana* var. *chorisiana*
E. franciscanum var. *crassifolium*	*A. diffusa*
Grossularia senilis	*Acanthomintha obovata* ssp. *duttonii*
Lupinus latifolius var. *dudleyi*	*Penstemon rattanii* ssp. *kleei*
Ceanothus cuneatus var. *dubius*	*Pedicularis dudleyi*
Malacothamnus arcuatus	*Grindelia maritima* f. *maritima*
Clarkia franciscana	*G. maritima* f. *anomala*
Arctostaphylos franciscana	*Lessingia ramulosa* var. *glabrata*
A. andersonii var. *andersonii*	*L. germanorum* var. *germanorum*
A. andersonii var. *imbricata*	*L. hololeuca* var. *arachnoidea*
Cirsium fontinale var. *fontinale*	

Some of the endemics are closely restricted to certain geological formations. Others occur in a number of different habitats. The largest group of endemics restricted to a certain soil type are those confined more or less to serpentine and are as follows: *Clarkia franciscana*, *Arctostaphylos franciscana*, *Acanthomintha obovata* ssp. *duttonii*, *Grindelia maritima*, *Lessingia hololeuca* var. *arachnoidea*, *L. ramulosa* var. *glabrata*, and *Cirsium fontinale* var. *fontinale*. The Santa Margarita formation in Santa Cruz County is the site of two endemics, *Erysimum teretifolium* and *Arctostaphylos silvicola*. *Arctostaphylos silvicola* does occur occasionally elsewhere, but is most common in the Santa Margarita formation. *Arctostaphylos glutinosa* is entirely restricted to Monterey shale and *Penstemon rattanii* ssp. *kleei*, although occurring at Aptos and Loma Prieta, is fairly common on soil derived from the Monterey shale. *Arctostaphylos crustacea* var. *rosei* and *Lessingia germanorum* var. *germanorum* are restricted to the coastal sand dunes in San Francisco and northern San Mateo County. *Cupressus abramsiana* is known from only four localities. In each case, the trees are restricted to sandstone or sands derived from the Butano sandstone or Vaqueros sandstone formations. The remaining endemics do not appear to be closely correlated with geological formations.

Numerous taxa have their southern limits of distribution in the Santa Cruz Mountains. No absolute correlation exists between floristic element and southern limit of distribution. *Tanacetum camphoratum* has its southern limit of distribution in San Francisco, but extends no farther north than Marin County. It is thus a narrowly restricted central California endemic. *Ceanothus foliosus*, so far as is known, does not occur farther south than Mount Madonna, but extends northward to Humboldt County and must thus be considered a California endemic. *Epilobium halleanum* is not known south of Camp Evers, Santa Cruz County, but ranges northward to British Columbia and eastward to Colorado.

Grove of *Cupressus abramsiana* at Eagle Rock, Santa Cruz County, in the distance and one of the vineyards of the Locatelli Ranch in the foreground.

Lysichiton americanum also has its southern limit of distribution in the Coast Ranges in Santa Cruz County, but ranges northward to Alaska; and a closely related species, *L. camtschatcense* (L.) Schott, occurs in eastern Asia. In general, however, taxa with southern limits of distribution do belong to a more northerly floristic element.

It is of interest to compare the number of taxa having southern limits of distribution in the Santa Cruz Mountains with the number reaching their southern limits in adjacent areas. In the Santa Cruz Mountains 181 taxa reach their southern limits of distribution. In Marin County the number is 89, on Mount Diablo six, and in the Mount Hamilton Range six. On the basis of the data in the taxonomic portion of this study, I would estimate that between 175 and 200 taxa reach their southern limits of distribution in the Coast Range in the Santa Lucia Mountains of Monterey County. The northern element becomes progressively less important in the flora from north to south, and this element is of less importance in the Inner Coast Ranges than in the Outer Coast Ranges.

The following is a list of species, subspecies, and varieties with southern limits of distribution in the Coast Range in the Santa Cruz Mountains together with a list of the southernmost stations at which these plants have been collected. In some cases two stations have been listed if they are at the same latitude but separated longitudinally.

Equisetum hyemale var. *affine*—Los Gatos, Boulder Creek
Botrychium silaifolium var. *californicum*—Camp Evers
Polypodium scouleri—Cahill Ridge
P. californicum—Soquel Valley
Dryopteris dilatata—Butano Creek
Blechnum spicant—Waddell Creek
Cheilanthes intertexta—Loma Prieta
Torreya californica—Loma Prieta, Little Basin

Poa kelloggii—Felton
Melica subulata—Congress Springs
Pleuropogon californicus—Millbrae
Glyceria leptostachys—San Francisco
G. occidentalis—Baden
G. pauciflora—Pilarcitos Creek
Puccinellia grandis—Near Palo Alto
Festuca subuliflora—Jamison Creek
F. idahoensis—Empire Grade
Hystrix californica—Santa Cruz
Elymus mollis—Waddell Creek

Elymus virescens—San Francisco
E. × *vancouverensis*—Santa Cruz
Trisetum cernuum—Glenwood
Phleum alpinum—San Francisco
Calamagrostis rubescens—Boulder Creek
Agrostis pallens—San Francisco
A. californica—Santa Cruz
Beckmannia syzigachne—Saratoga
Carex simulata—Camp Evers
C. densa—Morgan Hill
C. phyllomanica—Camp Evers
C. leptopoda—Big Basin
C. gracilior—Watsonville
C. salinaeformis—Camp Evers
C. nudata—Near Santa Cruz
C. amplifolia—Boulder Creek
C. exsiccata—Wrights, Big Basin
C. comosa—Santa Cruz County
Eleocharis obtusa—Pinto Lake
Eriophorum gracile—San Francisco
Lysichiton americanum—Near Felton
Calochortus tolmiei—Big Basin, Loma
Prieta
C. umbellatus—Kings Mountain
Scoliopus bigelovii—Los Gatos, Santa
Cruz
Maianthemum dilatum—Pescadero Creek
Disporum smithii—Woodside
Streptopus amplexifolius var. *denticulatus*
—Near Boulder Creek
Allium serratum—Morgan Hill
A. breweri—Loma Prieta
Brodiaea multiflora—San Juan Hills
B. congestum—Stevens Creek, Glenwood
B. appendiculata—New Almaden
Iris macrosiphon—Gilroy, Graham Hill
Cyripedium montanum—Corte de Madera
C. fasciculatum—Loma Prieta
Goodyera oblongifolia—Boulder Creek
Corylus californicus—Stevens Creek, Ben
Lomond Sand Hills
Urtica californica—Santa Cruz
Asarum caudatum—Los Gatos, Zayante
Chorizanthe cuspidata var. *cuspidata*—
Santa Cruz
Eriogonum hirtiflorum—Empire Grade
Rumex persicarioides—Moss Beach
Paronychia franciscana—San Bruno
Mountain
Montia sibirica—Mount Hermon
Spraguea umbellata—Ben Lomond Sand
Hills
Stellaria littoralis—San Francisco
Arenaria pusilla—San Francisco

Spergularia marina var. *tenuis*—Alviso
Silene scouleri ssp. *grandis*—Pescadero
Delphinium hesperium—San Jose, Big
Basin
Anemone quinquefolia var. *grayi*—Loma
Prieta, Aptos
Ranunculus lobbii—Coal Mine Ridge,
Aptos
R. pusillus—Coal Mine Ridge, Santa Cruz
R. bloomeri—Alviso
Meconella californica—Big Basin
Dicentra formosa—Big Basin
Erysimum franciscanum var. *franciscanum*
—Crystal Springs
Arabis blepharophylla—Boulder Creek
Sedum spathulifolium ssp. *spathulifolium*
—Santa Cruz County
Dudleya cymosa ssp. *cymosa*—Loma
Prieta, Eagle Rock
Tolmiea menziesii—Boulder Creek
Tiarella unifoliata—Felton
Grossularia leptosma—Loma Prieta,
Waddell Creek
Rubus leucodermis—Big Trees
R. spectabilis var. *franciscanus*—Liddell
Creek
Prunus subcordata—Permanente Creek
Lupinus affinis—Santa Cruz County
L. polyphyllus var. *grandifolius*—Mount
Hermon
L. arboreus var. *eximius*—Seal Cove
L. pachylobus—Los Gatos
Lotus torreyi—Santa Cruz
Psoralea strobilina—New Almaden
Trifolium depauperatum—Glenwood
T. cyathiferum—Edenvale
T. dichotomum—Gilroy
Astragalus pycnostachyus—San Gregorio
A. nuttallii var. *virgatus*—Pescadero
Lathyrus torreyi—Mount Hermon
Hesperolinon congestum—Woodside
Ptelea crenulata—Gilroy
Euonymus occidentalis—Felton
Ceanothus velutinus var. *laevigatus*—
Swanton
C. foliosus—Mount Madonna
Fremontodendron californicum ssp. *crassi-*
folium—Loma Prieta, Big Basin
Dirca occidentalis—Searsville
Epilobium halleanum—Camp Evers
Clarkia concinna—Loma Prieta
C. davyi—Santa Cruz
C. purpurea ssp. *viminea*—Los Gatos,
Locatelli Ranch

Ludwigia palustris var. *pacifica*—Camp
Evers
Perideridia kelloggii—Aptos
Ligusticum apiifolium—San Mateo
Angelica hendersonii—Pescadero
A. tomentosa—Santa Cruz
Conioselinum chinense—San Francisco
Pleuricospora fimbriolata—Big Basin
Ledum glandulosum var. *australe*—
Boulder Creek
Rhododendron macrophyllum—Scott
Valley
Arctostaphylos sensitiva—Mount Hermon
A. canescens—Loma Prieta
Vaccinium parvifolium—Big Basin
Dodecatheon hendersonii ssp. *cruciatum*—
Stanford, Jamison Creek
Fraxinus latifolia—San Francisquito
Creek
Centaurium floribundum—Waterman Gap
C. muhlenbergii—Palo Alto
C. trichanthum—San Jose
Menyanthes trifoliata—San Francisco
Convovulus malacophyllus ssp. *collinus*—
Madrone
Linanthus acicularis—Coal Mine Ridge
Polemonium carneum—Pilarcitos Canyon
Eriastrum abramsii—Black Mountain
Navarretia heterodoxa—Los Gatos
N. viscidula—Searsville
Gilia capitata ssp. *capitata*—Black
Mountain, Alba Grade
G. capitata ssp. *chamissonis*—San
Francisco
Nemophila menziesii var. *atomaria*—Loma
Prieta
Phacelia nemoralis—Loma Prieta, Ben
Lomond Mountain
P. californica—Santa Cruz
P. suaveolens—Loma Prieta, Eagle Rock
P. divaricata—Los Gatos
Romanzoffia suksdorfii—San Mateo
Canyon
Amsinckia lunaris—San Mateo County
Cryptantha torreyana var. *pumila*—
Stevens Creek, Swanton
Stachys chamissonis—Año Nuevo Point
Collinsia corymbosa—San Francisco
C. tinctoria—Laguna Honda, San
Francisco

Tonella tenella—Los Gatos
Mimulus rattanii—Near Santa Cruz
Gratiola ebracteata—Near Woodside
Cordylanthus pilosus—Los Gatos
Orthocarpus lithospermoides—Gilroy
O. floribundus—Belmont
Castilleja latifolia var. *wightii*—Santa
Cruz
Orobanche grayana var. *jepsonii*—Palo
Alto
O. pinorum—Boulder Creek
Plantago juncoides var. *juncoides*—
Mayfield
Galium triflorum—Near Santa Cruz
Sambucus callicarpa—Near Santa Cruz
Marah oreganus—Saratoga Summit,
Blooms Grade
Campanula angustiflora—Boulder Creek
C. californica—Camp Evers
Legenere limosa—Coal Mine Ridge
Grindelia humilis—Alviso
Chrysopsis villosa var. *bolanderi*—Portola
C. oregona var. *rudis*—Gilroy
Erigeron foliosus var. *hartwegii*—Ben
Lomond Mountain
E. inornatus var. *angustatus*—Crystal
Springs
Lessingia hololeuca var. *hololeuca*—Los
Gatos
Adenocaulon bicolor—Boulder Creek
Psilocarphus brevissimus var. *multiflorus*
—Near Gilroy
Stylocline amphibola—Near Santa Cruz
Helianthella californica—Black Mountain
Lagophylla congesta—Black Mountain
Hemizonia clevelandii—Near Boulder
Creek
Calycadenia multiglandulosa ssp.
cephalotes—Emerald Lake
Franseria chamissonis ssp. *chamissonis*—
Pajaro River
Eriophyllum lanatum var. *arachnoideum*—
Loma Prieta
Lasthenia glaberrima—San Jose
Tanacetum camphoratum—San Francisco
Senecio aronicoides—Ben Lomond Sand
Hills
Cirsium andrewsii—Northern San Mateo
County
C. remotifolium—San Francisco

Sixty-one taxa have their northern limits of distribution in the Coast Range in the Santa Cruz Mountains. For Marin County this figure is 34, for Mount Diablo 32, and for the Mount Hamilton Range 23. I have no figures available for the Santa Lucia Mountains of Monterey County, but it is probably a fairly high number. Plants with their northern limits of distribution in the Santa Cruz Mountains in general constitute a southern floristic element. The southern element becomes less important from south to north and is more important in the Inner Coast Ranges than in the Outer Coast Ranges. The following is a list of taxa with northern limits of distribution in the Coast Range in the Santa Cruz Mountains:

Selaginella bigelovii—Los Gatos
Cheilanthes cooperae—Near Felton
Pinus radiata—Año Nuevo Point
Bromus pseudolaevipes—Stanford
B. grandii—Eagle Rock
Elymus condensatus—Valencia Creek
Carex schottii—Sargents
Anemopsis californica—San Jose
Chorizanthe robusta—San Francisco
C. cuspidata var. *marginata*—San Francisco
C. pungens var. *hartwegii*—San Francisco
C. diffusa—Jasper Ridge
Brodiaea lutea—Woodside
Parietaria floridana—Near Santa Cruz
Eriogonum saxatile—Mount Umunhum
Delphinium parryi ssp. *seditosum*—Mount Hermon
Erysimum ammophilum—Sunset Beach
Descurainia pinnata ssp. *menziesii*—Stanford
Horkelia cuneata ssp. *cuneata*—San Francisco
H. bolanderi ssp. *parryi*—San Andreas Valley
Lupinus hirsutissimus—Swanton, Stevens Creek
L. truncatus—Santa Cruz
Lotus salsuginosus—Watsonville
Croton californicus—San Francisco
Ceanothus papillosus—Kings Mountain
C. dentatus—Between Soquel and Watsonville
Malacothamnus hallii—Near Madrone
Clarkia epilobioides—San Francisco
Oenothera hookeri ssp. *montereyensis*—San Francisco
Osmorhiza brachypoda—Castle Rock Ridge
Sanicula maritima—San Francisco

Tauschia hartwegii—Stanford
Lomatium parvifolium—Near Gilroy
Arctostaphylos hookeri—Near Watsonville
A. crustacea var. *crustacea*—Near San Andreas Lake
Dodecatheon clevelandii ssp. *sanctarum*—Año Nuevo Point
Linanthus ambiguus—Crystal Springs
Gilia tenuiflora—Ben Lomond Sand Hills
Pholistoma membranaceum—San Jose
Phacelia douglasii—San Francisco
Allocarya chorisiana var. *hickmanii*—Pescadero
Echidocarya californica—Eagle Rock
Salvia mellifera—Los Gatos
Stachys bullata—San Francisco
Antirrhinum multiflorum—Los Gatos
Collinsia bartsiaefolia var. *hirsuta*—San Francisco
C. multicolor—San Francisco
Cordylanthus rigidus—San Andreas Valley
Haplopappus venetus ssp. *vernonioides*—San Francisco
H. ericoides ssp. *blakei*—Bonny Doon
Chrysopsis villosa var. *camphorata*—Stevens Creek
Corethrogyne californica—San Francisco
C. leucophylla—Año Nuevo Point
C. filaginifolia var. *filaginifolia*—Soda Lake
C. filaginifolia var. *rigida*—Ben Lomond
Conyza coulteri—Loma Prieta, Felton
Lessingia germanorum var. *tenuipes*—Searsville
Gnaphalium bicolor—Waddell Creek
Monolopia gracilens—Sawyer Ridge
Hulsea heterochroma—Loma Prieta
Senecio breweri—San Mateo

A number of taxa, growing mainly on the eastern slope of the Sierra Azule, show definite Inner Coast Range affinities. Some taxa extend over large areas of the eastern slope of the Sierra Azule. Others are restricted and have been collected only once or twice.

The most common and one of the most conspicuous Inner Coast Range plants in the Santa Cruz Mountains is *Arctostaphylos glauca*. *Pinus sabiniana*, another typical Inner Coast Range and Sierra Nevada foothill species, occurs only in a few areas in the Santa Cruz Mountains, mostly in the vicinity of Loma Prieta and Mount Umunhum, usually between 1500 and 3000 feet in elevation. Some plants such as *Emmenanthe penduliflora* are more widespread in the Santa Cruz Mountains, but nevertheless belong to the Inner Coast Range–desert element.

Some of the other Inner Coast Range taxa in the Santa Cruz Mountains are presented in the subsequent list:

Selaginella bigelovii	*Rhamnus crocea* ssp.	*Microseris elegans*
Allium breweri	*ilicifolia*	*Erigeron petrophilus*
Arceuthobium campylo-	*Ceanothus ferrisae*	*Chrysothamnus nauseosus*
podum	*Malacothamnus hallii*	ssp. *occidentalis*
Eriogonum argillosum	*Fraxinus dipetala*	*Helianthella castanea*
E. saxatile	*Mentzelia dispersa*	*Layia hieracoides*
Calyptridium parryi var.	*M. lindleyi*	*Calycadenia multiglandu-*
hesseae	*Clarkia breweri*	*losa* var. *multiglandulosa*
Dudleya cymosa ssp.	*C. modesta*	*Eriophyllum lanatum* var.
setchellii	*Garrya fremontii*	*achillaeoides*
Lewisia rediviva	*Eriastrum abramsii*	*Chaenactis tanacetifolia*
Arabis breweri	*Allocarya glabra*	*Hulsea heterochroma*
Ptelea crenulata	*Galium andrewsii*	

Several other areas in the Santa Cruz Mountains are of interest with respect to plant distribution. The serpentine areas to the east of the crests support a varied flora, and a number of these species are either restricted to serpentine soil or are more common on serpentine soil than on nonserpentine soil. Some of the more obligate local serpentine plants are listed below:

Selaginella bigelovii	*Arenaria pusilla*	*Convolvulus malacophyllus*
Festuca pacifica	*Silene californica*	ssp. *collinus*
F. reflexa	*Streptanthus glandulosus*	*Linanthus liniflorus*
Koeleria gracilis	var. *glandulosus*	*L. ambiguus*
Calochortus venustus	*Dudleya cymosa* ssp.	*Acanthomintha obovata*
Allium lacunosum	*setchellii*	ssp. *duttonii*
A. amplectens	*Psoralea strobilina*	*Grindelia maritima*
A. serratum	*Astragalus gambellianus*	*Lessingia hololeuca* var.
A. breweri	*Hesperolinon micranthum*	*arachnoidea*
Quercus durata	*H. congestum*	*L. ramulosa* var. *glabrata*
Eriogonum saxatile	*Clarkia franciscana*	*Chrysothamnus nauseosus*
E. argillosum	*Sanicula bipinnatifida*	ssp. *occidentalis*
Lewisia rediviva	*Arctostaphylos franciscana*	*Rigiopappus leptocladus*
		Cirsium fontinale

The Ben Lomond Sand Hills and the sand deposits near Bonny Doon are of interest because of the endemics and because of several other distributional patterns. *Cardionema ramosissima, Horkelia cuneata* ssp. *cuneata,* and *Armeria maritima* var. *californica,* typical coastal plants, are found inland in the Ben Lomond Sand Hills. *Spraguea umbellata* reaches its southern limit of distribution in the Coast Range in this area. *Haplopappus ericoides* ssp. *blakei* reaches its northern limit of distribution near Bonny Doon, and *Gilia tenuiflora* reaches its northern limit in the Ben Lomond Sand Hills. The most distinctive feature is the stand of *Pinus ponderosa.* It is known locally only from near Bonny Doon and the Ben Lomond Sand Hills. To the north it occurs again in Napa County, to the south at Pico Blanco in the Santa Lucia Mountains, and to the east there is a small stand in the Mount Hamilton Range.

The marsh at Camp Evers is interesting because a large number of species of *Carex* are present and because several species of *Carex* and other genera have their southern limits of distribution in the Coast Range there. Unfortunately much of the marsh has been drained and other parts have been used as a source of peat. The most extensive stand of *Lupinus polyphyllus* var. *grandifolius* in the Santa Cruz Mountains occurred here. Among the plants with southern limits of distribution at Camp Evers are *Botrychium silaifolium* var. *californicum*, *Carex simulata*, *C. phyllomanica*, *C. salinaeformis*, and *Epilobium halleanum*.

Some of the marshes in San Francisco were of equal if not of greater interest. Most have been filled or drained. Behr has discussed the plants in some of them. Such species as *Eriophorum gracile*, *Castilleja miniata*, and *Menyanthes trifoliata* once grew in San Francisco marshes and in addition reached their southern limits of distribution there. Others like *Arenaria paludicola*, *Cornus nuttallii*, and *Lycopus americanus* once grew in the marshes of San Francisco. Unfortunately, specimens of a number of these no longer exist.

Several taxa with southern limits of distribution in the Santa Cruz Mountains are not encountered to the north for some distance. These taxa would appear to represent a relic element in the Santa Cruz Mountains. The following lists these species and the first county to the north in which they occur:

Carex salinaeformis—Mendocino County
Festuca subuliflora—Mendocino County
Streptopus amplexifolius var. *denticulatus*
—Mendocino County
Cypripedium fasciculatum—Del Norte County
Spraguea umbellata—Lake County

Tolmiea menziesii—Mendocino County
Rubus leucodermis—Sonoma County
Pleuricospora fimbriata—Humboldt County
Vaccinium parvifolium—Sonoma County
Orobanche pinorum—Siskiyou County

History of Botanical Collecting in the Santa Cruz Mountains

The history of botanical collecting in the Santa Cruz Mountains probably began in 1791. Until about 1850 the men who collected were botanists, naturalists, or the surgeons of scientific or diplomatic expeditions sent from Europe or from the eastern part of the United States. The specimens collected prior to 1850 were sent to such botanical centers as London, Paris, Prague, St. Petersburg, Cambridge, New York, and, eventually, Washington. Not until after 1850 were some residents in central California interested in the study of botany and natural history. The men who collected locally prior to 1850 number about 18. Quite likely small collections of plants were made by various travelers, but if so these collections never reached the leading botanists of the day and did not contribute to the development of California botany. The number of collectors since 1850 probably approaches 1000, but many of these were students and their period of collecting was short.

The distinction of being the first to collect in the Santa Cruz Mountains appears to belong to Thaddeus Haenke, a botanist on the Spanish Malaspina Expedition. Haenke spent from September 12 to 23, 1791, at Monterey, and quite possibly got as far north as Santa Cruz. Other early botanists or naturalists

Upper left, A. D. E. Elmer, 1927 (courtesy of his son, Dr. A. D. Elmer) ; *upper right,* Amos Arthur Heller, about 1921 (courtesy of his daughter, Mrs. Mildred Pritchett) ; *below,* LeRoy Abrams, about 1927 on a field trip to the Mojave Desert (photograph by G. T. Benson, in the Dudley Herbarium).

Upper left, Volney Rattan, about 1900 (courtesy of the Jepson Herbarium, University of California); *upper right*, W. R. Dudley, about 1900 (from a photograph in the Dudley Herbarium); *below*, C. L. Anderson, about 1885 (from *History of Santa Cruz County, California*).

include Archibald Menzies, Georg Heinrich von Langsdorff, Adalbert von Chamisso, Johann Friedrich Eschscholz, Alexander Collie, Paolo Emilio Botta, David Douglas, Thomas Coulter, Richard Brinsley Hinds, William Dunlop Brachenridge, Eugene Duflot de Mofras, William Gambel, John Charles Fremont, Karl Theodore Hartweg, and August Fitch.

Albert Kellogg (1813–1887) was the first resident botanist in central California. He arrived in California in August 1849 and collected extensively in western North America.

William Russel Dudley (1849–1911), the first professor of systematic botany at Stanford University, collected in many parts of the Santa Cruz Mountains from 1893 to 1910. Without doubt, Dudley must be considered the most important local collector. There were few plants he did not collect, or few localities he did not visit. His collections were not widely distributed until recently, despite the wealth of duplicate material. One of the most restricted local endemics, *Pedicularis dudleyi*, was named for him.

LeRoy Abrams (1874–1956) was a member of the faculty of Stanford University from 1906 until his retirement in 1940. He collected in many parts of the Santa Cruz Mountains, and his collections, made over a period of 42 years, are probably the most important local ones in the Dudley Herbarium, next to those of Dudley. Abrams is best known for his four-volume *Illustrated Flora of the Pacific States*.

Charles Lewis Anderson, M.D., (1827–1910) lived in Santa Cruz from 1864 until his death. Although he is better known for his Nevada and Sierra Nevada collections, he did collect extensively in Santa Cruz County. His grass collection is in the Dudley Herbarium. Many of his plants were sent to Asa Gray, and Dr. Gray named *Arctostaphylos andersonii* for him. Anderson published a *Catalogue of Flowering Plants and Ferns of Santa Cruz County*.

Adolph Daniel Edward Elmer (1870–1942) collected in many parts of the Santa Cruz Mountains during and prior to 1903, the year in which he received an M.A. from Stanford University. Elmer published a number of new species from central California in the early numbers of the *Botanical Gazette*. In 1904 he went to the Philippines and collected extensively in the far east.

Amos Arthur Heller (1867–1944) lived in Los Gatos from about 1904 to 1909. During this time he collected in the Santa Cruz Mountains and published the results of his explorations in his own hand-set and hand-printed journal, *Muhlenbergia*. Heller's specimens are among the most widely distributed that were ever collected in the Santa Cruz Mountains. Almost every major United States and European herbarium has a set of his California plants. His original herbarium is now part of that of the Brooklyn Botanic Garden.

Volney Rattan (1840–1915) taught high school in Santa Cruz for many years and later became a professor of botany at the California State Normal School at San Jose. He collected in many parts of central California, and in 1904 his herbarium of about 2000 sheets was acquired by Stanford University. Unfortunately many of his collections lack even the place and date of collection. Among his plants are a number of specimens collected by H. A. Bolander, but these, too, often lack sufficient data. Of Rattan's local collections, the most valuable are probably those from the serpentine hills south of San Jose. These hills are the type locality of *Gilia ambigua* Rattan. Rattan's name is commemorated in three

36   INTRODUCTION

species that occur locally, *Phacelia rattanii, Penstemon rattanii* ssp. *kleei*, and *Mimulus rattanii*. Rattan published several editions of a *Popular California Flora* between 1879 and 1894, which was used as a class textbook.

Among the other important collectors are Nils Johan Andersson, Elmer I. Applegate, William Sackston Atkinson, Rimo Charles Bacigalupi, Charles Fuller Baker, Margaret Alice Barry, Hans Herman Behr, John Milton Bigelow, Frederic Theodore Bioletti, Hiram G. Bloomer, Henry N. Bolander, Mary Katharine Brandegee, Townshend Stith Brandegee, William Henry Brewer, Thomas Bridges, Stewart Henry Burnham, Joseph Burtt-Davy, Evelina Cannon, Joseph Whipple Congdon, Horace Davis, Harry Arnold Dutton, Alice Eastwood, Roxana Judkins Stinchfield Ferris, Edward Lee Greene, Stella Duffield Halsey, Ida R. Hayward, Vesta F. Hesse, John Thomas Howell, Willis Linn Jepson, Beryl Schreiber Jespersen, Marcus Eugene Jones, Christian Ferdinand Leithold, Herbert Louis Mason, James Ira Wilson McMurphy, George Thomas Oberlander, Samuel Bonsall Parish, Charles Christopher Parry, R. L. Pendleton, Cyrus Guernsey Pringle, Josephine Louise Dows Randall, Peter Raven, C. A. Reed, Willis Horton Rich, Lewis Samuel Rose, Everett Winder Rust, Charles Piper Smith, Wilhelm Nikolaus Suksdorf, and Ira Loren Wiggins.

Keys to the Divisions, Classes, Subclasses, and Families

Plants herbaceous, reproducing by spores; sporangia borne adaxially (in *Selaginella* borne on very small sporophylls arranged in a 4-ranked strobilus; in *Isoetes* borne at the base of long, linear-subulate sporophylls); plants small, the stems not longitudinally grooved; plants not producing flowers or seeds. DIVISION LEPIDOPHYTA

Plants herbaceous, reproducing by spores; sporangia borne on sporangiophores, the sporangiophores aggregated into a terminal strobilus; plants small or to 1.5 m. tall, the stems longitudinally grooved and often with verticillate branches; plants not producing flowers or seeds. DIVISION CALAMOPHYTA

Plants herbaceous or woody, reproducing by spores or seeds (if by spores, the spores in sporangia on the abaxial surfaces of frond-like sporophylls or in sporocarps in *Azolla*, *Marsilea*, and *Pilularia*); plants small or very large, leaves or fronds usually present. DIVISION PTEROPHYTA

DIVISION LEPIDOPHYTA

Stems elongate, densely clothed with leaves; leaves ovate, under 1 cm. long.
1. Selaginellaceae, p. 53
Stems very short, corm-like; leaves appearing basal, linear, over 5 cm. long.
2. Isoetaceae, p. 53

DIVISION CALAMOPHYTA

One family. **1. Equisetaceae, p. 54**

DIVISION PTEROPHYTA

Plants herbaceous (the rhizomes sometimes woody), sometimes free-floating or submerged aquatics; not reproducing by seeds; without flowers and cones; fronds or sporophylls with abaxial sori or the sori borne in sporocarps. CLASS FILICINAE

Trees, shrubs, or herbs (often aquatic); reproducing by seeds; flowers or cones present.
Plants woody, trees or shrubs; flowers none; seeds borne on the adaxial surface of naked scales, the scales forming a cone (or in *Taxus* a fleshy, berry-like fruit); leaves linear, needle-like or decussate and scale-like. CLASS GYMNOSPERMAE

Plants woody or herbaceous; flowers present; seeds enclosed in an ovary; leaves various, rarely needle-like. CLASS ANGIOSPERMAE

CLASS FILICINAE

Plants small, free-floating, aquatic; leaves small, imbricate.

1. Salviniaceae, p. 55

Plants not small, free-floating and aquatic, if aquatic then rooting on the bottom.

Leaves filiform or quadrifoliolate; sori borne in sporocarps; plants of shallow water or muddy banks. **2. Marsiliaceae, p. 55**

Leaves neither filiform or quadrifoliolate; sori not borne in sporocarps; plants of moist or dry situations.

Fertile and sterile leaves seemingly dimorphic; plants fleshy; leaves not circinate in vernation; sporangia lacking an annulus.

3. Ophioglossaceae, p. 56

Fertile and sterile leaves not dimorphic (except in *Blechnum*); plants not fleshy; leaves circinate in vernation; sporangia with an annulus.

4. Polypodiaceae, p. 56

CLASS GYMNOSPERMAE

Leaves decussate, scale-like; cones globose, the scales more or less peltate.

3. Cupressaceae, p. 63

Leaves alternate, needle-like; cones ovoid or elongate, not globose, the scales usually distinctly flattened.

Leaves in bundles; cones usually over 7 cm. long, often asymmetrical at least at the base. **2. Pinaceae, p. 62**

Leaves solitary; cones usually under 7 cm. long, symmetrical.

Leaves not decurrent on the stems; cones 4–7 cm. long, the bracts subtending the ovuliferous scales conspicuous and exceeding the scales.

2. Pinaceae, p. 62

Leaves decurrent on the stems; cones under 3.5 cm. long; bracts not conspicuous.

Leaves falling separately; cone modified into a fleshy, 1-seeded, drupe-like "fruit." **1. Taxaceae, p. 62**

Small twigs deciduous with several leaves attached; cone many-seeded, not fleshy. **4. Taxodiaceae, p. 64**

CLASS ANGIOSPERMAE

Flowers usually 3-merous; leaves usually with parallel veins; plants herbaceous; cotyledon 1. Subclass Monocotyledoneae

Flowers usually 4- or 5-merous; leaves usually net-veined; plants herbaceous or woody; cotyledons usually 2. Subclass Dicotyledoneae

Subclass Monocotyledoneae

Trees. **13. Palmae, p. 111**

Herbs.

Plants aquatic.
Marine plants.   **5. Zosteraceae, p. 67**
Fresh- or brackish-water plants.
Plants minute, leafless, thallose, under 5 mm. long.
                                                            15. Lemnaceae, p. 112
Plants larger, leaves present, not thallose, over 5 mm. long.
Perianth present, conspicuously petaloid.
Petioles conspicuously inflated.   **17. Pontederiaceae, p. 113**
Petioles not inflated.
Perianth-segment 1; flowers in spikes; leaves basal, over 6 cm.
long.   **6. Aponogetonaceae, p. 67**
Perianth-segments 3 or 6; flowers solitary in the axils of the leaves;
leaves cauline, under 1.5 cm. long.
                                            10. Hydrocharitaceae, p. 70
Perianth absent or, if present, not conspicuous.
Flowers solitary in the leaf-axils; leaves minutely toothed.
                                                     4. Naiadaceae, p. 67
Flowers not solitary, rather in spikes or clusters; leaves not toothed.
Leaves fleshy, terete, linear-subulate.   **7. Lilaeaceae, p. 68**
Leaves not fleshy, not terete, usually with a blade (if linear, then
filiform).   **3. Potamogetonaceae, p. 65**
Plants terrestrial, often growing in water, but the major portion of the plant
aerial.
Perianth present, petaloid or fleshy in part, never scale-like or bract-like.
Perianth-segments small, fleshy, green, 2–3 mm. long. Plants of salt
marshes.   **8. Juncaginaceae, p. 68**
Perianth-segments usually larger, usually colored or white.
Gynoecium of many distinct carpels; fruit of achenes.
                                                  9. Alismataceae, p. 69
Gynoecium of united carpels; fruit not of achenes.
Ovary superior.
Inflorescence not umbellate.
Fruit a capsule; plants with bulbs or rootstocks.
Styles 3, distinct.   **19. Melanthaceae, p. 116**
Style 1, entire to 3-lobed or -parted.   **20. Liliaceae, p. 117**
Fruit usually a berry; plants with rootstocks.
                                           21. Convallariaceae, p. 120
Inflorescence umbellate.
Erect, usually scapose plants; perianth-segments all essentially
alike, not ephemeral or deliquescent.
                                           22. Amaryllidaceae, p. 123
Prostrate, creeping plants; inner perianth-segments petaloid,
ephemeral, deliquescent, outer perianth-segments green.
                                           16. Commelinaceae, p. 113
Ovary inferior.
Corollas actinomorphic; stamens 3.   **23. Iridaceae, p. 127**
Corollas zygomorphic; stamen usually 1.
                                           24. Orchidaceae, p. 128

Perianth present or absent, not petaloid, not fleshy, often scale-like or bract-like.
Plants with a large spathe subtending the spadix.   **14. Araceae, p. 111**
Plants without a spathe.
Inflorescence of two types, one of single, basal, sessile, pistillate flowers, the other scapose, of numerous staminate, pistillate, or perfect flowers; leaves terete, fleshy.   **7. Lilaeaceae, p. 68**
Inflorescence of one type; leaves various.
Flowers without conspicuous bractlets; inflorescence large, compact, the flowers unisexual.
Pistillate inflorescences elongate, cylindric.
1. Typhaceae, p. 64
Pistillate inflorescences of globose heads.
2. Sparganiaceae, p. 65
Flowers with conspicuous, often scarious bractlets; inflorescences large or small, usually not conspicuously compact, the flowers perfect or unisexual.
Flowers not in the axils of dry chaffy bracts; perianth of 6 parts; fruit a capsule.   **18. Juncaceae, p. 113**
Flowers in the axils of dry chaffy bracts; perianth none; fruit an achene or caryopsis.
Culms commonly round in cross section, hollow; sheaths open; flowers subtended by 2 bracts.   **11. Gramineae, p. 70**
Culms commonly triangular in cross section, solid; sheaths closed; flowers subtended by 1 bract.
12. Cyperaceae, p. 103

Subclass Dicotyledoneae

Calyx and corolla fused to form a woody, deciduous calyptra; stamens numerous; trees; leaves aromatic; fruit a woody or leathery capsule; ovary inferior.   **94. Myrtaceae, p. 244**
Plants not as above in all respects.

A. COROLLA ABSENT; CALYX PRESENT OR ABSENT

Either the staminate or the pistillate flowers or both in catkins; plants dioecious or monoecious; trees or shrubs.
Both staminate and pistillate flowers in catkins; fruit not a nut, less than 1 cm. long.
Leaves opposite. Chaparral plants.   **100. Garryaceae, p. 262**
Leaves alternate. Not chaparral plants.
Pistillate catkins herbaceous; leaves linear, lanceolate, elliptic, or deltoid, rarely ovate, rarely sharply serrate.
Plants dioecious; leaves deciduous; seeds comose; fruit not covered with a white waxy coating.   **26. Salicaceae, p. 132**

Plants monoecious; leaves evergreen; seeds not comose; fruit covered with a white waxy coating. **27. Myricaceae, p. 133**

Pistillate catkins woody; leaves ovate, sharply serrate. **28. Betulaceae, p. 134**

Only the staminate flowers in catkins; pistillate flowers sometimes clustered; fruit a nut, 1–4 cm. long.

Fruit surrounded by a scaly cup or a spiny bur; trees, or if shrubs, the mature leaves coriaceous. **30. Fagaceae, p. 135**

Fruit surrounded by an herbaceous, tubular involucre; shrubs, the leaves never coriaceous. **29. Corylaceae, p. 135**

Flowers not in catkins; plants dioecious or monoecious or the flowers perfect; herbs, occasionally trees or shrubs.

B. OVARY SUPERIOR

Flowers hypogynous.

Calyx lacking; pistil 1.

Plants with perfect flowers; inflorescences subtended by a petaloid involucre. **25. Saururaceae, p. 132**

Plants monoecious; involucre present or absent, if present not petaloid or only partly so.

Plants aquatic.

Leaves opposite, entire; ovary 4-celled, winged, smooth. **76. Callitrichaceae, p. 229**

Leaves whorled, finely dissected; fruit 1-celled, sometimes winged, usually spiny or tuberculate. **45. Ceratophyllaceae, p. 166**

Plants terrestrial.

Trees; juice not milky; flowers in capitate clusters arranged along a naked axis. **60. Platanaceae, p. 193**

Herbs; juice milky; flowers not in capitate clusters. **75. Euphorbiaceae, p. 227**

Calyx present; pistils 1 to many.

Pistils several to many, distinct. **46. Ranunculaceae, p. 167**

Pistil 1.

Ovary 1-celled.

Fruit a capsule, dehiscent.

Leaves alternate, stellate-pubescent. **75. Euphorbiaceae, p. 227**

Leaves opposite, not stellate-pubescent.

Calyx synsepalous.

Stipules lacking; plants erect. **105. Primulaceae, p. 268**

Stipules present; plants prostrate. **41. Aizoaceae, p. 156**

Calyx of distinct sepals. **43. Caryophyllaceae, p. 160**

Fruit various, indehiscent or occasionally circumscissile.

Fruit of achenes, utricles, or samaras, dry.

Trees. **32. Ulmaceae, p. 139**

Herbs, subshrubs, or vines.

Herbage often with stinging hairs; calyx 4-parted; stamens 4; stigmas often tufted.   **31. Urticaceae, p. 138**

Herbage never with stinging hairs; calyx variously lobed or rarely absent in pistillate flowers; stamens 1–9; stigmas not tufted.

Plants usually scurfy; flowers perfect or unisexual, when unisexual the pistillate lacking a calyx, but enclosed within 2 bracts.   **36. Chenopodiaceae, p. 148**

Plants not scurfy; flowers various.

Calyx not tubular, usually not corolla-like; leaves alternate, opposite, or basal; stipules present or not.

Fruit an achene, these trigonous or lenticular; calyx herbaceous or often colored; stipules present or lacking.   **35. Polygonaceae, p. 140**

Fruit an utricle, indehiscent, irregularly dehiscent, or circumscissile; calyx never colored, scarious; stipules lacking.   **37. Amaranthaceae, p. 154**

Calyx tubular, simulating the corolla; leaves opposite; stipules lacking.   **38. Nyctaginaceae, p. 155**

Fruit a berry or a drupe, fleshy.

Leaves compound; perennial herbs; fruit a berry.   **46. Ranunculaceae, p. 167**

Leaves simple; trees; fruit a drupe.   **48. Lauraceae, p. 172**

Ovary 2-, 5-, or many-celled.

Shrubs or trees.

Fruit a capsule; calyx showy; herbage stellate-pubescent; leaves simple.   **84. Sterculiaceae, p. 238**

Fruit a samara; calyx not showy; herbage not stellate-pubescent; leaves simple or compound.

Styles 2; fruit a double samara.   **80. Aceraceae, p. 232**

Style 1; fruit a single samara.   **107. Oleaceae, p. 271**

Herbs or occasionally somewhat woody perennial herbs.

Plants usually with milky juice; flowers monoecious.   **75. Euphorbiaceae, p. 227**

Plants without milky juice; flowers perfect.

Leaves opposite or whorled; low prostrate herbs.   **41. Aizoaceae, p. 156**

Leaves alternate; erect perennial herbs, 1–3 m. tall.   **39. Phytolaccaceae, p. 155**

Flowers perigynous.

Flowers in umbels.   **82. Rhamnaceae, p. 233**

Flowers solitary or clustered, but not in umbels.

Stipules present.

Fruit an achene; erect herbs or shrubs.   **61. Rosaceae, p. 194**

Fruit an utricle; prostrate herbs.   **40. Illecebraceae, p. 155**

Stipules absent.
Leaves alternate; shrubs; stamens 8.   **92. Thymelaeaceae, p. 243**
Leaves opposite; prostrate herbs; stamens 5.
                                            38. Nyctaginaceae, p. 155

BB. OVARY INFERIOR

Plants parasitic on various trees, mainly members of the Pinaceae and Fagaceae;
leaves or scales opposite.   **33. Loranthaceae, p. 139**
Plants not parasitic; leaves opposite or alternate.
Leaves alternate; plants terrestrial.
Leaves divided; ovary 1-celled.   **91. Datiscaceae, p. 243**
Leaves entire; ovary with more than 1 cell.
Leaves succulent, deltoid-ovate.   **41. Aizoaceae, p. 156**
Leaves not succulent, cordate at the base. **34. Aristolochiaceae, p. 140**
Leaves opposite or whorled; plants aquatic or of very wet habitats.
Leaves opposite, entire; fruit a 4-celled capsule.   **95. Onagraceae, p. 245**
Leaves whorled, entire or dissected; fruit of 1–4 nutlets.
                                            96. Haloragidaceae, p. 251

AA. COROLLA PRESENT; CALYX PRESENT

C. PETALS DISTINCT

D. OVARY SUPERIOR

Stamens hypogynous.
Stamens more than 10.
Pistils 2 or more (sometimes 1 in *Ranunculus ajacis*).
                                            46. Ranunculaceae, p. 167
Pistil 1.
Leaves opposite, simple.
Stamens united into groups; fruit a capsule, not separating into sepa-
rate carpels.   **85. Hypericaceae, p. 238**
Stamens all distinct; fruit separating into separate carpels.
                                            49. Papaveraceae, p. 173
Leaves alternate or basal, simple or compound.
Ovary 1-celled.
Sepals caducous.   **49. Papaveraceae, p. 173**
Sepals persistent.
Petals 13–15; leaves basal, succulent.
                                            42. Portulacaceae, p. 157
Petals 1, 2, or 5; leaves cauline, succulent or not.

Leaves compound.

Fruit a legume; shrubs or trees.   **64. Mimosaceae, p. 202**

Fruit a berry; perennial herbs.   **46. Ranunculaceae, p. 167**

Leaves simple.

Sepals 2; plants herbaceous.   **42 Portulacaceae, p. 157**

Sepals 5; plants woody.   **88. Cistaceae, p. 240**

Ovary with 5 or more cells.

Plants terrestrial; stamens united into a tube around the gynoecium; petals 5.   **83. Malvaceae, p. 236**

Plants aquatic; stamens distinct; petals 10–20.

                                           44. Nymphaeaceae, p. 166

Stamens 10 or less.

Pistils more than 1.

Pistils several times as many as the sepals; leaves usually dissected, not fleshy.   **46. Ranunculaceae, p. 167**

Pistils 4 or 5; leaves entire, fleshy.   **53. Crassulaceae, p. 188**

Pistil 1.

Flowers zygomorphic.

Flowers papilionaceous.   **66. Fabaceae, p. 203**

Flowers not papilionaceous.

Shrubs.   **65. Caesalpiniaceae, p. 202**

Herbs.

Leaves peltate; one sepal produced into a long spur; stamens 8.

                               69. Tropaeolaceae, p. 224

Leaves not peltate; sepals not produced into a long spur; stamens of various numbers.

One petal saccate or spurred; stamens regularly 5.

                               89. Violaceae, p. 240

Petals not spurred; stamens 6–40.

Leaves compound; stamens 6, united into 2 sets.

                               50. Fumariaceae, p. 174

Leaves simple; stamens 8–40.   **52. Resedaceae, p. 187**

Flowers actinomorphic.

Flowers cruciferous.   **51. Cruciferae, p. 176**

Flowers not cruciferous.

Leaves palmately compound or trifoliolate.

Sepals 5; leaves with an acid taste.   **68. Oxalidaceae, p. 223**

Sepals 6; leaves lacking an acid taste.

                               47. Berberidaceae, p. 171

Leaves not palmately compound.

Ovary 1-celled.

Leaves pinnately compound, prickly.

                               47. Berberidaceae, p. 171

Leaves simple, not prickly.

Calyx of 2 or 3 sepals.

Sepals 2; plants fleshy.   **42. Portulacaceae, p. 157**

Sepals usually 3; plants not fleshy.

                               49. Papaveraceae, p. 173

Calyx of 4 or more sepals, distinct or not.
 Leaves opposite.
  Leaves not succulent; plants of various habitats.
   43. Caryophyllaceae, p. 160
  Leaves succulent; plants of saline soils along the coast.
   87. Frankeniaceae, p. 240
 Leaves alternate or basal.
  Plants not saprophytic; leaves green, not reduced.
   Calyx synsepalous.   **106. Plumbaginaceae, p. 270**
   Calyx of separate sepals.   **55. Parnassiaceae, p. 192**
  Plants saprophytic; leaves not green, much reduced.
   102. Monotropaceae, p. 263
Ovary with 2 or more cells.
 Stipules present.
  Leaves simple, entire; plants aquatic or growing in very wet
   areas; stamens 2–5.   **86. Elatinaceae, p. 239**
  Leaves simple or compound, if simple, the margins not entire;
   plants terrestrial; stamens 10 (occasionally some ster-
   ile).
   Leaves simple or variously dissected, rarely compound;
    petals white to pink, or lavender; fruit not spiny;
    styles 5.   **67. Geraniaceae, p. 221**
   Leaves compound; petals yellow; fruit spiny; style 1.
    71. Zygophyllaceae, p. 226
 Stipules lacking.
  Leaves compound; fruit a samara; stamens 2.
   107. Oleaceae, p. 271
  Leaves simple; fruit not a samara; stamens 5 or more.
   Styles 2–5; flowers 5-merous; petals readily deciduous.
    70. Linaceae, p. 224
   Style 1; flowers 4- or 5-merous.
    Plants woody; leaves almost toothed; fruit a capsule.
     Leaves not reduced.
      Petals white.   **103. Ericaceae, p. 264**
      Petals dark red to purple.
       59. Pittosporaceae, p. 193
     Leaves reduced.   **101. Pyrolaceae, p. 263**
    Plants herbaceous; leaves dissected; fruit separating into
     carpels.   **77. Limnanthaceae, p. 230**
Stamens perigynous.
 Stamens borne on a disk lining the calyx at the base.
  Perennial herbs; leaves bipinnate or tripinnate.   **72 Rutaceae, p. 226**
  Trees or shrubs; leaves various.
   Leaves opposite.
    Fruit a samara.   **80. Aceraceae, p. 232**
    Fruit not a samara.
     Leaves palmately compound.   **81. Hippocastanaceae, p. 232**
     Leaves simple.

Petals hooded; stamens opposite the petals; seeds without an aril.
                82. Rhamnaceae, p. 233
Petals not hooded; stamens alternate with the petals; seeds with
     an aril.   **79. Celastraceae, p. 231**
Leaves alternate.
  Leaves simple.   **82. Rhamnaceae, p. 233**
  Leaves compound.
    Fruit a samara.
      Leaves 3–5 foliolate.   **72. Rutaceae, p. 226**
      Leaves 11–25-foliolate.   **73. Simaroubaceae, p. 226**
    Fruit dry, berry-like.   **78. Anacardiaceae, p. 230**
Stamens borne on the calyx or hypanthium or epipetalous.
  Calyx a caducous calyptra.   **49. Papaveraceae, p. 173**
  Calyx lobed or divided.
    Stipules lacking; leaves simple; pistil 1.
      Style and stigma 1.   **93. Lythraceae, p. 243**
      Styles and stigmas more than 1. **54. Saxifragaceae, p. 189**
    Stipules usually present; leaves simple or compound; pistils 1 to several.
      Fruit not a drupe.   **61. Rosaceae, p. 194**
      Fruit a drupe.   **62. Amygdalaceae, p. 200**

DD. OVARY INFERIOR OR PARTLY SO

Trees and shrubs.
  Stamens 8 to many.
    Stamens 8–20.
      Leaves alternate; fruit a pome. **63. Malaceae, p. 201**
      Leaves opposite; fruit a capsule. **56. Hydrangeaceae, p. 192**
    Stamens more than 20.   **94. Myrtaceae, p. 244**
  Stamens 4 or 5.
    Petals hooded; stamens opposite the petals. **82. Rhamnaceae, p. 233**
    Petals not hooded; stamens alternate with the petals.
      Leaves opposite; style 1.   **99. Cornaceae, p. 262**
      Leaves alternate; styles 1 or 2, sometimes partly united.
        Fruit a berry. Native plants. **58. Grossulariaceae, p. 192**
        Fruit a capsule. Introduced plants. **57. Escalloniaceae, p. 192**
Herbs.
  Petals numerous; plants very fleshy. **41. Aizoaceae, p. 156**
  Petals not more than 5; plants not fleshy (except in the Portulacaceae).
    Inflorescences umbellate.
      Fruit not a berry; breaking into 2, 1-seeded carpels; styles 2.
                98. Umbelliferae, p. 252
      Fruit a berry, black at maturity; styles 5. **97. Araliaceae, p. 252**
    Inflorescences not umbellate.
      Styles 2–5.   **54. Saxifragaceae, p. 189**
      Style 1 or none.
        Style none; plants aquatic; leaves whorled. **96. Haloragidaceae, p. 251**
        Style 1; plants rarely aquatic; leaves not whorled.

Sepals 2; plants fleshy. **42. Portulacaceae, p. 157**
Sepals 4 or 5; plants not fleshy.
Herbage rough-hairy (sticking readily to cloth); flowers 5-merous; fruit 1-celled. **90. Loasaceae, p. 242**
Herbage not rough-hairy; flowers usually 4-merous; fruit usually 4-celled. **95 Onagraceae, p. 245**

CC. PETALS UNITED AT LEAST AT THE BASE

Stamens more than 5 (5 in one species of *Rhododendron*).
Petals united at the base, not tubular, urceolate, or campanulate.
Pistils 4 or 5; herbage succulent. **53. Crassulaceae, p. 188**
Pistil 1; herbage not usually succulent.
Corollas zygomorphic.
Flowers papilionaceous; petals 3 or 5.
Leaves compound; fruit a legume; petals 5. **66. Fabaceae, p. 203**
Leaves simple; fruit a capsule; petals 3. **74. Polygalaceae, p. 226**
Flowers not papilionaceous; petals 4, in 2 dissimilar sets; stamens in 2 groups of 3.
50. Fumariaceae, p. 174

Corollas actinomorphic.
Leaves trifoliolate; stamens 10. **68. Oxalidaceae, p. 223**
Leaves simple; stamens numerous. **83. Malvaceae, p. 236**
Corollas tubular, urceolate, or campanulate.
Ovary superior. **103. Ericaceae, p. 264**
Ovary inferior. *stamens arise on petals* **104. Vacciniaceae, p. 268**
Stamens 5 or less, epipetalous; sometimes 6 or 7 in *Trientalis*.
Ovary superior.
Corollas actinomorphic.
Pistils 2, the stigmas united; plants with milky juice.
Stamens united with the stigmas; the column with hood-like appendages. **112. Asclepiadaceae, p. 274**
Stamens not united with the stigmas; hoods lacking.
111. Apocynaceae, p. 272
Pistil 1; plants without milky juice.
Stamens opposite the corolla lobes, equaling them in number.
Style 1; fruit a several-seeded capsule. **105. Primulaceae, p. 268**
Styles 5; fruit an utricle. **106. Plumbaginaceae, p. 270**
Stamens alternate with the corolla lobes, equaling them in number or fewer.
Corollas dry-scarious; inflorescences scapose, spike-like.
123. Plantaginaceae, p. 321
Corollas colored; inflorescences various.

48   *KEYS*

Ovary 4-celled, usually 4-lobed, splitting into 4 nutlets; inflores-
cence scorpioid.   **116. Boraginaceae, p. 287**
Ovary 1-, 2-, or 3-celled, usually not lobed nor splitting into 4 nut-
lets; inflorescence scorpioid or not.
Ovary 3-celled; style 3-cleft.   **114. Polemoniaceae, p. 276**
Ovary 1–2-celled; style not 3-cleft.
Sepals distinct or united only at the base.
Plants not trailing or twining; plants not parasitic; flowers
often in scorpioid inflorescences.
                                              115. Hydrophyllaceae, p. 283
Plants trailing or twining; plants sometimes parasitic;
flowers not in scorpioid inflorescences.
                                              113. Convolvulaceae, p. 274
Calyx 4–5-toothed.
Leaves simple; ovary 1–2-celled.
Leaves alternate; ovary 2-celled.
                                              119. Solanaceae, p. 301
Leaves opposite; ovary 1–2-celled.
Shrubs; ovary 2-celled.   **108. Loganiaceae, p. 271**
Herbs; ovary 1-celled.   **109. Gentianaceae, p. 272**
Leaves 3-foliolate; ovary 1-celled.
                                              110. Menyanthaceae, p. 272
Corollas zygomorphic, usually 2-lipped.
Fertile stamens 5.   **120. Scrophulariaceae, p. 306**
Fertile stamens 2 or 4.
Fruit of 2 or 4 nutlets; leaves opposite.   *new name*
Ovary 4-lobed; plants with a mint-like odor; stamens 4. *LAMIACEA*
                                              118. Labiatae, p. 295
Ovary not lobed; plants lacking a mint odor; stamens 2 or 4.
                                              117. Verbenaceae, p. 294
Fruit a capsule; leaves opposite or alternate.
Ovary 2-celled.   **120. Scrophulariaceae, p. 306**
Ovary 1-celled.
Plants aquatic, not parasitic; stamens 2; leaves green, finely dis-
sected.   **121. Lentibulariaceae, p. 320**
Plants terrestrial, parasitic; stamens 4; leaves not green, entire.
                                              122. Orobanchaceae, p. 320
Ovary inferior.
Stamens distinct.
Leaves alternate; flowers actinomorphic.   **129. Campanulaceae, p. 330**
Leaves opposite or whorled; flowers actinomorphic or zygomorphic.
Stamens 1–3; flowers zygomorphic; fruit 1-celled; often slender herbs.
                                              127. Valerianaceae, p. 328
Stamens 4–5; flowers zygomorphic or actinomorphic; fruit 1–5-celled;
herbs or shrubs.

Ovary 1-celled; flowers in dense involucrate heads.
126. **Dipsacaceae, p. 327**
Ovary 2–5-celled; flowers not in dense heads, not involucrate.
Flowers actinomorphic; leaves opposite or whorled; ovary 2-celled; plants herbaceous or woody; stems often with retrorse hairs.
124. **Rubiaceae, p. 323**
Flowers actinomorphic or zygomorphic; leaves opposite; ovary 2–5-celled; plants woody, shrubs, trees, or vines; stems without retrorse hairs.
125. **Caprifoliaceae, p. 325**
Stamens united into a tube around the style.
Flowers not in heads.
Stamens 3; plants with tendrils, vine-like; leaves palmately veined.
128. **Cucurbitaceae, p. 329**
Stamens 5; plants without tendrils, not vine-like; leaves pinnately veined.
130. **Lobeliaceae, p. 331**
Flowers in involucrate heads.
131. **Compositae, p. 332**

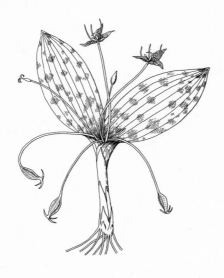

PART II

Annotated Catalogue
of Vascular Plants

Catalogue of Vascular Plants

) DIVISION LEPIDOPHYTA (

1. SELAGINELLACEAE. Selaginella Family

1. Selaginella Beauv.

1. S. bigelovii Underw. *Fig. 1.* Bigelow's Club Moss. Occasional in the vicinity of Mount Umunhum and Los Gatos, growing on serpentine and other rock outcroppings and often quite common under *Pinus sabiniana*.

2. ISOETACEAE. Quillwort Family

1. Isoetes L.

Both species of *Isoetes* which are listed below have been poorly collected, probably due to their habitats and the lack of conspicuous features.

Plants terrestrial; velum complete; corm 3-lobed.   1. *I. nuttallii*
Plants amphibious; velum incomplete; corm 2-lobed.   2. *I. howellii*

1. I. nuttallii A. Br. ex Engelm. Nuttall's Quillwort. Moist shaded areas along or near streams and in moist meadows; San Francisco, Blooms Grade, and near Boulder Creek.
2. I. howellii Engelm. *Fig. 2.* Howell's Quillwort. Occasional in ponds and ditches; Woodside Valley, Menlo Park, and Coal Mine Ridge.

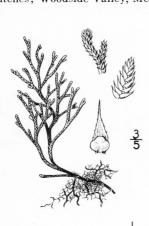

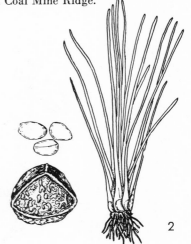

$\frac{3}{5}$

1

2

✱) DIVISION CALAMOPHYTA (✱

1. EQUISETACEAE. Horsetail Family

1. Equisetum L. Scouring Rush, Horsetail

Strobili blunt or slightly acute; stems annual.
Stems dimorphic, the fertile ones brown or pinkish, lacking chlorophyll, usually unbranched; stomata not in regular rows.
Sterile stems very slender, usually 4 mm. or less in diameter, often with the
lateral branches asymmetrically arranged; strobili 0.5–3.5 cm. long,
rarely found locally.   1. *E. arvense*
Sterile stems stout, usually over 5 mm. in diameter, usually with the lateral
branches symmetrically arranged; strobili 4–9.5 cm. long, usually
present.   2. *E. telmateia braunii*
Stems not dimorphic, the fertile ones green; stomata in regular rows.
3. *E. laevigatum*
Strobili distinctly apiculate.
Sheaths longer than broad, the lower ones sometimes with a dark band.
4. *E.* × *ferrissii*
Sheaths about as long as broad, all with two dark bands.   5. *E. hyemale affine*

1. E. arvense L. Common Horsetail. Occasional in moist, often sandy soil along
streams; San Francisco, Rancho del Oso, Boulder Creek, La Selva Beach, and
Pajaro River.
2. E. telmateia Ehrh. var. **braunii** (Milde) Milde. Giant Horsetail. Probably
the most common Horsetail locally, growing along streams in moist soil, often
with *Woodwardia fimbriata* and *Asarum caudatum* in redwood–Douglas fir forests; San Francisco, northern San Mateo County, San Francisco Watershed
Reserve, near La Honda, Coal Mine Ridge, Glenwood, Boulder Creek, near Santa
Cruz, and Smith Grade.

$\frac{3}{5}$

3

3. E. laevigatum A. Br. *Fig. 3.* California Horsetail. Usually in sandy river bottoms; near Gilroy, Felton, near Soquel, Pajaro River, and Watsonville. —*E. funstoni* Eaton, *E. kansanum* Schaffn.
4. E. × ferrissii Clute. Scouring Rush. Occasional in the Santa Cruz Mountains, usually in moist situations along streams; San Francisco, Los Gatos, near Ben Lomond, and Santa Cruz. —*E. laevigatum* of authors.
5. E. hyemale L. var. **affine** (Engelm.) Eaton. Western Scouring Rush. Occasional along creeks and streams; San Francisco, Pescadero Creek, Black Mountain, Los Gatos, Waddell Creek, Boulder Creek, and near Felton. —*E. hyemale* L. var. *californicum* Milde, *E. hyemale* L. var. *elatum* (Engelm.) Morton, *E. prealtum* Raf., *E. hyemale* L. var. *robustum* (A. Br.) Eaton.

✳) DIVISION PTEROPHYTA (✳

CLASS FILICINAE

1. SALVINIACEAE. Water Fern or Salvinia Family

1. Azolla Lam.

1. A. filiculoides Lam. American Water Fern, Fern-like Azolla. A floating aquatic, fairly common in small ponds, stagnant pools along streams, and in temporary puddles; San Francisco, Millbrae, Searsville Lake, Coal Mine Ridge, Davenport Landing, Camp Evers, and Aptos.

2. MARSILEACEAE. Pepperwort or Marsilea Family

Leaves with a 4-foliolate blade.   1. *Marsilea*
Leaves filiform.   2. *Pilularia*

1. Marsilea L.

1. M. vestita Hook. & Grev. Clover Fern, Hairy Pepperwort. Known locally from Lake Lagunita at Stanford University and reported from San Francisco and the San Francisco Watershed Reserve. Sporocarps of *M. vestita* were introduced into Lake Lagunita by Professor Douglas H. Campbell. The first specimen from Lake Lagunita of which we have record was collected in 1929 and *Marsilea* has thus maintained itself for at least 30 years. It grows in shallow water 4–6 inches deep along the shore of the lake. The lake is drained yearly during the latter part of May when it begins to become stagnant. A fluctuating water level is necessary for the ripening of the sporocarps.

56   *POLYPODIACEAE*

2. Pilularia L.

1. P. americana A. Br. *Fig. 4.* American Pillwort. Known only from a seasonal pond near Morgan Hill, but to be expected elsewhere in the Santa Clara Valley.

3. OPHIOGLOSSACEAE. Adder's Tongue Family

1. Botrychium Swartz. Grape Fern

1. B. silaifolium Presl var. **californicum** (Underw.) Jeps. California Grape Fern. Known locally from the bogs near Camp Evers in Santa Cruz County and reported from San Francisco. —*B. multifidum* (Gmelin) Rupr. var. *intermedium* (Eaton) Farw., *B. multifidum* (Gmelin) Rupr. ssp. *silaifolium* (Presl) Clausen.

4. POLYPODIACEAE. Fern Family

Sori not marginal, but rather on the abaxial surface of the frond.
  Indusium lacking.
    Fronds not golden-yellow on the abaxial surface; fronds once-pinnate.
                                                              1. *Polypodium*
    Fronds golden-yellow on the abaxial surface; at least the basal pinnae 2-
      pinnate.   2. *Pityrogramma*
  Indusium present.
    Sori linear, in two parallel rows, one on either side of the midvein; fronds
      1–3 m. tall.   3. *Woodwardia*
    Sori not linear, not in definite rows; fronds usually under 1.5 m. tall.
      Indusium peltate, orbicular.   4. *Polystichum*
      Indusium not peltate.
        Stipe scaly below the blade, occasionally above; fronds annual; indu-
          sium inconspicuous when sporangia are mature.
            Fronds commonly under 2.5 dm. long, 1–2-pinnate; indusium at-
              tached at the base.   5. *Cystopteris*
            Fronds commonly over 5 dm. long, 2–3-pinnate; indusium attached
              along one side.   6. *Athyrium*
        Stipe scaly both above and below the blade; fronds usually perennial;
          indusium conspicuous.   7. *Dryopteris*
Sori marginal or submarginal.
  Fronds dimorphic, segments of fertile fronds narrowly linear.   8. *Blechnum*
  Fronds not dimorphic.
    Pinnae very thin, fragile, finely veined; sporangia partly hidden by the re-
      flexed lobes.   9. *Adiantum*
    Pinnae thick, often dry and brittle.
      Large coarse ferns, often to 2 m. tall; fronds arising singly.
                                                              10. *Pteridium*
      Small ferns, usually under 2.5 dm. tall; fronds clustered.
        Abaxial surface of blade densely hairy and scaly, glandular-pubescent,
          or glabrous; indusium present.   11. *Cheilanthes*
        Abaxial surface of blade glabrous; indusium lacking. 12. *Pellaea*

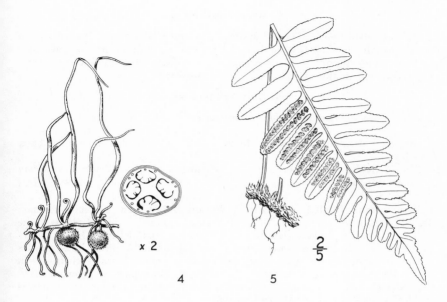

x 2

$\frac{2}{5}$

4   5

1. **Polypodium** L. Polypody

Fronds very coriaceous, segments rounded at the tips.   1. *P. scouleri*
Fronds not coriaceous, segments rounded to acute.
  Segments acute to obtuse or rounded. Common.   2. *P. californicum*
  Segments attenuate, acute. Rare.   3. *P. glycyrrhiza*

1. **P. scouleri** Hook. & Grev. Coast Polypody, Leather Fern. Rock crevices on exposed hills along the coast; San Francisco, San Bruno Hills, and Cahill Ridge.
2. **P. californicum** Kaulf. *Fig. 5*. California Polypody. The most common species of *Polypodium* locally, growing in a variety of habitats from wind-swept grassy summits to rocky slopes and shaded canyon areas; San Francisco, San Bruno Hills, Pedro Point, Pilarcitos Canyon, Butano Creek, Kings Mountain, Stanford, Stevens Creek, near Los Gatos, Waddell Creek, near Davenport, and Soquel Valley. Plants with smaller, thicker blades have been called *P. californicum* Kaulf. var. *kaulfussii* Eaton.
3. **P. glycyrrhiza** Eaton. Licorice Fern. Rare, known locally only from Kings Mountain.

2. **Pityrogramma** Link

1. **P. triangularis** (Kaulf.) Maxon. *Fig. 6*. Goldenback Fern. Shaded slopes in oak-madrone woods, at the edges of redwood forests, brushy slopes, and common along moist road banks; San Francisco, Pilarcitos Canyon, Coal Mine Ridge, Saratoga Canyon, Rancho del Oso, near Bielawski Lookout, Santa Cruz, Soquel Valley, and near Mount Madonna. —*Gymnogramme triangularis* Kaulf.

3. **Woodwardia** Smith

1. W. fimbriata Smith. *Fig. 7.* Western or Giant Chain Fern. Shaded stream banks, usually in the redwood–Douglas fir forests; San Francisco, Pilarcitos Canyon, Kings Mountain, Black Mountain, Los Gatos Canyon, Waterman Gap, Big Basin, and Boulder Creek. —*W. chamissoi* Brack.

4. **Polystichum** Roth

Blades simply pinnate, the pinnae not pinnately lobed or pinnatifid. Common.

1. *P. munitum*

Blades more than simply pinnate, the pinnae lobed or divided to near the base. Rare.

Pinnae divided to near the base, but not pinnate.   2. *P. californicum*
Pinnae pinnate.   3. *P. dudleyi*

1. P. munitum (Kaulf.) Presl. *Fig. 8.* Western Sword Fern. The most common *Polystichum* in the Santa Cruz Mountains, growing in moist shaded places in the redwood forests and along streams; San Francisco, Cahill Ridge, Kings Mountain, Coal Mine Ridge, Los Gatos Canyon, near Los Gatos, Waddell Creek, 6 miles north of Boulder Creek, Alba Road, and Casserly Creek. Plants with fronds in which the pinnae are short, imbricate, and ascending may be referred to *P. munitum* (Kaulf.) Presl var. *imbricans* (Eaton) Maxon. The typical variety is much the more common of the two locally.
2. P. californicum (Eaton) Underw. California Shield Fern. Fairly rare in moist shaded areas along creeks and streams; Pescadero Creek, Los Trancos Creek, near Los Gatos, Jamison Creek, and near Santa Cruz.
3. P. dudleyi Maxon. Dudley's Shield Fern. Rare in the Santa Cruz Mountains, growing in moist shaded areas along streams; Pilarcitos Canyon, San Mateo, Peters Creek, Kings Mountain, Stevens Creek, Los Gatos Canyon, Waddell Creek, Boulder Creek, Santa Cruz, and Aptos Creek. —*P. aculeatum* (Swartz) Roth var. *dudleyi* (Maxon) Jeps.

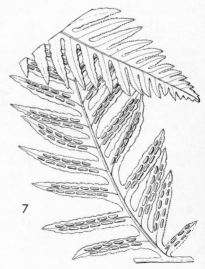

6   7

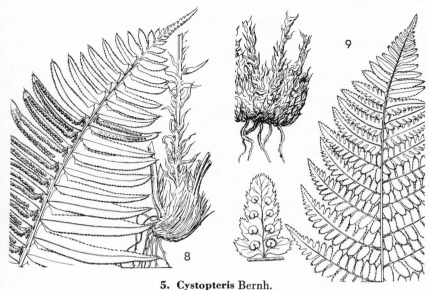

5. **Cystopteris** Bernh.

1. C. fragilis (L.) Bernh. Brittle Fern. Occasional, growing in partial shade on wooded slopes; Llagas Creek, Big Basin, and near Boulder Creek.

6. **Athyrium** Roth

1. A. filix-femina (L.) Roth. Western Lady Fern. Occasional in moist woods, commonly along streams; San Francisco, Pescadero Creek, Waddell Creek, and Big Basin. —*A. filix-femina* (L.) Roth var. *californicum* Butters, *A. filix-femina* (L.) Roth var. *cyclosorum* (Rupr.) Moore.

7. **Dryopteris** Adans.

Fronds 2-pinnate. Common.   1. *D. arguta*
Fronds 3-pinnate. Rare.   2. *D. dilatata*

1. D. arguta (Kaulf.) Watt. *Fig. 9.* Coastal Wood Fern. A common fern of wooded slopes, but rarely in the dense redwood forest; San Francisco, Kings Mountain, Coal Mine Ridge, Stanford, near Los Gatos, Llagas Creek, near Saratoga Summit, Waddell Creek, Ben Lomond Sand Hills, and Soquel Valley.
2. D. dilatata (Hoffm.) Gray. Spreading Wood Fern. Rare, mainly in moist areas in redwood–Douglas fir forests; San Francisco Watershed Reserve and Butano Creek.

8. **Blechnum** L.

1. B. spicant (L.) Smith. Deer Fern. Occasional in very moist places in the redwoods; Butano Creek, Gazos Creek, Waddell Creek, and near Boulder Creek. —*Struthiopteris spicant* (L.) Weis.

9. **Adiantum** L. Maidenhair Fern

Rachis of the frond continuous; ultimate pinnules symmetrical, attached at the center.   1. *A. jordani*
Rachis of the frond dichotomous; ultimate pinnules asymmetrical, attached laterally.   2. *A. pedatum aleuticum*

1. A. jordanii Muell. California Maidenhair. Rocky soil on moist shaded slopes and in shaded ravines, often under *Arbutus menziesii, Lithocarpus densiflora, Umbellularia californica,* and *Quercus* ssp.; San Francisco, Pilarcitos Canyon, Coal Mine Ridge, near Stanford, Bear Creek, Waddell Creek, Santa Cruz, and Soquel Valley.

2. A. pedatum L. var. **aleuticum** Rupr. *Fig. 10.* Western Maidenhair, Five-finger Fern. Moist shaded canyons and slopes, commonly growing under *Sequoia sempervirens, Pseudotsuga menziesii, Umbellularia californica,* and *Alnus rhombifolia*; Pilarcitos Canyon, Pescadero Creek, Los Gatos Canyon, Castle Rock Ridge, Waddell Creek, Boulder Creek, Santa Cruz, and Hester Creek.

10. **Pteridium** Gled. ex Scop.

1. P. aquilinum (L.) Kuhn var. **pubescens** Underw. *Fig. 11.* Bracken. A common fern of oak-madrone woods, margins of grasslands, roadsides, and fallow fields. San Francisco southward. —*P. aquilinum* (L.) Kuhn var. *lanuginosum* (Bong.) Fern.

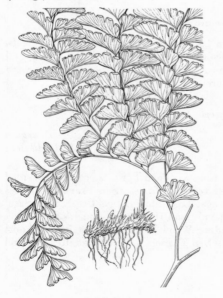 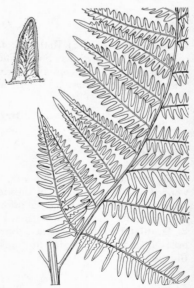

10   11

11. Cheilanthes Swartz. Lace Fern, Lip Fern

Plants not glandular-pubescent.
  Blades with scales.   1. *C. intertexta*
  Blades glabrous.   2. *C. californica*
Plants glandular-pubescent.   3. *C. cooperae*

1. C. intertexta (Maxon) Maxon. Coastal Lip Fern. Rocky slopes of Black Mountain and Mount Umunhum in Santa Clara County.
2. C. californica (Hook.) Mett. California Lace Fern. Occasional in Santa Cruz County; near Boulder Creek and Jamison Creek. —*Aspidotis californica* Nutt. ex Copeland.
3. C. cooperae Eaton. Cooper's Lip Fern. Known locally only from the limestone cliffs west of Felton in Santa Cruz County.

12. Pellaea Link. Cliffbrake

Ultimate pinnae obtuse to retuse; stipes flesh-colored.   1. *P. andromedaefolia*
Ultimate pinnae acute, apiculate; stipes purplish brown.   2. *P. mucronata*

1. P. andromedaefolia (Kaulf.) Fee. Coffee Fern. Dry, open or shaded habitats, usually among rock outcroppings, often in chaparral; San Francisco, San Andreas Valley, Coal Mine Ridge, Coyote Creek, Guadalupe Canyon, Loma Prieta, Barrocal Canyon, near Saratoga Summit, Rancho del Oso, and near Soquel.
2. P. mucronata (Eaton) Eaton. *Fig. 12.* Birds Foot Fern. Dry rock outcroppings, often on serpentine; San Francisco, Searsville, Stanford, Loma Prieta, near Boulder Creek, and Bonny Doon.

12

CLASS GYMNOSPERMAE

1. TAXACEAE. Yew Family

1. Torreya Arn.

1. T. californica Torr. California Nutmeg. Widespread in the Santa Cruz Mountains, but nowhere common, occupying a wide range of habitats from chaparral to the shade of redwood–Douglas fir forests; La Honda Grade, Stevens Creek, Alma Soda Springs, Mount Umunhum, Loma Prieta, Swanton, Big Basin, Little Basin, and near Saratoga Summit.

Several reports in the literature indicate that *Taxus brevifolia* Nutt. occurred along Laguna Creek in Santa Cruz County. No specimen of this species from the Santa Cruz Mountains exists in the herbaria of the following institutions: University of California, California Academy of Sciences, Harvard University, and Stanford University. It is possible that an isolated tree once grew in Santa Cruz County. To the north *T. brevifolia* is known from Marin County.

2. PINACEAE. Pine Family

Needle-leaves in fascicles, in ours usually in 3's, over 8 cm. long; bracts of mature cones not evident, fused to the ovuliferous scales.   1. *Pinus*
Leaves not in fascicles, separate, 2–3 cm. long; bracts of the mature cones evident, longer than and not fused to the scales.   2. *Pseudotsuga*

1. Pinus L. Pine

Cones very large, usually at least 20 cm. long, the scales with stout spines.
                                                                          1. *P. sabiniana*
Cones smaller, usually under 15 cm. long, the scales with slender prickles.
  Cones symmetrical, falling within 2–3 years; basal scales not distally tuberculate or mammillate; trunk of mature trees usually over 1 m. in diameter at breast height; bark separated into large plates.   2. *P. ponderosa*
  Cones asymmetrical, remaining closed and on the trees for many years; at least the basal scales distally either tuberculate or mammillate; trunk of mature trees under 1 m. in diameter; bark in narrow ridges.
    Basal cone-scales mammillate distally; cones broadly ovoid, deflexed, the axis of the cone not parallel to that of the branch.   3. *P. radiata*
    Basal cone-scales tuberculate distally; cones narrowly ovoid, sharply recurved, the axis of the cone parallel to that of the branch.
                                                                          4. *P. attenuata*

1. P. sabiniana Dougl. Digger Pine. Eastern slopes of the Santa Cruz Mountains, growing in chaparral, the trees usually widely spaced, rarely forming dense stands; near Los Gatos, Mount Umunhum, and Loma Prieta Ridge.
2. P. ponderosa Dougl. ex Lawson. Ponderosa or Western Yellow Pine. Known locally from the vicinity of Bonny Doon and Ben Lomond on inland marine sand deposits, commonly growing with *Arctostaphylos*, *Pinus attenuata*, and various oaks. In the Ben Lomond Sand Hills the trees are very large and in the past some have been cut for lumber.

3. P. radiata Don. Monterey Pine. Known locally from the vicinity of Año Nuevo Point south to Waddell Creek, commonly planted as an ornamental, and reproducing by seed in San Francisco.

4. P. attenuata Lemmon. *Fig. 13.* Knobcone Pine. Common on drv rocky outcroppings, in poor soil, and on the inland marine sand deposits, usually associated with various chaparral species; Butano Creek, Mount Umunhum, Loma Prieta, Waddell Creek, Bonny Doon, Ben Lomond Sand Hills, and near Mount Madonna.

Pinus × *attenuradiata* Stockwell & Righter is a hybrid between *P. attenuata* and *P. radiata* and occurs where these two species occur together in the vicinity of Año Nuevo Point and Waddell Creek near the coast. It is known in nature only from the above locality and is intermediate between the two parents in its morphological characters. The type was described from artificial hybrids grown at Placerville, California.

Dr. H. H. Behr recorded that a few small conifers grew on Lone Mountain in San Francisco about 1850. He did not indicate the genus or species of the conifers and now it is impossible to determine what they were.

2. Pseudotsuga Carr.

1. P. menziesii (Mirb.) Franco. *Fig. 14.* Douglas Fir. One of the most common trees in the Santa Cruz Mountains, growing from near sea level to about 3000 feet elevation, commonly associated with *Sequoia sempervirens, Lithocarpus densiflora, Arbutus menziesii,* and *Umbellularia californica;* Cahill Ridge, Pescadero Creek, Kings Mountain, near Saratoga Summit, Rancho del Oso, Big Basin, and near Mount Madonna. —*P. taxifolia* (Lamb.) Britt.

3. CUPRESSACEAE. Cypress Family

1. Cupressus L. Cypress

Mature ovulate cones 2–2.5 cm. long; seeds 3–5 mm. long; mature trees rarely over 10 m. tall.   1. *C. abramsiana*

Mature ovulate cones 2.5–3.5 cm. long; seeds 5–6 mm. long; mature trees 20–25 m. tall.   2. *C. macrocarpa*

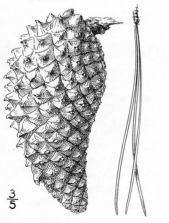

$\frac{3}{5}$ 13

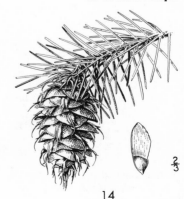

$\frac{2}{3}$

14

1. C. abramsiana Wolf. Abrams' or Santa Cruz Cypress. Endemic in the Santa Cruz Mountains, known from only four groves, growing on sandstone outcroppings or in marine sand deposits, commonly with *Pinus attenuata, Arctostaphylos, Adenostoma fasciculatum,* and *Vaccinium ovatum*; Butano Ridge, Eagle Rock, Brackenbrae, and Bonny Doon.

2. C. macrocarpa Hartw. Monterey Cypress. Widely planted as an ornamental and as a windbreak, occasionally producing seedlings; spontaneous seedlings are very common along the road to the west of Upper Crystal Springs Reservoir. San Francisco, Cahill Ridge, and Skyline Boulevard; native of the Monterey Peninsula.

4. TAXODIACEAE. Taxodium Family

1. Sequoia Endl.

1. S. sempervirens (Lamb.) Endl. *Fig. 15.* Coast Redwood. Canyon bottoms and moist slopes, central San Mateo County southward, more common on the western slopes of the Santa Cruz Mountains than on the eastern, rarely more than 20 miles from the coast, in general confined to the fog belt, commonly associated with *Pseudotsuga menziesii* or forming pure stands.

15

$\frac{2}{3}$

CLASS ANGIOSPERMAE

Subclass Monocotyledoneae

1. TYPHACEAE. Cat-tail Family

1. Typha L. Cat-tail

Staminate and pistillate portions of the inflorescence usually contiguous; compound pedicels bristle-like, over 1.5 mm. long; pollen-grains in 4's.

1. *T. latifolia*

Staminate and pistillate portions of the inflorescence not contiguous, separated; compound pedicels papillate, under 1.5 mm. long; pollen-grains single.
Inflorescence dark brown at maturity, overtopped by the upper leaves; leaves usually less than 10; plants to 1.5 m. tall.   2. *T. angustifolia*
Inflorescence light brown at maturity, exceeding the upper leaves; leaves usually more than 10; plants usually over 2.5 m. tall.   3. *T. domingensis*

1. T. latifolia L. *Fig. 16.* Broad-leaved Cat-tail, Soft Flag. Widespread in ditches, in quiet water along streams, in fresh or slightly brackish water marshes along the coast, and along the margins of ponds and lakes; San Francisco, San Andreas Lake, Searsville Lake, Stanford, Alviso, near Boulder Creek, and between Santa Cruz and Scott Valley. June–October.
2. T. angustifolia L. Nail Rod, Narrow-leaved Cat-tail. Fairly common in ponds, ditches, and brackish water sloughs; San Francisco, Emerald Lake, Pescadero Beach, Año Nuevo Point, Stevens Creek, and Alviso. June–August.
3. T. domingensis Pers. Lakes, roadside ditches, and somewhat brackish sloughs, often confused with the preceding species; San Francisco, Felt Lake, Alviso, near Mountain View, and near Camp Evers. July–November.

2. SPARGANIACEAE. Burreed Family

1. Sparganium L.

1. S. eurycarpum Engelm. *Fig. 17.* Broad-fruited Burreed. Marshes, drainage ditches, pot holes in drying river and stream channels, and lake margins; San Francisco, near Moss Beach, Searsville Lake, Lake Lagunita, Stanford, Alviso, Guadalupe Creek, Santa Cruz, and Watsonville. June–October. —*S. greenei* Morong.

3. POTAMOGETONACEAE. Pondweed Family

Flowers perfect; stamens more than 1; leaves alternate.
   Peduncles not coiled; stamens 4.   1. *Potamogeton*
   Peduncles coiled; stamens 2.   2. *Ruppia*
Flowers monoecious; stamen 1; leaves opposite.   3. *Zannichellia*

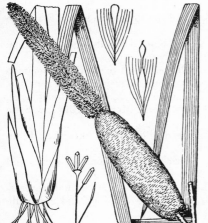

16

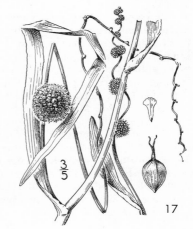

$\frac{3}{5}$

17

66   POTAMOGETONACEAE

1. Potamogeton L. Pondweed

Leaves dimorphic, the floating ones with a distinct, dilated blade, the submerged
    ones narrower.
  Submerged leaves less than 3 mm. wide, very narrowly linear, floating leaves
    ovate to elliptic or oblong, rounded or truncate at the base.   1. *P. natans*
  Submerged leaves usually over 5 mm. wide.
    Submerged leaves long-petiolate; fruit 3.5–4 mm. long, reddish, keel more
      or less muricate.   2. *P. nodosus*
    Submerged leaves short-petiolate; fruit 1.7–3.5 mm. long, greenish, keel
      usually not muricate.
      Submerged leaves usually over 1.5 cm. wide, when dry usually appearing
        distinctly rugose.
        Leaf-margins serrulate only at the tips.   3. *P. illinoensis*
        Leaf-margins serrulate throughout.   4. *P. crispus*
      Submerged leaves usually under 1.5 cm. wide, when dry not distinctly
        rugose.   5. *P. gramineus*
Leaves not dimorphic, all submerged, none with a distinct, dilated blade.
  Stipules adnate to the leaf-blades, free only at the tips.   6. *P. pectinatus*
  Stipules completely free from the leaf-blades.
    Peduncles usually more than 1 cm. long, filiform.
      Leaves 1–3 mm. wide.   7a. *P. pusillus pusillus*
      Leaves under 1 mm. wide.   7b. *P. pusillus minor*
    Peduncles usually under 1 cm. long, thick.   8. *P. foliosus macellus*

1. P. natans L. Common Floating or Broad-leaved Pondweed. Occasional in
fresh water ponds and sloughs; Coal Mine Ridge and near Watsonville. July–
September.
2. P. nodosus Poir. Long-leaved or Common American Pondweed. Probably
fairly common in shallow ponds, but poorly collected; San Francisco and Año
Nuevo Point. May–August. —*P. americanus* C. & S.
3. P. illinoensis Morong. Illinois Pondweed. The most common species of the
broad-leaved Potamogetons locally; San Francisco, Pilarcitos Lake, San Andreas
Lake, Felt Lake, College Lake, and Kelly Lake. May–October. —*P. lutescens* of
authors.
4. P. crispus L. Curl-leaved Pondweed. Known from Pine Lake in San Francisco
and to be expected elsewhere; native of Europe.
5. P. gramineus L. Various-leaved Pondweed. Known locally only from Coal
Mine Ridge. May–August. —*P. heterophyllus* Schreb.
6. P. pectinatus L. *Fig. 18.* Sego or Fennel-leaved Pondweed. Fairly common
in reservoirs, lakes, ponds, and quiet pools along streams; San Francisco,
Pilarcitos Lake, Belmont, Mussel Rock, Pescadero Creek, San Andreas Lake,
Año Nuevo Point, Menlo Park, Stanford, Raymonds, Morgan Hill, and Soda
Lake. May–October.
7a. P. pusillus L. var. **pusillus.** Small Pondweed. Ponds and lakes; San Francisco and Crystal Springs Lake. May–October.
7b. P. pusillus L. var. **minor** (Biv.) Fern. & Schub. Occasional in ponds and
reservoirs; San Francisco and Searsville. June–October.
8. P. foliosus Raf. var. **macellus** Fern. Leafy Pondweed. Occasional in reservoirs, ponds, and streams; San Francisco, near Half Moon Bay, and Searsville
Lake. May–August.

2. Ruppia L.

1. R. maritima L. Ditch Grass. Common in brackish water ponds and sloughs; San Francisco, Granada, near Redwood City, Pebble Beach, Alviso, near Chittenden, Santa Cruz, Seabright, and Soda Lake. June–November.

3. Zannichellia L.

1. Z. palustris L. Horned Pondweed. Ditches, slow-moving creeks and streams, and along the margins of lakes and reservoirs; San Francisco, Pilarcitos Lake, Redwood City, Crystal Springs Reservoir, and near Santa Clara. September–December.

4. NAIADACEAE. Naias or Water Nymph Family

1. Naias L.

1. N. guadalupensis (Spreng.) Morong. Guadalupe or Common Water Nymph. Occasional in fresh water ponds and lakes; San Francisco and near Watsonville.

5. ZOSTERACEAE. Eelgrass Family

Flowers monoecious; leaves usually over 5 mm. wide. Plants of muddy or sandy
    bottoms.   1. *Zostera*
Flowers dioecious; leaves usually under 2 mm. wide. Plants of rocks and reefs.
    2. *Phyllospadix*

1. Zostera L.

1. Z. marina L. Eelgrass, Wrack. To be expected along the coast where the bottom is muddy or sandy; so far known definitely only from San Francisco.

2. Phyllospadix Hook.

Spadices several, cauline; peduncle over 10 cm. long.   1. *P. torreyi*
Spadix usually 1, basal; peduncle usually under 6 cm. long.   2. *P. scouleri*

1. P. torreyi Wats. Torrey's Surfgrass. Rocks and reefs in the surf; San Francisco, Santa Cruz, and doubtless elsewhere.
2. P. scouleri Hook. Scouler's Surfgrass. This species has been collected on rocks in the surf in Marin County and in northern Monterey County and is to be expected along our coast.

6. APONOGETONACEAE. Water Hawthorn Family

1. Aponogeton L. f.

1. A. distachyus L. f. Cape Water Hawthorn. Known from a small pond near Skyline Boulevard in northern San Mateo County; native of South Africa, often used as an ornamental and occasionally escaping. February–April.

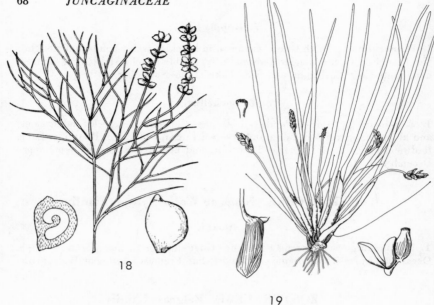

18

19

7. LILAEACEAE. Flowering Quillwort Family

1. Lilaea Humb. & Bonpl.

1. **L. scilloides** (Poir.) Haum. *Fig. 19.* Flowering Quillwort. Shallow ponds, along the edges of lakes, slow-moving streams, and irrigation ditches; San Francisco, Crystal Springs, Pigeon Point, Coal Mine Ridge, Stanford, Black Mountain, Morgan Hill, Camp Evers, and Watsonville. March–November. —*L. subulata* Humb. & Bonpl.

8. JUNCAGINACEAE. Arrow Grass or Arrow Weed Family

1. Triglochin L. Arrow Grass

Fertile carpels 3; fruit nearly spherical.   1. *T. striata*
Fertile carpels 6; fruit elongate.
    Plants usually over 3 dm. tall; rhizomes stout, short; ligules entire.
                                           2. *T. maritima*
    Plants usually under 3 dm. tall; rhizomes slender, elongate; ligules 2-parted.
                                           3. *T. concinna*

1. **T. striata** R. & P. Three-ribbed Arrow Grass. Occasional in salt marshes along the coast; Santa Cruz and Aptos. July–November.
2. **T. maritima** L. Seaside Arrow Grass. Salt marshes along San Francisco Bay and along the coast; San Francisco, Alviso, and Mountain View. April–October.
3. **T. concinna** Davy. Slender Arrow Grass. Salt marshes and areas of brackish water along San Francisco Bay; San Mateo, Palo Alto, and Mayfield. March–July.

9. ALISMATACEAE. Water Plantain Family

Carpels in one whorl.   1. *Alisma*
Carpels in several series.
 Flowers perfect; leaves truncate to cordate at the base.   2. *Echinodorus*
 Lower flowers pistillate, the upper usually staminate; leaves sagittate at the
  base.   3. *Sagittaria*

1. Alisma L.

1. A. plantago-aquatica L. Common Water Plantain. Lakes and ponds which
are regularly subjected to a fluctuating water level; San Francisco, Lake Lagunita, Searsville Lake, Coal Mine Ridge, and near Watsonville. June–September.

2. Echinodorus Rich. ex Engelm.

1. E. berteroi (Spreng.) Fassett. *Fig. 20.* Upright Burhead. Shores of dry
lake beds; Lake Lagunita and Felt Lake. June–September. —*E. cordifolius* (L.)
Griseb.

3. Sagittaria L.

1. S. latifolia Willd. Broad-leaved Arrowhead, Wapato. Shallow water of lakes,
ponds, and streams; Alviso, Campbell Creek, near Santa Cruz, and Harkins
Slough. July–October.

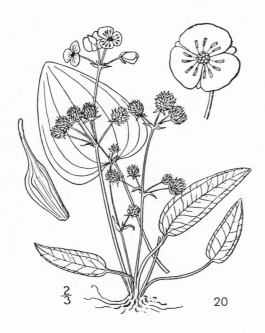

$\frac{2}{3}$ 20

10. HYDROCHARITACEAE. Tapegrass or Frogbit Family

1. **Elodea** Michx. Waterweed, Elodea

Leaves 0.5–1.5 cm. long, usually about 2 mm. wide; flowers 1 per spathe.
1. *E. canadensis*

Leaves 2–3 cm. long, 3–5 mm. wide; flowers 3 per spathe.
2. *E. densa*

1. E. canadensis Michx. Canadian Waterweed. Occasionally found in shallow ponds and in slowly moving water of irrigation ditches. Rarely flowering locally. —*Anacharis canadensis* (Michx.) Planch.

2. E. densa (Planch.) Casp. Brazilian Waterweed. Reported from San Francisco and elsewhere in California where it has been grown in aquaria and has become locally established; native of South America. July–August.

11. GRAMINEAE. Grass Family

Spikelets 1- to several-flowered; sterile or imperfect florets almost always above the fertile florets.

  Plants woody.
1. *Bambuseae*

  Plants herbaceous.

    Spikelets pedicellate, rarely subsessile.

      Spikelets 2- to several-flowered.

        Glumes shorter than the first florets; lemmas awned or not, if awned, usually from a bifid apex.
2. *Festuceae*

        Glumes usually longer than the first florets; lemmas awned or not, if awned, the awn usually inserted on the back of the lemma.
4. *Aveneae*

      Spikelets 1-flowered.
5. *Agrostideae*

    Spikelets sessile.

      Spikelets on opposite sides of the rachis; spikes terminal, solitary.
3. *Hordeae*

      Spikelets all on one side of the rachis; spikes 1 to several, racemosely or digitately arranged.
6. *Chlorideae*

Spikelets with 1 perfect terminal floret; sterile or staminate florets below the perfect floret, or rarely lacking.

  Spikelets not paired, or if paired both with fertile florets; fertile lemmas indurate.

    Glumes usually about equal in length.
7. *Phalarideae*

    Glumes not equal in length, the lower glume shorter than the upper, the upper about equaling the sterile lemma.
8. *Paniceae*

  Spikelets paired, the lower sessile and with a perfect floret, the upper pedicellate and with a sterile or staminate floret; fertile lemmas hyaline.
9. *Andropogoneae*

1. Bambuseae. Bamboo Tribe

One genus.
1. *Pseudosasa*

2. Festuceae. Fescue Tribe

Tall, reed-like grasses, usually 2 m. or more tall.
  Leaves mainly at the base of the culms.   2. *Cortaderia*
  Leaves more or less evenly distributed along the culms.   3. *Phragmites*
Grasses not reed-like, rarely over 1.5 m. tall.
  Plants dioecious. Plants of salt marshes and alkaline areas.   4. *Distichlis*
  Plants not dioecious (or if dioecious, not of salt marshes or alkaline areas);
    plants of various habitats.
    Inflorescences with fertile and sterile spikelets.
      Fertile spikelets 2–3-flowered.   5. *Cynosurus*
      Fertile spikelets 1-flowered.   6. *Lamarckia*
    Inflorescences with only fertile spikelets.
      Lemmas 3-nerved.   7. *Eragrostis*
      Lemmas 5- or more nerved.
        Lemmas as broad as long; florets spreading at right angles to the rachis,
        closely imbricate.   8. *Briza*
        Lemmas much longer than broad; florets not spreading at right angles
        to the rachis, closely imbricate or not.
          Lemmas keeled on the back.
            Spikelets in 1-sided clusters, the clusters scattered along the axis
            of the inflorescence or on short lateral branches.   9. *Dactylis*
            Spikelets not in 1-sided clusters.
              Lemmas awned from a bifid apex.   10. *Bromus*
              Lemmas awnless.   11. *Poa*
          Lemmas rounded on the back.
            Glumes papery; upper floret usually reduced to a clavate rudiment;
            spikelets commonly purple-tinged.   12. *Melica*
            Glumes not papery; upper florets reduced in size or not, but not
            reduced to a clavate rudiment; spikelets usually green.
              Nerves of the lemmas parallel.
                Spikelets in racemes.   13. *Pleuropogon*
                Spikelets in panicles.
                  Nerves of the lemmas conspicuous. Plants of very moist
                  fresh water habitats.   14. *Glyceria*
                Nerves of the lemmas very inconspicuous. Plants of saline
                  habitats.   15. *Puccinellia*
              Nerves of the lemmas converging at the apex.
                Lemmas awned from a bifid apex.   10. *Bromus*
                Lemmas entire, awned or not (rarely minutely bifid in *Festuca
                elmeri*).
                  Lemmas awned, long-attenuate, very gradually narrowed
                  distally.   16. *Festuca*
                  Lemmas not awned, rather abruptly narrowed distally.
                    Pedicels of the spikelets thick, short; plants annual;
                  panicles stiff.   17. *Scleropoa*
                    Pedicels of the spikelets slender, usually long; plants
                  annual or more usually perennial; panicles not stiff.
                  11. *Poa*

3. Hordeae. Barley Tribe

Spikelets 2 to several at each node of the rachis (occasionally only 1 in *Elymus*).
Rachis disarticulating at maturity (except in *H. vulgare*).

    Spikelets 3 per node, the central one fertile, the lateral ones sterile, often
       much reduced (in *H. vulgare* all the spikelets are fertile).

                                                                     18. *Hordeum*

    Spikelets 2 per node, both fertile.   **19. *Sitanion***

Rachis continuous, not disarticulating at maturity.

    Glumes much reduced or lacking; spikes not dense, the spikelets spreading.

                                                                 20. *Hystrix*

    Glumes present, conspicuous; spikes dense, the spikelets usually appressed
       or ascending.

        Plants annual, coarse, 6–12 dm. tall; spikelets always 3 per node, 1-
           flowered.   **18. *Hordeum***

        Plants annual or perennial (if annual, slender and usually under 5 dm.
           tall); spikelets 1 to many per node, 2- to several-flowered.

                                                          21. *Elymus*

Spikelets 1 per node of the rachis.

    Spikelets 2- to several-flowered; rachis not disarticulating at maturity; spike-
       lets not sunken into cavities in the rachis.

        Spikelets situated distinctly edgewise to the rachis; glumes 1 except in the
           terminal spikelet.   **22. *Lolium***

        Spikelets situated sidewise to the rachis (or sometimes somewhat edgewise);
           glumes 2.

           Plants perennial.

               Rachilla distorted at the base; spikelets more or less edgewise to the
                  rachis.   **21. *Elymus***

               Rachilla not distorted; spikelets situated sidewise to the rachis.

                                                 23. *Agropyron*

           Plants annual.

               Glumes ovate, 3-nerved.   **24. *Triticum***

              Glumes subulate, 1-nerved.   **25. *Secale***

    Spikelets 1-flowered; rachis disarticulating at maturity; spikelets sunken into
       cavities in the rachis.

        Lemmas awnless.

           Glume 1 per spikelet.   **26. *Monerma***

           Glumes 2 per spikelet.   **27. *Parapholis***

        Lemmas awned.   **28. *Scribneria***

4. Aveneae. Oat Tribe

Glumes 1 cm. long or longer.

    Plants annual; lemmas awned on the back or the awns occasionally lacking;
       awns round in cross section; panicles many-flowered.   **29. *Avena***

    Plants perennial; lemmas awned from between 2 terminal teeth; awns flat-
       tened; panicles few-flowered.   **30. *Danthonia***

Glumes less than 1 cm. long.
  Lemmas entire, not toothed or bifid at the apex.
    Spikelets 7–9 mm. long; awns about 2 times as long as the lemmas.
                                                31. *Arrhenatherum*
    Spikelets 3–6 mm. long; awns rarely longer than the lemmas or none.
      Glumes not exceeding the upper floret, scabrous, glabrous, or papillate-
        hirsute.   32. *Koeleria*
      Glumes exceeding the upper floret, scabrous to villous.   33. *Holcus*
  Lemmas variously toothed at the apex.
    Lemmas 4-toothed at the apex.   34. *Deschampsia*
    Lemmas bifid at the apex.
      Plants annual; spikelets 1.8–4 mm. long.   35. *Aria*
      Plants perennial; spikelets 6–8 mm. long.   36. *Trisetum*

5. Agrostideae. Timothy Tribe

Articulation below the glumes, the spikelets falling entire.
  Plants prostrate; inflorescence capitate, partly hidden by the bract-like upper
    leaves.   37. *Crypsis*
  Plants erect or ascending; inflorescence spike-like, narrowly cylindric or
    variously lobed, not capitate, not partly hidden by the upper leaves.
    Glumes not awned; spikelets strongly compressed.   38. *Alopecurus*
    Glumes long-awned; spikelets not strongly compressed.   39. *Polypogon*
Articulation above the glumes, the spikelets not falling entire.
  Fruit indurate; lemmas awned; callus conspicuous.
    Awns several times as long as the lemmas, persistent; glumes 6–20 mm. long.
                                                40. *Stipa*
    Awns about twice as long as the lemmas, deciduous; glumes 3 mm. long.
                                                41. *Oryzopsis*
  Fruit not indurate; lemmas awned or not; callus inconspicuous.
    Glumes longer than the lemmas.
      Glumes plumose.   42. *Lagurus*
      Glumes not plumose.
        Spikelets strongly compressed laterally; glumes keeled, the keels
          ciliate; inflorescences dense, cylindric or ellipsoid.   43. *Phleum*
        Spikelets not compressed laterally; glumes not keeled, not ciliate; in-
          florescences various.
          Glumes very unequal in length, saccate at the base; lemmas about ¼
            as long as the longest glume.   44. *Gastridium*
          Glumes subequal in length, not saccate at the base; lemmas at least
            ½ as long as the glumes.
            Spikelets 4–20 mm. long; paleae well developed; rachilla pro-
              longed behind the palea; plants perennial.
                                                45. *Calamagrostis*
            Spikelets rarely over 4 mm. long; paleae well developed or not;
              rachilla not prolonged behind the palea (except in *A. avena-
              cea*); plants annual or perennial.   46. *Agrostis*
    Glumes shorter than the lemmas or about equaling them.
      Spikelets 1–1.5 cm. long.   47. *Ammophila*
      Spikelets about 2 mm. long.   48. *Sporobolus*

6. Chlorideae. Grama Tribe

Plants annual; inflorescences paniculate.
  Spikelets nearly spherical, 1-flowered.   49. *Beckmannia*
  Spikelets elongate, 7–11-flowered.   50. *Leptochloa*
Plants perennial; inflorescences digitate or spike-like.
  Stems prostrate, rooting at the nodes; spikes usually 4–5, digitately spreading
    from the top of the culm; spikelets about 2 mm. long. Plants of disturbed
    areas.   51. *Cynodon*
  Stems erect, occasionally rooting from the lower nodes; spikes numerous, in
    spike-like racemes; spikelets about 12 mm. long. Plants of salt marshes.
          52. *Spartina*

7. Phalarideae. Canary Grass Tribe

Sterile lemmas awned; glumes unequal.   53. *Anthoxanthum*
Sterile lemmas not awned (lacking in one species of *Phalaris*); glumes subequal.
  Panicles open; sterile lemmas as long as or longer than the fertile ones; spike-
    lets not compressed laterally.   54. *Hierochloe*
  Panicles dense, spike-like; sterile lemmas at most ½ as long as the fertile ones
    or rarely lacking; spikelets compressed laterally.   55. *Phalaris*

8. Paniceae. Millet Tribe

Spikelets subtended by bristles or enclosed in a spiny bur.
  Spikelets enclosed in a spiny bur.   56. *Cenchrus*
  Spikelets subtended by bristles.
    Bristles persistent, not plumose; inflorescences not hidden.   57. *Setaria*
    Bristles not persistent, falling with the spikelets, plumose (or if not plumose
      the inflorescences hidden by the upper sheaths); inflorescences hidden
      or not.   58. *Pennisetum*
Spikelets not subtended by bristles or enclosed in a spiny bur.
  Spikelets pedicellate; inflorescences paniculate; spikelets not arranged on one
    side of the rachis.   59. *Panicum*
  Spikelets sessile or short-pedicellate; inflorescences various; spikelets arranged
    on one side of the rachis.
    Inflorescence simple, the rachis very broad and flattened.
          60. *Stenotaphrum*
    Inflorescences of 2 or more branches (occasionally 1 in *Paspalum*); rachis
      not broad and flattened.
      Rachis narrowly winged; inflorescences usually of 2 to several racemosely
        or digitately arranged branches.
        Branches of the inflorescence digitate.   61. *Digitaria*
        Branches of the inflorescence racemose.   62. *Paspalum*
      Rachis not winged; inflorescences paniculate.   63. *Echinochloa*

9. Andropogoneae. Sorghum Tribe

Outer glumes pectinate.   64. *Eremochloa*
Glumes not pectinate.   65. *Sorghum*

1. Bambuseae. Bamboo Tribe

1. Pseudosasa Makino

1. P. japonica (Sieb. & Zucc.) Makino. Metake. Well established in Golden Gate Park in San Francisco, spreading by runners, but also flowering; native of Asia. Summer months.

2. Festuceae. Fescue Tribe

2. Cortaderia Stapf

1. C. selloana (Schult.) Aschers. & Graebn. Pampas Grass. Commonly planted as an ornamental and becoming established in moist habitats; Crystal Springs Lake and Hilton Airport; native of South America. June–October. At several places between Saratoga Summit and Santa Cruz plants of this species occur on very steep, inaccessible road embankments, sites where they certainly were not planted.

3. Phragmites Trin.

1. P. communis Trin. Common Reed. Known locally only from San Francisco where it is probably now extinct, but to be expected elsewhere in marshy areas along streams and ponds. April–October. —*P. phragmites* (L.) Karst., *P. communis* Trin. var. *berlandieri* (Fourn.) Fern.

4. Distichlis Raf.

Spikelets congested; flowers uniformly 5–9 per spikelet; culms prostrate to decumbent; plants stoloniferous; leaf-blades 1–2 dm. long.
1a. D. spicata stolonifera
Spikelets not conspicuously congested; flowers 3–14 per spikelet, variable in number; culms erect; plants not stoloniferous; leaf-blades under 0.5 dm. long.
1b. D. spicata nana

1a. D. spicata (L.) Greene var. **stolonifera** Beetle. Salt Grass. Salt marshes along the coast and along the shores of San Francisco Bay; San Francisco, near Alviso, Alviso, Davenport Landing, Santa Cruz, and Pajaro River. May–October.
1b. D. spicata (L.) Greene var. **nana** Beetle. Known locally only from near Soda Lake in southern Santa Cruz County. May–July.

5. Cynosurus L.

Plants perennial; awns of the lemmas inconspicuous; panicles narrow, more or less linear. *1. C. cristatus*
Plants annual; awns of the lemmas about equaling the spikelets; panicles ovoid to subcapitate. *2. C. echinatus*

1. C. cristatus L. Crested Dog's Tail Grass. Known locally as a lawn weed from San Francisco; native of Europe. June–July.
2. C. echinatus L. Known from roadsides in Santa Cruz County, but to be expected occasionally elsewhere; near Big Basin and near Boulder Creek; native of Europe. June–August.

6. Lamarckia Moench

1. L. aurea (L.) Moench. Goldentop. Occasional, but apparently spreading, growing in dry or rocky soils in disturbed areas and occasionally as a weed in gardens; San Francisco, near Redwood City, near Stanford, Stanford, Almaden, Guadalupe Mines, and Edenvale; native of the Mediterranean region. March–May.

7. Eragrostis Host. Love Grass

Midveins of the lemmas with conspicuous glands; spikelets about 3 mm. wide.
1. *E. megastachya*
Midveins of the lemmas without glands.
Plants low, rooting at the nodes and forming mats.   2. *E. hypnoides*
Plants erect, not rooting at the nodes and forming mats.
Spikelets about 1 mm. wide, linear.
Spikelets 3–5 mm. long; lemmas 1–1.5 mm. long; plants 1–5 dm. tall; the lower branches of the inflorescence pilose in the axils.   3. *E. pilosa*
Spikelets 5–7 mm. long; lemmas 1.7–2 mm. long; plants 6–10 dm. tall; axils not pilose.   4. *E. orcuttiana*
Spikelets about 1.5 mm. wide, linear to ovoid.   5. *E. diffusa*

1. E. megastachya (Koel.) Link. Candy Grass. Known as a garden weed from San Francisco and Stanford; native of Eurasia. June–July. —*E. cilianensis* of authors.
2. E. hypnoides (Lam.) BSP. Creeping Eragrostis. Known from Pinto Lake in southern Santa Cruz County and to be expected elsewhere on moist ground near lakes and ponds. September–November.
3. E. pilosa (L.) Beauv. Pilose Eragrostis. Known locally as a weed from Stanford and Boulder Creek; native of Europe. July–September.
4. E. orcuttiana Vasey. *Fig. 21.* Orcutt's Eragrostis. Occasional as a weed in gardens and in planted areas; San Francisco, Los Gatos, near Raymonds, Gilroy, and Boulder Creek; native of the southwestern part of the United States. August–September.
5. E. diffusa Buckl. Fairly common as a weed in gardens, lawns, and disturbed areas, but rarely collected; San Francisco, Stanford, and Boulder Creek; native in much of North America, but introduced locally. June–October. —*E. caroliniana* (Spreng.) Scribn.

8. Briza L. Quaking Grass

Spikelets at least 10 mm. across at anthesis; panicles drooping.   1. *B. maxima*
Spikelets under 5 mm. across at anthesis; panicles erect.   2. *B. minor*

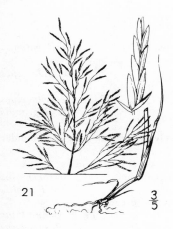

21    $\frac{3}{5}$

22    $\frac{3}{5}$

1. B. maxima L. *Fig. 22.* Rattlesnake or Big Quaking Grass. Disturbed habitats, often in shade. throughout the Santa Cruz Mountains; native of Europe. April–May.

2. B. minor L. Little Quaking Grass. Widely distributed in grasslands and often in disturbed habitats; native of Europe. January–June.

9. Dactylis L.

1. D. glomerata L. Orchard Grass. Fairly common in disturbed areas, along roads. and in overgrazed pastures; San Francisco, near Baden, Cahill Ridge, Palo Alto, Ben Lomond Mountain. and Glenwood; native of Europe. May–August.

10. Bromus L. Brome Grass

Awns of the lemmas distinctly geniculate.   1. *B. trinii*
Awns of the lemmas straight or lacking.
  Lemmas distinctly keel-compressed.
    Lemmas awnless or the awns under 2.5 mm. long.   2. *B. willdenowii*
    Lemmas awned. the awns over 3 mm. long.
      Leaf-blades about 2 mm. wide, often involute, densely canescent.
                                    3. *B. breviaristatus*
      Leaf-blades usually over 5 mm. wide, usually flat, glabrous or variously pubescent.
        Awns 7 mm. long or longer.   4. *B. carinatus*
        Awns under 7 mm. long.
          Panicles very narrow. the branches short. erect. Plants of coastal sand dunes.   5. *B. maritimus*
          Panicles broad. the branches longer, spreading. Plants of various habitats. but not found on coastal sand dunes.
            Lower glumes usually 5-nerved.   6. *B. stamineus*
            Lower glumes 1- or 3-nerved.   7. *B. marginatus*

Lemmas not keeled-compressed.
  Plants perennial.
    Lower glumes 1-nerved; plants dark green.   8. *B. vulgaris*
    Lower glumes 3-nerved; plants pale green.
      Glumes glabrous.   9. *B. laevipes*
      Glumes pubescent.
        Upper glume 5-nerved.   10. *B. pseudolaevipes*
        Upper glume 3-nerved.   11. *B. grandis*
  Plants annual.
    Lemmas rounded at the top, the teeth rarely over 1 mm. long; awns under
      15 mm. long.
      Panicles dense, contracted, the branches erect to ascending.
        Lemmas glabrous or rarely minutely scabrous.   12. *B. racemosus*
        Lemmas distinctly pubescent.   13. *B. mollis*
      Panicles open, not contracted, the branches spreading.
        Lemmas glabrous.   14. *B. commutatus*
        Lemmas pubescent.   15. *B. arenarius*
    Lemmas acute at the top, the teeth usually 2–5 mm. long; awns 12–50 mm.
      long.
      Panicles contracted, the pedicels short, erect to ascending; awns of the
        lemmas 16–22 mm. long.
        Culms densely puberulent below the panicles.   16. *B. rubens*
        Culms glabrous or sparsely scabrous below the panicles.
                                                   17. *B. madritensis*
      Panicles open, the pedicels often long, spreading; awns of the lemmas
        12–50 mm. long.
        Upper glumes under 1 cm. long; spikelets pubescent or glabrous;
          awns of the lemmas 12–14 mm. long.   18. *B. tectorum*
        Upper glumes 2.5–3 cm. long; spikelets glabrous; awns of the lemmas
          30–50 mm. long.   19. *B. rigidus*

1. B. trinii Desv. Chilean Brome Grass. Occasional on serpentine areas and as
a weed elsewhere; San Francisco, Woodside, and Stanford; native of Chile.
March–May.
2. B. willdenowii Kunth. Rescue Grass. Occasional as a weed in disturbed
areas; San Francisco, near San Bruno, and Santa Clara; native of South America. March–September. —*B. catharticus* Vahl, nomen confusum, *B. unioloides*
(Willd.) Rasp.
3. B. breviaristatus Buckl. Occasional in grasslands; near San Mateo, near
Stanford, and Palo Alto. May–June.
4. B. carinatus H. & A. *Fig. 23.* California Brome. Common on grassy slopes
and in brush-covered areas; San Francisco, Montara Mountain, Moss Beach,
Crystal Springs, La Honda, Coal Mine Ridge, Black Mountain, Permanente
Creek, near Los Gatos, near Saratoga Summit, Glenwood, Big Basin, and Boulder
Creek. March–July.
5. B. maritimus (Piper) Hitchc. Seaside Brome Grass. Stabilized coastal sand
dunes; San Francisco and Año Nuevo Point. April–May.
6. B. stamineus Desv. Chilean Brome. Occasional as a weed in disturbed areas;
San Francisco, Belmont, and Menlo Park; native of Chile. April–May.

23

$\frac{3}{5}$

7. B. marginatus Nees. Large Mountain Brome Grass. Open slopes and coastal bluffs; San Francisco, near Colma, Crystal Springs Lake, Pigeon Point, near Woodside, and near Santa Cruz. May–October.

8. B. vulgaris (Hook.) Shear. Narrow-flowered Brome Grass. Shaded areas in redwoods and redwood–Douglas fir forests, often along streams; La Honda, near Pescadero Creek, Kings Mountain, Saratoga–Big Basin Road, Black Mountain and Waddell Creek. May–July.

9. B. laevipes Shear. Woodland Brome Grass. Occasional in shaded woods; near San Mateo, Portola Valley, Coal Mine Ridge, Black Mountain, near Los Gatos, and near Madrone. May–July.

10. B. pseudolaevipes Wagnon. Occasional in woods in Santa Clara and Santa Cruz counties; Stanford, Permanente Creek, and near Boulder Creek. March–July.

11. B. grandis (Shear) Hitchc. Tall Brome Grass. Known locally from near Eagle Rock. June–July.

12. B. racemosus L. Smooth-flowered Soft Cheat. Occasional in grasslands and in disturbed areas; San Francisco, San Andreas Valley, near Saratoga Summit, and Hilton Airport; native of Europe. April–June.

13. B. mollis L. Soft Chess. The most common annual *Bromus* locally, growing in open grassland, edges of chaparral, in woods, and as a weed in disturbed areas; native of Europe. March–June. —*B. hordeaceus* of authors.

14. B. commutatus Schrad. Hairy Chess, Downy-sheathed Cheat. Occasional in disturbed areas, along roadsides, and in grassland; San Francisco, Ben Lomond, and near Boulder Creek; native of Europe. May–June.

15. B. arenarius Labill. Australian Chess, Australian Brome Grass. Known locally from Boulder Creek and to be expected elsewhere; native of Australia. May–June.

16. B. rubens L. Foxtail Chess. Common as a weed in disturbed areas and occasionally on dry grassy slopes; introduced from Europe. March–May.

17. B. madritensis L. Spanish Brome Grass. Disturbed areas and dry open slopes, San Francisco southward; introduced from Europe. April–May.

18. B. tectorum L. Cheat Grass, Downy Chess, Downy Cheat. Occasional in disturbed areas and in sandy soils; San Francisco, Palo Alto, Ben Lomond Sand Hills, and Graham Hill; native of Europe. April–June. —*B. tectorum* L. var. *glabratus* Spenner.
19. B. rigidus Roth. Ripgut Grass. Disturbed areas, chaparral, and grasslands, a very common introduced grass; native of the Mediterranean region. April–July.

11. Poa L. Bluegrass

Plants annual.
    Plants rarely more than 2 dm. tall; sheaths of the leaves smooth; lemmas not
        webbed at the base.   1. *P. annua*
    Plants over 2 dm. tall; sheaths roughened; lemmas webbed at the base.
                                                         2. *P. howellii*
Plants perennial.
    Rhizomes present.
        Culms flattened, 2-edged.   3. *P. compressa*
        Culms terete.
        Plants dioecious; panicles very dense; lemmas not conspicuously webbed.
                                             4. *P. douglasii*
        Plants not dioecious; panicles open; lemmas conspicuously webbed.
             Lemmas glabrous except for the web at the base.   5. *P. kelloggii*
             Lemmas pubescent above the web.   6. *P. pratensis*
    Rhizomes absent.
        Web conspicuous.
             Lemmas glabrous.   7. *P. trivialis*
             Lemmas pubescent on the keels and nerves.   8. *P. nemoralis*
        Web lacking.
             Lemmas distinctly keeled, usually also compressed.   9. *P. unilateralis*
             Lemmas obscurely keeled, rounded on the back.   10. *P. scabrella*

1. P. annua L. Annual Bluegrass. A very common grass of gardens, moist disturbed areas, and cultivated fields; native of Europe. February–September.
2. P. howellii Vasey & Scribn. *Fig. 24.* Howell's Bluegrass. Wooded areas from San Mateo County southward; Cahill Ridge, Seal Cove, near Pescadero Creek, Coal Mine Ridge, Page Mill Road, Permanente Creek, Saratoga–Big Basin Road, near Eagle Rock, near Boulder Creek, Ben Lomond Sand Hills, and Santa Cruz. March–June. —*P. bolanderi* Vasey ssp. *howellii* (Vasey & Scribn.) Keck.
3. P. compressa L. Canada Bluegrass. Occasional as a weed; San Francisco and San Bruno Mountain; native of Europe. May–June.
4. P. douglasii Nees. Douglas' or Dune Bluegrass. Coastal sand dunes; San Francisco, San Pedro, and Santa Cruz. March–May.
5. P. kelloggii Vasey. Kellogg's Bluegrass. Occasional in shaded woods; Pescadero Creek, Boulder Creek, and Felton. April–May.
6. P. pratensis L. Kentucky Bluegrass. Lawns, disturbed areas, and open hills; San Francisco, Cahill Ridge, Millbrae, Stanford, Palo Alto, and Santa Cruz; widespread in North America and Eurasia, local collections from near habitations may be escapes from introduced European strains. March–June.

7. P. trivialis L. Rough-stalked Meadow Grass. Occasional as garden and lawn weeds in San Francisco; native of Europe. May–June.

8. P. nemoralis L. Rare as a garden weed in Golden Gate Park in San Francisco; native of Europe. June.

9. P. unilateralis Scribn. San Francisco Bluegrass. Fairly common in rocky areas in exposed locations; San Francisco, Montara Mountain, near Millbrae, near Pescadero, and Santa Cruz. March–May.

10. P. scabrella (Thurb.) Benth. ex Vasey. Pine or Malpais Bluegrass. Rocky grassland and serpentine soils; San Francisco, San Andreas Lake, Cahill Ridge, Woodside, Searsville, Page Mill Road, Stevens Creek, San Jose, Guadalupe Creek, Almaden Canyon, Loma Prieta, Edenvale, near Bielawski Lookout, and Little Basin. February–May.

12. Melica L. Melic Grass

Culms bulbous at the base.
  Lemmas tapering gradually, acuminate, pubescent on the nerves.
                                                  1. *M. subulata*
  Lemmas tapering abruptly, acute to obtuse, glabrous.
    Spikelets usually over 15 mm. long; first and second florets about 2.5 mm. apart at the base; glumes much shorter than the spikelets.
                                                  2. *M. geyeri*
    Spikelets usually under 15 mm. long; first and second florets about 2 mm. apart; glumes nearly as long as the spikelets.   3. *M. californica*
Culms not bulbous at the base.
  Fertile florets 1 or rarely 2 per spikelet.
    Rudimentary florets about 1 mm. long, on stipes about 2–2.5 mm. long; lemmas pubescent near the apex.   4. *M. torreyana*
    Rudimentary florets about 2 mm. long, on stipes about 0.5 mm. long; lemmas at most scabrous.   5. *M. imperfecta*
  Fertile florets 2 or more.
    Glumes much shorter than the spikelets; lemmas pilose-ciliate on the margins basally.   6. *M. harfordii*
    Glumes nearly as long as the spikelets; lemmas at most scabrous.
                                                  3. *M. californica*

1. M. subulata (Griseb.) Scribn. Alaska Onion Grass. Occasional on open wooded slopes; near La Honda, Kings Mountain, El Corte Madera, Congress Springs, and Boulder Creek. March–July.

2. M. geyeri Munro ex Bolander. Geyer's Onion Grass. Usually growing in redwoods or in redwood–Douglas fir forests, less frequently in oak-woodlands; San Francisco Watershed Reserve, near Stanford, and Stevens Creek. March–May.

3. M. californica Scribn. Western Melica. Fairly common in rocky areas in grasslands and occasionally in open areas in oak-woodland; San Francisco, Sawyer Ridge, Crystal Springs, near San Mateo, Woodside, Stanford, Permanente Creek, Black Mountain, Mount Umunhum, Mount Charlie Road, and Santa Cruz. March–June. —*M. bulbosa* Geyer, in part.

4. M. torreyana Scribn. Torrey's Melica. Fairly common in a wide range of habitats, rocky serpentine, grasslands, oak-woodlands, brush-covered slopes, and redwood forests, San Francisco southward. March–July.

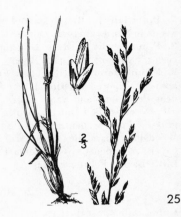

24   25

5. M. imperfecta Trin. *Fig. 25.* Small-flowered Melica. Fairly common on brush-covered hillsides, open woods, sandy coastal bluffs, and along the edges of chaparral; San Francisco, Montara Mountain, Searsville Ridge, Stanford, Stevens Creek, near Los Gatos, Loma Prieta, near Saratoga Summit, Ben Lomond Sand Hills, and Santa Cruz. March–June. Hybrids between *M. imperfecta* and *M. californica* occur occasionally. Plants with spreading panicles have been called *M. imperfecta* Trin. var. *refracta* Thurb.
6. M. harfordii Bolander. Harford's Melica. Rare in the Santa Cruz Mountains; near Lexington, near Saratoga Summit, and Boulder Creek. May–June.

13. Pleuropogon R. Br.

1. P. californicus (Nees) Benth. ex Vasey. California Pleuropogon, Semaphore Grass. Marshy areas in San Francisco and in the vicinity of Crystal Springs Lakes. May–June.

14. Glyceria R. Br. Manna Grass

Spikelets linear, 1–1.5 cm. long; branches of the panicles rigid, erect.
    Lemmas 2.5–4 mm. long.   1. *G. leptostachya*
    Lemmas 4–5.5 mm. long.   2. *G. occidentalis*
Spikelets ovate to oblong, about 0.5 cm. long; branches of the panicles nodding,
    not rigid.   3. *G. pauciflora*

1. G. leptostachya Buckl. Davy's Manna Grass. Known locally only from San Francisco, where it was collected in 1852. May–June.
2. G. occidentalis (Piper) Nels. Water Manna Grass. Collected in 1908 near Baden in northern San Mateo County. June–July. —*G. fluitans* of authors.
3. G. pauciflora Presl. Few-flowered Manna Grass. Occasional in very wet areas, swampy ground, and along the edges of reservoirs; San Francisco and Pilarcitos Dam. July. —*Puccinellia pauciflora* (Presl) Munz, *Torreyochloa pauciflora* (Presl) Church.

15. **Puccinellia** Parl. Alkali Grass

Anthers 0.7–0.8 mm. long; lemmas 1.5–3 mm. long; spikelets 4–7 mm. long,
   3–6-flowered.
  Lemmas 2–3 mm. long; lower branches of the panicles ascending.
1. P. airoides

  Lemmas 1.5–2 mm. long; lower branches of the panicles drooping or eventu-
    ally reflexed. 2. P. distans
Anthers 1.3–1.5 mm. long; lemmas 3–4 mm. long; spikelets 8–15 mm. long,
   5–12-flowered. 3. P. grandis

1. P. airoides (Nutt.) Wats. & Coult. Nuttall's Alkali Grass. Known only from
Soda Lake. July. —*P. nuttalliana* (Schult.) Hitchc.
2. P. distans (L.) Parl. Little Alkali Grass. Occasional as a weed in San Francisco and to be expected elsewhere; native of Europe. June.
3. P. grandis Swallen. Pacific Alkali Grass. Levees and salt marshes along San
Francisco Bay. May–August. —*P. nutkaensis* of authors.

16. **Festuca** L. Fescue

Plants annual.
  Glumes pubescent.
    Lemmas glabrous. 1. F. confusa
    Lemmas pubescent.
      Pedicels closely appressed to the rachis. 2. F. grayi
      Pedicels at least eventually all spreading or reflexed. 3. F. eastwoodae
  Glumes glabrous.
    Glumes very unequal, the lower one 0.5–2 mm. long, less than ½ as long
      as the upper.
      Lemmas ciliate. 4. F. megalura
      Lemmas not ciliate. 5. F. myuros
    Glumes not very unequal, the lower one over 2 mm. long, at least ½ as long
      as the upper.
      Lemmas glabrous or slightly scabrous.
        Lower branches of the inflorescence erect or ascending, not reflexed;
          florets often 6 or more per spikelet.
          Florets usually 6 or more per spikelet; plants low, much tufted;
            lemmas 3–5 mm. long, the awns rarely over 5 mm. long, usually
            shorter than the lemmas. 6. F. octoflora
          Florets usually less than 6 per spikelet; plants tall, usually not
            tufted; lemmas 5–7 mm. long, the awns over 5 mm. long.
7. F. dertonensis
        Lower branches of the inflorescence reflexed at maturity; florets 5 or
          fewer per spikelet.
          Florets 3–5 per spikelet; upper pedicels remaining appressed.
8. F. pacifica
          Florets usually 1–2 per spikelet; all the pedicels at length reflexed.
9. F. reflexa

84   GRAMINEAE

Lemmas pubescent.
Lower branches of the inflorescence erect or ascending; plants usually
    low, tufted.   6. *F. octoflora*
Lower branches of the inflorescence reflexed at maturity; plants usually
    tall, not much tufted.
    Pedicels appressed.   2. *F. grayi*
    Pedicels eventually spreading or reflexed.   10. *F. microstachys*
Plants perennial.
Florets long-stipitate.   11. *F. subuliflora*
Florets sessile.
Leaf-blades over 3 mm. wide, not involute.
    Lemmas with awns 2–8 mm. long.   12. *F. elmeri*
    Lemmas without awns, rarely slightly attenuate.
    Lemmas 5–7 mm. long; plants with rhizomes.   13. *F. elatior*
    Lemmas 7–10 mm. long; plants without rhizomes. 14. *F. arundinacea*
Leaf-blades involute (rarely flat in *F. californica*).
Leaves villous at the junction between the sheath and the blade.
    15. *F. californica*
Leaves glabrous at the junction between the sheath and the blade.
    Awns longer than the lemmas.   16. *F. occidentalis*
    Awns shorter than the lemmas or nearly lacking.
    Culms decumbent at the base, loosely tufted; leaf-blades smooth;
      basal sheaths brownish.   17. *F. rubra*
    Culms erect, densely tufted; leaf-blades usually scabrous; basal
      sheaths straw-colored.   18. *F. idahoensis*

1. F. confusa Piper. Hairy-leaved Fescue. Rare on the eastern slopes of the Santa Cruz Mountains; near Stanford, Permanente Creek, and near Madrone. April–May.
2. F. grayi (Abrams) Piper. Gray's Fescue. Known locally from only a few localities; near San Jose, near Saratoga Summit, and Coyote. April–May.
3. F. eastwoodae Piper. Eastwood's Fescue. Occasional in grasslands; Coyote, near Madrone, and Eagle Rock. April–May.
4. F. megalura Nutt. Foxtail or Western Six-weeks Fescue. A common annual grass of open slopes, coastal bluffs, inland sand deposits, chaparral, and disturbed areas; San Francisco southward. March–May.
5. F. myuros L. Rattail Fescue. A very common grass of open slopes, clearings in chaparral, and occasionally in disturbed areas; native of Europe. March–May.
6. F. octoflora Walt. Slender Fescue. Chaparral, fire burns, and in sandy soils, probably fairly common, but rarely collected; San Francisco, Butano Ridge, Big Basin, Boulder Creek, and Ben Lomond Sand Hills. April–May. —*F. octoflora* Walt. var. *hirtella* Piper.
7. F. dertonensis (All.) Aschers. & Graebn. Six-weeks Fescue. A common grass of open slopes, chaparral, brush-covered areas, coastal and inland sands, and disturbed habitats; native of Europe. March–May. —*F. bromoides* of authors.
8. F. pacifica Piper. *Fig. 26.* Pacific Fescue. Common in rocky grassland and on serpentine; San Francisco, Sawyer Ridge, Jasper Ridge, Alamitos Creek, Monte Bello Ridge, Almaden Canyon, near San Jose, Boulder Creek, and Santa Cruz. April–May.

9. F. reflexa Buckl. Few-flowered Fescue. Serpentine soils and outcroppings; San Francisco, San Mateo, Woodside, Stanford, Permanente Creek, and Boulder Creek. April–May.

10. F. microstachys Nutt. Nuttall's Fescue. Rare, usually on grassy slopes or in chaparral; Stanford, near Madrone, near Eagle Rock, and Blooms Grade. March–May.

11. F. subuliflora Scribn. Coast Range Fescue. Occasional in the Santa Cruz Mountains; Jamison Creek and Boulder Creek. May–July.

12. F. elmeri Scribn. & Merr. Elmer's Fescue. Occasional in wooded areas; Stanford, Black Mountain, near Los Gatos, and near Boulder Creek. May–June. —*F. elmeri* Scribn. & Merr. var. *luxurians* Piper.

13. F. elatior L. Meadow Fescue. Disturbed areas, lawns and gardens, probably common, but poorly collected locally; San Francisco, Palo Alto, and near Boulder Creek; native of Eurasia. June–July.

14. F. arundinacea Schreb. Alta or Reed Fescue. Occasional in disturbed areas; San Francisco and Crystal Springs Lake; native of Europe. May–June.

15. F. californica Vasey. California Fescue. Open woods and brush-covered slopes, most common to the east of the crests of the Santa Cruz Mountains; San Francisco, Seal Cove, Crystal Springs, Kings Mountain, and Permanente Creek. March–May.

16. F. occidentalis Hook. Western Fescue. Moist areas in redwoods and red-wood–Douglas fir forests; Pescadero Creek, near La Honda, Searsville, Big Basin, and near Boulder Creek. April–July.

17. F. rubra L. Red Fescue. Common on grassy hills, brushy areas, sand dunes, and inland sand deposits; San Francisco, San Bruno Mountain, Boulder Creek, and Ben Lomond Sand Hills. May–June.

18. F. idahoensis Elmer. Blue Bunch Grass. Open meadows, grassy slopes, and occasionally in brush-covered areas; San Francisco, San Francisco Watershed Reserve, Crystal Springs, and Empire Grade. April–June.

17. Scleropoa Griseb.

1. S. rigida (L.) Griseb. Occasional as a weed in San Francisco and to be expected elsewhere; native of Europe. April–May.

3. Hordeae. Barley Tribe

18. Hordeum L. Barley

Plants perennial; auricles lacking at the base of the leaf-blades.
    Awns 2–5 cm. long.   1. *H. jubatum caespitosum*
    Awns under 1 cm. long.
        Leaf-blades 3–8 mm. wide; spikes 6–10 mm. in diameter; anthers usually
        1–1.5 mm. long.   2. *H. brachyantherum*
        Leaf-blades 2–3 mm. wide; spikes about 5 mm. in diameter; anthers usually
        1.5–3 mm. long.   3. *H. californicum*
Plants annual; auricles present or not.
    Rachis of the spike continuous, not disarticulating; spikelets all sessile;
        auricles present.   4. *H. vulgare*

Rachis of the spike disarticulating; lateral spikelets pedicellate; auricles present or not.
 Glumes of the central spikelets ciliate; auricles present.
  Internodes of the rachis 2–3 mm. long; florets of the lateral spikelets longer than the floret of the central spikelet.   5. *H. leporinum*
  Internodes of the rachis 1.8–2 mm. long; florets of the lateral spikelets not longer than the floret of the central spikelet.   6. *H. glaucum*
 Glumes of the central spikelets not ciliate; auricles lacking.
  Glumes thick, rigid; florets of the lateral spikelets awned.   7. *H. hystrix*
  Glumes slender, not rigid; florets of the lateral spikelets awnless.
                                                                   8. *H. depressum*

1. H. jubatum L. var. **caespitosum** (Scribn.) Hitchc. Tufted Barley. Rare locally, known as a weed only from San Francisco, but to be expected elsewhere occasionally; probably introduced locally from other parts of North America. June–August.

2. H. brachyantherum Nevski. Meadow Barley. Grassy slopes, edges of streams, and in partly wooded areas; San Francisco, San Bruno Mountain, Crystal Springs Lake, Pescadero Beach, Stanford, near Gilroy, Swanton, and near Boulder Creek. April–June.

3. H. californicum Covas & Stebbins. California Barley. Rare locally; Spring Valley, near San Mateo, Mountain View, and Saratoga. May–June.

4. H. vulgare L. Common Barley. Occasionally persisting in disturbed areas and in fallow fields, widely planted; San Francisco, Crystal Springs Lake, Pescadero, Stanford, La Selva Beach, and Pajaro River; native of Eurasia. January–June.

5. H. leporinum Link. Farmer's Foxtail. Common on grassy slopes and in disturbed areas, San Francisco southward; native of southern Europe. February–July. —*H. murinum* of authors, in part.

6. H. glaucum Steud. Wall Barley. Occasional as a weed in disturbed areas and in grasslands; San Francisco, Woodside, and Palo Alto; native of Eurasia. April. —*H. stebbinsii* Covas, *H. murinum* of authors, in part.

7. H. hystrix Roth. Mediterranean Barley. Common on grassy slopes and in disturbed areas; San Francisco, Cahill Ridge, near Woodside, Jasper Ridge, Cooleys Landing, Stanford, near Gilroy, Hilton Airport, and Santa Cruz; native of Europe. May–July. —*H. gussoneanum* Parl.

8. H. depressum (Scribn. & Smith) Rydb. Low Barley. Low ground near tidal sloughs; Redwood City and Palo Alto. May–June.

19. Sitanion Raf. Squirreltail

Glumes entire or occasionally 2–3-lobed; spikes narrow, several times as long as broad.
                                                                   1. *S. hanseni*
Glumes divided into 3 or more awns; spikes about as long as broad.
                                                                   2. *S. jubatum*

1. S. hanseni (Scribn.) Smith. Hansen's Squirreltail. Occasional in open exposed grasslands; San Francisco and near Redwood City. May–June. It has been shown that *S. hanseni* is a sterile hybrid between *Elymus glaucus* and *S. jubatum*.

2. S. jubatum Smith. *Fig. 27.* Big Squirreltail. Occasional in rocky soils and on open grassy slopes; San Francisco, San Bruno Hills, San Andreas Lake, Woodside, Black Mountain, and near Eagle Rock. April–July.

20. Hystrix Moench

1. H. californica (Bolander) Kuntze. California Bottle Brush. Wooded slopes near the coast, usually in redwood or redwood–Douglas fir forests; near Crystal Springs Lake, San Mateo, and near Santa Cruz. May–July.

21. Elymus L. Rye Grass

Plants annual; lemmas awned, the awns at least 10 times as long as the lemmas.
                1. *E. caput-medusae*
Plants perennial; lemmas awned or not, if awned the awns less than 3 times as long as the lemmas.
 Lemmas awned, the awns 1–2 times as long as the lemmas; glumes usually strongly 3–4-nerved, scabrous abaxially.   2. *E. glaucus*
 Lemmas not awned or with only very short awns; glumes nerved or not.
  Glumes distinctly nerved, villous; rhizomes present. Plants of coastal sand dunes.   3. *E. mollis*
  Glumes nerved or not, glabrous to sparsely pubescent, but not villous; rhizomes present or not. Plants of various habitats.
   Spikelets usually 3, 5, or more per node; inflorescences commonly branched; leaf-blades 1.5–3.5 cm. broad; glumes subulate, obscurely nerved.   4. *E. condensatus*
   Spikelets 1–2 per node; inflorescences not branched; leaf-blades under 1.5 cm. broad; glumes lanceolate to subulate, nerved or not.
    Plants without rhizomes; glumes more or less flat, strongly nerved, the nerves usually 5–7; spikelets 2 per node. 5. *E. virescens*
    Plants with rhizomes; glumes usually somewhat folded, nerveless or the nerves 1–3; spikelets 1–2 per node.
     Spikelets usually 2 per node; culms usually at least 6 dm. tall.
      Culms finely pubescent below the spike. Plants of dunes and sandy beaches.   6. *E.* × *vancouverensis*
      Culms glabrous below the spike or at most minutely scabrous. Plants of inland habitats.   7. *E. triticoides*
     Spikelets usually 1 per node; culms usually 1–2 dm. tall.
                8. *E. pacificus*

1. E. caput-medusae L. Medusa Head. Known locally only as a weed at Los Gatos; native of Europe. June.

2. E. glaucus Buckl. Western Rye Grass, Blue Wild Rye. A common grass of wooded and brush-covered slopes; San Francisco, Pilarcitos Canyon, Montara Mountain, near Redwood City, Coal Mine Ridge, Stanford, Permanente Creek, near Almaden, Mount Charlie Road, Glenwood, near Big Basin, Empire Grade, and Santa Cruz. April–June. Plants with pubescent leaf-blades and sheaths have been called *E. glaucus* Buckl. var. *jepsonii* Davy. Considerable variation with respect to the amount of pubescence exists, even among the plants of one collection. Occasional plants have awnless lemmas.

3. E. mollis Trin. Sea Lyme or American Dune Grass. Sand dunes along the coast; San Francisco, near Pescadero, and Waddell Creek. May–October. Local plants are referable to *E. mollis* Trin. ssp. *mollis* var. *mollis* f. *mollis*.

4. E. condensatus Presl. Giant Rye Grass. Occasional on slopes near the coast; San Francisco, near Año Nuevo Point, near Waddell Creek, Watsonville, and Chittenden. May–October.

5. E. virescens Piper. Pacific Rye Grass. Occasional on grassy or brush-covered slopes; San Francisco and probably elsewhere. June–August. —*E. glaucus* Buckl. ssp. *virescens* (Piper) Gould.

6. E. × vancouverensis Vasey. Vancouver's Rye Grass. Coastal sand dunes and sandy flats; Santa Cruz. June–August. *Elymus × vancouverensis* is the hybrid between *E. mollis* and *E. triticoides*.

7. E. triticoides Buckl. *Fig. 28.* Alkali Rye Grass. Woods, brush-covered areas, and in weedy places; San Francisco, near Woodside, Palo Alto, Glenwood, near Boulder Creek, Santa Cruz, and Watsonville Slough. May–July. Plants with several spikelets at the node have been called *E. triticoides* Buckl. var. *multiflorus* Gould.

8. E. pacificus Gould. Gould's Rye Grass. Sandy soils in San Francisco. May–June. —*Agropyron arenicola* Davy.

22. **Lolium** L. Darnel, Ryegrass

Glumes as long as or longer than the spikelets; lemmas awned (except in the variety); awns usually about 2–3 times as long as the lemmas; plants annual.   1. *L. temulentum*
Glumes shorter than the spikelets; lemmas awned or not; if awned, the awns usually about as long as the lemmas; plants annual or perennial.
    Plants perennial; lemmas usually awnless; culms subcompressed.
                                                                        2. *L. perenne*
    Plants annual or short-lived perennials; lemmas awned (except in one variety); culms cylindric.   3. *L. multiflorum*

1. L. temulentum L. Darnel. Occasional in disturbed areas and on overgrazed slopes; near Baden, near La Honda, Stanford, Palo Alto, near San Jose, Blooms Grade, and Santa Cruz; native of Europe. April–June. Plants in which the lemmas are awnless or have awns which do not exceed the lemmas may be referred to *L. temulentum* L. var. *leptochaeton* A. Br.
2. L. perenne L. English or Perennial Ryegrass. Fairly common as a weed in urban areas and occasionally on grassy slopes; San Francisco, Montara Mountain, Sawyer Ridge, Stanford, and Hilton Airport; native of Europe. March–July. Plants in which the spikelets are crowded together to form an ovoid spike have been called *L. perenne* L. var. *cristatum* Pers. The lemmas of *L. perenne* are usually awnless, but occasional individuals are encountered in which short awns are present.
3. L. multiflorum Lam. Italian or Australian Ryegrass. A very common grass of disturbed areas, grassy slopes, and along the edges of salt marshes; native of Europe. April–October. Plants with branched panicles have been called *L. multiflorum* Lam. var. *ramosum* Guss. ex Arcang. Awnless plants may be called *L. multiflorum* Lam. var. *muticum* DC.

23. Agropyron Gaertn. Wheat Grass

Plants with creeping rhizomes.
  Spikelets 1–1.5 cm. long; lemmas minutely scabrous on the abaxial surface, 8–10 mm. long; glumes 3–7-nerved.   1. *A. repens*
  Spikelets 1.5–2 cm. long; lemmas smooth on the abaxial surface, 10–15 mm. long; glumes 9-nerved.   2. *A. junceum*
Plants without creeping rhizomes.
  Glumes acute; lemmas acute or awned; spikelets to 1.5 cm. long.
     3. *A. trachycaulum*
  Glumes obtuse; lemmas obtuse, not awned; spikelets 1.5–2.5 cm. long.
     4. *A. elongatum*

1. A. repens (L.) Beauv. Quack Grass. Occasional as a weed in gardens and in disturbed areas in San Francisco; native of Eurasia. June–August.
2. A. junceum (L.) Beauv. Rush-like Wheat Grass. Known as a weed on sand dunes in San Francisco; native of Europe. May–September.
3. A. trachycaulum (Link) Malte. Slender Wheat Grass. Known locally from near Crystal Springs. April–May. —*A. tenerum* Vasey, *A. pauciflorum* (Schwein.) Hitchc.
4. A. elongatum (Host.) Beauv. Elongate Wheat Grass. Known locally from along the edge of a small salt marsh stream near the Palo Alto Yacht Harbor; native of Europe. July–August.

24. Triticum L.

1. T. aestivum L. Wheat. Widely grown locally and occasionally spontaneous in disturbed areas and in fallow fields; not known outside of cultivation or as an escape from cultivation, presumably this species originated in Asia Minor. April–July.

25. Secale L.

1. S. cereale L. Cereal Rye. Occasional as an escape from cultivation or growing from seed which has been scattered about during the transportation of hay; San Francisco, Stanford, and near Santa Cruz; probably derived from wild populations which are native in southwestern Asia. May.

26. Monerma Beauv.

1. M. cylindrica (Willd.) Coss. & Dur. Thin Tail. Salt marshes, coastal bluffs, and spreading into grasslands and serpentine areas; San Francisco, San Pedro Point, and Stanford; native of Eurasia. May–June. —*Lepturus cylindricus* (Willd.) Trin.

27. Parapholis Hubb.

1. P. incurva (L.) Hubb. Sickle Grass. Sea beaches, sand dunes, and along the edges of salt marshes; San Francisco, Palo Alto, Santa Cruz, and near Capitola; native of Europe. May–June. —*Pholiurus incurvus* (L.) Schinz. & Thell.

28. Scribneria Hack.

1. S. bolanderi (Thurb.) Hack. *Fig. 29.* Scribner's Grass. Rare in the Santa Cruz Mountains, known locally from the vicinity of Jamison Creek in Santa Cruz County, but to be expected elsewhere. March–April.

4. Aveneae. Oat Tribe

29. Avena L. Oat

Lemmas hairy.
 Lemmas with awn-tipped teeth, the awns usually about 4 mm. long; pedicels
  capillary.   1. *A. barbata*
 Lemmas without awn-tipped teeth; pedicels stout.   2a. *A. fatua fatua*
Lemmas glabrous.
 Awns well developed, strongly geniculate; spikelets usually 3-flowered.
                                                          2b. *A. fatua glabrata*
 Awns usually lacking, if present more or less straight; spikelets usually 2-
  flowered.   3. *A. sativa*

1. A. barbata Brot. Slender Wild Oat. A common annual grass of fields, orchards, open grassy slopes, and disturbed areas, San Francisco southward; native of the Mediterranean region. February–June.

2a. A. fatua L. var. **fatua.** Wild Oat. A common grass of fields, pastures, open slopes, and disturbed areas, San Francisco southward; native of Europe. February–July.

 Avena brevis Roth has been collected as a weed in Belmont. It is not known outside of cultivation or as an occasional escape from cultivation. The glumes of *A. brevis* are less than 1.5 cm. long, while those of *A. fatua* var. *fatua* are 2.5 cm. long.

2b. A. fatua L. var. **glabrata** Peterm. Occasional as a weed near Crystal Springs Lake; native of Europe. March–June.

3. A. sativa L. Cultivated Oat. Occasional as an escape from cultivation; San Francisco and near Edenvale; native of Eurasia. April–June.

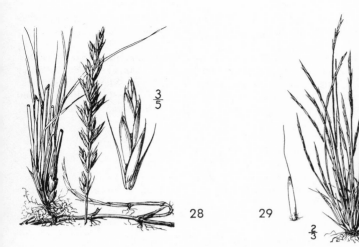

30. Danthonia Lam. & DC. Oat Grass

Lemmas 7–8 mm. long, exclusive of the awns; glumes 15–18 mm. long.
    Sheaths glabrous, pilose only at the throat.   1*a. D. californica californica*
    Sheaths pilose throughout.   1*b. D. californica americana*
Lemmas 5–6 mm. long, exclusive of the awns; glumes 8–10 mm. long.
                                               2. *D. pilosa*

1*a.* **D. californica** Bolander var. **californica.** California Wild Oat Grass. Fairly common on coastal bluffs, in grasslands, and occasionally in serpentine soils; San Francisco, San Francisco Watershed Reserve, Crystal Springs Lake, near Woodside, Moss Beach, near Menlo Park, Hilton Airport, near Eagle Rock, and Boulder Creek. April–June.
1*b.* **D. californica** Bolander var. **americana** (Scribn.) Hitchc. Coastal bluffs and grasslands; San Francisco, San Bruno Mountain, near Pescadero, Pigeon Point, and Santa Cruz. April–May.
2. **D. pilosa** R. Br. Purple-awned Wallaby Grass. Known locally as a weed in Palo Alto; native of Australia. June–July.

31. Arrhenatherum Beauv.

1. **A. elatius** (L.) Presl. Tall Oat Grass. Occasional as a weed in disturbed areas and on grassy slopes; near Boulder Creek and Loma Prieta; native of Europe. June–July.

32. Koeleria Pers.

Plants annual, introduced; spikelets 2–4 mm. long; lemmas short awned.
                                      1. *K. gerardi*
Plants perennial, native; spikelets 4–6 mm. long; lemmas usually awnless.
                                  2. *K. macrantha*

1. K. gerardi (Vill.) Shinners. Bristly Koeleria. Known from dry rocky areas in grasslands in San Francisco and to be expected in similar habitats elsewhere; native of Europe. April–May.
2. K. macrantha (Ledeb.) Spreng. June Grass. Occasional on dry grassy slopes, serpentine outcroppings, and on brush-covered slopes; San Francisco, Sweeney Ridge, Crystal Springs Lake, Stanford, Permanente Creek, Stevens Creek, near Los Gatos, Mount Umunhum, near Saratoga Summit, Boulder Creek, Ben Lomond Sand Hills, and Santa Cruz. April–June. —*K. gracilis* Pers., *K. cristata* (L.) Pers.

33. Holcus L.

1. H. lanatus L. Velvet Grass. Common in moist areas such as drainage ditches, along seeps and springs, and on grassy slopes; San Francisco southward; native of Europe. March–October.

34. Deschampsia Beauv. Hair Grass

Plants annual; glumes usually 6–7 mm. long; leaves few, not tufted at the base
of the culms.   1. *D. danthonioides*
Plants perennial; glumes 3.5–6 mm. long; leaves numerous, tufted at the base
of the culms.
  Leaf blades filiform; panicles narrow, the branches usually appressed; glumes
     usually 3.5–4 mm. long.   2. *D. elongata*
  Leaf blades not filiform; panicles wider, the branches ascending; glumes usu-
     ally 4–6 mm. long.
    Spikelets 4–5 mm. long; panicle open, lax.   3a. *D. caespitosa caespitosa*
    Spikelets 6–8 mm. long; panicle narrow, not lax.
                           3b. *D. caespitosa holciformis*

1. D. danthonioides (Trin.) Munro ex Benth. Annual Hair Grass. Low moist areas where water has stood during the winter and early spring, occasionally from San Francisco southward; San Francisco, Woodside, Jasper Ridge, near Coyote, Camp Evers, and Santa Cruz. April–June. —*D. danthonioides* (Trin.) Munro ex Benth. var. *gracilis* (Vasey) Munz, *Aira danthonioides* Trin.
2. D. elongata (Hook.) Munro ex Benth. *Fig. 30.* Slender Hair Grass. Moist, shaded areas, usually in woods or on brush-covered slopes; San Francisco, Pescadero Creek, Jasper Ridge, Coal Mine Ridge, Stevens Creek, near Los Gatos, Big Basin, San Vicente Road, near Felton, and Santa Cruz. May–July. —*Aira elongata* Hook.
3a. D. caespitosa (L.) Beauv. ssp. **caespitosa.** Tufted Hair Grass. Reported from San Francisco and known from near Boulder Creek. —*Aira caespitosa* L.
3b. D. caespitosa (L.) Beauv. ssp. **holciformis** (Presl) Lawr. California Hair Grass. Most common along the coast in marshy areas or on coastal bluffs, occasionally inland on grassy slopes or under Douglas firs; San Francisco, near San Bruno, Cahill Ridge, Pebble Beach, near Burlingame, and Santa Cruz. April–July. —*Aira holciformis* (Presl) Steud.

35. Aira L.

1. A. caryophyllea L. Silvery Hair Grass. A very common grass of open slopes, sandy soils, in chaparral, and occasionally in disturbed habitats; native of Europe. March–June. —*Aspris caryophyllea* (L.) Nash.

36. Trisetum Pers.

Panicle loose, the lower branches filiform; spikelets usually with 3 flowers, the florets distant; leaf-blades 6–12 mm. wide.   1. *T. cernuum*
Panicle rather dense, the lower branches not filiform; spikelets 2–3-flowered, the flowers usually close together; leaf-blades 2–7 mm. wide.   2. *T. canescens*

1. T. cernuum Trin. *Fig. 31.* Nodding Trisetum. Moist areas in woods and forests; Cahill Ridge, Kings Mountain, Coal Mine Ridge, Stanford, Stevens Creek, Big Basin, and Glenwood. April–June.
2. T. canescens Buckl. Tall Trisetum. Redwood forests, stream banks, and in moist wooded areas; San Francisco, San Francisco Watershed Reserve, near Woodside, Kings Mountain, Alamitos School, Black Mountain, near Boulder Creek, and Santa Cruz. May–August. —*T. cernuum* Trin. var. *canescens* (Buckl.) Beal.

5. Agrostideae. Timothy Tribe

37. Crypsis Ait.

1. C. nilaca Fig. & DeNot. Prickle Grass, Sharp-leaved Crypsis. Occasional in dry lake beds and to be expected in vernal pools; Stanford and between Gilroy and Morgan Hill; native of Egypt and southwestern Asia. June–October.

38. Alopecurus L. Foxtail

Spikelets 5–6 mm. long; plants perennial.   1. *A. pratensis*
Spikelets 3–3.5 mm. long; plants annual.   2. *A. howellii*

1. A. pratensis L. Known only from Golden Gate Park in San Francisco, but to be expected elsewhere occasionally; native of Eurasia. March–April.
2. A. howellii Vasey. Howell's Meadow Foxtail. Occasional along the edges of ponds; Crystal Springs Lake, San Mateo County, and Coal Mine Ridge. May–June.

39. Polypogon Desf. Beard Grass

Panicles spikelike; plants annual; glumes both scabrous and pubescent.
                                                  1. *P. monspeliensis*
Panicles with distinct branches; plants perennial; glumes scabrous.
                                                  2. *P. interruptus*

1. P. monspeliensis (L.) Desf. Annual Beard Grass, Rabbit's Foot Grass. A common weed of wet areas, drainage ditches, creek bottoms, and along the edges of puddles and ponds; native of Europe. April–November.
2. P. interruptus HBK. Beard Grass. Common in ditches, low wet areas, and depressions where water stood during the spring, San Francisco southward. April–August. —*P. lutosus* of authors, *P. littoralis* Sm.

40. Stipa L. Spear or Needle Grass

Spikelets (excluding the awns) 6–10 mm. long.   1. *S. lepida*
Spikelets (excluding the awns) 15–25 mm. long.
    Awns about 7–9 times as long as the lemmas; leaves green.   2. *S. pulchra*
    Awns 9–12 times as long as the lemmas; leaves somewhat glaucous.
                                                              3. *S. cernua*

1. S. lepida Hitchc. *Fig. 32.* Small-flowered Stipa. Oak woods, edges of chaparral, grasslands, and occasionally in serpentine and rocky outcroppings; San Francisco, San Andreas Valley, near Stanford, near Los Altos, Saratoga–Big Basin Road, San Vicente Road, and Santa Cruz. April–June. —*S. lepida* Hitchc. var. *andersonii* (Vasey) Hitchc.
2. S. pulchra Hitchc. Nodding Stipa. Rocky outcroppings, grassy slopes, and sometimes in serpentine soils; San Francisco, San Bruno Mountain, Montara Mountain, Crystal Springs Lake, Redwood City, near Woodside, Jasper Ridge, Stanford, Almaden Canyon, and Santa Cruz. March–June.
3. S. cernua Stebbins & Love. Grasslands and rocky outcroppings; San Francisco and perhaps elsewhere. April–May.

41. Oryzopsis Michx. Rice Grass

1. O. miliacea (L.) B. & H. ex Aschers. & Schwein. Smilo Grass, Millet Mountain Rice. Occasional as a weed in disturbed areas, especially urban ones; San Francisco, Palo Alto, and Santa Cruz; native of the Mediterranean region. April–November.

42. Lagurus L.

1. L. ovatus L. Hare's Tail Grass. Well established on sand dunes and as a weed in San Francisco; native of the Mediterranean region. May–July.

43. Phleum L. Timothy

Spikes linear-cylindric.	1. *P. pratense*
Spikes ovate-cylindric.	2. *P. alpinum*

1. P. pratense L. Cultivated Timothy. Occasional as a weed in lawns and disturbed areas; San Francisco, Stanford, and Santa Cruz; native of Eurasia. June–July.
2. P. alpinum L. Mountain Timothy. Known from one specimen collected in San Francisco in 1862 and not subsequently recollected.

44. Gastridium Beauv.

1. G. ventricosum (Gouan) Schinz & Thell. Nit Grass. Fairly common as a garden weed and in grasslands; San Francisco, San Francisco Watershed Reserve, La Honda, Palo Alto, near Los Gatos, near Hilton Airport, near Eccles, and Santa Cruz; native of Europe. May–August.

45. Calamagrostis Adans. Reed Grass

Glumes 3–5 mm. long; sheaths pubescent on the collars; plants with long, creeping rhizomes.   1. *C. rubescens*
Glumes 5–7 mm. long; sheaths glabrous on the collars; plants without long, creeping rhizomes.
    Plants 10–15 dm. tall; awns of the lemmas not equaling the glumes; callus hairs about ½ as long as the lemmas.   2. *C. nutkaensis*
    Plants 4–8 dm. tall; awns of the lemmas about equaling the glumes; callus hairs about ¼–⅓ as long as the lemmas.   3. *C. koelerioides*

1. C. rubescens Buckl. *Fig. 33.* Pine Grass. Usually growing in redwood forests, occasionally in shaded woods; Tunitas Creek Road, Kings Mountain, Big Basin, near Boulder Creek, and Boulder Creek. June–September.

$\frac{3}{5}$ 32 $\frac{3}{5}$ 33

2. C. nutkaensis (Presl) Steud. Pacific Reed Grass. Usually along or near the coast at the edges of marshes and in moist areas farther inland; San Francisco, San Pedro, Arroyo de Los Frijoles, and Camp Evers. May–June.

3. C. koelerioides Vasey. Tufted Pine Grass. Open areas in redwood–Douglas fir forests, edges of brush-covered slopes, and in rocky grasslands; La Honda, Saratoga–Big Basin Road, near Big Basin, Big Basin, between Felton and Empire Grade, and near Davenport. June–September.

46. Agrostis L. Bent Grass

Many specimens of *Agrostis* in herbaria lack the basal portions of the plant, thus making it impossible to tell whether the plant is annual or perennial or has stolons or rhizomes. It is essential that the entire plant be available for accurate identification.

Paleae nearly as long as the lemmas or at least ½ as long. Introduced grasses.
  Rachilla prolonged behind the paleae; lemmas pubescent, awned from the
    back.   1. *A. avenacea*
  Rachilla not prolonged; lemmas not pubescent, rarely awned.
    Glumes scabrous both on the midribs and on the abaxial surface.
                                                        2. *A. semiverticillata*
    Glumes scabrous only on the midribs, glabrous on the rest of the abaxial
      surface.
      Ligules 1–2 mm. long.   3. *A. tenuis*
      Ligules 3–6 mm. long.
        Rhizomes lacking; culms more or less decumbent at the base.

                                                              4. *A. stolonifera*
        Rhizomes present; culms usually erect.   5. *A. alba*
Paleae never more than ½ as long as the lemmas or lacking. Native or introduced
  grasses.
  Plants with long creeping rhizomes.
    Panicles spikelike, contracted.   6. *A. pallens*
    Panicles open.
      Lemmas about 3 mm. long, the hairs at the base 1 mm. long or longer;
        spikelets 3.5–4.5 mm. long.   7. *A. hallii*
      Lemmas about 2–2.5 mm. long, the hairs at the base less than 1 mm. long;
        spikelets usually under 3.5 mm. long.   8. *A. diegoensis*
  Plants without long creeping rhizomes.
    Panicles very diffuse, with long filiform branches; the branches branched
      distally.   9. *A. scabra*
    Panicles contracted, often spikelike to moderately diffuse, the branches not
      filiform; the branches branched below the middle.
      Lemmas not awned, or if awned the awns not exceeding the glumes.
        Panicles spikelike, usually partly hidden by the uppermost sheaths;
          paleae about 0.5 mm. long.   10. *A. californica*
        Panicles spikelike or diffuse, much exserted beyond the uppermost
          sheaths; paleae less than 0.5 mm. long.   11a. *A. exarata exarata*
      Lemmas awned, the awn exserted beyond the glumes.
        Plants perennial, 2–12 dm. tall.   11b. *A. exarata pacifica*
        Plants annual, 1–4 dm. tall.   12. *A. microphylla*

1. A. avenacea Gmel. Hairy-flowered Bent Grass. Occasional as a weed; San Francisco, near Burlingame, Crystal Springs Lake, and near Woodside; native of the Hawaiian Islands and Polynesia. May–July. —*A. retrofracta* Willd.
2. A. semiverticillata (Forsk.) Christ. Water Bent Grass. A common weed of drainage ditches, moist areas in gardens, and in other wet disturbed habitats; San Francisco, Crystal Springs Lake, near Stanford, near Alviso, Boulder Creek, and Watsonville; native of Europe. June–September. —*A. verticillata* Vill.
3. A. tenuis Sibth. Colonial or Common Bent. Occasional as a weed in gardens and in moist disturbed areas; San Francisco and Stanford; native of Europe. May–July.
4. A. stolonifera L. Creeping Bent. A weed of disturbed areas and brush-covered slopes; San Francisco and probably elsewhere; native of Europe.
5. A. alba L. Redtop. Occasional as a weed in gardens and lawns and in other moist areas; San Francisco, Redwood City, Stanford, and Loma Prieta; native of Europe. June–September. —*A. palustris* of authors.
6. A. pallens Trin. Seashore Bent Grass. Known locally only from San Francisco. June–July.
7. A. hallii Vasey. Hall's Bent Grass. Occasional in wooded areas; San Francisco, Coal Mine Ridge, and Alpine Creek. June–July.
8. A. diegoensis Vasey. Leafy Bent Grass. Fairly common on open grassy slopes, brush-covered areas, and along the margins of chaparral; San Francisco, near Hillsborough, Alpine Creek, Saratoga–Big Basin Road, Glenwood, Ben Lomond Sand Hills, and Santa Cruz. June–August.
9. A. scabra Willd. Rough Hair Grass. Known locally only from near Ben Lomond and Mill Creek in Santa Cruz County, but to be expected elsewhere in moist open woods. July–August. —*A. hiemalis* of authors.
10. A. californica Trin. California Bent Grass. Occasional on coastal bluffs; San Francisco, near Waddell Creek, and Santa Cruz. June–July.
11a. A. exarata Trin. var. **exarata.** Western Bent Grass. Occasional in low moist areas where water has stood; San Francisco, Searsville Lake, near Boulder Creek, and Watsonville. July–August.
11b. A. exarata Trin. var. **pacifica** Vasey. Pacific Bent Grass. Fairly common along streams, occasionally in bogs, and in other moist areas; San Francisco, Pescadero, Stanford, near Boulder Creek, near Felton, Camp Evers, and Santa Cruz. May–August.
12. A. microphylla Steud. Small-leaved Bent Grass. Occasional in serpentine areas, along the coast, and in very wet places where water has stood during the spring; San Francisco, Crystal Springs Lake, Jasper Ridge, and Santa Cruz. May–July.

<h3 style="text-align:center">47. Ammophila Host</h3>

Ligules 10–30 mm. long, thin. 1. *A. arenaria*
Ligules 1–3 mm. long, firm. 2. *A. breviligulata*

1. A. arenaria (L.) Link. Beach Grass. Sand dunes along the coast, once planted to stabilize the dunes, now well established; San Francisco, Waddell Creek, Santa Cruz, and Pajaro River; native of Europe. June–October. The sand dunes which once covered much of the western portion of San Francisco and what is now Golden Gate Park were planted with *A. arenaria* prior to other reclamation efforts.

2. A. breviligulata Fernald. Known locally from sand dunes near the mouth of Gazos Creek; native of eastern North America. June–August.

48. Sporobolus R. Br. Dropseed

1. S. poiretii (Roem. & Schult.) Hitchc. Smutgrass. Known as a lawn weed in San Francisco and to be expected elsewhere; native of tropical Asia. May–December.

6. Chlorideae. Grama Tribe

49. Beckmannia Host

1. B. syzigachne (Steud.) Fern. Slough Grass. Known definitely from Cahill Ridge and reported from Saratoga. June–July. *—B. erucaeformis* of authors.

50. Leptochloa Beauv.

1. L. fascicularis (Lam.) Gray. Sprangletop Grass. Rare as a weed in San Francisco; native throughout much of the United States, southward to South America, undoubtedly introduced locally. August–October.

51. Cynodon Rich.

1. C. dactylon (L.) Pers. Bermuda Grass. A very common weed of drainage ditches, lawns, roadsides, and other disturbed habitats; native of Europe. May–October.

52. Spartina Schreb.

1. S. foliosa Trin. *Fig. 34.* California or Pacific Cordgrass. Salt marshes bordering San Francisco Bay; San Francisco, Dumbarton Bridge, Cooleys Landing, and Palo Alto. October–November.

7. Phalarideae. Canary Grass Tribe

53. Anthoxanthum L.

1. A. odoratum L. Sweet Velvet Grass, Sweet Vernal Grass. Occasional as a weed in lawns and in moist disturbed areas; San Francisco and probably elsewhere; native of Eurasia. March–May.

54. Hierochloe R. Br.

1. H. occidentalis Buckl. *Fig. 35.* California or Western Vanilla Grass. A common grass in redwoods and redwood–Douglas fir forests, northern San Mateo County southward; Pescadero Creek, San Gregorio Creek, La Honda, Kings Mountain, Searsville, near Stanford, Wrights, near Felton, and near Mount Madonna. January–May. *—H. macrophylla* Thurb. ex Bolander.

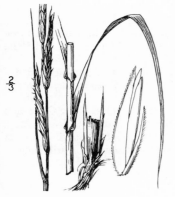

34

35

55. **Phalaris** L. Canary Grass

Each fertile spikelet subtended by 6 sterile spikelets; sterile lemmas none.
    Upper sterile spikelets long pedicellate.   1*a. P. paradoxa paradoxa*
    All spikelets short pedicellate.   1*b. P. paradoxa praemorsa*
All spikelets usually fertile; sterile lemmas 1 or 2.
    Plants perennial.
        Sterile lemmas 2 per spikelet; glumes 6–8 mm. long; panicles 2–5 cm. long,
            2–2.5 cm. in diameter.   2. *P. californica*
        Sterile lemma 1 per spikelet; glumes 5–6 mm. long; panicles 5–15 cm. long,
            about 1.5 cm. in diameter.   3. *P. tuberosa stenoptera*
    Plants annual.
        Sterile lemma 1 per spikelet.   4. *P. minor*
        Sterile lemmas 2 per spikelet.
            Glumes conspicuously and broadly winged, 7–8 mm. long; sterile lemmas
                about ½ the length of the fertile ones.   5. *P. canariensis*
            Glumes narrowly winged or not, 3.5–4 mm. long; sterile lemmas about ⅓
                the length of the fertile ones.
                Glumes not winged.   6. *P. lemmoni*
                Glumes narrowly winged.   7. *P. angusta*

1*a*. **P. paradoxa** L. var. **paradoxa.** Paradox Canary Grass. Occasional as a
weed in cultivated fields and disturbed areas; San Francisco and Stanford; native of the Mediterranean region. April–June.
1*b*. **P. paradoxa** L. var. **praemorsa** (Lam.) Coss. & Dur. Occasional as a weed
in disturbed areas; San Francisco and Palo Alto; native of the Mediterranean
region. June–July.
2. **P. californica** H. & A. *Fig. 36.* California Canary Grass. Along streams,
coastal bluffs, and other moist areas; San Francisco, Crystal Springs, Cahill
Ridge, Moss Beach, near San Mateo–Santa Cruz county line, Black Mountain,
Saratoga, near Los Gatos, and near Gilroy. April–June.
3. **P. tuberosa** L. var. **stenoptera** (Hack.) Hitchc. Harding Grass. Known
locally from disturbed areas and as an escape from grain fields along the coast;
San Francisco, near San Bruno, and San Gregorio Creek; native of Eurasia.
June–August.

4. P. minor Retz. Mediterranean Canary Grass. Disturbed areas throughout the Santa Cruz Mountains, probably fairly common, but poorly collected; San Francisco, Crystal Springs, Redwood City, Palo Alto, Gilroy, Watsonville, and Palm Beach; native of the Mediterranean region. April–October.
5. P. canariensis L. Canary Grass. Occasional as a weed in gardens and disturbed areas; San Francisco, Palo Alto, Santa Cruz, and Watsonville; native of the Mediterranean region. June–August.
6. P. lemmoni Vasey. Lemmon's Canary Grass. Occasional in Santa Clara and Santa Cruz counties, usually in low moist areas; Saratoga and Camp Evers. May.
7. P. angusta Nees ex Trin. Narrow Canary Grass. Occasional in low wet ground or along the edges of ponds; Santa Cruz and Watsonville. April–June.

8. Paniceae. Millet Tribe

56. Cenchrus L. Burgrass

1. C. pauciflorus Benth. Small-flowered Sandbur. Known locally only from Stanford and Gilroy; perhaps introduced locally from other parts of the western hemisphere. July–September.

57. Setaria Beauv. Bristle Grass

Each spikelet subtended by 5 or more bristles; plants annual or perennial.
    Plants annual; spikelets 3 mm. long.   1. *S. glauca*
    Plants perennial; spikelets 2–2.5 mm. long.   2. *S. geniculata*
Each spikelet subtended by 1–3 bristles; plants annual.   3. *S. viridis*

1. S. glauca (L.) Beauv. Yellow Bristle Grass. Lawn weeds and in disturbed areas; San Francisco and Boulder Creek; native of Europe. July–October. —*Chaetochloa lutescens* (Wiegel) Stuntz.
2. S. geniculata (Lam.) Beauv. Perennial Foxtail, Knotroot Bristle Grass. Known locally so far only as a lawn weed in San Francisco; native of eastern North America and South America. June–September. —*Chaetochloa geniculata* (Lam.) Millsp. & Chase.
3. S. viridis (L.) Beauv. Green Bristle Grass. Occasional in disturbed areas; San Francisco, Stanford, San Jose, and Boulder Creek; native of Europe. July–August. —*Chaetochloa viridis* (L.) Scribn.

58. Pennisetum Rich.

Culms erect; panicles long-exserted beyond the upper sheaths.   1. *P. villosum*
Culms creeping; panicles partly hidden by the upper sheaths.
                                         2. *P. clandestinum*

1. P. villosum R. Br. Feathertop. Occasional as weeds; San Francisco and Santa Cruz; native of Africa. August–October.
2. P. clandestinum Hochst. ex Chiov. Kikuyu Grass. A weed of lawns, gardens, and disturbed areas in San Francisco and probably elsewhere; native of Africa. August–October.

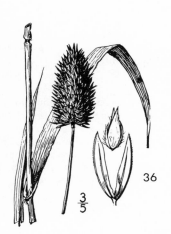

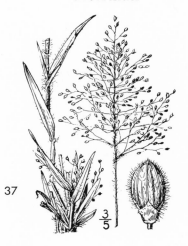

59. Panicum L.

Spikelets 4.5–5 mm. long; plants annual.   1. *P. miliaceum*
Spikelets about 2–4 mm. long; plants annual or perennial.
   Plants annual; mature panicles usually at least 20 cm. long.   2. *P. capillare*
   Plants perennial; mature panicles rarely over 8 cm. long.   3. *P. pacificum*

1. P. miliaceum L. Broomcorn Millet. Occasional as a weed in San Francisco and near Boulder Creek; native of Eurasia. July–October.
2. P. capillare L. Old Witch Grass. Rare as a weed in San Francisco and Boulder Creek; native of the eastern part of the United States. July–September. —*P. capillare* L. var. *occidentale* Rydb.
3. P. pacificum Hitchc. & Chase. *Fig. 37.* Pacific Panicum. Occasional in meadows where water seeps in the spring and early summer; Hilton Airport, near Felton, and Graham Hill Road. May–July.

60. Stenotaphrum Trin.

1. S. secundatum (Walt.) Kuntze. Saint Augustine Grass. Occasional as a weed in Golden Gate Park in San Francisco; native of the southeastern part of the United States and the American tropics. July–August.

61. Digitaria Heist. ex Adans.

Sheaths glabrous; spikelets about 2 mm. long.   1. *D. ischaemum*
Sheaths pilose; spikelets about 3 mm. long.   2. *D. sanguinalis*

1. D. ischaemum (Schreb.) Schreb. ex Muhl. Common as a lawn weed and in disturbed areas; native of Eurasia. September–October. —*Syntherisma ischaemum* (Schreb.) Nash.
2. D. sanguinalis (L.) Scop. Crab Grass. A very common weed of lawns, gardens, and disturbed areas, rarely collected; native of Europe. May–September. —*Syntherisma sanguinalis* (L.) Dulac.

62. Paspalum L.

Culms tufted, erect or geniculate at the base; racemes usually 3–5.
                                                        1. *P. dilatatum*
Culms from creeping rhizomes, erect; racemes commonly 2 (rarely 1 or 3).
                                                        2. *P. distichum*

1. P. dilatatum Poir. Dallis Grass. Occasional as lawn weeds and along the edges of sloughs; San Francisco, Stanford, and Watsonville; native of South America. July–September.

2. P. distichum L. Knot Grass. Edges of salt marshes, along the edges of ponds, streams, and lakes, and in lawns; often growing in water to about 1½ feet deep; San Francisco, Searsville Lake, Palo Alto, near Alviso, Campbell Creek, near Coyote, Santa Cruz, and near Watsonville. July–October.

63. Echinochloa Beauv.

1. E. crusgalli (L.) Beauv. Barnyard Grass. Common as a weed in moist soil, often along drainage ditches. July–December. —*E. crusgalli* (L.) Beauv. var. *zelayensis* (HBK.) Hitchc., *E. crusgalli* (L.) Beauv. var. *mitis* (Pursh) Peterm., *E. crusgalli* (L.) Beauv. var. *frumentacea* (Roxb.) Wight.

9. Andropogoneae. Sorghum Tribe

64. Eremochloa Buese

1. E. ciliaris (L.) Merrill. This species was collected in San Francisco by Bolander and has not been recollected; native of southeastern Asia.

65. Sorghum Moench

Plants perennial; stout creeping rhizomes present, culms 0.5–1.5 m. tall.
                                                        1. *S. halepense*
Plants annual; lacking rhizomes.
    Culms 2–3 m. tall; leaf-blades 0.8–1.2 cm. wide; inflorescence loose.
                                                        2. *S. sudanense*
    Culms usually less than 0.5 m. tall; leaf-blades 4–5 cm. wide; inflorescence
        very dense.   3. *S. vulgare*

1. S. halepense (L.) Pers. Johnson Grass. Occasional as a weed, usually in ditches and cultivated fields; San Francisco, Santa Cruz, and Watsonville; native of the Mediterranean region. June–October. —*Holcus halepensis* L.

2. S. sudanense (Piper) Stapf. Sudan Grass. Known from the waterfront in San Francisco; native of the Anglo-Egyptian Sudan. May–July.

3. S. vulgare Pers. Sorghum. Occasional as an escape from cultivation, known locally from Palo Alto; native of Eurasia. September–November. —*S. bicolor* (L.) Moench.

12.   CYPERACEAE. Sedge Family

Each pistillate flower enclosed within a sac-like bract, the perigynium. 1. *Carex*
Each pistillate flower not enclosed within a perigynium.
  Scales of the spikelets 2-ranked; spikelets laterally flattened; bristles none.
                                                        2. *Cyperus*
  Scales of the spikelets spirally arranged; spikelets not laterally flattened;
    bristles present or not.
    Spikes terminal, solitary, bractless; base of the style persistent as a tubercle.
                                                     3. *Eleocharis*
    Spikes terminal or lateral, solitary or not, subtended by one or more bracts;
      base of the style not persistent as a tubercle.
      Base of the style swollen; bristles lacking.   4. *Fimbristylis*
      Base of the style not swollen; bristles usually present.
        Bristles many, several times as long as the achene.   5. *Eriophorum*
        Bristles few, usually about as long as the achene.   6. *Scirpus*

1. Carex L. Sedge, Carex

Stigmas 2; achenes lenticular.
  Lateral spikelets sessile.
    Culms solitary or a few from long-creeping rhizomes.
      Beaks of the perigynia ⅕–⅓ the length of the bodies; perigynia chest-
        nut-colored; lower sheaths light brown.   1. *C. simulata*
      Beaks of the perigynia ⅓–½ as long as the bodies; perigynia blackish
        in age; lower sheaths dark brown to black.
        Beaks of the perigynia hyaline at the orifice; scales light brown.
                                          2. *C. praegracilis*
        Beaks of the perigynia not hyaline at the orifice; scales dark brown.
                                          3. *C. pansa*
    Culms several to many, more or less caespitose; rhizomes if present not
      long-creeping.
      Spikes androgynous.
        Spikes few, few-flowered; inflorescences slender, not congested.
                                         4. *C. tumulicola*
        Spikes numerous, many-flowered; inflorescences usually not slender,
          usually congested.
          Perigynia blackish-brown at maturity, more or less biconvex; inflores-
            cences loosely branched at least below. Rare.   5. *C. cusickii*
          Perigynia yellowish or tinged with brown in age, plano-convex; inflo-
            rescences congested. Common.
            Scales shorter than the perigynia, acute to cuspidate, usually not
              conspicuously awned; perigynia ovate, 3.5–4.5 mm. long.
                                          6. *C. densa*
           Scales longer than the perigynia, conspicuously awned; perigynia
              ovate-lanceolate, 2.5–3 mm. long.   7. *C. dudleyi*

At least the upper spikelet gynaecandrous.
Perigynia essentially wingless, the lower portion spongy-thickened.
    Perigynia spreading at maturity.   8. *C. phyllomanica*
    Perigynia appressed at maturity.
        Beaks of the perigynia deeply bidentate; bodies of the perigynia
            hidden by the scales.   9. *C. bolanderi*
        Beaks of the perigynia shallowly bidentate; upper part of the
            bodies of the perigynia exposed.   10. *C. leptopoda*
Mature perigynia winged, the lower portion not spongy-thickened.
    Perigynia 3.5 mm. long or less.
        The adaxial surface of the perigynia distinctly nerved; perigynia
            about 3 mm. long.   11. *C. montereyensis*
        The adaxial surface of the perigynia not nerved or obscurely so;
            perigynia 2.7–3.5 mm. long.
            Perigynia appressed, 3–3.5 mm. long; culms stiff; leaf-blades
                2–3.5 mm. wide.   12. *C. subfusca*
            Perigynia more or less spreading, 2.7–3.3 mm. long; culms
                slender, not stiff; leaf-blades 1–2.5 mm. wide.
                        13. *C. teneraeformis*
    Perigynia 3.5–5 mm. long.
        The adaxial surface of the perigynia not nerved or obscurely so
            toward the base.
            Spikelets with 10–20 perigynia, congested; inflorescences usually
                capitate; leaf-blades 2.5–4 mm. wide.   14. *C. subbracteata*
            Spikelets with 4–12 perigynia, not conspicuously congested; in-
                florescences not strongly capitate; leaf-blades 1–2 mm. wide.
                        15. *C. gracilior*
        The adaxial surface of the perigynia definitely nerved.
                        16. *C. harfordii*
Lateral spikelets pedicellate.
    Lowest bract of the inflorescence sheathing; perigynia essentially beakless;
        pistillate spikes usually few-flowered.
        Scales with a distinct awn.   17. *C. salinaeformis*
        Scales awnless.   18. *C. hassei*
    Lowest bract of the inflorescence not sheathing; perigynia at least short-
        beaked; pistillate spikes many-flowered.
        Scales 1–2 times as long as the mature perigynia.   19. *C. obnupta*
        Scales about equaling or shorter than the perigynia.
            Flowering culms arising laterally, not surrounded by the bases of the
                leaves of the previous year; pistillate spikes 1–4 cm. long.
                        20. *C. nudata*
            Flowering culms not arising laterally, surrounded at the base by the
                leaves of the previous year; pistillate spikes 2.5–20 cm. long.
            Beaks of the perigynia bidentate.   21. *C. barbarae*
            Beaks of the perigynia not bidentate.
                 Leaf-blades 6–12 mm. wide; lower sheaths keeled; plants 10–15
                    dm. tall.   22. *C. schottii*
                 Leaf-blades 3–5 mm. wide; lower sheaths not keeled; plants 3–10
                    dm. tall.   23. *C. senta*

Stigmas 3; achenes trigonous.
  Bodies of the perigynia pubescent.
    Lowest pistillate spikelet arising from among the basal leaves.
      Perigynia with several nerves between the lateral ribs.   24. *C. globosa*
      Perigynia without nerves between the lateral ribs.   25. *C. brevicaulis*
    Lowest pistillate spikelet not arising from among the basal leaves.
      Perigynia densely pubescent; beaks prominently bidentate; scales acute,
        prolonged into a short awn.   26. *C. lanuginosa*
      Perigynia sparingly pubescent; beaks minutely bidentate; scales rounded,
        rarely apiculate.   27. *C. gynodynama*
  Bodies of the perigynia glabrous (the beaks rarely pubescent).
    Perigynia rarely over 4 mm. long including the beaks.
      Leaves usually about 1.5 cm. wide.   28. *C. amplifolia*
      Leaves about 0.5 cm. wide.   29. *C. serratodens*
    Perigynia 6–10 mm. long including the beaks.
      Beaks of the perigynia about 1.5 mm. long, the teeth about 0.5 mm. long.
        30. *C. rostrata*
      Beaks of the perigynia 2–3 mm. long, the teeth 1–1.5 mm. long.
        Perigynia 7–10 mm. long, 2–3 mm. wide at the widest part.
          31. *C. exsiccata*
        Perigynia 6–7 mm. long, 1.5 mm. wide at the widest part. 32. *C. comosa*

1. C. simulata MacKenzie. Short-beaked Sedge. Known from bogs in Santa Cruz County; Camp Evers and near Ben Lomond. April–August.
2. C. praegracilis Boott. Clustered Field Sedge. Boggy areas along the edges of salt marshes; San Francisco, Woodside, Mayfield, and Camp Evers. April–July.
3. C. pansa Bailey. Sand Dune Sedge. Known locally from sand dunes in San Francisco where it is now probably extinct. March–June.
4. C. tumulicola MacKenzie. Foothill Sedge. Open grassy areas and occasionally in marshes; San Francisco, San Bruno Hills, Pilarcitos Lake, Crystal Springs, near San Mateo, Coal Mine Ridge, Stanford, Permanente Ravine, near Los Gatos, Boulder Creek, and Soquel Valley. March–June.
5. C. cusickii MacKenzie. Cusick's Sedge. Occasional in swamps and bogs; San Francisco, Camp Evers, and near Ben Lomond. May–July.
6. C. densa (Bailey) Bailey. Dense Sedge. One of the more common species of *Carex* locally, usually in marshes, moist grassland, boggy ground, and occasionally in roadside ditches, San Francisco southward. March–June. Occasional specimens approach *C. vicaria* Bailey in having perigynia that are weakly nerved adaxially. *Carex densa* is definitely nerved adaxially.
7. C. dudleyi MacKenzie. Dudley's Sedge. Occasional in moist situations in northern San Mateo County; near South San Francisco, Brisbane, and east of Burlingame. April–June.
8. C. phyllomanica Boott. Coastal Stellate Sedge. Known so far only from boggy areas at Camp Evers in central Santa Cruz County. May–July.
9. C. bolanderi Olney. Bolander's Sedge. Along streams and in marshes; Pilarcitos Creek, San Mateo Creek, Kings Mountain, near Holy City, Wrights, Llagas Post Office, Glenwood, near Boulder Creek, Ben Lomond, and Felton. May–July.

10. C. leptopoda MacKenzie. Short-scaled Sedge. Known only from Big Basin. June.

11. C. montereyensis MacKenzie. Monterey Sedge. Open slopes and in shaded areas in the redwoods; San Francisco, Cahill Ridge, Año Nuevo Point, near Pescadero, near Boulder Creek, Jamison Creek, Big Basin Highway, and near Camp Evers. April–July.

12. C. subfusca Boott. Rusty Sedge. Rare locally in moist places; San Francisco, Pescadero Creek, and near Big Basin. April–July.

13. C. teneraeformis MacKenzie. Sierra Slender Sedge. Moist areas in the redwoods; near Basin Way, near Boulder Creek, and near Felton. May–June.

14. C. subbracteata Mackenzie. Small-bracted Sedge. Fairly common in marshy ground, wet meadows, and in moist woods; San Francisco, South San Francisco, Cahill Ridge, Crystal Springs Lake, Half Moon Bay, Pescadero Creek, Coal Mine Ridge, Wrights, near Camp Evers, New Brighton, and Santa Cruz. April–June.

15. C. gracilior MacKenzie. Slender Sedge. Occasional, growing in boggy ground, along streams, along the edges of ponds, and on wet grassy slopes; San Francisco, Seal Cove, Pescadero, Corte Madera Creek, near Camp Evers, and Watsonville. March–July.

16. C. harfordii MacKenzie. Harford's Sedge. Bogs, edges of ponds, and in sandy flats near the coast; San Francisco, South San Francisco, Cahill Ridge, Pigeon Point, near Pescadero, Boulder Creek, Graham Hill, and Camp Evers. May–August.

17. C. salinaeformis MacKenzie. Deceiving Sedge. Known only from boggy ground at Camp Evers. May–June.

18. C. hassei Bailey. Hasse's Sedge. Known locally from Loma Prieta. June.

19. C. obnupta Bailey. Slough Sedge. Salt marshes and low wet ground, usually near the coast; San Francisco, Pilarcitos Lake, Pescadero, Butano Ridge, mouth of Waddell Creek, near Boulder Creek, near Ben Lomond, near Felton, and Santa Cruz. April–November.

20. C. nudata Boott. Torrent Sedge. Stream beds, often growing in crevices between rocks; Peters Creek, Stanford, San Jose, Wrights, Big Basin, and San Lorenzo River. March–May.

21. C. barbarae Dewey. Santa Barbara Sedge. Common in wet meadows and on moist, brush-covered slopes, San Francisco southward. April–August.

22. C. schottii Dewey. Schott's Sedge. Rare locally; near Pescadero, Sargents, and Watsonville. March–July.

23. C. senta Boott. Rough Sedge. Occasional along creeks and streams from northern Santa Clara County southward; Stanford and Coyote Creek. April–May.

24. C. globosa Boott. *Fig. 38.* Round-fruited Sedge. In redwoods and on wooded slopes; San Francisco, Big Basin, near Glenwood, Boulder Creek, near Felton, Ben Lomond Sand Hills, and near Santa Cruz. May–August.

25. C. brevicaulis MacKenzie. Short-stemmed Sedge. San Francisco south to central Santa Cruz County, growing in exposed areas, often in sandy soil; San Francisco, San Bruno Hills, near Seal Cove, San Pedro, Pigeon Point, Ben Lomond Mountain, Ben Lomond Sand Hills, and near Felton. March–June.

26. C. lanuginosa Michx. Woolly Sedge. Occasional in moist areas in central Santa Cruz County; near Big Basin and Ben Lomond. April–May.

27. C. gynodynama Olney. Olney's Hairy Sedge. Rare in moist places near the coast; Pescadero, Boulder Creek, and Empire Grade. April–May.

28. C. amplifolia Boott. Ample-leaved Sedge. Marshy areas and along streams,

San Mateo and Santa Cruz counties; Pilarcitos Lake, Pilarcitos Canyon, Kings Mountain, Squealer Gulch, and Boulder Creek. May–September.
29. C. serratodens Boott. Bifid Sedge. Most common in serpentine soils along springs and seeps, San Mateo County southward; Spring Valley, near Woodside, Emerald Lake, Kings Mountain, Little Los Gatos Creek, near Los Gatos, Loma Prieta, and between Madrone Springs and Llagas Creek. April–August. —*C. bifida* of authors.
30. C. rostrata Stokes. Beaked Sedge. Rare locally, known from bogs at Camp Evers in Santa Cruz County and from near San Jose. June–July.
31. C. exsiccata Bailey. *Fig. 39.* Western Inflated Sedge. Occasional in low marshy ground along ponds; between Woodside and Crystal Springs, Wrights, and Big Basin. April–July.
32. C. comosa Boott. Bristly Sedge. Known locally only from San Francisco and Santa Cruz County. June–July.

2. Cyperus L.

Spikelets 1-flowered.	1. *C. brevifolius*
Spikelets several-flowered.	
Styles bifid; achenes lenticular.	
Achenes flattened dorsally.	2. *C. laevigatus*
Achenes flattened laterally.	3. *C. niger capitatus*
Styles trifid; achenes trigonous.	

Spikelets deciduous above the basal pair of scales; inflorescence at length consisting only of the rachis of the spike and the basal scales of each spikelet. 4. *C. strigosus*
Spikelets persistent; scales deciduous, the inflorescence at length consisting of the rachis of the spike and the rachillae of the spikelets.
Rachillae of the spikelets not winged.
Plants perennial, rhizomatous; scales about 2 mm. long.
Stamen 1; bracts 5–8, very unequal; culm-leaves with blades. 5. *C. eragrostis*
Stamens 3; bracts 18–20, equal in length; culm-leaves reduced to sheaths. 6. *C. alternifolius*
Plants annual, not rhizomatous; scales about 0.7 mm. long. 7. *C. difformis*
Rachillae of the spikelets winged with hyaline appendages at each node.
Scales 1–1.5 mm. long; plants annual, not stoloniferous. 8. *C. erythrorhizos*
Scales 2.5–3 mm. long; plants perennial, stoloniferous. 9. *C. esculentus*

1. C. brevifolius (Rottb.) Hassk. Occasional as a weed of lawns and moist disturbed areas; San Francisco and Boulder Creek; native of tropical regions. July.
2. C. laevigatus L. Smooth Cyperus. Occasional in southern Santa Cruz County in wet alkaline soils; near Chittenden and Soda Lake. June–October.
3. C. niger R. & P. var. **capitatus** (Britt.) O'Neill. Brown Cyperus. Wet areas in marshes and along the edges of ponds; San Francisco, Alviso, Camp Evers, and between Aptos and Watsonville. July–October. —*C. melanostachyus* HBK.

4. C. strigosus L. Straw-colored Cyperus. Known locally only from Boulder Creek but to be expected elsewhere. September.

5. C. eragrostis Lam. Tall Cyperus. Common in low wet depressions, drainage ditches, river bottoms, and marshes; San Francisco southward. May–October. —*C. vegetus* Willd.

6. C. alternifolius L. Umbrella Plant. Occasional as an escape from cultivation; San Francisco, Palo Alto, and Santa Cruz; native of Madagascar. March–August.

7. C. difformis L. Known locally only from San Francisco where it has been collected in wet areas near San Francisco State College; native of Asia. July–October.

8. C. erythrorhizos Muhl. Red-rooted Cyperus. Occasional along the edges of ponds and lakes in southern Santa Cruz County. October–November.

9. C. esculentus L. Chufa, Yellow Nutgrass. Sloughs, moist areas, and occasionally as a lawn weed; San Francisco, near Boulder Creek, and Watsonville. July–November.

3. Eleocharis R. Br.   Spike-rush

Styles trifid; achenes trigonous.
    Achenes with several longitudinal ridges and with fine horizontal lines.
        Stamens 2; scales straw-colored; culms spongy, striate.   1. *E. radicans*
        Stamens 3; scales brown; culms capillary, furrowed.   2. *E. acicularis*
    Achenes not longitudinally ridged, the surface finely reticulate.
        Tubercles subulate, continuous with the apex of the achene.   3. *E. rostellata*
        Tubercles conic, constricted at the base.   4. *E. montevidensis*
Styles bifid; achenes lenticular.
    Plants perennial, rhizomatous; tubercles pyramidal, constricted at the base.
                        5. *E. macrostachya*
    Plants annual, roots fibrous; tubercles lamelliform.   6. *E. obtusa*

38   39

1. E. radicans (Poir.) Kunth. Occasional along the edges of lakes and ponds; San Francisco, Pinto Lake, and near Watsonville. August–November. —*E. acicularis* (L.) R. & S. var. *radicans* (Poir.) Britton.
2. E. acicularis (L.) R. & S. Needle Spike-rush. Marshes and along the edges of lakes and ponds; San Francisco, Pilarcitos Canyon, Coal Mine Ridge, Stanford, near Mountain View, and Watsonville. May–August.
3. E. rostellata (Torr.) Torr. Walking Sedge, Beaked Spike-rush. Occasional in fresh or brackish water marshes; San Francisco and Watsonville Slough. August. —*E. rostellata* (Torr.) Torr. var. *congdonii* Jeps.
4. E. montevidensis Kunth. Dombey's Spike-rush. Occasional, sandy creek bottoms, marshes, and springs; Emerald Lake, near Mayfield, Alviso, Loma Azule, Llagas Creek, and Pajaro River. May–November. —*E. montana* of authors.
5. E. macrostachya Britton. Wire Grass, Common or Creeping Spike-rush. Fairly common along the edges of lakes and marshes; San Francisco, San Pedro, Emerald Lake, Coal Mine Ridge, Stanford, Palo Alto, near Alviso, Santa Cruz County, and Watsonville. April–October. —*E. palustris* of authors.
6. E. obtusa (Willd.) Schult. Blunt Spike-rush. Known locally so far only from Pinto Lake in southern Santa Cruz County. September–November.

4. Fimbristylis Vahl

1. F. miliacea (Thunb.) Vahl. Grass-like Fimbristylis. Collected in San Francisco nearly one hundred years ago and not subsequently recollected; native of the tropics, occasional as a weed elsewhere.

5. Eriophorum L.

1. E. gracile Koch. Slender Cotton Grass, Cotton Sedge. Recorded from swamps in San Francisco, now probably extinct.

6. Scirpus L. Bulrush, Tule, Club Rush

Involucral bracts not foliaceous.
  Plants annual, rarely over 1 dm. tall; culms filiform.
    Scales obtuse, only slightly keeled near the tip.   1. *S. cernuus californicus*
    Scales acute, strongly keeled on the back.   2. *S. koilolepis*
  Plants perennial, usually at least 2 dm. tall; culms stout.
    Culms leafy, triangular to subterete in cross section; spikelets 1–12, in a dense capitate cluster.
      Involucral bracts 3–10 cm. long; leaf-blades convolute; plants usually under 1 m. tall.   3. *S. americanus*
      Involucral bracts 1–3 cm. long; leaf-blades flat; plants 0.5–2.2 m. tall.   4. *S. olneyi*
    Culms with leaves only at the base, terete; spikelets numerous, in irregular umbels.
      Bristles ciliate to plumose; culms triangular in cross section at the top.   5. *S. californicus*

110   CYPERACEAE

Bristles with retrorse barbs; culms terete in cross section.
Spikelets ovoid; roots fibrous.   6. *S. validus*
Spikelets subcylindric; roots fleshy.
Scales at least 1.5 times as long as the mature achenes; styles trifid.
                                                                             7. *S. acutus*
Scales about as long as the mature achenes; styles trifid or bifid.
                                                                             8. *S. rubiginosus*
Involucral bracts foliaceous.
Spikelets 0.3–0.6 cm. long.   9. *S. microcarpus*
Spikelets 1 cm. long or longer.
Inflorescence umbellate; achenes trigonous.   10. *S. fluviatilis*
Inflorescence a dense capitate cluster; achenes lenticular.   11. *S. robustus*

1. S. cernuus Vahl var. **californicus** (Torr.) Beetle. *Fig. 40.* Low Club Rush.
Moist depressions in stabilized sand dunes, marshes, seeps, and occasionally on
moist coastal bluffs; San Francisco, Mussel Rock, Montara Point, Crystal
Springs Lakes, Pebble Beach, Año Nuevo Point, near Palo Alto, Waddell Creek,
Santa Cruz, and Watsonville. April–October.
2. S. koilolepis (Steud.) Gleason. Dwarf or Keeled Club Rush. Rare in marshes
and low wet areas; San Francisco, San Andreas Lake, and near Boulder Creek.
April–May. —*S. carinatus* (H. & A.) Gray.
3. S. americanus Pers. Three Square. Common in fresh and brackish water
marshes; San Francisco, Moss Beach, Año Nuevo Point, Redwood City, Alviso,
Waddell Creek, and near Watsonville. April–October. —*S. americanus* Pers. var.
polyphyllus (Boeckl.) Beetle.
4. S. olneyi Gray. Olney's Bulrush. Occasional in marshes; San Francisco,
Alviso, Chittenden, and Watsonville. July–October.
5. S. californicus (C. A. Mey.) Steud. California Tule or Bulrush. Common in
fresh and salt water marshes; San Francisco, San Andreas Lake, Alviso, Waddell
Creek, Santa Cruz, between Santa Cruz and Watsonville, and Watsonville. May–
November.
6. S. validus Vahl. American Great Bulrush, Tule. Known locally only from
San Francisco. July–August.
7. S. acutus Muhl. ex Bigel. Common Tule, Viscid Bulrush. Occasional along
the edges of ponds and in marshes; Searsville Lake, Felt Lake, Alviso, and Big
Basin. July–November.
8. S. rubiginosus Beetle. Rare locally in ponds and lakes; San Francisco,
Emerald Lake, and Felt Lake. May–August.
9. S. microcarpus Presl. Small-fruited or Panicled Bulrush. A common plant
of wet areas, springs, and marshes; San Francisco, Pilarcitos Canyon, San Antonio Creek, Soda Springs, Loma Prieta, Glenwood, Big Basin, Blooms Grade,
near Olympia, Mount Hermon, and Watsonville Slough. May–October.
10. S. fluviatilis (Torr.) Gray. River Bulrush. Occasional in marshes and in
marshy areas along rivers and streams; Crystal Springs Lake and Pajaro River.
June–July.
11. S. robustus Pursh. Prairie Bulrush. Salt and fresh water marshes and in
moist roadside ditches; San Francisco, Granada, Crystal Springs Lake, Palo
Alto, Alviso, Santa Cruz, Pajaro River, and Watsonville. May–October. —*S.
paludosus* Nels.

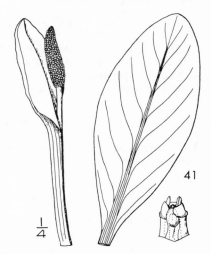

40

41

13. PALMAE. Palm Family

1. Phoenix L.

1. P. canariensis Chaband. Canary Island Date Palm. Seedlings common in the vicinity of mature trees and occasionally sprouting on dump heaps; Stanford and probably elsewhere; native of the Canary Islands.

In addition to those of *Phoenix canariensis*, the seeds of the Guadalupe Island Palm, *Erythea edulis* Watson, occasionally germinate and produce seedlings on the Stanford University Campus. *Erythea* is a fan-leaved palm, while *Phoenix* has pinnate fronds.

14. ARACEAE. Calla or Arum Family

Spathe yellowish.	1. *Lysichiton*
Spathe white.	2. *Zantedeschia*

1. Lysichiton Schott

1. L. americanum Hult. & St. John. *Fig. 41.* Yellow Skunk Cabbage. Known from only a few localities in the Santa Cruz Mountains, growing in wet ground near perennial springs in the redwoods; Butano Creek, Butano Canyon, near Ben Lomond, near Alba School, and near Felton. March–June. —*L. kamtschatcensis* of authors.

2. Zantedeschia Spreng.

1. Z. aethiopica (L.) Spreng. Common Calla, Calla Lily. Occasional as an escape from cultivation, on dump heaps, where the rhizomes have been discarded and in wet or marshy ground; San Francisco, near Davenport, and near Watsonville; native of South Africa. Spring.

15. LEMNACEAE. Duckweed Family

Rootlets present on the underside of the thallus.
    Rootlets more than 1 per thallus.   1. *Spirodela*
    Rootlets 1 per thallus.   2. *Lemna*
Rootlets absent.   3. *Wolffiella*

1. Spirodela Schleid.

1. S. polyrhiza (L.) Schleid. Greater Duckweed. Occasional in freshwater ponds and lakes; Coal Mine Ridge and Watsonville. June–October.

2. Lemna L. Duckweed

Thallus usually less than 5 mm. long, floating, sessile or very short stipitate.
    Thallus thick, obviously fleshy even in dried specimens.   1. *L. gibba*
    Thallus thin, not obviously fleshy in dried specimens.
        Thallus elongate, pale green, more or less translucent, asymmetrical at the
            base.   2. *L. valdiviana*
        Thallus more or less round in outline, dark green, not translucent, usually
            not asymmetrical at the base.
            Thallus 3–5 mm. long, lacking papillae.   3. *L. minor*
            Thallus 1–3 mm. long, with papillae.   4. *L. minima*
Thallus usually 6–12 mm. long, more or less submerged, long stipitate.
                                                        5. *L. trisulca*

1. L. gibba L. Gibbous Duckweed. Pools, ponds, and temporary puddles; San Francisco, Mussel Rock, San Francisquito Creek, Santa Cruz, and Watsonville. June–December.
2. L. valdiviana Philippi. Valdivia Duckweed. Occasional in ponds and temporary puddles; San Francisco, Coal Mine Ridge, Camp Evers, Santa Cruz, and near Watsonville. August–September. —*L. cyclostasa* (Ell.) Chev.
3. L. minor L. Smaller Duckweed. Occasional in ponds, in isolated puddles along streams, and in small lakes; San Francisco, near Woodside, between Pescadero and Swanton, Mountain View–Alviso Road, and Camp Evers. June–January.
4. L. minima Philippi. Least Duckweed. Swampy areas, moist rocks, and borders of ponds; San Francisco, 10 miles south of San Francisco, Alpine Road, Sargents, Santa Cruz, and Watsonville. March–October.
5. L. trisulca L. Ivy-leaved Duckweed. Known locally only from San Francisco.

3. Wolffiella Hegelm.

1. W. lingulata (Hegelm.) Hegelm. Tongue-shaped Wolffiella. Occasional in fresh water ponds and puddles; San Francisco, Mussel Rock, near Freedom, and near Watsonville Junction. June–October.

16.   COMMELINACEAE.   Spiderwort Family

1.   Tradescantia L.

1. T. fluminensis Vell. Wandering Jew, Spiderwort. Becoming well established in San Francisco, where it is widely grown as a ground cover; native of South America.

17.   PONTEDERIACEAE.   Pickerel Weed Family

1.   Eichornia Kunth

1. E. crassipes (Mart.) Solms. Water Hyacinth. Well established at Kelly Lake near Santa Cruz; native of tropical America. Rarely flowering locally. —*Piaropus crassipes* (Mart.) Britton.

18.   JUNCACEAE.   Rush Family

Seeds many; leaves rarely grass-like; sheaths open.   1. *Juncus*
Seeds 3; leaves grass-like; sheaths closed.   2. *Luzula*

1.   Juncus L.   Rush, Wiregrass

Inflorescence appearing lateral, the lower bract of the inflorescence appearing as a continuation of the stem; each flower subtended by 2 scarious bracts.
  Plants caespitose; rhizomes poorly developed; anthers shorter than, to as long as, the filaments; perianth-segments 2–3 mm. long.
    Stamens 6.   1. *J. patens*
    Stamens 3.
      Perianth-segments not very stiff, 2–2.5 mm. long, brown. Plants mainly coastal.   2a. *J. effusus brunneus*
      Perianth-segments stiff, 2.5–3 mm. long, straw-colored. Inland plants.
                                                                2b. *J. effusus pacificus*
  Plants not caespitose; rhizomes well developed; anthers longer than the filaments; perianth-segments 5–6 mm. long.
    Some basal bracts with blades; stems flattened.   3. *J. mexicanus*
    Basal bracts lacking blades; stems terete.
      Perianth about 6 mm. long. Plants of coastal salt marshes and sand dunes.
                                                                         4. *J. leseurii*
      Perianth about 5 mm. long. Plants usually of inland habitats.
                                                                         5. *J. balticus*
Inflorescence appearing terminal, the lowest bract of the inflorenscence not appearing as a continuation of the stem.
  Each flower subtended by 2 scarious bracts; flowers not in heads.
    Plants perennial; plants commonly over 2 dm. tall.
      Leaves more than ½ the length of the stems; perianth-segments with brown scarious margins. Common.   6. *J. occidentalis*

Leaves less than ½ the length of the stems; perianth-segments with white
scarious margins. Rare.   7. *J. tenuis*
Plants annual; plants usually under 2 dm. tall (occasionally to 3 dm.).
                                                                     8. *J. bufonius*
Each flower not subtended by 2 scarious bracts; flowers in heads, the heads 1-
many.
Plants annual, less than 1 dm. tall.
Basal bract of the inflorescence at least twice as long as an individual
flower; flowers usually several per head; stems not filiform.
                                                                     9. *J. capitatus*
Basal bract of the inflorescence only as long as an individual flower;
flowers often solitary; stems filiform.   10. *J. kelloggii*
Plants perennial, over 1 dm. tall.
Leaf-blades septate, terete.
Septa of the leaf-blades complete, *i.e.*, continuous across the blade;
stamens 3 or 6.
Stamens 6.   11. *J. dubius*
Stamens 3.
Heads few, many-flowered.   12. *J. bolanderi*
Heads many, relatively few-flowered.   13. *J. acuminatus*
Septa of the leaf-blades not complete; stamens 6.
Perianth about 3 mm. long; anthers up to as long as the filaments.
                                                                    14. *J. xiphioides*
Perianth about 4–6 mm. long; anthers longer than the filaments.
                                                                    15. *J. phaeocephalus*
Leaves lacking septa, grass-like.   16. *J. falcatus*

1. J. patens Meyer. *Fig. 42.* Spreading or Common Rush. A common rush in
the Santa Cruz Mountains, along streams, in grassland, and in areas that are
wet in the spring; San Francisco, Crystal Springs Lake, Black Mountain, Alviso,
Palo Alto, Loma Prieta, Big Basin, near Boulder Creek, Mount Hermon, and
Watsonville. May–November.
2a. J. effusus L. var. **brunneus** Engelm. Bog Rush. Usually along or near the
coast in bogs and along the borders of lagoons; San Francisco, near Half Moon
Bay, Palo Alto, near Alviso, Glenwood, Santa Cruz, and Seabright. June–
November.
2b. J. effusus L. var. **pacificus** Fern. & Wieg. Usually inland in marshes and
along streams; San Andreas Lake, Crystal Springs Lake, Kings Mountain, Black
Mountain, Palo Alto, Los Gatos Creek, Loma Prieta, near Boulder Creek, and
Watsonville. June–September.
3. J. mexicanus Willd. Mexican Rush. Occasional near lakes and sloughs; San
Francisco, Crystal Springs Lake, and near Watsonville. June–July.
4. J. leseurii Bolander. Salt Rush. Coastal salt marshes, dunes along the
coast, and occasionally inland; San Francisco, Año Nuevo Point, Pilarcitos
Lake, and mouth of Pajaro River. April–November.
5. J. balticus Willd. Baltic Rush. Along ponds, moist places in meadows, and
occasionally in salt marshes; San Francisco, Crystal Springs Lake, Woodside,

Coal Mine Ridge, Palo Alto, Loma Prieta, near Boulder Creek, and Pajaro River. May–October. —*J. balticus* Willd. var. *montanus* Engelm.

6. J. occidentalis (Cov.) Wieg. Western Rush. Fairly common in grassy slopes and meadows throughout the Santa Cruz Mountains from San Francisco southward. April–July. —*J. tenuis* Willd. var. *congesta* Engelm.

7. J. tenuis Willd. Slender Rush. Known locally only from San Francisco, where it was probably introduced; widespread in North America.

8. J. bufonius L. Toad Rush. Common throughout the area covered by this flora in sandy soils, along roadside ditches, in depressions made by cattle in overgrazed areas, and sometimes in marshes, often forming pure stands. April–September. —*J. bufonius* L. var. *fasciculatus* Koch, *J. bufonius* L. var. *congestus* Wahl.

9. J. capitatus Wieg. Rare, known locally only from between Felton and Santa Cruz. May–June.

10. J. kelloggii Engelm. Kellogg's Rush. Occasional in the Santa Cruz Mountains; San Francisco, Jamison Creek Road, and Graham Hill Road. April.

11. J. dubius Engelm. Mariposa Rush. Known from only a few localities; San Andreas Lake and near Los Gatos. June–July.

12. J. bolanderi Engelm. Bolander's Rush. Known from the vicinity of Stanford and to be expected occasionally elsewhere. May–June.

13. J. acuminatus Michx. Sharp-fruited Rush. Occurring along the edges of the pond behind Mill Creek Dam in Santa Cruz County. May–July. —*J. bolanderi* Engelm. var. *riparius* Jeps.

14. J. xiphioides Meyer. Iris-leaved Rush. Along streams and creeks and occasionally in sloughs; San Francisco, Pilarcitos Canyon, Palo Alto, near Alviso, Saratoga, Loma Prieta, Big Basin, Watsonville, and Pajaro River. June–October. —*J. xiphioides* Meyer var. *auratus* Engelm.

15. J. phaeocephalus Engelm. Brown-headed Rush. Widely distributed, growing along the coast in sand dunes, marshes, and sloughs and inland in wet grassy meadows, bogs, and along lakes and streams, San Francisco southward. March–August. *Juncus phaeocephalus* is a variable species in which several subspecific entities have been described, mainly on the basis of the branching patterns of the inflorescences. Plants with few, many-flowered heads have been called *J. phaeocephalus* Engelm. var. *phaeocephalus*. Plants with many, few-flowered heads have been called *J. phaeocephalus* Engelm. var. *paniculatus* Engelm. Plants with many, many-flowered heads have been referred to *J. phaeocephalus* Engelm. var. *glomeratus* Engelm.

16. J. falcatus Meyer. Sickle-leaved Rush. Occasional in moist ground in bogs or in sandy soil; San Francisco, Santa Cruz Mountains, and Camp Evers. May–July.

2. Luzula DC. Wood Rush

1. L. multiflora (Retz.) Lejeune. *Fig. 43.* Common Wood Rush. Common in grasslands and on shaded slopes; San Francisco, San Andreas Lake, Coal Mine Ridge, Stevens Creek, Guadalupe Creek, near Waddell Creek, Empire Grade, near Felton, and Santa Cruz. February–June. —*L. subsessilis* (Wats.) Buch., *L. campestris* DC. var. *congesta* Buch.

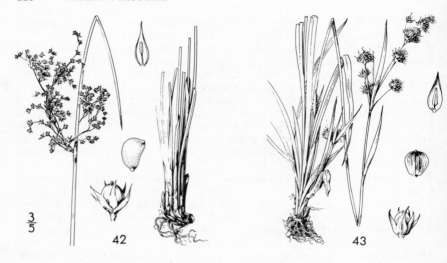

$\frac{3}{5}$

42   43

19.   MELANTHACEAE.   Bunch Flower Family

Plants with short woody rootstalks; racemes very dense.   1. *Xerophyllum*
Plants with membranous-coated bulbs; racemes lax.   2. *Zygadenus*

1.   Xerophyllum Michx.

1. X. tenax (Pursh) Nutt. Elk Grass, Fire Lily, Bear Grass, Indian Basket
Grass, Western Turkey Beard. Known from only two localities in the Santa Cruz
Mountains; near Bonny Doon and Big Basin. May–June.

2.   Zygadenus Michx. Zygadene

Stamens about 1/2 as long as the perianth-segments.
  Plants usually over 3 dm. tall; inflorescences many-flowered.
                                                              1*a*. *Z. fremontii fremontii*
  Plants usually under 3 dm. tall; inflorescences few-flowered.
                                                              1*b*. *Z. fremontii minor*
Stamens as long as or longer than the perianth-segments.   2. *Z. venenosus*

1*a*. Z. fremontii (Torr.) Torr. ex Wats. var. **fremontii.** *Fig. 44.* Fremont's Star
Lily, Fremont's Zygadene. Fairly common in the Santa Cruz Mountains in
woods and on brush-covered slopes; San Francisco, Crystal Springs Lake, Woodside, Pigeon Point, Kings Mountain, Coal Mine Ridge, Stanford, near Los Gatos,
Mount Umunhum, Swanton, Glenwood, and Empire Grade. February–September.
1*b*. Z. fremontii (Torr.) Torr. ex Wats. var. **minor** (H. & A.) Jeps. Occasional
from San Francisco southward, usually on open wind-swept grassy hills; San
Francisco, San Bruno Hills, South San Francisco, near Bonny Doon, and near
Santa Cruz. February–April.
2. Z. venenosus Wats. Death Camas, Deadly Zygadene. Known locally only
from Crystal Springs Lake. April.

20. LILIACEAE. Lily Family

Perianth-segments essentially all alike.
    Leaves basal or nearly so.
        Perianth of separate segments; inflorescence paniculate. Native or intro-
            duced plants.
            Perianth-segments 15–20 mm. long; leaves not stiff, undulate-margined.
                Native plants.   1. *Chlorogalum*
            Perianth-segments 7–10 mm. long; leaves stiff, not undulate-margined.
                Introduced plants.   2. *Chlorophytum*
        Perianth of united segments; inflorescence racemose. Introduced plants.
            Perianth blue, globular-urceolate, 4–5 mm. long; lobes less than 1 mm.
                long, not reflexed.   3. *Muscari*
            Perianth of various colors, swollen at the base, about 25 mm. long; lobes
                about equalling the tube, reflexed.   4. *Hyacinthus*
    Leaves cauline.
        Leaves over 3 dm. long; perianth tubular; inflorescence a dense raceme.
            Introduced plants.   5. *Kniphofia*
        Leaves under 1.5 dm. long; perianth of separate segments; inflorescence
            not a dense raceme. Native plants.
            Perianth-segments 5–8 cm. long, orange; plants usually over 10 dm. tall.
                                                                         6. *Lilium*
            Perianth-segments 1–3.5 cm. long, never orange; plants usually under
                6 dm. tall.   7. *Fritillaria*
Perianth-segments of 2 dissimilar sets.   8. *Calochortus*

1. Chlorogalum Kunth

1. C. pomeridianum (DC.) Kunth. *Fig. 45.* Amole, Soap Plant. Grasslands,
serpentine slopes, and borders of woodland; San Francisco, San Andreas Lake,
near Redwood City, Woodside, Kings Mountain, Stanford, Black Mountain,
Swanton, near Santa Cruz, and near Pajaro River. May–June. —*C. pomeridianum* (DC.) Kunth var. *divaricatum* (Lindl.) Hoover.

2. Chlorophytum Ker

1. C. capense (L.) Druce. Occasionally persisting in San Francisco as an escape
from cultivation; native of South Africa.

3. Muscari Mill.

1. M. botryoides (L.) Mill. Grape Hyacinth. Commonly grown as a garden
ornamental, frequently persisting and spreading into disturbed habitats; San
Francisco and Menlo Park; native of Europe. March–May.

4. Hyacinthus L.

1. H. orientalis L. Common Hyacinth. Occasionally found in disturbed areas
in San Francisco; native of the Mediterranean Region.

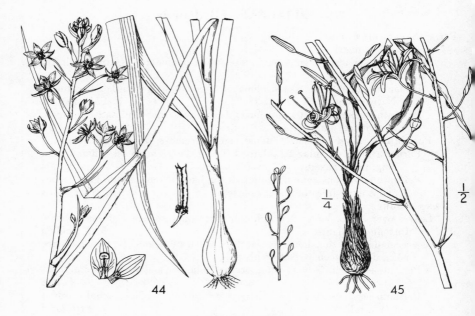

44   45

5. Kniphofia Moench

1. K. uvaria (L.) Hook. f. Red-Hot-Poker Plant, Torch Flower. Widely cultivated as an ornamental, and becoming established in San Francisco and perhaps
elsewhere; native of South Africa.

6. Lilium L.

1. L. pardalinum Kell. Panther, Leopard, or Tiger Lily, California Tiger Lily.
Wet meadows, along springs, and near streams, mainly on the western slopes of
the Santa Cruz Mountains; San Francisco, San Mateo Creek, near La Honda,
Pescadero Creek, Loma Prieta, Big Basin, and Santa Cruz. June–September.

7. Fritillaria L. Fritillary

Leaves in verticils, usually on the upper half of the stem; bulbs with rice-grain
    bulblets.   1. *F. lanceolata*
Leaves usually alternate, on the lower half.of the stem; bulbs lacking rice-grain
    bulblets, composed of several scales.
    Perianth-segments white.   2. *F. liliacea*
    Perianth-segments not white.
        Perianth-segments greenish-yellow.   3. *F. agrestis*
        Perianth-segments brown or purple-greenish.   4. *F. biflora*

1. F. lanceolata Pursh. *Fig. 46.* Checker Lily, Mission Bells, Purple Rice-
bulbed Fritillary. The most common species of Fritillary in the Santa Cruz

Mountains, usually growing on wooded slopes, occasionally in exposed places; San Francisco, Pillar Point, Pilarcitos Canyon, Crystal Springs Lake, Año Nuevo Point, Coal Mine Ridge, Monte Bello Ridge, near Los Gatos, Swanton, San Lorenzo Road, and near Eccles. March–May. —*F. lanceolata* Pursh var. *floribunda* Benth., *F. mutica* Lindl.

2. F. liliacea Lindl. Fragrant or White Fritillary. San Francisco southward to the vicinity of Palo Alto, usually on open grassy hills; San Francisco, South San Francisco, Spring Valley Lakes, Crystal Springs Lake, Redwood City, and Stanford. February–April.

3. F. agrestis Greene. Ill-scented Fritillary, Stink Bells. Known locally only from along the coast at Año Nuevo Point and between Santa Cruz and Soquel. March.

4. F. biflora Lindl. Black Fritillary, Mission Bells, Chocolate Lily. Occasional in the Santa Cruz Mountains; near Hillsborough and San Jose. April–May. —*F. biflora* Lindl. var. *inflexa* Jeps.

8. Calochortus Pursh. Calochortus, Mariposa Lily

Fruits 3-winged, oblong; flowers not yellow.
  Flowers globose, nodding; some leaves usually at least 1 cm. wide.   1. *C. albus*
  Flowers campanulate, erect or spreading; leaves usually under 1 cm. wide.
    Petals conspicuously ciliate, clawed. Common.   2. *C. tolmiei*
    Petals not conspicuously ciliate (except at the base on the adaxial surface),
      cuneate. Rare.
      Stems branched, the internodes long. San Mateo County.
          3. *C. umbellatus*
      Stems unbranched, the internodes very short. Santa Cruz County.
          4. *C. uniflorus*
Fruits 3-angled, linear; flowers sometimes yellow.
  Flowers always yellow; glands of the petals lunate.   5. *C. luteus*
  Flowers usually white, pink, or rose; glands of the petals quadrate.
          6. *C. venustus*

1. C. albus Dougl. ex Benth. *Fig. 47.* White Globe Lily, White Fairy Lantern. The most common species of *Calochortus* in the Santa Cruz Mountains, commonly growing in shade in oak-woodland or in the open, northern San Mateo County southward. April–June. Plants with rose-colored flowers have been called *C. albus* Dougl. ex Benth. var. *rubellus* Greene.

2. C. tolmiei H. & A. Tolmie's or Hairy Star Tulip, Pussy Ears. Usually in oak-madrone woods, occasionally in chaparral, most common on the western slopes of the Santa Cruz Mountains; near Pescadero, La Honda Grade, Kings Mountain, Mount Umunhum, Uvas Canyon, near Big Basin, Big Basin, Little Basin, and Loma Prieta. April–June.

3. C. umbellatus Wood. Oakland Star Tulip. Known locally only from a few localities, usually on wooded slopes; Belmont and Kings Mountain. April–June.

4. C. uniflorus H. & A. Large-flowered Star Tulip. Occasional in Santa Cruz County; Empire Grade, near Santa Cruz, and Soquel. April.

5. C. luteus Dougl. ex Lindl. Yellow Mariposa. Open grassy slopes and in adobe, clay, or rocky soils, San Francisco southward; San Francisco, near Crystal Springs Lake, Kings Mountain, San Francisquito Creek, Searsville, Stanford, near San Jose, near Los Gatos, Wrights, Bielawski Lookout, and Santa Cruz. April–June.

6. C. venustus Dougl. ex Benth. White or Butterfly Mariposa. Southern San Mateo County southward on serpentine and on open slopes; near Redwood City, near Woodside, Portola, Jasper Ridge, Palo Alto, San Jose, and near Mount Hermon. May–July.

21. CONVALLARIACEAE. Lily-of-the-Valley Family

Leaves usually 3, whorled; flowers 1; sepals 2.5 cm. long or longer.   1. *Trillium*
Leaves 2 or more, not whorled; flowers more than 1; sepals less than 2 cm. long.
  Leaves reduced to scales; branches filiform. Introduced plants.   2. *Asparagus*
  Leaves not reduced to scales; branches not filiform. Native plants.
    Leaves basal, 2–6 in number.
      Flowering pedicels over 5 cm. long; stamens 3; flowers mottled with green
        and purple.   3. *Scoliopus*
      Flowering pedicels less than 2.5 cm. long; stamens 6; flowers rose-purple.
                                                   4. *Clintonia*
    Leaves cauline, usually more than 10 in number.
      Flowers terminal in racemes or panicles; stems not branched.
                                        5. *Smilacina*
      Flowers solitary or in few-flowered umbels, pendant; stems branching.
        Flowers terminal.   6. *Disporum*
        Flowers axillary, peduncles adnate to the internode above its origin.
                                      7. *Streptopus*

46   47

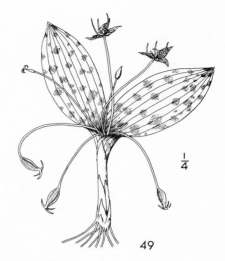

1. **Trillium** L. Wake Robin

Flowers sessile; petals over 3.5 cm. long, usually maroon.   1. *T. chloropetalum*
Flowers pedicellate; petals under 3 cm. long, white to pink, lavender upon drying.
                                                                   2. *T. ovatum*

1. T. chloropetalum (Torr.) Howell. Giant Wake Robin, Common Trillium.
Wooded slopes and thickets, rarely in coniferous forests; San Francisco, Pilar-
citos Canyon, Montara Mountain, Pescadero Beach, near Portola, Coal Mine
Ridge, Stevens Creek, and near Los Gatos. February–April. —*T. sessile* L. var.
giganteum H. & A.
2. T. ovatum Pursh. *Fig. 48.* Western Wake Robin, Coast or Western Trillium.
Redwood–Douglas fir forests, most common to the west of the crests of the Santa
Cruz Mountains; Pilarcitos Canyon, near Pescadero, Woodside, near Saratoga,
Swanton, Ben Lomond, and near Soquel. March–May.

2. **Asparagus** L.

1. A. officinalis L. Garden Asparagus. Occasional as an escape from cultivation
near Gilroy; native of Europe.

3. **Scoliopus** Torr.

1. S. bigelovii Torr. *Fig. 49.* Slink Pod, Brownies, California Fetid Adder's
Tongue. Moist slopes and along creek banks in redwood forests; Pilarcitos Lake,
La Honda, Kings Mountain, Woodside Hills, Los Gatos, Waddell Creek, and near
Santa Cruz. February–March. This species is one of the earliest to flower in our
flora and with its glossy green foliage, often mottled with purplish pigment, is
an attractive element in the herbaceous understory of the redwood forest. By the
end of March the long reflexed pedicels bear maturing capsules. Usually plants
of this species bear only two leaves, but occasionally individuals with three leaves
are found.

4. Clintonia Raf.

1. C. andrewsiana Torr. Red Clintonia. Occasional as a member of the herbaceous understory in moist coniferous or mixed forests; Pescadero, Kings Mountain, Swanton, Ben Lomond, Glenwood, near Felton, Santa Cruz, and Loma Prieta. April–June.

5. Smilacina Desf. False Solomon's Seal

Inflorescence paniculate.	1. *S. racemosa amplexicaulis*
Inflorescence racemose.	2. *S. stellata sessilifolia*

1. S. racemosa (L.) Desf. var. **amplexicaulis** (Nutt.) Wats. Fat Solomon, Western Solomon's Seal. On slopes in coniferous and mixed evergreen forests, usually in shade and in rich soil; Pilarcitos Canyon, Palo Alto, Coal Mine Ridge, Saratoga, near Los Gatos, Glenwood, and Soquel. March–May.
2. S. stellata (L.) Desf. var. **sessilifolia** (Baker) Henders. Slim Solomon. Foothills and mountains, usually on wooded slopes in partial shade; San Francisco, Pilarcitos Canyon, San Andreas Lake, Belmont, Coal Mine Ridge, Stanford, near Los Gatos, Loma Prieta, and Swanton. February–March.
Maianthemum dilatum (Wood) Nels. & Macbr. has been reported from Pescadero Creek by Mrs. E. J. Rebert of San Francisco, but at present no herbarium specimens are available to confirm this record. *Maianthemum dilatum* ranges northward to Alaska and east to Idaho and may be distinguished from *Smilacina* by its cordate-ovate leaves.

6. Disporum Salisb. Fairy Bells

Flowers cylindrical, truncate at the base; styles pubescent; stigma 3-cleft; seeds
    usually more than 6; leaf-margins glabrous or sparsely ciliate.   1. *D. smithii*
Flowers turbinate; styles glabrous or the lower half pubescent; stigma usually
    entire; seeds usually 6 or less; leaf-margins with short, pointed hairs pointing forward.   2. *D. hookeri*

1. D. smithii (Hook.) Piper. Large-flowered Fairy Bell, Fairy Lantern. Redwood–Douglas fir forests in southern San Mateo County and adjacent Santa Clara County; Butano Creek, Kings Mountain, La Honda Road, and Woodside. March–May.
2. D. hookeri (Torr.) Nichols. *Fig. 50.* Hooker's Fairy Bell. Shaded places in redwood–Douglas fir forests, mixed forests, and brush-covered hills; San Francisco, Crystal Springs Lake, La Honda, Kings Mountain, Stevens Creek, Los Gatos, Uvas Canyon, Swanton, and Soquel. March–May.

7. Streptopus Michx. Twisted-Stalk

1. S. amplexifolius (L.) DC. var. **denticulatus** Fassett. Clasping-leaved Twisted-Stalk. Known locally only from the vicinity of Boulder Creek in Santa Cruz County. April–May.

22. AMARYLLIDACEAE. Amaryllis Family

Perianth-segments free or only slightly united at the base, not forming a definite tube.
    Perianth-segments rose, lavender, or purplish, rarely white; plants with the odor of onion.   1. *Allium*
    Perianth-segments yellowish; plants lacking an onion odor.   2. *Muilla*
Perianth-segments united, forming a definite tube.
    Perianth-segments blue to violet, if white then over 10 mm. long; pedicels jointed beneath the flowers. Native plants.   3. *Brodiaea*
    Perianth-segments about 10 mm. long, white, tinged with pink; pedicels not jointed below the flowers. Introduced plants.   4. *Nothoscordum*

1. Allium L. Wild Onion

Scapes terete or triquetrous.
    Outer bulb-coats lacking "herringbone" reticulations.
        Leaves usually 5 mm. wide or wider; pedicels usually 1.5–3 cm. long; perianth-segments about 10 mm. long, white. Introduced plants.
            Perianth-segments broadly elliptic to ovate; scape round in cross section below.   1. *A. neapolitanum*
            Perianth-segments narrowly ovate-lanceolate; scape triquetrous.
                2. *A. triquetrum*
        Leaves about 2 mm. wide or less; pedicels usually under 1 cm. long; perianth-segments 6–8 mm. long, rose-colored. Native plants.   3. *A. lacunosum*
    Outer bulb-coats with "herringbone" reticulations.
        Ovary crested.
            Bracts of the inflorescence about 1 cm. long; perianth-segments 6–8 mm. long.   4. *A. amplectens*
            Bracts of the inflorescence over 1.5 cm. long; perianth-segments over 8 mm. long.
                Inflorescence usually not congested; perianth-segments 10 mm. or more long. Growing in both serpentine and nonserpentine soils.   5. *A. dichlamydeum*
                Inflorescence more or less congested; perianth-segments 8–10 mm. long. Growing in serpentine soils.   6. *A. serratum*
        Ovary not crested.
            Perianth-segments becoming hyaline, under 9 mm. long; scapes coming directly from the bulbs.   7. *A. hyalinum*
            Perianth-segments not becoming hyaline, over 10 mm. long; scapes coming from lateral offshoots.   8. *A. unifolium*
Scapes flattened.   9. *A. breweri*

1. **A. neapolitanum** Cyr. Neapolitan Onion. Occasional as an escape from cultivation in San Francisco and Boulder Creek; native of southern Europe. April–June.
2. **A. triquetrum** L. Becoming naturalized in moist areas in San Francisco; native of Eurasia.

50   51

3. A. lacunosum Wats. Pitted Onion. Serpentine outcroppings from southern San Mateo County southward; Emerald Lake, Redwood City, near Menlo Park, and Jasper Ridge. May–June.

4. A. amplectens Torr. Narrow-leaved Onion. Occasional on serpentine outcroppings; Stanford, Jasper Ridge, and Coyote Creek. May.

5. A. dichlamydeum Greene. Coastal Onion. Rocky and serpentine soils, San Francisco southward; San Francisco, San Bruno Hills, Crystal Springs, San Andreas Lake, Half Moon Bay Road, near Redwood City, Woodside, Jasper Ridge, Stanford, and Page Mill Road. April–June.

6. A. serratum Wats. Serrated Onion. Occasional on serpentine outcroppings; San Mateo, Evergreen, Guadalupe Canyon, Guadalupe Reservoir, Almaden, Edenvale, and Morgan Hill. March–April.

7. A. hyalinum Curran. Paper-flowered Onion. Known locally only from the vicinity of Uvas Creek and Almaden. April–May.

8. A. unifolium Kell. One-leaved Onion. Moist areas along the coast and along ponds and streams, often in shaded areas; San Francisco, Cahill Ridge, near Woodside, Emerald Lake, Año Nuevo Point, and Los Gatos. April–June.

9. A. breweri Wats. *Fig. 51.* Brewer's Onion. Serpentine slopes, Crystal Springs Lake southward to Loma Prieta; Crystal Springs Lake, Redwood City Hills, Emerald Lake, Jasper Ridge, and Loma Prieta. April–May.

2. **Muilla** Wats. Muilla

1. M. maritima (Torr.) Wats. Common Muilla. Serpentine, inland sand deposits, and grassy slopes, San Francisco southward; San Francisco, San Andreas Dam, Woodside, Stanford, near San Jose, near Coyote, and Quail Hollow. February–April.

3. Brodiaea Smith

Inflorescence capitate, congested; pedicels usually much shorter than the flowers.
  Stamens with anthers 6.   1. *B. pulchella*
  Stamens with anthers 3.
    Staminodia obtuse; pedicels completely free.   2. *B. multiflora*
    Staminodia bifid; pedicels more or less united toward the base.
                                   3. *B. congesta*
Inflorescence open, not congested; pedicels usually as long as the flowers or
  longer.
  Stamens 6.
    Perianth-segments 25–40 mm. long, blue; stamens inserted at two levels.
                                   4. *B. laxa*
    Perianth-segments 10–20 mm. long, white, yellow, rarely lilac; stamens in-
      serted at one level.
      Perianth-segments 15–20 mm. long, yellow; filaments 2-toothed at the
        apex.   5. *B. lutea*
      Perianth-segments 10–15 mm. long, white, rarely lilac; filaments not 2-
        toothed at the apex.   6. *B. hyacinthina*
  Stamens 3.
    Anthers each with two appendages on the abaxial surface.
                             7. *B. appendiculata*
    Anthers lacking appendages.
      Scape mainly subterranean, usually no more than 5 cm. of it above ground.
                               8. *B. terrestris*
      Scape mainly above ground, more than 5 cm. of it above ground.
        Staminodia usually shorter than the stamens, acute.   9. *B. elegans*
        Staminodia usually longer than the stamens, usually obtuse and often
          3-lobed at the apex.   10. *B. coronaria*

1. B. pulchella (Salisb.) Greene. *Fig. 52.* Common Saitas, Common Brodiaea, Blue Dicks, Wild Hyacinth. Common on grassy slopes, occasionally on sand dunes, and in brushy places or wooded areas, San Francisco southward. February–May. —*B. capitata* Benth., *Dichelostemma pulchellum* (Salisb.) Heller.
2. B. multiflora Benth. Many-flowered Saitas, Many-flowered Brodiaea. Rare and isolated in southern San Mateo County and near San Jose; San Francisquito Creek, Lake Lagunita, and near San Jose. May. —*Dichelostemma multiflorum* (Benth.) Heller.
3. B. congesta Smith. Narrow Saitas, Ookow. Occasional on grassy slopes and in open wooded areas; Stanford, Page Mill Creek, Stevens Creek, and near Glenwood. April–May. —*Dichelostemma congestum* (Smith) Kunth.
4. B. laxa (Benth.) Wats. *Fig. 53.* Common Triteleia, Grass Nut, Ithuriel's Spear, Triplet Lily. Common, especially on the eastern slopes of the Santa Cruz Mountains in grasslands and on serpentine slopes; San Francisco, Cahill Ridge, Redwood City, Woodside, Stanford, near San Jose, Edenvale, Loma Prieta, near Big Basin, and Santa Cruz. April–June. —*Triteleia laxa* Benth.
5. B. lutea (Lindl.) Mort. Common Calliprora, Pretty Face, Golden Brodiaea. Oak-woodland from northern San Mateo County southward; Woodside to Crystal

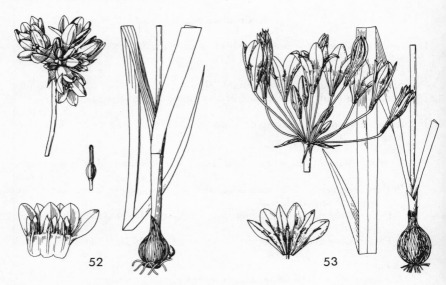

52   53

Springs Lake, La Honda Grade, Kings Mountain, Palo Alto, near San Jose, between Gilroy and Watsonville, and near Boulder Creek. April–June. —*Triteleia ixioides* (Ait. f.) Greene, *Calliprora ixioides* (Ait. f.) Greene.

6. B. hyacinthina (Lindl.) Baker. Wild Hyacinth, White Brodiaea. Meadows and low areas which are rather wet during the early spring; San Francisco, San Andreas Lake, near Woodside, Lake Lagunita, Matadero Creek, Hilton Airport, and Graham Hill. April–June. —*Triteleia hyacinthina* (Lindl.) Greene, *Hesperoscordum hyacinthinum* Lindl.

7. B. appendiculata Hoover. Known from only a few localities in the Santa Cruz Mountains; San Francisquito Creek, Stanford, and near Almaden. May.

8. B. terrestris Kell. Dwarf Brodiaea. Serpentine and sandy soils; San Francisco, Sawyer Ridge, Crystal Springs Lake, near San Mateo, Woodside, Searsville, Stanford, Swanton, Big Basin, Camp Evers, and Graham Hill Road. March–June. —*B. coronaria* (Salisb.) Engler var. *macropoda* (Torr.) Hoover.

9. B. elegans Hoover. Harvest Brodiaea. Occasional in grasslands and on serpentine; San Francisco, near Redwood City, Searsville, Stanford, near Los Gatos, and Eccles. May–June.

10. B. coronaria (Salisb.) Engler. Harvest Brodiaea. Rare in the Santa Cruz Mountains; Hillsborough, Woodside, and Watsonville Junction. May–June.

4. Nothoscordum Kunth. Grace or False Garlic

1. N. inodorum (Ait.) Aschers. & Graebn. Occasional as an escape from cultivation; San Francisco and Palo Alto; probably native of Africa and well established in the southeastern part of the United States. May–June. —*N. fragrans* (Vent.) Kunth.

23. IRIDACEAE. Iris Family

Perianth-segments essentially free, not united into a tube.
Flowers white, in a dense cluster; filaments free. Introduced plants.
                                                          1. *Libertia*
Flowers blue or yellow, not in dense clusters; filaments united into a tube.
    Native plants.   2. *Sisyrinchium*
Perianth-segments united into a tube at the base.
Flowers irregular.   3. *Chasmanthe*
Flowers regular.
    Perianth-segments in 2 dissimilar sets, the outer spreading or reflexed, the
        inner narrower and erect.   4. *Iris*
    Perianth-segments all more or less similar.
    Flowers in an erect spike; perianth-segments longer than the tube.
                                                             5. *Ixia*
    Flowers in a more or less one-sided horizontal spike; perianth-segments
        shorter than the tube.   6. *Freesia*

1. Libertia Spreng.

1. L. formosa Graham. Showy Libertia. Occasionally naturalized on ocean
bluffs in San Francisco; native of Chile.

2. Sisyrinchium L. Blue-eyed Grass

Perianth-segments blue; inflorescences both terminal and lateral; seeds flattened.
                                                          1. *S. bellum*
Perianth-segments yellow; inflorescences terminal; seeds cup-shaped.
                                                      2. *S. californicum*

1. S. bellum Wats. *Fig. 54.* Nigger Babies, California Blue-eyed Grass. Very
common throughout the Santa Cruz Mountains, usually in grasslands, often suffi-
ciently abundant locally to form blue patches among the vegetation; San Francisco, San Bruno Hills, near Belmont, near Half Moon Bay, Woodside, Stanford,
near Los Gatos, Swanton, and near Santa Cruz. February–May.
2. S. californicum (Ker) Dryand. Yellow-eyed Grass, California Golden-eyed
Grass. Usually along or near the coast in boggy areas, occasionally found along
the margins of San Francisco Bay and inland; San Francisco, Colma, Millbrae,
Crystal Springs, near Pescadero, and Camp Evers. March–May. —*Hydastylus
californicus* (Ker) Salisb.

3. Chasmanthe N. E. Brown

1. C. aethiopica (L.) N. E. Brown. Escaping from cultivation in San Francisco;
native of South Africa.

4. Iris L. Iris, Flag

Standards 10 mm. or more wide; flowers white to purple, sometimes yellow.
  Growing on dry land.
  Perianth-tube under 1 cm. long.   1. *I. longipetala*
  Perianth-tube over 1.5 cm. long.
    Perianth-tube 1.5–2 cm. long.   2. *I. douglasiana*
    Perianth-tube more than 3 cm. long.   3. *I. macrosiphon*
Standards about 2–8 mm. wide; flowers yellow. Growing along the edges of
  ponds and lakes.   4. *I. pseudacorus*

1. I. longipetala Herbert. Long-petaled, Coast, or Field Iris. Open grasslands
and slopes, mainly along the coast; San Francisco, South San Francisco, San
Bruno Mountain, San Andreas Lake, Half Moon Bay, and Año Nuevo Point.
March–May. Colonies of *I. longipetala* on the east slope of San Bruno Mountain
are very extensive and, despite overgrazing in that area, manage to survive very
well.
2. I. douglasiana Herbert. *Fig. 55.* Douglas' or Mountain Iris. Open woods,
grasslands, and coastal bluffs, mainly on the west slopes of the Santa Cruz Mountains; San Francisco, near San Bruno, near Millbrae. Upper Pescadero Creek,
Woodside, Rancho del Oso, Swanton, and Big Basin. February–June.
3. I. macrosiphon Torr. Ground or Slender-tubed Iris. Grassy slopes in oak-
madrone woods, edges of redwood–Douglas fir forests, and on brush-covered
slopes; Pescadero Creek, Portola, near Los Gatos, Wrights, Loma Prieta, Gilroy,
near Saratoga Summit, Eagle Rock, Glenwood, Big Basin, and near Santa Cruz.
March–June. —*I. californica* Leichtl., *I. fernaldii* Foster. Flower color is exceedingly variable in this species, which hybridizes with *I. douglasiana*. The
specimen from near Eagle Rock came from a population in which the color of the
flowers varied from white through various shades of lavender and yellow to deep
purple.

54

55

4. I. pseudacorus L. Yellow Iris. Occasional along the edges of lakes and lagoons; Searsville Lake and Neary Lagoon, Santa Cruz. April–June. Native of Europe.

5. Ixia L.

1. I. maculata L. African Ixia. Occasional as an escape from cultivation in San Francisco and becoming established on sand dunes; native of South Africa.

6. Freesia Klatt

1. F. refracta Klatt. Common Freesia. Becoming well established in old gardens in San Francisco; native of South Africa.

24. ORCHIDACEAE. Orchid Family

Flowers large, generally few; inflorescence racemose.
 Lip saccate and inflated.   1. *Cypripedium*
 Lip neither saccate nor inflated.   2. *Epipactis*
Flowers small, many; inflorescence spicate.
 Spur present and free from the ovary.   3. *Habenaria*
 Spur lacking (or in *Corallorhiza maculata* adnate to the ovary).
  Plants with green foliage, not saprophytic.
   Leaves both basal and cauline, linear to linear-lanceolate or oblanceolate, 5–35 cm. long, not marked with white; spike spirally twisted.
                4. *Spiranthes*
   Leaves basal, ovate-lanceolate, 3–6 cm. long, often marked with white along the veins; spike somewhat one-sided but not spirally twisted.
                5. *Goodyera*
  Plants lacking green foliage, plant either white or brownish.
   Entire plant white; sepals and petals 12–15 mm. long.   6. *Cephalanthera*
   Entire plant brownish; sepals and petals 6–10 mm. long.   7. *Corallorhiza*

1. Cypripedium L. Lady's Slipper

Leaves alternate, several.   1. *C. montanum*
Leaves opposite, two.   2. *C. fasciculatum*

1. C. montanum Dougl. ex Lindl. Mountain Lady's Slipper. Very rare in the Santa Cruz Mountains; La Honda Creek, Corte de Madera Creek, and Santa Cruz Mountains. April–May.
2. C. fasciculatum Kell. ex Wats. *Fig. 56.* Clustered Lady's Slipper. Rare in moist woods; Kings Mountain, Stevens Creek, Black Mountain, Loma Prieta, and Glenwood. April–May. To the north, *C. fasciculatum* is not encountered again until Del Norte County, a distance of about 300 miles.

2. Epipactis Sw.

1. E. gigantea Dougl. ex Hook. Stream Orchid, Giant Helleborine. Occasional in moist places among rocks and under Douglas fir; San Francisco, Cahill Ridge, Pescadero Creek, Loma Prieta, and Powder Mill Gulch. May–August.

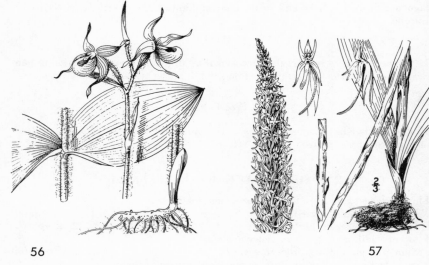

56   57

3. **Habenaria** Willd. Rein Orchis

Stem-leaves several.
   Lip linear, not rhombic at base, 6 mm. long; flowers not white. Inland plants.
                                                          1. *H. sparsiflora*
   Lip rhombic at base, about 8 mm. long; flowers white. Coastal plants.
                                                    2. *H. dilatata leucostachys*
Stem-leaves usually 2–4, commonly near the base.
   Spur equal to or slightly longer than the lip; inflorescence slender, open,
      usually about 1 cm. in diameter.   3a. *H. unalascensis unalascensis*
   Spur at least twice as long as the lip; inflorescence more congested, usually
      over 1.5 cm. in diameter.
      Inflorescence very congested, about 2 cm. in diameter. Coastal plants.
                                                    3b. *H. unalascensis maritima*
      Inflorescence only slightly congested, about 1.5 cm. in diameter. Inland
         plants.   3c. *H. unalascensis elata*

1. **H. sparsiflora** Wats. Sparsely-flowered Bog Orchid. Rare in northern Santa
Clara County, growing near springs; near Stanford, and Black Mountain. June–
July. —*Limnorchis sparsiflora* (Wats.) Rydb.
2. **H. dilatata** (Pursh) Hook. var. **leucostachys** (Lindl.) Ames. White Rein
Orchis, White-flowered Bog Orchid. Marshy areas along the coast; San Fran-
cisco, Pebble Beach, Arroyo de Los Frijoles, and Watsonville. May–June.
—*Limnorchis leucostachys* (Lindl.) Rydb.
3a. **H. unalascensis** (Spreng.) Wats. var. **unalascensis**. Alaska Piperia. Fairly
rare locally; Pescadero Creek, Black Mountain, and Boulder Creek. June–
August. —*Piperia unalascensis* (Spreng.) Rydb.
3b. **H. unalascensis** (Spreng.) Wats. var. **maritima** (Greene) Correll. Coast
Piperia. Occasional on brushy slopes along the coast; San Francisco, Edgemar,

and Pebble Beach. August. —*H. greenei* Jeps., *H. elegans* (Lindl.) Bolander var. *maritima* (Greene) Ames, *Piperia maritima* (Greene) Rydb.
3c. H. unalascensis (Spreng.) Wats. var. **elata** (Jeps.) Correll. *Fig. 57*. Elegant Piperia. The most common variety of this species in the Santa Cruz Mountains usually growing in oak-madrone woods; San Francisco, San Francisco Watershed Reserve, Pescadero Creek, Kings Mountain, Coal Mine Ridge, Los Gatos, Big Basin, Waddell Creek, Empire Grade, and Santa Cruz. April–September. —*H. elegans* (Lindl.) Bolander, *H. michaelii* Greene, *Piperia elegans* (Lindl.) Rydb.

4. **Spiranthes** Rich. Lady's Tresses

Lip lacking two mammillate callosities at the base, definitely dilated or expanded at the tip.   1*a. S. romanzoffiana romanzoffiana*
Lip with two mammillate callosities at the base, slightly or not dilated at the apex.   1*b. S. romanzoffiana porrifolia*

1a. S. romanzoffiana C. & S. var. **romanzoffiana**. Hooded Lady's Tresses. Rare, usually growing in wet meadows near the coast; San Francisco, Swanton, Graham Hill, and Santa Cruz. May–July. —*Ibidium romanzoffianum* (C. & S.) House.
1b. S. romanzoffiana C. & S. var. **porrifolia** (Lindl.) Ames & Correll. Western Lady's Tresses. Rare, known definitely only from a few localities in the Santa Cruz Mountains; San Andreas Valley and near Boulder Creek. May–July. —*Ibidium porrifolium* (Lindl.) Rydb.

5. **Goodyera** R. Br. Rattlesnake Plantain

1. G. oblongifolia Raf. Menzies Rattlesnake Plantain. Known locally only from near Boulder Creek. August–September. —*G. decipiens* (Hook.) Hubbard, *Peramium decipiens* (Hook.) Piper.

6. **Cephalanthera** Rich.

1. C. austinae (Gray) Heller. Phantom Orchid. Rare locally, known only from Big Basin but perhaps to be expected occasionally elsewhere. June. —*Eburophyton austinae* (Gray) Heller.

7. **Corallorhiza** Chatelain. Coral Root

Perianth-segments not distinctly striate-veined; lip white, purple-spotted; ovary with a spur adnate to it.   1. *C. maculata*
Perianth-segments distinctly striate-veined; lip not white, striped with purple; ovary without a spur.   2. *C. striata*

1. C. maculata Raf. Spotted Coral Root. Occasional in fairly dense shade, growing with *Lithocarpus densiflora*, *Pseudotsuga menziesii*, *Arbutus menziesii*, and *Quercus agrifolia*; Sawyer Ridge, La Honda Grade, Pescadero Creek, Mayfield, Castle Rock, near Big Basin Park, and Big Basin. April–July.
2. C. striata Lindl. Striped Coral Root. Occasional in oak-madrone woods and on brushy slopes; San Francisco, Cahill Ridge, San Mateo Creek, Burlingame, La Honda Grade, Peters Creek, Searsville, Adobe Creek, Permanente Creek, Black Mountain, Saratoga Springs, and Big Basin. April–June.

2. Subclass Dicotyledoneae

25. SAURURACEAE. Lizard Tail Family

1. Anemopsis H. & A.

1. A. californica (Nutt.) H. & A. Yerba Mansa. To be expected occasionally in the Santa Clara Valley in moist ditches and along stream banks; near San Jose. April–June.

26. SALICACEAE. Willow Family

Leaves broad, commonly deltoid; the margins of the floral bracts laciniate; ovary globose.   1. *Populus*
Leaves narrow, elliptic to lanceolate or narrower; the margins of the floral bracts not laciniate; ovary ovoid.   2. *Salix*

1. Populus L. Poplar, Cottonwood

Petioles terete; leaves ovate, distinctly lighter beneath. The most common species.   1. *P. trichocarpa*
Petioles flattened; leaves usually broadly triangular-ovate, not bicolored.
    Branches appressed-ascending; plants reproducing vegetatively.
                                                              2. *P. nigra italica*
    Branches ascending or spreading; plants reproducing sexually.
                                                              3. *P. fremontii*

1. P. trichocarpa T. & G. ex Hook. *Fig. 58.* Black Cottonwood. Along the lower portions of stream courses from southern San Mateo County southward; near Crystal Springs Lake, Pescadero, San Francisquito Creek, Los Gatos Reservoir, Felton, Santa Cruz, and Pajaro River. February–April.
2. P. nigra L. var. **italica** Muenchh. Lombardy Poplar. Occasional along stream courses, roads, and moist areas, spreading vegetatively and escaping from cultivation; near San Andreas Dam and Carnadero Creek.
3. P. fremontii Wats. Fremont's Cottonwood. Known only from one specimen along San Francisquito Creek near Stanford. March–April.

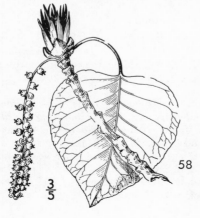

58

59

2. Salix L. Willow

Leaves essentially alike on both surfaces with respect to color and/or pubescence, narrowly linear to linear-lanceolate, silky-villous at least when young; stamens 2; capsules pubescent. 1. *S. hindsiana*
Leaves not alike on both surfaces, variously shaped, pubescent or glabrous; stamens 1–10; capsules pubescent or glabrous.
Leaves densely silky-villous abaxially; stamen 1; capsules 4–4.5 mm. long, pubescent. 2. *S. coulteri*
Leaves not densely silky-villous abaxially; stamens 2–10; capsules pubescent or glabrous.
Leaves usually broadest near the middle or above; stamens 2; capsules glabrous or pubescent; pedicels 0.5–1.5 mm. long.
Capsules 4–5.5 mm. long, glabrous; styles 0.5 mm. long; leaves usually oblong to oblanceolate. 3. *S. lasiolepis*
Capsules 7–9 mm. long, pubescent; stigmas sessile; leaves broadly oblanceolate to obovate. 4. *S. scouleriana*
Leaves usually broadest in the lower half; stamens 3–10; capsules glabrous; pedicels 1.5–3 mm. long.
Petioles with wart-like glands; capsules 5–6 mm. long. 5. *S. lasiandra*
Petioles without wart-like glands; capsules 3.5–5 mm. long. 6. *S. laevigata*

1. S. hindsiana Benth. Valley or Hinds' Willow. Along streams and creeks in the Santa Cruz Mountains; San Francisco, Permanente Creek, Saratoga, Pajaro River, and Watsonville. March–May.
2. S. coulteri Anders. Coulter's Willow. Occasional along streams and creeks, more common on the eastern slopes of the Santa Cruz Mountains than on the western, San Francisco southward. March–May.
3. S. lasiolepis Benth. Arroyo Willow. The most common species of *Salix* in the Santa Cruz Mountains, growing along streams and the edges of marshes, San Francisco southward. February–April. —*S. lasiolepis* Benth. var. *bigelovii* (Torr.) Bebb.
4. S. scouleriana Barratt ex Hook. Scouler's Willow. Moist places in woods, on brush covered slopes, and along streams; Baden, Pescadero, Portola Valley, Black Mountain, Stevens Creek, near Saratoga Summit, Los Gatos Creek, Ben Lomond Mountain, Big Basin, Boulder Creek, and Big Trees. February–April.
5. S. lasiandra Benth. Yellow Willow. A common willow along the edges of streams, lakes, and on very moist slopes, San Francisco southward. February–April.
6. S. laevigata Bebb. *Fig. 59.* Red Willow. Along the edges of lakes, ponds, and streams; San Andreas Lake, near Portola, Stanford, Guadalupe Creek, Wrights, near Felton, and Aptos. February–May.

27. MYRICACEAE. Wax Myrtle, Sweet Gale, or Bayberry Family

1. Myrica L.

1. M. californica C. & S. *Fig. 60.* California Wax Myrtle. Along or near the coast, often on sand dunes, and usually within 10 miles of the ocean; San Francisco, Arroyo de los Frijoles, near Pigeon Point, Swanton, Big Basin, and Santa

Cruz. April–June. Near the coast *M. californica* is a shrub, whereas in protected valleys it becomes a small tree to about 30 feet tall and grows with *Sequoia sempervirens, Pseudotsuga menziesii, Lithocarpus densiflora, Vaccinium ovatum, Rhododendron occidentale, Gaultheria shallon, Arbutus menziesii, Acer macrophyllum,* and *Alnus oregona.*

28. BETULACEAE. Birch Family

1. Alnus Hill. Alder

Leaves with narrow, revolute margins, the margins coarsely toothed. Plants of
    the coastal areas.   1. *A. oregona*
Leaves lacking revolute margins, the margins finely toothed. Inland plants.
                                              2. *A. rhombifolia*

1. A. oregona Nutt. *Fig. 61.* Red or Oregon Alder. Confined to the western side of the Santa Cruz Mountains, usually near the coast along streams; Pilarcitos Canyon, Pescadero Creek, San Gregorio Creek, Año Nuevo Point, Gazos Creek, Waddell Creek, Big Basin, Liddell Creek, and Aptos. Staminate catkins appear in September, October, or November and shed pollen in February and March; the pistillate ones are evident in January, February, or March and produce mature seeds by fall.—*A rubra* Bong.

2. A. rhombifolia Nutt. White Alder. Along streams on the eastern slopes of the Santa Cruz Mountains and occasionally on the western, but always away from the immediate coast; San Mateo Creek, Coal Mine Canyon, Stanford, Los Gatos Creek, Los Gatos Reservoir, San Jose, and Boulder Creek. Staminate catkins produce pollen in January and February; mature seeds are to be found in late summer and early fall.

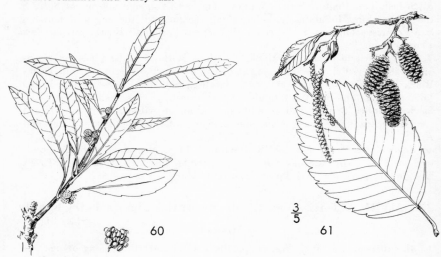

$\frac{3}{5}$

60   61

29. CORYLACEAE. Hazel Family

1. Corylus L. Hazel

1. C. californica (A. DC.) Rose. *Fig. 62.* California Hazel. A fairly common shrub of oak-madrone woods and along creeks, usually growing in at least partial shade; San Francisco, Sawyer Ridge, Pescadero, Kings Mountain, Black Mountain, Stevens Creek, Swanton, near Glenwood, Alba Grade, and Ben Lomond Sand Hills. The staminate catkins appear in October and November and shed pollen from January through March; mature nuts are to be found as early as May. —*C. cornuta* Marsh var. *californica* (A. DC.) Sharp, *C. rostrata* Ait. var. *californica* A. DC.

30. FAGACEAE. Oak or Beech Family

Fruit surrounded by a spiny involucre; abaxial leaf-surfaces and young twigs
    covered with a dense golden-colored pubescence.   1. *Castanopsis*
Fruit surrounded basally by a scaly cup-like involucre; abaxial leaf-surfaces
    glabrous to golden pubescent.
Staminate catkins stiff, erect; trees evergreen; leaves oblong, serrate, the
    lateral veins parallel; cup with stiff, spreading scales.   2. *Lithocarpus*
Staminate catkins not stiff, usually drooping; trees or shrubs, evergreen or
    deciduous; leaves variously shaped, toothed, lobed, or divided, the lateral
    veins usually not parallel; cup without stiff, spreading scale.

3. *Quercus*

1. Castanopsis (D. Don) Spach. Chinquapin

1. C. chrysophylla (Dougl.) A. DC. var. **minor** (Benth.) A. DC. Golden Chinquapin. Dry open ridges in chaparral, usually at higher elevations, northern San Mateo County southward; Cahill Ridge, Kings Mountain, between Loma Prieta and Mount Umunhum, Loma Prieta, and Chalks Lookout. July–August.

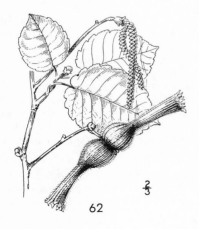

62

63

2. Lithocarpus Blume

1. L. densiflora (H. & A.) Rehd. Tan Oak, Tan Bark Oak. A common tree in the Santa Cruz Mountains, commonly growing with *Umbellularia californica* and *Arbutus menziesii* along the edges of coniferous forests or forming pure, dense stands; La Honda, Coal Mine Ridge, Black Mountain, Los Gatos Creek, Loma Prieta, Big Basin, Waddell Creek, Opal Creek, Blooms Grade, and Santa Cruz. May–July. —*Pasania densiflora* (H. & A.) Oersted.

3. Quercus L. Oak

Scales of the involucre thin; abaxial leaf-surfaces not conspicuously paler than the adaxial; stigmas on slender styles.

Tree deciduous; leaves thin, deeply to shallowly sinuately lobed.

Lobes of the leaves usually with several long, bristle-tipped teeth, leaves usually lobed about ¾ of the way to the midrib.   1. *Q. kelloggii*

Lobes of the leaves usually with only 1 bristle-like tooth, leaves usually lobed less than ⅓ of the way to the midrib.   2. *Q.* × *morehus*

Trees evergreen; leaves thick, toothed.

Leaves usually convex, the teeth always spine-tipped, usually with tufts of hair in the axils of the veins abaxially, veins not conspicuously parallel.   3. *Q. agrifolia*

Leaves usually plane, often entire, lacking tufts of hair, the lateral veins conspicuously parallel.   4. *Q. wislizeni*

Scales of the involucre usually warty or thick at the base, or often obscured by dense tomentum; abaxial leaf-surfaces usually conspicuously paler than the adaxial; stigmas more or less sessile.

Trees, deciduous; leaves relatively thin, usually 5–15 cm. long.

Leaves blue-green, shallowly lobed, usually less than ⅓ of the way to the midrib.   5. *Q. douglasii*

Leaves not blue-green, distinctly lobed, usually at least ½ of the way to the midrib.

Acorns 3–5 cm. long, gradually tapering to the apex; involucres deep.   6. *Q. lobata*

Acorns 2–3 cm. long, rounded at the apex; involucres shallow.   7. *Q. garryana*

Trees or shrubs, evergreen; leaves thick, leathery, usually 1.5–5 cm. long.

Abaxial surfaces of the young leaves densely golden-pubescent; acorns maturing in 2 years; trees or chaparral shrubs.   8. *Q. chrysolepis*

Abaxial surfaces of young leaves lacking dense golden-pubescence; shrubs.

Leaves more or less plane or undulate, usually glossy adaxially. Not growing on serpentine.   9. *Q. dumosa*

Leaves convex, the margins revolute, not glossy adaxially, stellate-pubescent. Growing on serpentine.   10. *Q. durata*

1. Q. kelloggii Newb. Kellogg's or California Black Oak. Growing in oak-grassland, usually at higher elevations, and on slopes with other oaks and *Arbutus menziesii*, most common to the east of the crests of the Santa Cruz Mountains;

near Woodside, Searsville, Coal Mine Ridge, Black Mountain, near Los Gatos, Madrone Station, and near Burrell. March–May.

2. Q. × morehus Kell. Oracle Oak. Occasional in the Santa Cruz Mountains; Black Mountain, near Saratoga Summit, Mount Charlie Road, near Gilroy, and near Bonny Doon. March–May. *Quercus × morehus* is a hybrid between *O. kelloggii* and *Q. wislizenii*. The leaves are usually larger than those of *Q. wislizenii*, rarely lobed more than ⅓ of the way to the midrib or merely coarsely toothed, thin or leathery, and usually have only one bristle-tip per lobe.

3. Q. agrifolia Nee. *Fig. 63.* Encina, California Live Oak. A very common tree of valleys and of rolling hills, often growing with *Q. lobata, Arbutus menziesii, Umbellularia californica,* and *Aesculus californica;* San Francisco, Sawyer Ridge, Jasper Ridge, Stanford, Los Gatos, Rancho del Oso, near Saratoga Summit, near Glenwood, Tuxedo, and Ben Lomond Sand Hills. February–April.

4. Q. wislizeni A. DC. Sierra or Interior Live Oak. Woodlands and chaparral, central San Mateo County southward; Peters Creek, Butano Creek, Kings Mountain, Jasper Ridge, Black Mountain, Mount Umunhum, Loma Prieta, near Eagle Rock, Big Basin, near Boulder Creek, and Santa Cruz. April–May. In chaparral *Q. wislizenii* is a shrub, usually three to seven feet tall, and has been called *Q. wislizenii* A. DC. var. *frutescens* Engelm. *Quercus wislizenii* is a medium- or large-sized tree on wooded slopes and in lowlands.

5. Q. douglasii H. & A. Douglas' or Blue Oak. Wooded, rolling hills, southern San Mateo County southward along the eastern slopes of the Santa Cruz Mountains, often growing with *Q. agrifolia, Q. lobata, Photinia arbutifolia,* and *Aesculus californica.* March–May. *Quercus douglasii* apparently hybridizes with *Q. lobata,* and hybrids are to be found where both species occur together.

6. Q. lobata Nee. *Fig. 64.* California White or Valley Oak, Roble. Mainly in the valleys and on low, rolling foothills, rare west of the crests of the Santa Cruz Mountains; near Woodside, Alpine Road, Page Mill Road, Stanford, Black Mountain Road, Guadalupe Creek, and Glenwood. March–May.

7. Q. garryana Dougl. Garry's, Oregon, or Post Oak. Rare in the Santa Cruz Mountains, usually occurring along the summits as individual trees; Black Mountain and 7 miles north of Saratoga Summit. April–June.

8. Q. chrysolepis Liebm. Maul, Canyon, or Gold Cup Oak. Open woods, growing with *Q. kelloggii, Q. agrifolia, Lithocarpus densiflora,* and *Arbutus menziesii,* or growing as a shrub or a small tree in chaparral; San Francisco, Portola, Coal Mine Ridge, Black Mountain, Stevens Creek Road, Castle Rock Ridge, near Los Gatos, Loma Prieta, Chalks Lookout, Big Basin, and near Boulder Creek. May–June. Dwarf specimens may be called *Q. chrysolepis* Liebm. var. *nana* Jeps.

9. Q. dumosa Nutt. California Scrub Oak. Chaparral from southern San Mateo County southward; Jasper Ridge, Page Mill Road, near Los Gatos, Loma Prieta, Llagas Post Office, Rancho del Oso, and Mill Creek. April–June.

10. Q. durata Jeps. Leather Oak. Chaparral, usually restricted to serpentine soils, rocks, and ridges, Crystal Springs Lakes south to southern Santa Clara County on the eastern slopes of the Santa Cruz Mountains; San Francisco Watershed Reserve, between Woodside and Crystal Springs Lake, Woodside, Jasper Ridge, Black Mountain, Loma Prieta, and near Mount Madonna. May–July.

31. URTICACEAE. Nettle Family

Plants with stinging hairs; leaves opposite.
 Pistillate flowers with distinct sepals.   1. *Urtica*
 Pistillate flowers with united sepals.   2. *Hesperocnide*
Plants without stinging hairs; leaves alternate.
 Plants more or less erect, not rooting at the nodes; leaves over 5 mm. long,
  not shiny.   3. *Parietaria*
 Plants prostrate, rooting at the nodes; leaves under 5 mm. long, usually shiny.
                4. *Helxine*

1. Urtica L. Nettle

Plants annual; staminate and pistillate flowers in the same cluster; plants usually under 5 dm. tall.   1. *U. urens*
Plants perennial; staminate and pistillate flowers in different clusters; plants
 usually over 10 dm. tall.
 Stems more or less glabrous; adaxial leaf-surfaces nearly glabrous; leaf-blade
  broadly ovate.   2. *U. californica*
 Stems densely white-pubescent; adaxial leaf-surfaces short-pubescent; leaf-
  blade lanceolate to ovate.   3. *U. holosericea*

1. U. urens L. Dwarf Nettle. Common, but poorly collected, weeds of disturbed
areas, cultivated fields, and gardens; introduced from Europe. March–June.
2. U. californica Greene. *Fig. 65.* California or Coast Nettle. Occasional along
or near the coast from San Francisco south to central Santa Cruz County. April–
September.
3. U. holosericea Nutt. Hoary Nettle. Moist areas in redwood forests and along
streams; Lake Merced, between La Honda and Pescadero, Menlo Park, near
Holy City, near Felton, Boulder Creek, and Pajaro River. May–October.

2. Hesperocnide Torr.

1. H. tenella Torr. Western Nettle. Fairly common in shaded areas, San Francisco southward. March–May.

3. Parietaria L. Pellitory

Leaves commonly 1.5 cm. long or longer, broadly ovate; plants not weak-
 stemmed, perennial. Introduced.   1. *P. judaica*
Leaves commonly 0.5–1 cm. long, suborbicular to deltoid; plants weak-stemmed,
 annual. Native.   2. *P. floridana*

1. P. judaica L. Pellitory. Occasional as a weed in San Francisco, Santa Cruz,
and to be expected elsewhere; native of Europe. May–October.
2. P. floridana Nutt. Florida Pellitory. Known locally only from near the coast
along San Vicente Road in Santa Cruz County. March–May.

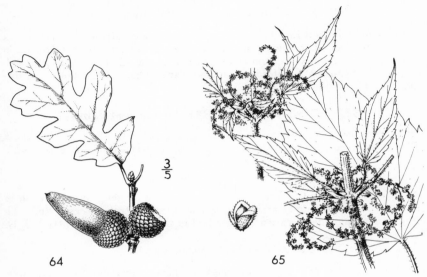

64   65

4. Helxine Req.

1. H. soleirollii Req. Mind-your-own-business, Mother-of-Thousands, Baby's Tears. Occasional in gardens, lawns, and disturbed areas about habitations; San Francisco, Palo Alto, and Rancho del Oso; native of Corsica and Sardinia. To be found in flower at all times of the year.

32. ULMACEAE. Elm Family

1. Ulmus L. Elm

1. U. procera Salisb. English Elm. Reproducing vegetatively and forming thickets in Golden Gate Park and probably elsewhere; native of Europe.

33. LORANTHACEAE. Mistletoe Family

Leaves conspicuous. green, fleshy. Parasitic on Angiosperms.   1. *Phoradendron*
Leaves scale-like, yellowish. Parasitic on Gymnosperms.   2. *Arceuthobium*

1. Phoradendron Nutt. Mistletoe.

1. P. villosum Nutt. Hairy Mistletoe, Oak Mistletoe. Parasitic on various species of *Quercus*, on *Castanopsis*, occasionally on *Adenostoma*, and to be expected on *Alnus* and *Umbellularia*. To be found with flowers throughout most of the year. —*P. flavescens* (Pursh) Nutt. var. *villosum* (Nutt.) Engelm.

2. Arceuthobium M. Bieb. Pine Mistletoe

1. A. campylopodum Engelm. *Fig. 66.* Western Dwarf Mistletoe. Known as a native in the Santa Cruz Mountains only on *Pinus attenuata* and *P. sabiniana* in the vicinity of Mount Umunhum and Loma Prieta, but this species has been collected on *P. radiata* at Stanford and in other parts of Santa Clara County. July–October.

34. ARISTOLOCHIACEAE. Birthwort Family

Plants with creeping rootstocks; flowers regular; stamens 12.   1. *Asarum*
Plants caulescent; flowers irregular; calyx saccate; stamens 6.   2. *Aristolochia*

1. Asarum L. Wild Ginger

1. A. caudatum Lindl. *Fig. 67.* Long-tailed Wild Ginger. Growing in deep shade in redwood-Douglas fir forests, commonly with *Oxalis oregana* and with it often forming a dense carpet; naturalized in Golden Gate Park. March–July.

2. Aristolochia L. Dutchman's Pipe

1. A. californica Torr. California Pipe Vine. Known definitely from Lake Merced in San Francisco and reported from other parts of the Santa Cruz Mountains. February–May.

35. POLYGONACEAE. Buckwheat Family

Stipules none; cauline leaves when present opposite or alternate.
  Cauline leaves opposite, usually 2-lobed; flowers not enclosed within an involucre; stems weak, prostrate to decumbent.   1. *Pterostegia*
  Cauline leaves opposite or alternate, entire; flowers enclosed within an involucre; stems not weak, usually more or less erect.
    Lobes of the involucre spiny; plants annual.   2. *Chorizanthe*
    Lobes of the involucre not spiny; plants annual or perennial.   3. *Eriogonum*
Stipules present, often scarious and sheathing; leaves alternate.
  Leaves not suborbicular; plants usually not vine-like.
    Perianth 6-parted.
      Perianth becoming bur-like in fruit, spinescent.   4. *Emex*
      Perianth herbaceous or papery in fruit, not spinescent.   5. *Rumex*
    Perianth usually 5-parted (occasionally 4-parted).   6. *Polygonum*
  Leaves suborbicular to elliptic; plants much-branched vines.   7. *Muhlenbeckia*

1. Pterostegia F. & M.

1. P. drymarioides F. & M. Pterostegia. A common prostrate herb of open woods, brush-covered slopes, coastal sand dunes, and occasionally on serpentine; San Francisco, Montara Mountain, Coal Mine Ridge, Stanford, Guadalupe Creek, San Jose, near Edenvale, Waddell Creek, near Boulder Creek, Santa Cruz, and Sunset Beach State Park. March–June.

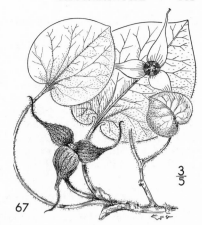

66   67

2. Chorizanthe R. Br. ex Benth. Spine-flower, Chorizanthe

Involucres 3-ribbed, the sides folded together, transversely wrinkled; teeth of the involucre lacking membranous margins.   1. *C. polygonoides*
Involucres 6-ribbed, the sides not folded together, usually not transversely wrinkled; teeth of the involucre with scarious membranous margins (except in *C. cuspidata* var. *cuspidata*).
Membranous margins continuous across the sinuses between adjacent teeth.
                                                    2. *C. membranacea*
Membranous margins not continuous across the sinuses.
  Involucres over 4 mm. long; plants usually erect.   3. *C. robusta*
  Involucres less than 4 mm. long; plants erect to diffuse.
    Lobes of the perianth pubescent abaxially, cuspidate; tube of the involucre with transverse wrinkles. Sand dune plants.
      Teeth of the involucre lacking membranous margins.
                                            4a. *C. cuspidata cuspidata*
      Teeth of the involucre with membranous margins.
                                            4b. *C. cuspidata marginata*
    Lobes of the perianth usually glabrous abaxially, not cuspidate. Inland plants.
      Membranous margins of the involucral teeth rose-colored; lobes of the perianth usually obovate, erose.   5. *C. pungens hartwegii*
      Membranous margins of the involucral teeth usually white; lobes of the perianth oblong to lanceolate, entire.   6. *C. diffusa*

1. C. polygonoides T. & G. Knotweed Chorizanthe. Known locally only from northern Santa Clara County; Page Mill Road. May–June.
2. C. membranacea Benth. Pink Chorizanthe. Dry rocky slopes along the crests of the Santa Cruz Mountains; Crystal Springs, Castle Rock Ridge, and Mount Umunhum. May–July.
3. C. robusta Parry. *Fig. 68.* Robust Spine-flower. Coastal sand dunes, rarely inland, San Francisco southward; San Francisco, Colma, Santa Cruz, Sunset Beach, and near Watsonville. April–July.

4a. C. cuspidata Wats. var. **cuspidata.** San Francisco Chorizanthe. Coastal and inland sand dunes, San Francisco southward to the vicinity of Santa Cruz; San Francisco, San Pedro, and Santa Cruz. April–July.
4b. C. cuspidata Wats. var. **marginata** Goodman. Coastal sand dunes and on sandy coastal bluffs; San Francisco, Colma, and San Pedro. May–June.
5. C. pungens Benth. var. **hartwegii** (Benth.) Goodman. Hartweg's Spine-flower. Inland marine sand deposits and elsewhere in sandy soil; San Francisco, Saratoga to Big Basin, Big Basin, Hilton Airport, near Boulder Creek, Glenwood, Ben Lomond, Felton, Mount Hermon, and Santa Cruz. April–June.
6. C. diffusa Benth. Diffuse Spine-flower. Dry sandy and rocky areas from southern San Mateo County southward; Jasper Ridge, China Grade, Eagle Rock, Big Basin, Ben Lomond Sand Hills, near Mount Hermon, and Santa Cruz. May–June.

3. Eriogonum Michx. Wild Buckwheat

Plants perennial; stout, woody caudex present.
  Leaves not borne in fascicles at the nodes.
    Involucres aggregated into heads, rarely solitary, the heads terminal or axil-
      lary; leaves green adaxially, densely white-tomentose abaxially; stems
      glabrous or tomentose.
        Stems white-tomentose, glabrate in age; heads usually 1 per peduncle-
          like stem, terminal. Coastal.   1. *E. latifolium*
        Stems glabrous; heads usually several per stem, the stems branched above.
          Inland.   2. *E. nudum*
    Involucres scattered along the stems; leaves densely white-tomentose on
      both surfaces; stems tomentose, glabrous in age.   3. *E. saxatile*
  Leaves borne in fascicles at the nodes.   4. *E. fasciculatum foliolosum*
Plants annual; roots slender.
  Perianth with uncinate bristles; involucres about 1 mm. long; plants diffusely
    branched above.   5. *E. hirtiflorum*
  Perianth glabrous; involucres 3–5 mm. long; plants not diffusely branched
    above.
      Involucres broadly campanulate; branches of the inflorescence short, not
        virgate; involucres mainly borne terminally.   6. *E. argillosum*
      Involucres cylindric; branches of the inflorescence long, virgate; involucres
        borne at the nodes of the virgate branches.
          Plants usually densely tomentose or floccose throughout; involucres 4–5
            mm. long, uniformly tomentose.   7. *E. virgatum*
          Plants usually tomentose or floccose only below, the distal internodes usu-
            ally glabrous, reddish; involucres 2.5–3 mm. long, not uniformly
            tomentose, the ribs usually glabrous.   8. *E. vimineum*

1. E. latifolium Smith. *Fig. 69.* Coast Eriogonum, Tibinagua. Coastal bluffs, often growing with *Baccharis pilularis, Fragaria chiloensis, Eriophyllum staechadifolium, Dudleya farinosa,* and *Artemisia pycnocephala;* San Francisco, Ocean View, Half Moon Bay, Pescadero, Swanton, Davenport, Santa Cruz, and Capitola. June–October.
2. E. nudum Dougl. ex Benth. Tibinagua, Naked-stemmed Eriogonum. Fairly common on dry open slopes, inland marine sand deposits, rock outcroppings,

and along the edges of chaparral; San Francisco, Sweeney Ridge, Crystal Springs Lake, Peters Creek, Coal Mine Ridge, Stanford, near Los Gatos, Loma Prieta, near Saratoga Summit, Glenwood, near Felton, Mount Hermon, and Soda Lake. June–August. —*E. latifolium* Smith ssp. *nudum* (Dougl. ex Benth.) Stokes, *E. latifolium* Smith ssp. *decurrens* Stokes. *Eriogonum nudum* is a variable species in which several subspecific entities have been proposed.

3. E. saxatile Wats. Rock Eriogonum. Known locally only from the eastern slope of Mount Umunhum, growing in rocky outcroppings and on serpentine. June–August.

4. E. fasciculatum Benth. ssp. **foliolosum** (Nutt.) Stokes. Flat-top, Wild Buckwheat, California Buckwheat Brush. Occasional along the coast; San Francisco and near Watsonville; native from Monterey County southward. July–September.

5. E. hirtiflorum Gray ex Wats. Hairy-flowered Eriogonum. Known locally only from Empire Grade near Eagle Rock. July–September.

6. E. argillosum Howell. Clay-loving Eriogonum. Occasional in the serpentine hills south of San Jose and to be expected elsewhere in the Santa Clara Valley and adjacent foothills. March–May.

7. E. virgatum Benth. Virgate Eriogonum. Occasional on the eastern slopes of the Santa Cruz Mountains in Santa Clara County and in dry creek bottoms in the Santa Clara Valley; Black Mountain and Carnadero Creek. July–August.

8. E. vimineum Dougl. ex Benth. Wicker Eriogonum. Fairly common on rocky slopes, serpentine outcroppings, and in chaparral, central San Mateo County southward, mainly on the eastern slopes of the Santa Cruz Mountains; Crystal Springs Lake, Belmont, Emerald Lake, Jasper Ridge, Stanford, Black Mountain, Mount Umunhum, Loma Prieta, and New Almaden. July–September. *Eriogonum vimineum* is variable and has been divided into a number of subspecific and varietal units. Our plants may for the most part be referred to *E. vimineum* Dougl. ex Benth. var. *caninum* Greene.

4. Emex Neck.

1. E. australis Steinh. Occasional as a weed in disturbed areas; San Francisco; native of Australia. June–July.

68

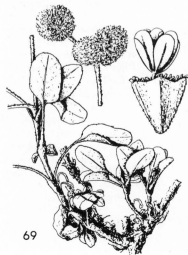

69

5. **Rumex** L. Dock, Sorrel

Leaves hastate at the base; plants dioecious, perennial; inner perianth-segments
entire, lacking callous grains.   1. *R. acetosella*
Leaves at most cordate at the base; plants monoecious, annual or perennial;
inner perianth-segments entire or not, callous grains present or not.
Stems branched below the inflorescence, often prostrate or decumbent; plants
perennial; inner perianth-segments entire; callous grains usually present
on at least one perianth-segment in each flower.
Callous grains present.
Leaf-blades ovate-lanceolate or oblong-lanceolate, usually 2–3 times as
long as wide; inner perianth-segments 4–5 mm. long; callous grains
about as wide as the inner perianth-segment; branches of the inflorescence densely flowered.   2. *R. crassus*
Leaf-blades linear-oblong to lanceolate, usually at least 3 times as long
as broad; inner perianth-segments 2–4 mm. long; callous grains narrower than the inner perianth-segments; branches of the inflorescence not densely flowered.
Callous grains usually 1 per flower, rarely lacking; achenes about 2
mm. long.   3. *R. salicifolius*
Callous grains usually 3 per flower; achenes about 2.5 mm. long.
4. *R. transitorius*
Callous grains lacking.   5. *R. californicus*
Stems not branched below the inflorescence, always erect; plants annual or
perennial; inner perianth-segments entire or not.
Callous grains lacking.   6. *R. fenestratus*
Callous grains present on at least one inner perianth-segment per flower.
Mature perianth-segments entire.
Inner perianth-segments 4–6 mm. long; callous grains narrower than
the margins of the inner perianth-segments; verticillate-like glom-
erules approximate, obscuring the stem; branches of the inflor-
escence strict.   7. *R. crispus*
Inner perianth-segments 2.5–3 mm. long; callous grains wider than the
margins of the inner perianth-segments; verticillate-like glom-
erules remote, not obscuring the stem; branches of the inflorescence spreading.   8. *R. conglomeratus*
Mature perianth-segments variously toothed or divided.
Plants annual; leaves 3 or more times as long as broad.
Callous grains ⅓–½ the width of the body of the inner perianth-
segments, the body 1.7–2 mm. wide.   9. *R. fueginus*
Callous grains about as wide as the body of the inner perianth-seg-
ments, the body about 1 mm. wide.   10. *R. persicarioides*
Plants perennial; leaves usually less than 2.5 times as long as wide.
Some branches of the inflorescence at right angles to the stem;
achenes 3–4 mm. long; all inner perianth-segments usually with
callous grains and with 5–10 teeth on each margin.
11. *R. pulcher*
Branches of the inflorescence rarely divergent at more than 45° from
the stem; achenes about 2 mm. long; usually only one inner
perianth-segment with a callous grain, segments with 3–5 nar-
row teeth on each margin.   12. *R. obtusifolius*

1. R. acetosella L. Sheep Sorrel. A common species of *Rumex* of disturbed areas, fallow fields, and in poor soils of open slopes, San Francisco southward; native of Europe. March–September.

2. R. crassus Rech. f. Most common along the coast in sandy soil, but occasionally inland; San Francisco, Montara, Mussel Rock, Pescadero, Waddell Creek, Santa Cruz, and near Seabright. June–August. —*R. salicifolius* Weinm. var. *crassus* (Rech. f.) Howell. The group of taxa represented by *R. crassus*, *R. salicifolius*, *R. transitorius*, and *R. californicus* is a difficult one taxonomically, is in need of careful study, and is perhaps only one polymorphic species.

3. R. salicifolius Weinm. Willow-leaved Dock. Occasional in wet areas, along lakes, ponds, streams, and drainage ditches; San Francisco, Crystal Springs Reservoir, Stevens Creek Road, Uvas Creek, Glenwood, Blooms Grade, Bear Creek Road, and Felton. May–August.

4. R. transitorius Rech. f. Edges of lakes and ponds or in moist disturbed areas; San Francisco, San Pedro, Lake Lagunita, Stanford, Palo Alto, and near Moffett Field. —*R. salicifolius* Weinm. f. *transitorius* (Rech. f.) Howell.

5. R. californicus Rech. f. California Dock. Occasional in the Santa Cruz Mountains; San Francisco and the Santa Clara Valley. May–August. —*R. salicifolius* Weinm. f. *ecallosus* Howell.

6. R. fenestratus Greene. Known from marshes in San Francisco and northern San Mateo County. March–August. —*R. occidentalis* Wats. var. *procerus* (Greene) Howell.

7. R. crispus L. Curly or Yellow Dock. The most common species of *Rumex* locally, growing along streams, drainage ditches, wet fields, and along the edges of ponds, San Francisco southward; native of Eurasia. April–July.

8. R. conglomeratus Murr. Green Dock. Fairly common in low moist areas where water stands in the spring and along the edges of marshes, ponds, and streams; San Francisco, Crystal Springs Lake, San Mateo, Stanford, Alviso, Camp Evers, Santa Cruz, and Watsonville; native of Europe. April–October.

9. R. fueginus Philippi. Golden Dock. Borders of lakes, ponds, and marshes; San Francisco, San Andreas Lake, Crystal Springs Lake, near Pescadero, Kelly Lake, Pinto Lake, and Pajaro River. June–October.

10. R. persicarioides L. *Fig. 70.* Golden Dock. Occasional in marshes near the coast and in ponds; San Francisco, Granada, Pilarcitos Creek, and Moss Beach. June–August.

11. R. pulcher L. Fiddle Dock. Common in moist pastures and disturbed areas; San Francisco, Pilarcitos Dam, Crystal Springs Lake, Millbrae, Coal Mine Ridge, Stanford, and Seabright; native of Europe. May–July.

12. R. obtusifolius L. Bitter Dock. Occasional along streams and the shores of lakes; San Francisco and probably elsewhere; native of Europe. May–September. —*R. obtusifolius* L. ssp. *agrestis* (Fries) Danser.

6. Polygonum L. Knotweed, Smartweed

Plants climbing vines; leaves hastate to cordate at the base.   1. *P. convolvulus*
Plants not vines; leaves never hastate, rarely cordate at the base.
  Leaves articulate with the sheaths, dehiscent at the point of articulation; flowers scattered along the stems, solitary or few in the axils of the leaves.
    Plants perennial; leaf-margins strongly revolute; midribs of the leaves keeled, prominent.   2. *P. paronychia*

Plants annual; leaf-margins revolute or not; midribs of the leaves not keeled, usually not prominent.
Achenes smooth.
Achenes 2.5–3 mm. long, exserted. Usually growing in salt marshes.
3. *P. patulum*
Achenes about 1.5 mm. long, not exserted. Growing in weedy areas, but usually not in salt marshes.   4. *P. argyrocoleon*
Achenes striate.   5. *P. aviculare*
Leaves not articulate with the sheaths, rarely dehiscent, usually withering; flowers in definite, spike-like racemes.
Inflorescences 1 or rarely 2 per stem or branch, terminal; peduncles stout, 1–2 mm. in diameter; perianth rose-red.
Inflorescences 5–10 cm. long; leaves lanceolate to ovate-lanceolate; sheaths never with herbaceous margins.   6. *P. coccineum*
Inflorescences 1–4 cm. long; leaves elliptic to oblong; sheaths of terrestrial plants with herbaceous margins.   7. *P. natans*
Inflorescences numerous, terminal or in the axils of the leaves; peduncles slender, under ¾ mm. in diameter; perianth pink, greenish, or white.
Perianth glandular-punctate.   8. *P. punctatum*
Perianth glabrous, not glandular-punctate.
Sheaths bristly-ciliate; perianth usually 5-parted.
Spikes oblong, 1–3 cm. long; plants annual.   9. *P. persicaria*
Spikes slender, linear, 3–8 cm. long; plants perennial.
10. *P. hydropiperoides*
Sheaths not bristly-ciliate; perianth often 4-parted.
11. *P. lapathifolium*

1. P. convolvulus L. Black Bindweed. Occasional in disturbed areas, in gardens, and along roadsides; San Francisco, near Half Moon Bay, and Stanford; native of Europe. July–September.
2. P. paronychia C. & S. Dune Knotweed. Coastal sand dunes and sandy bluffs along the coast; San Francisco, Pescadero, Pebble Beach, Año Nuevo Point, and Palm Beach. May–July.
3. P. patulum Bieb. Known from salt marshes bordering San Francisco and' as an occasional urban weed; Redwood City, Cooleys Landing, and between Palo Alto and Sunnyvale; native of Europe. June–October.
4. P. argyrocoleon Steud. ex Kunze. Silversheath Knotweed. Occasional in disturbed areas in San Francisco and probably elsewhere; native of Asia. June–October.
5. P. aviculare L. Common or Dooryard Knotweed. A common species of dry land, packed soils, and disturbed areas; native of Eurasia. May–October. *Polygonum aviculare* is exceedingly variable with respect to internode length, length of the branches, position of the branches, and length of the perianth. Plants with erect stems have been called *P. aviculare* L. var. *erectum* (L.) Roth. Fleshy-leaved plants with prostrate stems and very short internodes have been called *P. aviculare* L. var. *littorale* (Link) Martens & Koch. *Polygonum aviculare* L. var. *aviculare*, itself, has prostrate stems and internodes which are about as long as the leaves.

70   71

6. P. coccineum Muhl. ex Willd. Swamp Knotweed. A common species of *Polygonum* in ponds, lakes, and along the edges of slow-flowing streams, becoming terrestrial as the water dries up, San Francisco southward. July–October. —*P. coccineum* Muhl. ex Willd. f. *natans* (Wieg.) Stanford, *P. coccineum* Muhl. ex Willd. f. *terrestris* (Willd.) Stanford.

7. P. natans (Michx.) Eat. Water Smartweed. Occasional in ponds in northern San Mateo County and probably elsewhere. July–October. —*P. amphibium* L. var. *stipulaceum* Coleman.

8. P. punctatum Ell. *Fig. 71.* Water Smartweed. Fairly common along the edges of ponds, lakes, and slow-moving streams, San Francisco southward. June–October.

9. P. persicaria L. Lady's Thumb. Marshes, edges of ponds, and drainage ditches, occasional from San Francisco southward; San Francisco, San Andreas Lake, near Boulder Creek, and Watsonville; native of Europe. June–October.

10. P. hydropiperoides Michx. Known locally from Pinto Lake in southern Santa Cruz County and to be expected occasionally elsewhere. June–October. —*P. hydropiperoides* Michx. var. *asperifolium* Stanford.

11. P. lapathifolium L. Willow Weed. Common in streams, marshes, ditches, ponds, and lakes; San Francisco, Pillar Point, Felt Lake, Searsville Lake, Campbell Creek, Wrights, near Coyote, Aptos, Santa Cruz–Watsonville Road, and Watsonville; native of Europe. August–October. Plants in which the abaxial leaf-surfaces are white-tomentose have been called *P. lapathifolium* L. var. *salicifolium* Sibth.

Polygonum pennsylvanicum L., Pinkweed, has been reported from San Francisco in the older literature and is known from other parts of California as an introduced weed from the eastern part of the United States. It may be distinguished from the previous species by the presence of stalked glands on the peduncles. *Polygonum lapathifolium* either lacks glands or has sessile ones.

7. Muhlenbeckia Meissn.

1. M. complexa Meissn. Mattress Vine. Occasional as an escape from gardens in San Francisco; native of New Zealand. January–November.

36.   CHENOPODIACEAE.   Goosefoot or Saltbush Family

Leaves reduced to opposite scales; stems and branches succulent, terete, the basal
ones becoming woody.   1. *Salicornia*
Leaves not reduced to scales, alternate; stems often somewhat succulent.
  Fruit dehiscent; calyx indurate at the base.   2. *Beta*
  Fruit commonly indehiscent; calyx not indurate at the base.
    At least some utricles in each plant enclosed by 2 separate or partly united
      bracts.
      Leaf-blades under 5 cm. long, variously shaped; plants commonly scurfy,
        occasionally glabrous; bracts not becoming spinescent.   3. *Atriplex*
      Leaf-blades 5–20 cm. long, deltoid to hastate; plants glabrous; bracts with
        2–3 spines at maturity.   4. *Spinacia*
    Utricles subtended by a 3- or 5-merous calyx.
      Fruiting calyx winged or armed with hooks on the abaxial surface.
        Calyx-lobes each with a hook.   5. *Echinopsilon*
        Calyx-lobes horizontally winged.
          Leaves pungent; wings broad.   6. *Salsola*
          Leaves not pungent; wings small.   7. *Kochia*
      Fruiting calyx sometimes keeled, but usually not with wings or hooks.
        Stamens and calyx-lobes commonly 1.   8. *Monolepis*
        Stamens and calyx-lobes commonly 3 or 5.
          Leaves narrowly linear, semiterete, fleshy.   9. *Suaeda*
          Leaves with a flattened blade, fleshy or not.
            Fruiting calyx herbaceous or fleshy, not reticulate.
                                                                  10. *Chenopodium*
            Fruiting calyx dry, reticulate.   11. *Roubieva*

1.   Salicornia L.

Plants perennial, commonly rooting at the nodes. Common.   1. *S. pacifica*
Plants annual. Rare.   2. *S. depressa*

1. S. pacifica Standley. *Fig. 72.* Woody Glasswort, Pickleweed. Salt marshes
along the ocean and along San Francisco Bay, occasionally inland along the
borders of sloughs; San Francisco, San Mateo, Palo Alto, near Mountain View,
Alviso, near Santa Cruz, Capitola, near Chittenden Station, mouth of Pajaro
River, and near Watsonville. July–November. —*S. virginica* L. in part.
2. S. depressa Standley. Glasswort. Occasionally in the salt marshes along San
Francisco Bay; Redwood City and Palo Alto. August–November. —*S. rubra*
Nels.

2.   Beta L.

1. B. vulgaris L. Garden Beet. Common as an escape from cultivation in the
lower portion of Santa Clara Valley in low alkaline areas, fallow fields, and salt
marshes, less frequent elsewhere; San Francisco, Redwood City, Palo Alto, near
Moffett Field, Alviso, near Santa Clara, and near mouth of Pajaro River; native
of Europe. April–November.

3. Atriplex L. Saltbush

Plants perennial; herbage usually at least somewhat furfuraceous.
Erect shrubs; plants dioecious; mature bracts 2–3 mm. long.
                                         1. *A. lentiformis breweri*
Prostrate to decumbent herbaceous perennials, often woody at the base; plants
  monoecious; mature bracts 3–7 mm. long.
Mature bracts united to the middle, 5–7 mm. long, tuberculate on the surfaces; leaves oblong to orbicular.   2. *A. leucophylla*
Mature bracts distinct or united only at the base, 3–5(7) mm. long, the sur-
  faces not tuberculate; leaves lanceolate to elliptic or oblong.
Mature bracts united at the base, 4–5(7) mm. long; leaves oblong to
  oblong-ovate, about 3.5 cm. long, the lower usually toothed, usually
  glabrous and green adaxially.   3. *A. semibaccata*
Mature bracts distinct, about 3 mm. long; leaves elliptic to lanceolate or
  oblanceolate, 1–2 cm. long, entire, densely furfuraceous adaxially.
                                      4. *A. californica*
Plants annual; herbage furfuraceous to glabrous.
Pistillate flowers of 2 kinds, some with 2 bracts, others with a 5-parted calyx;
  bracts broadly ovate to orbicular, entire.   5. *A. hortensis*
Pistillate flowers all 2-bracteate; bracts usually more or less deltoid, toothed
  or divided.
Mature bracts variously toothed or truncate.
Mature bracts with several sharp teeth at the apex, 2–2.5 mm. long, usu-
  ally not tuberculate on the surfaces.   6. *A. serenana*
Mature bracts toothed on the margins, but the terminal portion of the
  bract entire or more or less truncate, 2–12 mm. long, usually more or
  less tuberculate on the surfaces.
Mature leaves densely furfuraceous; mature bracts 4–12 mm. long.
Mature bracts truncate or shallowly 3-toothed terminally; leaves usu-
  ally cordate at the base.   7. *A. expansa mohavensis*
Mature bracts acute terminally; leaves cuneate or rounded at the
  base.   8. *A. rosea*
Mature leaves more or less glabrous.
Lower cauline leaves lanceolate to oblong, not hastate.
                           9a. *A. patula patula*
Lower cauline leaves deltoid to rhomboidal, hastate.
                    9b. *A. patula hastata*
Mature bracts entire.   10. *A. joaquiniana*

1. A. lentiformis (Torr.) Wats. var. **breweri** (Wats.) McMinn. Brewer's Saltbush. Occasional in saline soils near the coast, known locally from Watsonville.
August–October. This variety is occasionally used as an ornamental hedge plant.
2. A. leucophylla (Moq.) Dietr. Beach Saltbush, Sea-scale. Coastal sand dunes
and beaches; San Francisco, Princeton, Moss Beach, Swanton, Waddell Creek,
Davenport, Santa Cruz, near Capitola, and Palm Beach. June–September.
3. A. semibaccata R. Br. Australian Saltbush. Occasional as a weed, usually in
saline soils; San Francisco and Alviso; native of Australia. June–September.

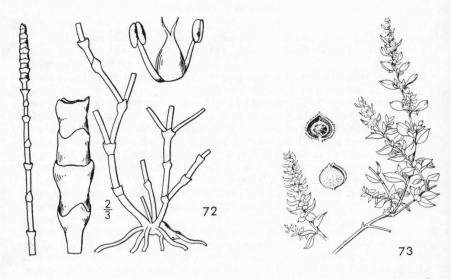

$\frac{2}{3}$ 72

73

4. A. californica Moq. *Fig. 73.* California Saltbush. Sand dunes along the coast; San Francisco, Pillar Point, Santa Cruz, Seabright, and Palm Beach. June–July.

5. A. hortensis L. Sea Purselane, Garden Orache. Occasional as a weed in disturbed places and in salt marshes along San Francisco Bay; Redwood City, Palo Alto, and near Moffett Field; native of Asia. July–September.

6. A. serenana Nels. Bracted Saltbush, Bract-scale. Known locally only from San Francisco and near Watsonville. July–October.

7. A. expansa Wats. var. **mohavensis** Jones. Mohave Fog Weed. Rare as a weed in San Francisco; introduced from central California. May–July.

8. A. rosea L. Redscale, Red Saltbush, Red Orache. Occasional as a weed in disturbed areas and along the edges of cultivated fields; San Francisco, Palo Alto, and Palo Alto Yacht Harbor; native of Eurasia. August–September.

9a. A. patula L. var. **patula.** Spear Orache, Spear Saltbush. Occasional in salt marshes; San Francisco and Santa Cruz. July–September.

9b. A. patula L. var. **hastata** (L.) Gray. Fat Hen, Halberd-leaved Orache. Common in salt marshes along the coast and San Francisco Bay, occasionally inland; San Francisco, near Colma, Moss Beach, San Mateo, Cooleys Landing, Searsville Lake, Palo Alto, Alviso, Santa Cruz, Watsonville Slough, and mouth of Pajaro River. July–September. —*A. hastata* L.

10. A. joaquiniana Nels. San Joaquin Saltbush.. Occasional as a weed in San Francisco; perhaps introduced locally from the Central Valley. June–September. —*A. patula* L. ssp. *spicata* (Wats.) Hall & Clem.

4. Spinacia L.

1. S. oleracea L. Spinach. Occasional as an escape from cultivation in low wet areas, especially near the coast; northwestern San Mateo County. March–May.

5. Echinopsilon Moq.

1. E. hyssopifolium (Pall.) Moq. Salt marshes and disturbed areas in San Francisco and in the Santa Clara Valley; San Francisco, Palo Alto, Alviso, and San Jose; native of Europe. July–October. —*Bassia hyssopifolia* (Pall.) Kuntze.

6. Salsola L.

1. S. kali L. var. **tenuifolia** Tausch. Russian Thistle. Occasional in disturbed areas; San Francisco, Palo Alto, Coyote, Watsonville, and doubtless elsewhere; native of Eurasia. July–October.

7. Kochia Roth

1. K. scoparia (L.) Schrad. Summer Cypress. Disturbed areas near the water front in San Francisco; native of Eurasia. June–September. —*K. scoparia* (L.) Schrad. var. *subvillosa* Moq.

8. Monolepis Schrad.

1. M. nuttalliana (R. & S.) Greene. Nuttall's Monolepis. Known locally only as a weed from San Francisco. May–September.

9. Suaeda Forsk. Sea-Blite, Seep-Weed

Plants annual, erect; calyx-lobes with horn-like appendages.
1. *S. depressa erecta*
Plants perennial, prostrate to decumbent; calyx-lobes without appendages.
2. *S. californica*

1. S. depressa (Pursh) Wats. var. **erecta** Wats. Sea-Blite. Known locally only from Soda Lake in southern Santa Cruz County. May–August.
2. S. californica Wats. California Sea-Blite. Occasional in salt marshes along San Francisco Bay; San Francisco and Palo Alto. July–October.

10. Chenopodium L. Goosefoot, Pigweed

Leaves glandular-pubescent, glandular-punctate, or glandular-vesiculate at least abaxially.
Seeds situated vertically, 0.5–0.7 mm. broad; leaves with gold-colored glandular vesicles abaxially.
1. *C. pumilio*
Seeds usually situated horizontally, 0.7–1.0 mm. broad; leaves lacking gold-colored glandular vesicles abaxially.
Leaves glandular-pubescent at least abaxially; flowers in small axillary cymules.
2. *C. botrys*
Leaves glandular-punctate abaxially; flowers in axillary, spike-like or capitate clusters.
Inflorescence bracteate; leaves sinuate-dentate.
3a. *C. ambrosioides ambrosioides*

Inflorescence ebracteate; leaves laciniate-serrate.
Stems densely woolly.   3b. *C. ambrosioides vagans*
Stems more or less glabrous, not densely woolly.
                                                         3c. *C. ambrosioides anthelminticum*
Leaves glabrous or scurfy abaxially, never glandular.
Most seeds situated vertically; calyx 3–5-parted.
Plants perennial; flowers in a terminal spike-like inflorescence; calyx-lobes
    usually 4–5, the margins erose-dentate.   4. *C. californicum*
Plants annual; flowers in axillary clusters and spikes; calyx-lobes usually
    3–4, usually entire.
Calyx-lobes in those flowers with vertically situated seeds united nearly
    to the tip; leaves densely white-farinose abaxially at least when
    young.   5. *C. macrospermum farinosum*
Calyx-lobes in those flowers with vertically situated seeds separate or
    united to about the middle; leaves usually not densely farinose ab-
    axially.   6. *C. humile*
Seeds all situated horizontally; calyx 5-parted.
Inflorescences usually shorter than the leaves; leaves strongly deltoid, sinu-
    ate-dentate, shining on the adaxial surface; seeds with a narrow rim,
    dull; stigmas short and stout; calyx-lobes not or only obscurely keeled.
                                                         7. *C. murale*
Inflorescences usually longer than the leaves; leaves rhombic-ovate, vari-
    ously toothed, dull on the adaxial surface; seeds usually without a nar-
    row rim, dull or shiny; stigmas not stout; calyx-lobes keeled or not.
Seeds essentially smooth; pericarp smooth or inconspicuously irregu-
    larly roughened; leaves chiefly membranaceous in texture; calyx-
    lobes strongly keeled or not.
Seeds chiefly 1.1–1.5 mm. broad; calyx-lobes largely covering fruit.
                                                         8. *C. album*
Seeds 0.9–1.2 mm. broad; calyx-lobes exposing fruit at maturity.
                                                         9. *C. strictum glaucophyllum*
Pericarp and surface of seeds foveolate-reticulate; leaves thin-membrana-
    ceous, membranaceous, or nearly coriaceous in texture; calyx-lobes
    with a prominent winged keel often equaling half the width of the
    sepal.   10. *C. berlandieri*

1. **C. pumilio** R. Br. Tasmanian Goosefoot. Occasional in disturbed areas; near
Coyote and Camp Evers; native of Australia and Tasmania. June–September.
—*C. carinatum* of authors.

2. **C. botrys** L. Jerusalem Oak, Feather Geranium. Occasional along highways,
in stream beds, and other disturbed areas; San Francisco, Wrights, Lexington,
and Saratoga Summit; native of Eurasia. April–September.

3a. **C. ambrosioides** L. var. **ambrosioides.** Mexican Tea. Fairly common as a
weed in disturbed areas; San Francisco, Crystal Springs, Menlo Park, Stanford,
Wrights, near Coyote, near Eccles, Santa Cruz, and Pajaro River; native of tropical America. July–December.

3b. **C. ambrosioides** L. var. **vagans** (Standley) Howell. Occasional as a weed;
San Francisco and near Boulder Creek; native of South America. July–December. —*C. ambrosioides* L. ssp. *chilense* Aellen.

3c. C. ambrosioides L. var. **anthelminticum** (L.) Gray. Wormseed. Occasionally in disturbed areas; San Francisco and near Boulder Creek; native of tropical America. July–December.

4. C. californicum (Wats.) Wats. *Fig. 74.* California Goosefoot, Soap Plant. Widespread on moist slopes, occasionally in sandy soils, and becoming a weed in disturbed areas, poorly collected locally; San Francisco, Stanford, and Sunset Beach. February–June.

5. C. macrospermum Hook. f. var. **farinosum** (Wats.) Howell. Coast Goosefoot. Edges of salt marshes along the coast and San Francisco Bay and along the margins of stagnant pools; Salada, Menlo Park, near Santa Cruz, and Chittenden. July–November. —*C. farinosum* (Wats.) Standley.

6. C. humile Hook. Known locally so far only from Pinto Lake in southern Santa Cruz County, but to be expected elsewhere. September–November. —*C. rubrum* L. var. *humile* (Hook.) Wats.

7. C. murale L. Nettle-leaved or Wall Goosefoot. A common weed of disturbed areas, fallow fields, and orchards, San Francisco southward; native of Europe. June–October.

8. C. album L. White Goosefoot, Lamb's Quarter. Common as a weed of disturbed areas, fields, and vineyards; native of Europe. July–October.

9. C. strictum Roth var. **glaucophyllum** (Aellen) Wahl. Occasional as a weed in San Francisco and probably elsewhere; native of much of the United States, but introduced locally. August–October.

10. C. berlandieri Moq. Berlandier's Goosefoot. Disturbed areas in San Francisco and probably elsewhere; native throughout most of the United States, but introduced locally.

11. Roubieva Moq.

1. R. multifida (L.) Moq. Cut-leaved Goosefoot. Occasional in disturbed areas, quite common in San Francisco, less frequent elsewhere; San Francisco, Daly City, San Jose, and near Wrights; native of Chile. July–October. —*Chenopodium multifidum* L.

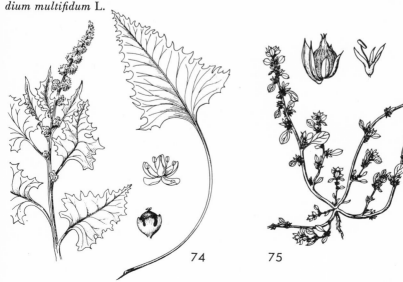

74   75

37. AMARANTHACEAE. Amaranth Family

1. Amaranthus L. Amaranth, Pigweed

Weeds generally, and *Amaranthus* particularly, are rarely collected. The paucity of cited localities is due not to the absence of plants in the field, but in the herbarium.

Plants with dense terminal spikes, these usually at least 2 cm. long, smaller axillary spikes also usually present.
  Stems prostrate to decumbent; utricles indehiscent.
    Utricles nearly smooth.   1. *A. deflexus*
    Utricles deeply wrinkled at maturity.   2. *A. gracilis*
  Stems erect; utricles circumscissile.
    Spikes slender, not stiff, not crowded; pistillate sepals erect, shorter than to longer than the utricles, gradually tapering to a terminal bristle.
      Seeds circular in outline, less than 1 mm. in diameter, black; stamens 5; calyx of the staminate flowers 2 mm. long.   3. *A. hybridus*
      Seeds elliptic to ovate in outline, 1.1–1.3 mm. long, brownish-black; stamens 3–5; calyx of the staminate flowers 3 mm. long.
                                     4. *A. powellii*
    Spikes thick, in a stiff glomerate panicle; pistillate sepals spreading at maturity, usually longer than the utricle, obtuse, truncate, or emarginate.   5. *A. retroflexus*
Plants with only axillary spikes.
  Stems erect; sepals of the pistillate flowers 3.   6. *A. albus*
  Stems prostrate or decumbent; sepals of the pistillate flowers 3–5.
    Sepals of the pistillate flowers 4–5, about 3 mm. long, enclosing the lower part of the utricle; bracts attenuate to spinose-tipped.   7. *A. graecizans*
    Sepals of the pistillate flowers 3, two often scale-like, usually less than 2 mm. long, not enclosing the lower part of the utricle; bracts narrowly subulate.   8. *A. californicus*

1. A. deflexus L. Low Amaranth. A common weed of disturbed areas, railroad rights of way, along sidewalks, roads, and vacant lots; San Francisco, San Carlos, Redwood City, Menlo Park, Palo Alto, Alviso, Santa Clara, and Pajaro River; native of tropical America. May–October.
2. A. gracilis Desf. Known locally only from San Francisco, where it grows as a sidewalk weed; native of the Old World.
3. A. hybridus L. Slender Pigweed, Spleen Amaranth. Occasional as a weed; San Francisco; native of tropical America. June–December.
4. A. powellii Wats. Powell's Amaranth. Weeds of disturbed areas; San Francisco, near Spanishtown, Stanford, and Santa Cruz. August–October.
5. A. retroflexus L. Rough Pigweed, Green Amaranth. Fairly common in disturbed areas and fallow fields; San Francisco, Spanishtown, Stanford, San Antonio Creek, and near Glenwood; native of eastern North America. June–December.
6. A. albus L. Tumbleweed. Weeds of disturbed areas, fields, and vineyards; San Francisco, Stanford, near San Jose, and Los Gatos; native of eastern North America. July–October. —*A. graecizans* of authors.

7. A. graecizans L. Prostrate Amaranth. Disturbed areas, usually in dry, hard-packed soils; San Francisco, Menlo Park, Stanford, and Wrights; native of tropical America. June–October. —*A. blitoides* of authors.

8. A. californicus (Moq.) Wats. *Fig. 75.* California Amaranth. Shores of lakes and ponds after the water has receded; San Francisco, Crystal Springs Lake, and Lower Crystal Springs Lake. July–October.

38. NYCTAGINACEAE. Four-O'clock Family

Fruits winged; leaves succulent. Sand dune plants. 1. *Abronia*
Fruits not winged; leaves not succulent. Garden escapes. 2. *Mirabilis*

1. Abronia Juss. Sand Verbena, Abronia

Flowers yellow; leaves orbicular-reniform to ovate. 1. *A. latifolia*
Flowers purplish; leaves elliptic to lanceolate. 2. *A. umbellata*

1. A. latifolia Esch. *Fig. 76.* Yellow Sand Verbena. Coastal sand dunes and sea beaches above the high water mark. February–November.

2. A. umbellata Lam. Beach or Pink Sand Verbena. Sand dunes along the coast. February–November.

2. Mirabilis L. Four-O'clock

1. M. jalapa L. Four-O'clock, Marvel-of-Peru. Occasional as an escape from cultivation in San Francisco, Los Altos, and probably elsewhere; native of tropical America. August–October.

39. PHYTOLACCACEAE. Pokeweed Family

1. Phytolacca L. Pokeweed or Pokeberry

1. P. heterotepala Walter. Mexican Pokeberry. Occasional as a weed in San Francisco in disturbed areas, and to be expected elsewhere; native of Mexico. July–August. *Phytolacca heterotepala* is closely related to the common neotropical *P. icosandra* L. and may actually only be a minor variant thereof.

40. ILLECEBRACEAE. Knotwort Family

Leaves pubescent; sepals similar.
 Stipules conspicuous; calyx 2 mm. long. 1. *Paronychia*
 Stipules minute; calyx 1 mm. long. 2. *Herniaria*
Leaves glabrous; sepals unequal. 3. *Cardionema*

1. Paronychia Adans.

1. P. franciscana Eastw. California Whitlowwort. Rocky hilltops and slopes, San Francisco and northern San Mateo counties. April–June.

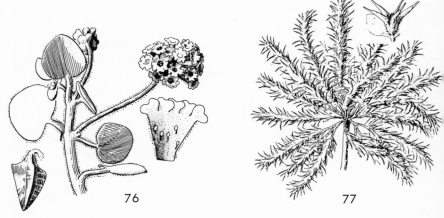

76   77

2. Herniaria L.

1. H. cinerea DC. Gray Herniaria. Known definitely only from near the Pajaro River east of Watsonville, but to be expected elsewhere in dry sandy localities; native of southern Europe. April–June.

3. Cardionema DC.

1. C. ramosissimum (Weinm.) Nels. & Macbr. *Fig. 77*. Sand Mat. Most common on coastal and inland sand dunes, but occasionally on rocky slopes; San Francisco south to the Pajaro River. April–August.

41.   AIZOACEAE.   Carpet Weed Family

Ovary superior; petals lacking; leaves not succulent.
   Leaves whorled; stipules lacking or inconspicuous.   1. *Mollugo*
   Leaves opposite; stipules present, conspicuous.   2. *Cypselea*
Ovary inferior; petals present or absent; leaves succulent.
   Leaves triangular-ovate in outline; petals lacking.   3. *Tetragonia*
   Leaves not triangular in outline; petals present.   4. *Mesembryanthemum*

1. Mollugo L. Carpet Weed

1. M. verticillata L. Indian Chickweed. Occasional in low areas where water has stood in the winter and early spring; known definitely from Stanford and San Jose; native of tropical America. June–October.

2. Cypselea Turp.

1. C. humifusa Turp. Cypselea. Low ground where water has stood during the winter and spring; Crystal Springs Lake and Aptos; native of the West Indies. July–August.

3. Tetragonia L. Sea Spinach

1. T. tetragonioides (Pall.) Ktze. New Zealand Spinach. Sandy soils and salt marshes along the coast and San Francisco Bay; native of Asia and New Zealand. April–September. —*T. expansa* Murr.

4. Mesembryanthemum L. Ice Plant

Leaves triangular in cross section, at least 8 mm. thick.
  Leaves usually less than 5 cm. long, obtusely angled; flowers 3–5 cm. in diameter.   1. *M. chilense*
  Leaves usually over 5 cm. long, sharply angled; flowers 8–10 cm. in diameter.   2. *M. edule*
Leaves more or less terete or with flat blades, less than 2 mm. thick.
  Leaves with a distinct petiole and flat blade, opposite.   3. *M. cordifolium*
  Leaves more or less terete, lacking a distinct petiole, opposite or alternate.
    Leaves 1–2 cm. long; calyx-tube under 0.5 cm. long.
      Leaves mainly alternate; plants annual.   4. *M. nodiflorum*
      Leaves mainly opposite; plants perennial.   5. *M. floribundum*
    Leaves 2.5–15 cm. long; calyx-tube over 1 cm. long.
      Leaves 2.5–3.5 cm. long, mainly opposite.   6. *M. crassifolium*
      Leaves 4–15 cm. long, mainly alternate.   7. *M. elongatum*

1. M. chilense Molina. *Fig. 78.* Sea Fig. Dunes along the coast, San Francisco southward. April–September.
2. M. edule L. Hottentot Fig. Occasional on sand dunes and bluffs along the coast; native of South Africa, widely introduced into California as an ornamental and as a plant to check erosion on highway cuts and earth fills, occasionally naturalized. April–September.
3. M. cordifolium L. f. Occasional as an escape from cultivation in San Francisco; native of South Africa.
4. M. nodiflorum L. Slender-leaved Ice Plant. Known from one specimen collected in the salt marshes east of Redwood City (*Nobs and Smith 1623*); native of Africa. July–September.
5. M. floribundum Haw. Naturalized on sea cliffs at San Francisco and Santa Cruz; native of South Africa. June–October.
6. M. crassifolium L. Common Ice Plant. Occasional as an escape from cultivation in San Francisco; native of South Africa.
7. M. elongatum Haw. Locally established along the cliffs in San Francisco from the Cliff House south to Lake Merced; native of South Africa. September–October.

42. PORTULACACEAE. Purselane Family

Capsules valvate.
  Capsules splitting into 3 valves.
    Stem leaves alternate; seeds numerous.   1. *Calandrinia*
    Stem leaves usually opposite; seeds few.   2. *Montia*

Capsules splitting into 2 valves.
　Styles long; drying petals twisting around the style; inflorescence umbellate-
　　cymose.　　　　　　　　　　　　　　　　　　　　　3. *Spraguea*
　Styles very short; drying petals forming a cap around the capsules; inflo-
　　rescence a spike-like panicle.　　　　　　　　　4. *Calyptridium*
Capsules circumscissile.
　Plants perennial; roots and caudex stout; sepals 4–8; petals 2 or more cm.
　　long, rose-colored or white.　　　　　　　　　　5. *Lewisia*
　Plants annual; roots slender; sepals 2; petals less than 1 cm. long, yellow.
　　　　　　　　　　　　　　　　　　　　　　　　6. *Portulaca*

1. **Calandrinia** HBK.

Mature capsules about as long as the calyx. Plants common.
　　　　　　　　　　　　　　　　　　　　1. *C. ciliata menziesii*
Mature capsules about twice as long as the calyx. Plants rare.　2. *C. breweri*

1. C. ciliata (R. & P.) DC. var. **menziesii** (Hook.) Macbr. *Fig. 79.* Red Maids.
Meadows, open areas, and frequently in cultivated fields and orchards. February–June. This variety is extremely variable with respect to flower size, and
numerous segregates have been described. The typical variety is native of Peru.
Calandrinia ciliata var. *menziesii* is one of the first plants to flower each year.
2. C. breweri Wats. Brewer's Calandrinia. Occasional on grassy slopes and cultivated fields, but most common in burned or disturbed chaparral; Stevens Creek
Reservoir, near Loma Prieta, and China Grade. April–June.

2. **Montia** L.

Cauline leaves alternate.　　　　　　　　　　　　1. *M. parvifolia*
Cauline leaves opposite.
　Stems with several pairs of leaves, often rooting at the nodes, prostrate to more
　　or less erect.
　　Seeds 1–1.4 mm. in diameter; sepals about 1.5 mm. long; lower leaves more
　　　or less spatulate.　　　　　　　　　　　　　2. *M. verna*
　　Seeds 0.6–0.9 mm. in diameter; sepals about 1 mm. long; lower leaves
　　　essentially linear.　　　　　　　　　　　　　3. *M. hallii*
　Stems with 1 pair of leaves, not rooting at the nodes, more or less erect.
　　Cauline leaves not connate, usually at least 2 cm. long; flowers bracteate.
　　　　　　　　　　　　　　　　　　　　　　　4. *M. sibirica*
　　Cauline leaves connate at least on one side; inflorescences with bracts only
　　　at the base.
　　　Cauline leaves connate on both sides, forming a conspicuous disk; basal
　　　　leaf blades often spatulate to rhomboidal.　　5. *M. perfoliata*
　　　Cauline leaves connate on one side, not forming a disk; basal leaf blades
　　　　linear to linear-spatulate.
　　　　Petals 2–4 times as long as the sepals; racemes over 3 cm. long, long
　　　　　pedunculate.　　　　　　　　　　　　　6. *M. gypsophiloides*
　　　　Petals about twice as long as the sepals; racemes 0.5–3 cm. long, short
　　　　　pedunculate.　　　　　　　　　　　　　7. *M. spathulata*

78

79

1. M. parvifolia (Moq.) Greene. Small-leaved Montia. Moist shaded areas along streams in redwood–Douglas fir forests, mainly on the western slopes of the Santa Cruz Mountains; San Mateo Canyon, Pescadero Creek, Waddell Creek, and Big Basin. April–July.

2. M. verna Neck. Vernal Montia. Moist areas in grasslands, along the edges of small puddles, and on muddy banks. March–June. —*M. fontana* of authors, in part.

3. M. hallii (Gray) Greene. Hall's Montia. Occasional on very moist grassy areas and in small local puddles, San Francisco southward. March–June. —*M. fontana* of authors, in part. *Montia verna* and *M. hallii* often grow together and care must be taken to keep the collections separate for purposes of identification.

4. M. sibirica (L.) Howell. Siberian Montia, Candy Flower. Known locally only from Santa Cruz County, growing along the shaded banks of streams and creeks that do not dry up completely in the late summer; Swanton, Boulder Creek, Zayante Creek, and near Mount Hermon. March–August.

5. M. perfoliata (Donn ex Willd.) Howell. *Fig. 80.* Miner's Lettuce. A common, variable species occurring in moist habitats throughout the Santa Cruz Mountains, often becoming weedy. February–May. Many segregates of *M. perfoliata* have been described. Much, if not most, of the variation is probably ecological. Until such time as this species is studied carefully and grown in a uniform environment, it seems pointless to accord each minor variation a name.

6. M. gypsophiloides (F. & M.) Howell. Coast Range Montia. Known only from serpentine slopes south of San Jose and near Edenvale in the Santa Clara Valley. February–May.

7. M. spathulata (Dougl.) Howell. Common Montia. Grasslands, shaded slopes, and edges of chaparral, San Francisco southward; San Francisco, San Francisco Watershed Reserve, Pescadero Creek, Emerald Lake, Searsville Ridge, Stanford,

Black Mountain, Monte Bello Ridge, Castle Rock, Loma Prieta, Edenvale, and near Eagle Rock. January–May. —*M. tenuifolia* (T. & G.) Howell, *M. exigua* (T. & G.) Jeps.

3. Spraguea Torr.

1. S. umbellata Torr. Pussy Paws. Known locally only from the Ben Lomond Sand Hills. June–July.

4. Calyptridium Nutt.

1. C. parryi Gray var. **hesseae** Thomas. Hesse's Calyptridium. Known from only a few localities in the Santa Cruz Mountains; Loma Prieta, Ben Lomond Mountain, and the Ben Lomond Sand Hills. June–August.

$\frac{1}{4}$

80

5. Lewisia Pursh

1. L. rediviva Pursh. Bitterroot. So far known from only one specimen collected on the serpentine hills near the mouth of Almaden Canyon (*Bacigalupi 1340*). March–June.

6. Portulaca L.

1. P. oleracea L. Purselane. Common throughout the Santa Cruz Mountains, especially as a weed of gardens and disturbed areas; native of Europe. May–October.

Boussingaultia gracilis Miers var. *pseudo-baselloides* (Haum.) Bailey, a member of the Basellaceae commonly known as Madeira Vine, has been reported as persisting in old gardens and in disturbed areas in San Francisco. This species is a perennial vine, native of tropical America, and its flowers have 2 sepals. None of the members of the Portulacaceae listed above are vines.

43. CARYOPHYLLACEAE. Pink Family

Sepals distinct.
    Stipules none.
        Capsules cylindric.   1. *Cerastium*
        Capsules globose, ovate, or oblong.
            Styles as many as the sepals and alternate with them.   2. *Sagina*
            Styles fewer than the sepals.
                Petals deeply 2-parted or -cleft, rarely none.   3. *Stellaria*
                Petals entire to emarginate.   4. *Arenaria*
    Stipules present, scarious.
        Leaves obovate.   5. *Polycarpon*
        Leaves linear to subulate, often fleshy.
            Leaves subulate-setaceous, 4–6 mm. long.   6. *Loeflingia*
            Leaves linear, usually somewhat fleshy, (6)10–50 mm. long.
                Styles and valves of the capsule 5; leaves appearing verticillate.
                    7. *Spergula*

Styles and valves of the capsule 3; leaves not appearing verticillate.
8. *Spergularia*
Calyx tubular, the tube at least ½ the length of the calyx.
Styles 3–5.
Plants densely white-woolly; styles usually 5.   9. *Lychnis*
Plants variously pubescent or glabrous, but not densely white-woolly; styles
usually 3.   10. *Silene*
Styles 2.
Leaves sessile, clasping, and often auriculate; calyx 10–14 mm. long.
11. *Vaccaria*
Leaves short-petiolate, not clasping; mature calyx 18–20 mm. long.
12. *Saponaria*

1. **Cerastium** L. Mouse-ear Chickweed

Petals 2–3 times as long as the sepals; mature capsules about equaling the
sepals.   1. *C. arvense*
Petals about equaling the sepals; mature capsules longer than the sepals.
Plants perennial; sepals shorter than the pedicels.   2. *C. vulgatum*
Plants annual; sepals longer than the pedicels.   3. *C. viscosum*

1. C. arvense L. Field or Meadow Chickweed. Open grassy hills, San Francisco
and northern San Mateo counties. March–May. —*C. arvense* L. var. *maximum*
Hollick and Britton.
2. C. vulgatum L. Large Mouse-ear Chickweed. Occasional as a lawn weed
and on grassy slopes; introduced from Europe. March–July. —*C. holosteoides*
Fries.
3. C. viscosum L. Mouse-ear Chickweed. A common weed of gardens, open
grassy slopes, cultivated fields, and disturbed areas; native of Europe. March–
June. —*C. glomeratum* Thuill.

2. **Sagina** L. Pearlwort

The members of this genus occurring locally are very poorly represented in
herbaria. The four species listed below are probably much more common than
the number of localities would indicate.

Sepals 4; petals present or absent; bases of the leaves glabrous or ciliolate.
Petals none; bases of the leaves ciliolate; plants glandular-pubescent.
1. *S. apetala barbata*
Petals present; bases of the leaves not ciliolate; plants glabrous.
2. *S. procumbens*
Sepals 5; petals present; bases of the leaves glabrous.
Sepals about 2 mm. long; plants annual; stems very slender.
3. *S. occidentalis*
Sepals about 3 mm. long; plants perennial; stems somewhat succulent.
4. *S. crassicaulis*

1. S. apetala Ard. var. **barbata** Fenzl ex Ledeb. Sticky Pearlwort. A diminutive
weed of paths, graveled areas, and other disturbed habitats; San Francisco,
Burlingame, Stanford, and Boulder Creek; native of Eurasia. April–May.

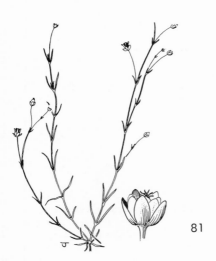

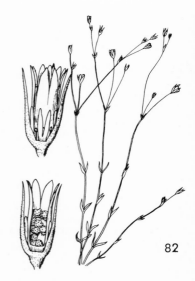

81

82

2. S. procumbens L. Procumbent Pearlwort. A weed of gardens and disturbed areas in San Francisco; native of Eurasia. May–September.
3. S. occidentalis Wats. *Fig. 81.* Western Pearlwort. Occasional in moist ground; San Francisco, Seal Cove, Stanford, and Jamison Creek Road. February–April.
4. S. crassicaulis Wats. Beach Pearlwort. Sea cliffs along the coast; San Francisco and Santa Cruz. May–December.

3. Stellaria L. Chickweed, Starwort

Plants with pubescent lines on the stems, annual.   1. *S. media*
Plants, if pubescent, not with hairs in distinct lines on the stems; annual or perennial.
  Plants with filiform stems, usually glabrous; sepals about 3 mm. long; annual.
                                        2. *S. nitens*
  Plants without filiform stems, glabrous to villous-pubescent; sepals 4–6 mm. long; perennial.
    Plants villous-pubescent; sepals 5–6 mm. long.   3. *S. littoralis*
    Plants essentially glabrous; sepals 4–4.5 mm. long.   4. *S. graminea*

1. S. media (L.) Cyrill. Common Chickweed. A very common species of moist disturbed areas, pastures, cultivated fields, grasslands, and gardens; native of Europe. February–October.
2. S. nitens Nutt. *Fig. 82.* Shiny Chickweed. Grasslands, sepentine soils, and inland sand deposits, probably quite common but inconspicuous, San Francisco southward. February–April.
3. S. littoralis Torr. Beach Starwort. Known locally only from along the coast at Lands End and Point Lobos in San Francisco. May–June.
4. S. graminea L. Lesser Starwort. Known from lawns in San Francisco and Stanford; native of Europe. April–June.

4. **Arenaria** L. Sandwort

Plants perennial; leaves linear-lanceolate or broader.
  Plants puberulent; leaves lanceolate to ovate-lanceolate. Not palustrine.
        1. *A. macrophylla*
  Plants glabrous; leaves linear-lanceolate. Palustrine.   2. *A. paludicola*
Plants annual; leaves linear-subulate or filiform.
  Leaves linear-subulate to lanceolate, 2–5 mm. long.
    Sepals 3-nerved, 3–4 mm. long.   3. *A. californica*
    Sepals 1-nerved, 2–2.5 mm. long.   4. *A. pusilla diffusa*
  Leaves filiform, 8–20 mm. long.   5. *A. douglasii*

1. A. macrophylla Hook. Large-leaved Sandwort. Known locally only from the peak of Loma Prieta, growing in rocky soil. May–June.
2. A. paludicola Robins. Swamp Sandwort. Swampy ground, known from San Francisco, where now probably extinct, and from Camp Evers. May–August.
3. A. californica (Gray) Brewer. California Sandwort. Shallow soil in rocky areas and inland sand deposits, San Francisco southward; fairly rare but locally abundant where it does occur. February–April.
4. A. pusilla Wats. var. **diffusa** Maguire. Dwarf Sandwort. Known only from San Francisco and San Mateo counties. April–May.
5. A. douglasii Fenzl ex T. & G. *Fig. 83.* Douglas' Sandwort. Serpentine and open grassy slopes, mainly to the east of the crests of the Santa Cruz Mountains; San Bruno Hills, Sawyer Ridge, Woodside serpentine, Waddell Creek, Los Gatos, Almaden Canyon, and Jamison Creek. March–May.

5. **Polycarpon** L.

1. P. tetraphyllum (L.) L. Four-leaved Polycarp. Occasional as a weed in clay and sandy soils and in pavement cracks; San Francisco and Stanford; native of Europe. April–September.

6. **Loeflingia** L.

1. L. squarrosa Nutt. California Loeflingia. Known locally only from the Ben Lomond Sand Hills, but to be expected elsewhere in similar habitats. April–May.

7. **Spergula** L. Spurry

1. S. arvensis L. Spurry. A very common weed of orchards, cultivated fields, and disturbed areas; native of Europe. To be found in flower at all times of the year, but most commonly so March–July.

8. **Spergularia** (Pers.) J. & C. Presl. Sand Spurry

Plants annual or short-lived perennials; seeds not winged.
  Leaves distinctly fascicled; stipules triangular-acuminate, conspicuous.
        1. *S. rubra*
  Leaves not fascicled or rarely so; stipules broadly deltoid, 2–4 mm. long, not very conspicuous.
    Seeds usually less than 0.5 mm. long; stamens 6–10.   2. *S. bocconii*

Seeds usually over 0.5 mm. long; stamens usually 2–5.
Inflorescence lax, not crowded; mature capsules 3.5–6.5 mm. long.
                                                                    3*a. S. marina marina*
Inflorescence crowded; mature capsules 3–4.5 mm. long.
                                                                    3*b. S. marina tenuis*
Plants usually perennial; seeds winged or not.
Plants usually glabrous; leaves not fascicled; annual to perennial; seeds about
    1 mm. long. Rare.   4. *S. media*
Plants usually pubescent and glandular, conspicuously so in the inflorescence;
    leaves fascicled or not; perennial with a stout taproot; seeds 0.5–1 mm.
    long.
Petals pink; seeds usually about 1 mm. long. Common.   5. *S. macrotheca*
Petals white; seeds usually about 0.5 mm. long. Fairly rare.   6. *S. villosa*

1. S. rubra (L.) J. & C. Presl. Purple Sand Spurry. Common, but rarely collected, growing in disturbed areas, hard-packed soils, and pastures; introduced from Europe. April–September.

2. S. bocconii (Scheele) Foucard ex Merino. Boccone's Sand Spurry. Disturbed areas throughout the Santa Cruz Mountains and adjacent lowlands, but nowhere common; introduced from southern Europe. April–July.

3a. S. marina (L.) Griseb. var. **marina.** Salt-marsh Sand Spurry. Along the coast and occasionally inland in more or less alkaline areas; San Francisco, Mussel Rock, San Mateo, near Palo Alto, Stanford, and Searsville. April–October.

3b. S. marina (L.) Griseb. var. **tenuis** (Greene) Rossbach. Known locally only from the salt marshes near Alviso. April–November.

4. S. media (L.) Presl. Middle-sized Sand Spurry. Known locally from only one specimen collected in 1908 near Redwood City; native of Europe. April–October.

5. S. macrotheca (Hornem. ex C. & S.) Heynh. Large-flowered Sand Spurry. Along or near the ocean and the San Francisco Bay in saline soils of marshes and bluffs. March–October.

6. S. villosa (Pers.) Camb. Villous Sand Spurry. Along the coast and inland in clay soils; San Francisco, Stanford, and Santa Clara; native of South America. May–July.

9. Lychnis L. Campion

1. L. coronaria (L.) Desr. Mullein Pink, Rose Campion. Occasional as an escape from cultivation in the Richmond District of San Francisco and probably elsewhere in the Santa Cruz Mountains; native of Europe. June–August.

10. Silene L. Catchfly, Campion

Calyx with 15 or more nerves.   1. *S. cucubalus*
Calyx with 10 nerves.
    Annuals; petals less than 5 mm. longer than the calyx.
        Calyx glabrous, or with very few hairs; upper internodes with a purplish,
            viscid-pubescent band between the nodes; inflorescence paniculate.
                                                                    2. *S. antirrhina*

Calyx hirsute; whole plant hirsute, lacking viscid-pubescent bands; inflorescence a one-sided raceme.   3. *S. gallica*
Perennials; petals at least 7 mm. longer than the calyx.
Flowers large, the calyx at least 1.5 cm. long; flowers not glomerate.
                                                   4. *S. californica*
Flowers small, the calyx less than 1.5 cm. long; flowers somewhat glomerate.
Stamens and styles long-exserted; flowers nodding at anthesis.
                                             5. *S. lemmonii*
Stamens and styles not long-exserted; flowers erect at anthesis.
Flowers often glomerate; stems few from a taproot.
                                             6. *S. scouleri grandis*
Flowers not glomerate; stems many from a branched rootcrown.
Calyx long-pubescent, the trichomes about 0.5 mm. long; leaves usually less than 5 cm. long. San Francisco south to northern Santa Cruz County.   7a. *S. verecunda verecunda*
Calyx densely pubescent, the trichomes much less than 0.5 mm. long; leaves usually over 5 cm. long. Central Santa Cruz County.
                                             7b. *S. verecunda platyota*

1. S. cucubalus Wibel. Bladder Campion. Occasional as an escape from cultivation and common as a weed near Moffett Field; native of Eurasia. June–July. —*S. latifolia* (Mill.) Britt. & Rendle.

2. S. antirrhina L. Sleepy or Snapdragon Catchfly. Occasional in disturbed and cultivated areas; known definitely from near San Jose and Boulder Creek. May–June.

3. S. gallica L. Common Catchfly, Windmill Pink. A very common weed of disturbed areas, along roads, and on open grassy slopes; the most common species of *Silene* locally; native of Europe. April–October.

4. S. californica Durand. California Indian Pink. Eastern slopes of the Santa Cruz Mountains from Crystal Springs Lake south to Stevens Creek and Black Mountain, commonly growing on serpentine outcroppings. April–July.

83   84

Silene laciniata Cav. ssp. *major* Hitchc. & Maguire var. *angustifolia* Hitchc. & Maguire has been reported from Santa Cruz County; however, it has been impossible to locate the specimen or specimens upon which this record was based. *Silene laciniata* and *S. californica* may be distinguished by the following key:

Plants usually over 2.5 dm. tall; leaves ovate, elliptic, to obovate; auricles about
    1 mm. long.   *S. laciniata*
Plants rarely 2.5 cm. tall; leaves linear to linear-lanceolate; auricles about 2
    mm. long.   *S. californica*

5. S. lemmonii Wats. *Fig. 84.* Lemmon's Campion. Rare in the Santa Cruz Mountains; known only from Saratoga and Boulder Creek. May–June.
6. S. scouleri Hook. ssp. **grandis** (Eastw.) Hitchc. & Maguire. Scouler's Large Campion. Mainly along the coast, San Francisco, Pescadero, Millbrae, but to be expected elsewhere. July–August. —*S. pacifica* Eastw.
7a. S. verecunda Wats. ssp. **verecunda.** Dolores Campion. A narrowly restricted endemic, occurring from San Francisco south to northern Santa Cruz County. March–October.
7b. S. verecunda Wats. ssp. **platyota** (Wats.) Hitchc. & Maguire. Usually in sandy soil of chaparral and often growing with *Pinus attenuata* and *P. ponderosa*, central Santa Cruz County. May–September.

11. Vaccaria Medic.

1. V. segetalis (Neck.) Garcke ex Asch. Known from San Francisco as a weed; introduced from Europe. May–August. —*V. vulgaris* Host.

12. Saponaria L. Bouncing Bet

1. S. officinalis L. Bouncing Bet, Soapwort. Common in the southern part of the Santa Clara Valley and adjacent foothills, growing in disturbed areas and in vineyards, where it is sometimes a serious weed; native of Europe. June–August.

44. NYMPHAEACEAE. Water Lily Family

1. Nymphaea L. Cow Lily

1. N. polysepala (Engelm.) Greene. Indian Pond Lily, Yellow Water Lily. Occasional in ponds near the coast, known from San Francisco, Pescadero, Big Basin, and Watsonville. April–September. —*Nuphar polysepalum* Engelm.

45. CERATOPHYLLACEAE. Hornwort Family

1. Ceratophyllum L. Hornwort

1. C. demersum L. Hornwort. Submerged aquatic plants of ponds, reservoirs, lakes, and slow-moving streams. June–August.

46. RANUNCULACEAE. Buttercup or Crowfoot Family

Carpels with more than 1 seed.
    Fruit a berry, usually red.   1. *Actaea*
    Fruit of separate follicles.
        Neither the petals nor sepals spurred.   2. *Isopyrum*
        Petals and/or sepals spurred.
            Flowers regular; petals 5, spurred; sepals not spurred.   3. *Aquilegia*
            Flowers irregular; petals 2 or 4, the upper 2 spurred; upper sepals pro-
                longed into a spur.   4. *Delphinium*
Carpels 1-seeded.
    Cauline leaves opposite or whorled; petals none.
        Perennial herbs.   5. *Anemone*
        Vines.   6. *Clematis*
    Cauline leaves alternate; petals absent or present.
        Petals usually present; plants not dioecious; leaves entire to once-pinnate.
            Receptacle cylindrical, several times as long as broad; sepals spurred.
                              7. *Myosurus*
            Receptacle conical, about as long as broad; sepals lacking spurs.
                            8. *Ranunculus*
        Petals absent; plants dioecious; leaves tripinnately compound.
                            9. *Thalictrum*

1. **Actaea** L. Baneberry

1. A. arguta Nutt. ex T. & G. Western Red Baneberry. Northern San Mateo
County southward, in woods and along streams, nowhere very abundant. March–
May. —*A. rubra* (Ait.) Willd. ssp. *arguta* (Nutt. ex T. & G.) Hult.

2. **Isopyrum** L.

1. I. occidentale H. & A. Western Rue Anemone. Known locally only from Big
Basin. April.

3. **Aquilegia** L. Columbine

Stems not densely glandular-pubescent or viscid, varying from glabrous to puber-
    ulent.   1. *A. formosa truncata*
Stems densely glandular-pubescent or viscid.   2. *A. eximia*

1. A. formosa Fisch. var. **truncata** (F. & M.) Jones. Northwest Crimson Columbine. Fairly common in mixed evergreen forests, along the margins of redwood
forests, and on brush-covered slopes. April–June.
2. A. eximia Van Houtte ex Planch. *Fig. 85.* Van Houtte's Columbine. Usually
growing on or near serpentine outcroppings; Crystal Springs Lake, Los Gatos
Creek, Los Gatos, and between Soquel and Aptos. June–August.

4. **Delphinium** L. Larkspur

Pistil 1; plants annual.   1. *D. ajacis*
Pistils 3; plants perennial.
  Flowers red.   2. *D. nudicaule*
  Flowers blue, purple, or occasionally white.
    Plants 1 m. or more tall; flowers usually over 50 in mature inflorescences.
                                                                                        3. *D. californicum*
    Plants under 1 m. tall; flowers under 50 in mature inflorescences.
      Pistils glabrous; stems slender, the diameter of the lowest inch of the
        stem smaller than the diameter of the rest of the stem; stems separat-
        ing easily from the tuberous roots.
        Rachis of the inflorescence puberulent. Plants of the immediate coast.
                                                                                        4. *D. decorum*
        Rachis of the inflorescence glabrous. Inland plants.   5. *D. patens*
      Pistils pubescent; stems stout, of about equal diameter throughout; stems
        not separating easily from the fusiform or fibrous roots.
        Stems spreading-pubescent, especially in the lower portion; inflores-
          cences spreading, paniculate; flowers usually few; sepals 12–24
          mm. long, 10–17 mm. wide.   6. *D. variegatum*
        Stems puberulent with incurved trichomes; inflorescences narrow, rac-
          emose; flowers numerous; sepals 8–12 mm. long, 5–7 mm. wide.
          Divisions of the leaves 3 mm. or more broad. Common on grassy
            slopes and in serpentine.   7. *D. hesperium*
          Divisions of the leaves 2 mm. or less broad. Rare, on inland marine
            sand deposits.   8. *D. parryi seditosum*

1. D. ajacis L. Rocket Larkspur. Occasional as an escape from cultivation;
native of Europe. June–July.
2. D. nudicaule T. & G. *Fig. 86.* Red Larkspur. Occasional in mixed evergreen forests; San Mateo Creek, Stevens Creek, Alba Grade, Boulder Creek, San
Lorenzo River. April–June.

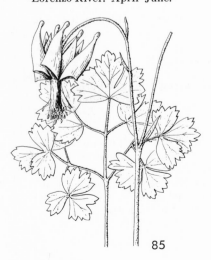
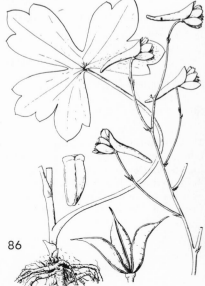

85   86

3. D. californicum T. & G. Coast Larkspur. San Francisco southward, growing in thickets and in chaparral, often forming extensive local colonies. April–July.

4. D. decorum F. & M. Coast or Blue Larkspur. Grassy slopes, on bluffs, and on rocky slopes near the coast; San Bruno Mountain, Montara Mountain, Pebble Beach, Pigeon Point, and Scott Creek. March–April.

5. D. patens Benth. Coast Larkspur. Grasslands and open woods from central San Mateo County southward, often growing in the shade of *Quercus, Acer,* and *Aesculus.* March–May. —*D. decorum* F. & M. var. *patens* (Benth.) Gray.

6. D. variegatum T. & G. Royal Larkspur. Grasslands, serpentine areas, and in the partial shade of oaks, Crystal Springs Lake region south to northern Santa Clara County. February–May. —*D. variegatum* T. & G. f. *superbum* Ewan.

7. D. hesperium Gray. Western Larkspur. Open grassy slopes and meadows, central San Mateo County southward. April–July.

8. D. parryi Gray ssp. **seditosum** (Jeps.) Ewan. Parry's Larkspur. Rare in the Santa Cruz Mountains, known locally only from Mount Hermon and from between Ben Lomond and Zayante in central Santa Cruz County. May. —*D. hesperium* Gray var. *seditosum* Jeps.

5. **Anemone** L. Anemone, Wind Flower

1. A. quinquefolia L. var. **grayi** (Behr & Kell.) Jeps. Western Wood Anemone, Wind Flower. Moist soil, often along streams, in deep shade in redwood–Douglas fir forests. March–May.

6. **Clematis** L. Clematis, Virgin's Bower

Leaflets usually 3–5; sepals about 2 cm. long; flowers usually 1–3. 1. *C. lasiantha*
Leaflets usually 5–7; sepals under 1 cm. long; flowers numerous, inflorescence
  paniculate.
  Plants dioecious. Native.   2. *C. ligusticifolia*
  Plants with perfect flowers. Introduced.   3. *C. vitalba*

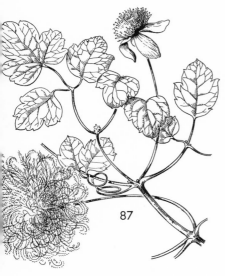

87

1. C. lasiantha Nutt. *Fig. 87.* Chaparral Clematis, Virgin's Bower. Growing on dry, brush-covered slopes and in chaparral, usually with *Adenostoma fasciculatum, Quercus dumosa, Ceanothus thyrsiflorus, Lepechina calycina, Dendromecon rigida,* and *Arctostaphylos.* March–May.

2. C. ligusticifolia Nutt. Western Virgin's Bower. Creek bottoms and along their banks, growing up through and over the other shrubby vegetation, more common on the eastern slopes of the Santa Cruz Mountains than on the western. June–September.

3. C. vitalba L. Traveler's Joy. Occasional as an escape from cultivation in San Francisco; native of Eurasia.

7. Myosurus L. Mouse Tail

1. M. minimus L. Common Mouse Tail. Occasional in heavy clay soils in San Francisco, the Santa Clara Valley, and probably elsewhere, but rapidly becoming extinct locally. April. —*M. minimus* L. var. *filiformis* Greene.

8. Ranunculus L. Buttercup

Plants aquatic; petals white, not glossy.
  Receptacles glabrous; styles 2–3 times as long as the ovaries; leaves usually
    floating.   1. *R. lobbii*
  Receptacles hispid; styles ½ as long as the ovaries; leaves floating or sub-
    merged.
    Floating leaves absent.   2a. *R. aquatilis capillaceus*
    Floating leaves present.   2b. *R. aquatilis hispidulus*
Plants not aquatic; petals yellow, glossy.
  Leaves entire to crenate, not lobed or divided.   3. *R. pusillus*
  Leaves lobed or divided.
    Achenes hairy or muricate; plants annual.
      Achenes muricate on the surface; petals 5–8 mm. long; plants coarse.
                                                                 4. *R. muricatus*
      Achenes not muricate, but with uncinate bristles; petals about 1.5 mm.
        long; plants slender.   5. *R. hebecarpus*
    Achenes glabrous (if pubescent then neither muricate nor with uncinate
      bristles); plants perennial.
      Stems creeping, rooting at the nodes.   6. *R. repens*
      Stems erect or ascending, not rooting at the nodes.
        Beak as long as the achene, not hooked.   7. *R. bloomeri*
        Beak much shorter than the achene, hooked.
          Basal leaf-blades usually compound; petals 9–16 or more; beak of
            the achene stout, under 0.8 mm. long.
                                                  8a. *R. californicus californicus*
          Basal leaf-blades usually simple; petals 5–16; beak of the achene
            stout or slender, 0.5–1.5 mm. long.
            Petals 5–9; beak of the achene slender, 1–1.5 mm. long; stems
              erect, 3–7 dm. long. Inland plants.   8b. *R. californicus gratus*
            Petals 9–16 or sometimes more; beak of the achene stout, under
              0.8 mm. long; stems more or less prostrate, 1–2.5 dm. long.
              Plants of coastal bluffs.   8c. *R. californicus cuneatus*

1. R. lobbii (Hiern) Gray. Lobb's Water Buttercup. Ponds, lakes, pools, and fresh water marshes, northern San Mateo County south to central Santa Cruz County. March–May.
2a. R. aquatilis L. var. **capillaceus** (Thuill.) DC. Water Buttercup. Lakes and ponds, San Francisco southward. March–July.
2b. R. aquatilis L. var. **hispidulus** Drew. Known locally only from a pond on Cahill Ridge, but to be expected elsewhere. June–July.
3. R. pusillus Poir. Low Buttercup, Spearwort. Growing in wet areas and along the margins of pools; Woodside, Coal Mine Ridge, and Santa Cruz. May–June.

4. R. muricatus L. Prickle-fruited Buttercup. Moist ground, seeps, and around springs, San Francisco southward; native of Europe. February–June.

5. R. hebecarpus H. & A. *Fig. 88.* Pubescent-fruited Buttercup. Wooded slopes, commonly under oaks and madrones in very moist situations; San Francisco, Sawyer Ridge, Coal Mine Ridge, Permanente Creek, Los Gatos, and Bear Creek Canyon. March–May.

6. R. repens L. Crowfoot, Creeping Buttercup. Common as a weed in lawns, in marshes, and along seeps and springs; native of Europe. To be found in flower at all seasons of the year. —*R. repens* L. var. *erectus* DC.

7. R. bloomeri Wats. Bloomer's Buttercup. Moist heavy soils, in marshland, and often in low fields near San Francisco Bay; San Francisco, Baden, Montara Mountain, San Andreas Lake, near Woodside, Alviso, and Mayfield. February–June.

8a. R. californicus Benth. var. **californicus.** California Buttercup. A very common species of open fields, meadows, and grasslands, sometimes in serpentine soil. February–April. Teratological forms in which all the flower parts are sepaloid are fairly common and have been collected between Woodside and Half Moon Bay and at Stanford.

8b. R. californicus Benth. var. **gratus** Jeps. Southern San Mateo County southward on the eastern slopes of the Santa Cruz Mountains, in ravines and deep canyons. March–June.

8c. R. californicus Benth. var. **cuneatus** Greene. Coastal bluffs and occasionally inland on open, wind-swept summits; San Francisco, San Gregorio Creek, Pigeon Point, and near Swanton. March–June.

9. Thalictrum L. Meadow Rue

1. T. polycarpum (Torr.) Wats. *Fig. 89.* Brush-covered slopes and occasionally along the edges of chaparral in moist areas, more common on the eastern slopes of the Santa Cruz Mountains than on the western. March–May.

47. BERBERIDACEAE. Barberry Family

Plants shrubby; leaves or the leaflets spinose, thick.   1. *Berberis*
Plants herbaceous; leaflets not spinose, thin.   2. *Vancouveria*

1. Berberis L. Barberry

Leaves simple.   1. *B. darwinii*
Leaves compound.
    Leaves borne in a terminal tuft; veins of the leaflets palmate from the base; bud scales persistent.   2. *B. nervosa*
    Leaves borne along the stem; veins of the leaflets pinnate throughout; bud scales not persistent.
        Leaflets ovate, regularly toothed, the basal leaflets usually remote from the base of the petiole. Introduced plants.   3. *B. aquifolium*
        Leaflets ovate-lanceolate, irregularly toothed, the basal leaflets near the base of the petiole. Native plants.   4. *B. pinnata*

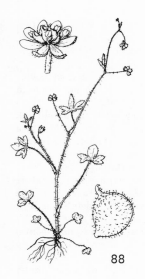

88   89

1. **B. darwinii** Hook. Darwin's Barberry. Known as an escape from cultivation in San Francisco; native of Chile and Patagonia.
2. **B. nervosa** Pursh. Oregon Grape. Occasional in coniferous woods, known from Butano Creek, Waddell Creek, Laguna Creek, near Bonny Doon, and Rincon Canyon. April–June. —*Mahonia nervosa* (Pursh) Nutt.
3. **B. aquifolium** Pursh. Holly-leaved Berberis. Known locally only as an escape from cultivation at Boulder Creek; native from British Columbia to Oregon and Idaho. February–March. —*Mahonia aquifolium* (Pursh) Nutt.
4. **B. pinnata** Lag. California Berberis, Coast Barberry. Rocky outcroppings throughout the Santa Cruz Mountains, but nowhere common; San Francisco, San Bruno Hills, San Andreas Reservoir, Pescadero Creek, Jasper Ridge, Waddell Creek, Bonny Doon Road, and Los Gatos Creek. March–May. —*Mahonia pinnata* (Lag.) Fedde.

2. Vancouveria Morr. & Des. Vancouveria

1. **V. planipetala** Calloni. *Fig. 90.* Small-flowered Vancouveria, Inside-out Flower. Most common on the western slopes of the Santa Cruz Mountains, usually growing in redwood forests; Kings Mountain, Gazos Creek, Glenwood, Ben Lomond, and Big Trees. April–September.

48. LAURACEAE. Laurel Family

Perfect stamens 9; calyx 6-parted.   1. *Umbellularia*
Perfect stamens 12 or more; calyx 4-parted.   2. *Laurus*

1. Umbellularia Nutt.

1. U. californica (H. & A.) Nutt. *Fig. 91.* California Laurel, California Bay, Oregon Myrtle, Oregon Pepperwood. A common tree of canyons, mountain slopes, valleys, and occasionally in chaparral, San Francisco southward. December–April.

2. Laurus L. Laurel

1. L. nobilis L. Laurel, Sweet Bay. Occasional as an escape from cultivation in San Francisco and probably elsewhere; native of Eurasia.

49. PAPAVERACEAE. Poppy Family

Plants woody shrubs.   1. *Dendromecon*
Plants annual or perennial herbs.
  Leaves entire, at least some usually opposite.
    Carpels remaining united, smooth, usually 3; stamens under 12.
                                          2. *Meconella*
    Carpels separating, 9–18, moniliform when mature; stamens numerous.
                                          3. *Platystemon*
  Leaves not entire, alternate.
    Sepals united to form a calyptra; petals orange or yellow.   4. *Eschscholzia*
    Sepals not united to form a calyptra; petals yellow or scarlet.
      Petals yellow; capsules 2-celled, bicarpellate.   5. *Glaucium*
      Petals scarlet; capsules 1-celled, several-carpellate.   6. *Stylomecon*

1. Dendromecon Benth.

1. D. rigida Benth. *Fig. 92.* Tree or Bush Poppy. Chaparral and closed-cone pine forests, commonly in sandy or rocky soils; Cahill Ridge, Peters Creek, Butano Ridge, Los Gatos Hills, Black Mountain, Loma Prieta, Eagle Rock, and Ben Lomond. Most commonly in flower April–July, but flowering at all seasons.

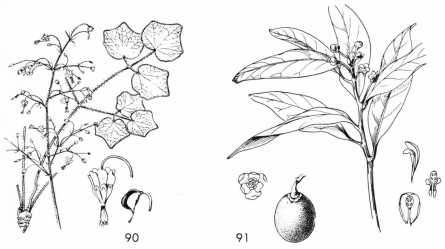

90   91

2. Meconella Nutt.

Capsules obovoid; basal leaves linear; scapes spreading-pilose.   1. *M. linearis*
Capsules linear; basal leaves obovate to spatulate; scapes glabrous.
2. *M. californica*

1. M. linearis (Benth.) Nels. & Macbr. Narrow-leaved Meconella. Occasional in the Santa Cruz Mountains, commonly in sandy soils; San Francisco, San Bruno Hills, Zayante Creek, and near Camp Evers. February–June. —*Hesperomecon linearis* (Benth.) Greene.
2. M. californica Torr. California Meconella. Fairly rare, wet rocky slopes and grasslands, often growing with *Mimulus guttatus, Montia hallii, Lithophragma affinis,* and *Alchemilla occidentalis;* San Francisco, San Bruno Mountain, near Crystal Springs, Big Basin, and Jamison Creek. February–April.

3. Platystemon Benth. Cream Cups

1. P. californicus Benth. *Fig. 93.* California Cream Cups. Widely distributed in the Santa Cruz Mountains and adjacent lowlands in rocky soils, in serpentine, on grassy slopes, and on coastal sand dunes; San Francisco, Crystal Springs Lake, Redwood City, Palo Alto, Stevens Creek, Los Gatos, San Jose, near Gilroy, Swanton, Scotts Valley, and Sunset Beach. March–June. Coastal populations usually have much wider leaves than inland ones.

4. Eschscholzia Cham.

1. E. californica Cham. *Fig. 94.* California Poppy. Common on grassy and rocky slopes, in fields, on coastal sand dunes and bluffs, and in disturbed habitats. March–October and occasionally in flower at other times of the year.

5. Glaucium Mill. Sea Poppy

1. G. flavum Crantz. Yellow-horned Poppy. Occasional as an escape from cultivation. June–September.

6. Stylomecon Taylor

1. S. heterophylla (Benth.) Taylor. Wind Poppy. Moist shaded slopes and occasionally on serpentine; Spanishtown, near San Mateo, Stanford, San Jose, Los Gatos, and Loma Azule. April–May. —*Meconopsis heterophylla* Benth., *Papaver heterophyllum* (Benth.) Greene.

50. FUMARIACEAE. Fumewort or Fumitory Family

Flowers 2–3 cm. long, yellow or lavender; outer petals similar; capsules ovate to
cylindrical, 1.5–2 cm. long. Native plants.   1. *Dicentra*
Flowers less than 7 mm. long, white or tinged with pink or purple; outer petals
dissimilar; capsules nearly globose, 2–3 mm. in diameter. Introduced plants.
2. *Fumaria*

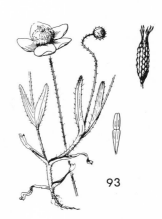

92

93

1. **Dicentra** Bernh. Bleeding Heart

Stems scapose; primary leaf-divisions ternate; flowers lavender; outer petals reflexed from the tip.   1. *D. formosa*

Stems leafy; primary leaf-divisions pinnate; flowers yellow; outer petals reflexed from the middle.   2. *D. chrysantha*

1. D. formosa (Haw.) Walp. *Fig. 95.* Pacific Bleeding Heart. Moist areas along stream banks in shade, abundant in local colonies, but not widespread; Crystal Springs, Waddell Creek, Scott Creek, Ben Lomond Ridge, Black Mountain, Big Basin, and as an escape from cultivation in San Francisco. April–June.

2. D. chrysantha (H. & A.) Walp. Golden Dicentra, Golden Ear-drops. Occasional on the eastern slopes of the Santa Cruz Mountains from about 400 to 2000 feet elevation, usually in dry rocky soil and often in chaparral; near Los Altos, near Black Mountain, Saratoga Summit, Wrights Station, and Alma-Sierra Azule Road. May–August.

2. **Fumaria** L. Fumitory

Flowers 3–4 mm. long, cream-colored, the tips purple; leaf-segments narrowly linear.   1. *F. parviflora*

Flowers 4–7 mm. long, purple; leaf-segments linear.   2. *F. officinalis*

1. F. parviflora Lam. Small-flowered Fumitory. Known locally as an introduced weed in disturbed areas at Moffett Field and in Mountain View, but to be expected elsewhere; native of Europe. April–June.

2. F. officinalis L. Common Fumitory. Known from San Francisco as an escape from cultivation; native of Europe. April–June.

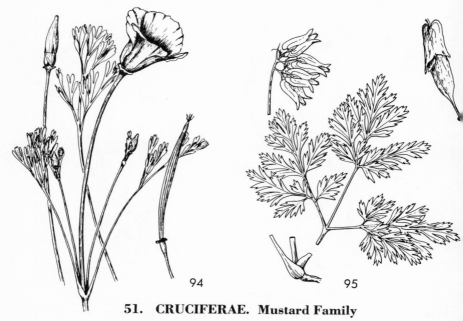

94   95

51. CRUCIFERAE. Mustard Family

Pods usually at least three times as long as broad.
  Pods indehiscent, but often breaking into 1-seeded joints.
    Leaves fleshy; pods 1-seeded. Sand dune and beach plants.   **1.** *Cakile*
    Leaves not fleshy; pods several-seeded. Plants of various habitats.
                                                    2. *Raphanus*
  Pods dehiscent by valves.
    Pods with a prominent beak; petals yellow.
      Upper cauline leaves more or less entire to pinnately lobed or divided.
                                              3. *Brassica*
      Upper cauline leaves bipinnately lobed.   **4.** *Erucastrum*
    Pods lacking a prominent beak; petals of various colors.
      Cauline leaves entire or dentate.
        Pods 2-jointed, the upper joint 5-ribbed; petals yellow. **5.** *Rapistrum*
        Pods not 2-jointed; petals of various colors.
          Pods tipped by the persistent, corniculate stigmas, the pods often to
            12 cm. long; petals purple; herbage stellate-pubescent.
                                            6. *Matthiola*
          Pods not tipped by persistent stigmas, pods usually less than 10 cm.
            long; petals yellow, white, or purplish; herbage if pubescent
            usually with simple or 2-forked hairs.
            Petals with undulate margins, purple or white, pubescence of
              simple hairs.   **7.** *Streptanthus*
            Petals plain, yellow, rose, or purplish; pubescence usually of forked
              hairs.
              Petals yellow to yellow-orange, over 10 mm. long. **8.** *Erysimum*
              Petals rose, purple, or if yellow, then 3 mm. long. **9.** *Arabis*

Cauline leaves usually deeply lobed or pinnatifid.
  Stems from underground tubers. 10. *Dentaria*
  Stems lacking tubers.
    Pods oblong, the seeds in 2 rows. Plants of very wet areas.
      Petals yellow. 11. *Rorippa*
      Petals white. 12. *Nasturtium*
    Pods linear, the seeds in 1 row. Plants of fairly dry areas.
      Pods flattened.
        Pods flattened parallel to the partition; leaves with ovate to orbicular leaflets. 13. *Cardamine*
        Pods flattened contrary to the partition; leaves deeply pinnately divided into linear lanceolate lobes. 14. *Tropidocarpum*
      Pods terete or quadrangular in cross section.
        Pods more or less 4-angled in cross section; midveins of the valves prominent; stems glabrous. 15. *Barbarea*
        Pods terete; veins of the valves prominent or not; stems variously pubescent.
          Petals yellowish or greenish, occasionally fading to white; pods erect or ascending, 0.5–10 cm. long.
            Pubescence of simple hairs; pods over 1 cm. long. 16. *Sisymbrium*
            Pubescence of forked hairs; pods 0.6–0.8 cm. long. 17. *Descurainia*
          Petals white; pods usually reflexed, 3–6 cm. long. 18. *Thelypodium*
Pods about as broad as long, at most twice as long as broad.
  Pods 1-seeded, indehiscent.
    Pods with hooked hairs (these rarely lacking), the margins never perforate. 19. *Athysanus*
    Pods glabrous or pubescent, the hairs not hooked, the margins often perforate. 20. *Thysanocarpus*
  Pods 2- to many-seeded, usually dehiscent.
    Pods flattened at right angles to the partition, the partition thus narrower than the pods.
      One pair of petals larger than the other. 21. *Iberis*
      Petals all about equal.
        Pods wrinkled or tuberculate. 22. *Coronopus*
        Pods smooth.
          Seeds several in each cell; pods obdeltoid to deltoid-obcordate or elliptic to oval.
            Pods obdeltoid to deltoid-obcordate. 23. *Capsella*
            Pods elliptic to oval. 24. *Hutchinsia*
          Seeds 1 per cell; pods variously shaped, but not obdeltoid to obcordate.
            Pods indehiscent, not notched or winged at the apex. 25. *Cardaria*
            Pods dehiscent, notched or winged at the apex. 26. *Lepidium*
    Pods flattened parallel to the partition or more or less terete, the partition thus about as wide as the pods.
      Petals purple; pods flattened, thin, over 15 mm. wide. 27. *Lunaria*

Petals white or yellow; pods less than 5 mm. wide.
Leaves pinnately divided, glabrous.
  Petals yellow.   11. *Rorippa*
  Petals white.   12. *Nasturtium*
Leaves entire, pubescence of stellate or forked hairs.
  Petals white; pubescence of 2-forked hairs.   28. *Lobularia*
  Petals yellow; pubescence of stellate hairs.   29. *Alyssum*

1. Cakile Mill. Sea Rocket

Leaves entire or sinuate-dentate; pods lacking horns at the joints.
    1. *C. edentula californica*
Leaves pinnately lobed; pods with horns at the joints. 2. *C. maritima*

1. **C. edentula** (Bigel.) Hook. ssp. **californica** (Heller) Hulten. *Fig. 96.* Pacific or California Sea Rocket. Coastal sand dunes and beaches; San Francisco, Waddell Creek, and Swanton. May–November.
2. **C. maritima** Scop. Sea Rocket. Coastal sand dunes and beaches, and becoming a weed in disturbed areas near San Francisco Bay; San Francisco, Half Moon Bay, Pescadero, and Davenport Landing; native of Europe and the Mediterranean region. March–October.

2. Raphanus L. Radish

Pods grooved lengthwise, constricted between the seeds when mature; petals yellow. 1. *R. raphanistrum*
Pods not grooved lengthwise, not constricted between the seeds when mature; petals white to purple or yellow, usually purple-veined. 2. *R. sativus*

1. **R. raphanistrum** L. Jointed Charlock. Occasional in disturbed areas; San Francisco, Menlo Park, Stanford, and Santa Cruz; native of Eurasia. May–July.
2. **R. sativus** L. Wild Radish. One of the most common introduced weeds in the Santa Cruz Mountains, growing in fallow fields, coastal sand dunes, road embankments, and disturbed areas; native of Europe. February–July.

3. Brassica L. Mustard

Beak at least ⅓ the length of the pod, usually somewhat flattened; valves usually with 3 equally distinct longitudinal veins; cauline leaves not clasping.
  Pods bristly, spreading at maturity; seeds usually 3 per cell. 1. *B. hirta*
  Pods not bristly, spreading or not at maturity; seeds usually 5 or more per cell.
    Pods 3–4 mm. wide at maturity, spreading; leaves glabrous or slightly hispid.
    2. *B. kaber*
    Pods about 1 mm. wide at maturity, appressed; leaves hirsute.
    3. *B. geniculata*
Beak less than ⅓ the length of the pod, usually round; valves with one prominent median longitudinal vein; cauline leaves sometimes clasping.
  Upper cauline leaves auriculate-clasping, sessile, glabrous, glaucous.
    Buds exceeding the flowers; basal leaves glabrous. 4. *B. oleracea*
    Buds not exceeding the flowers; basal leaves somewhat hirsute.
    5. *B. campestris*

Upper cauline leaves not clasping, sessile or petiolate, occasionally hirsute, usually not glaucous.

Pedicels of mature pods less than 5 mm. long; pods appressed to the rachis, 1–2 cm. long, to 2 mm. in diameter, somewhat quadrangular.

6. *B. nigra*

Pedicels of mature pods more than 5 mm. long; pods not appressed to the rachis, usually 2–3 cm. long or longer, at least 2 mm. in diameter.

Plants annual; pods not stipitate. 7. *B. juncea*

Plants biennial or perennial; pods short-stipitate, the stipes 1–3 mm. long.

8. *B. fruticulosa*

1. B. hirta Moench. White Mustard or Charlock. Known definitely from an old collection made in San Francisco, but to be expected occasionally in orchards, cultivated fields, and disturbed areas; native of Eurasia. March–August. —*Sinapis alba* L.

2. B. kaber (DC.) Wheeler. Charlock. Common in fields and disturbed areas; San Francisco, San Mateo County, Stanford, Edenvale, and Laurel; native of Europe. March–October. —*Sinapis arvensis* L.

3. B. geniculata (Desf.) Ball. Mediterranean or Summer Mustard. Occasional as a weed; San Francisco, Stanford, near Coyote, and Watsonville; native of the Mediterranean region. June–September. *B. adpressa* Boiss., *Sinapis incana* L.

4. B. oleracea L. Cabbage. Occasional as an escape from cultivation, persisting for a short time in neglected fields, and becoming established on maritime slopes; native of Europe. Probably to be found in flower at all seasons.

5. B. campestris L. *Fig. 97.* Common or Field Mustard. The most common species of *Brassica* locally, especially common during the spring in fields, orchards, and disturbed areas; native of Europe. To be found in flower at all seasons of the year.

96

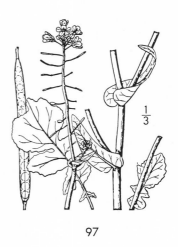

$\frac{1}{3}$

97

6. B. nigra (L.) Koch. Black Mustard. Disturbed areas; San Francisco, Crystal Springs Lake, Stanford, and Pajaro River; native of Europe. April–September.
7. B. juncea (L.) Cosson. Chinese, Indian, or Oriental Mustard. Occasional as a weed; San Francisco; native of Asia. June–September.
8. B. fruticulosa Cyr. Known locally only as a weed in San Francisco; native of the Mediterranean region. May–July.

4. Erucastrum Presl

1. E. gallicum (Willd.) Schulz. Occasional in the Santa Clara Valley; native of Europe. March–September.

5. Rapistrum Crantz

1. R. rugosum (L.) All. Known only from disturbed areas in San Francisco; native of the Mediterranean region. July.

6. Matthiola R. Br. Stock

1. M. incana (L.) R. Br. Stock. Occasional as an escape from cultivation along the sea cliffs in San Francisco and probably elsewhere; native of Europe. April–May.

7. Streptanthus Nutt.

Sepals and petals usually purple; at least the lower leaves coarsely dentate.
1*a. S. glandulosus glandulosus*
Sepals and petals white; lower leaves entire to dentate.
1*b. S. glandulosus albidus*

1a. S. glandulosus Hook. var. **glandulosus.** *Fig. 98.* Common Jewel Flower. Rocky slopes, usually serpentine, from the Los Altos Hills southward on the eastern slopes of the Santa Cruz Mountains; Los Altos Hills, Campbell Creek, Permanente Creek, Alma Soda Springs, near Los Gatos, and Loma Prieta. April–June.
1b. S. glandulosus Hook. var. **albidus** (Greene) Jeps. San Juan Bautista Hills south of San Jose. April–May.

8. Erysimum L. Wallflower

Leaves less than 1 mm. broad. Inland sand deposits, Santa Cruz County.
1. *E. teretifolium*
Leaves over 2 mm. broad.
  Seeds wingless; plants never woody; upper cauline leaves usually not toothed; mature pods usually about 2 mm. wide, somewhat quadrangular.
2. *E. capitatum*
  Seeds winged; plants woody or not; upper cauline leaves toothed or not; mature pods usually over 2 mm. wide, somewhat flattened.
    Pods usually ascending; pedicels not at right angles to the axis of the inflorescence.
    Leaves less than 6 mm. wide, entire to remotely denticulate.
3*a. E. franciscanum franciscanum*

Leaves 9–15 mm. wide, commonly sharply toothed.

3*b*. *E. franciscanum crassifolium*

Pods usually divaricate; pedicels more or less at right angles to the axis of the inflorescence.   4. *E. ammophilum*

1. E. teretifolium Eastw. Terete-leaved Erysimum. Endemic, known only from the inland marine sand deposits in Santa Cruz County; Glenwood, Ben Lomond Mountain, 2 miles southeast of Ben Lomond, and Ben Lomond Sand Hills. April–September. —*E. filifolium* Eastw.

2. E. capitatum (Dougl.) Greene. *Fig. 99.* Douglas' Wallflower. Chaparral, oak woods, and in disturbed areas in redwood–Douglas fir forests; near San Gregorio, La Honda, Castle Rock Ridge, Loma Prieta, and near Boulder Creek. April–June.

3*a***. E. franciscanum** Rossb. var. **franciscanum.** Franciscan Wallflower. Rocky slopes and summits, often on serpentine outcroppings, grassy slopes, and coastal sand dunes; Presidio, Lake Merced, San Bruno Hills, near Sharp Park, Montara Mountain, and Crystal Springs Lake. March–May.

3*b***. E. franciscanum** Rossb. var. **crassifolium** Rossb. Coarse-leaved Wallflower. Endemic in the Santa Cruz Mountains, growing on sandy or rocky slopes, in ravines, and in oak woods; San Bruno Hills, Salada, Montara Mountain, Sawyer Ridge, San Mateo Creek, Greyhound Rock, and Scott Creek. March–May.

4. E. ammophilum Heller. Coast Wallflower. Known locally only from the sand dunes and sand hills along the coast in southern Santa Cruz County. March–May.

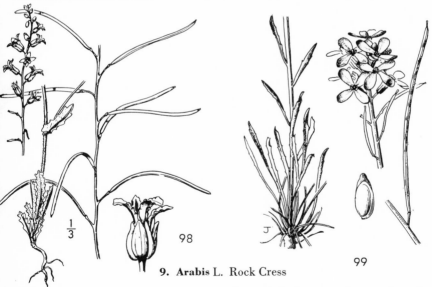

9. Arabis L. Rock Cress

Petals rose or purple; cauline leaves sessile or auriculate, not glaucous but glabrous or pubescent.

Petals 10–15 mm. long; pods straight, erect.   1. *A. blepharophylla*

Petals 6–8 mm. long; pods arcuate, spreading.   2. *A. breweri*

Petals yellow-white; cauline leaves auriculate, glaucous, glabrous. 3. *A. glabra*

1. A. blepharophylla H. & A. Coast Rock Cress. Rocky coastal bluffs, bare granitic soils, and open grassy slopes on the coastal side of the Santa Cruz Mountains; San Francisco, San Bruno Hills, Montara Mountain, and near Boulder Creek. February–May.

2. A. breweri Wats. Brewer's Rock Cress. Known definitely only from the vicinity of Mount Umunhum. May–June.

3. A. glabra (L.) Bernh. Tower Mustard. Occasional on coastal sand dunes and on wooded slopes; San Francisco, Montara Mountain, Coal Mine Ridge, Permanente Creek, between Scott and Mill creeks, and Sunset Beach. February–June. —*Turritis glabra* L.

10. Dentaria L. Toothwort

Leaflets of cauline leaves 3–5, ovate, toothed, thin. 1*a*. *D. californica californica*
Leaflets of cauline leaves usually 3, oblanceolate, usually entire, thick.
1*b*. *D. californica integrifolia*

1*a*. D. californica Nutt. var. **californica.** *Fig. 100.* California Toothwort. Common along moist stream banks and slopes in redwood–Douglas fir forests; San Francisco, San Mateo, La Honda, Coal Mine Ridge, Los Gatos foothills, Swanton, Jamison Creek, and Santa Cruz. December–May.

1*b*. D. californica Nutt. var. **integrifolia** (Nutt.) Detling. Milkmaids, Rainbells. Fields, open slopes, and occasionally in shade; South San Francisco, Arroyo de los Frijoles, Aptos, and Mount Madonna Road. November–April. *Dentaria californica* is a variable species with respect to shape, lobing, and thickness of both the cauline and rhizomal leaves. Intermediate specimens, which are difficult to assign to one variety or another, are often found. The initial flowering dates listed above for both varieties are the exception rather than the rule. In general, plants do not flower in any great number until February.

11. Rorippa Scop. Yellow Cress

Stems much branched from the base; pedicels usually 2–4 mm. long; segments
    of the leaves sharply toothed.   1. *R. curvisiliqua*
Stems not branched from the base; pedicels usually over 6 mm. long; segments
    of the leaves with rounded teeth or the margin more or less entire.
2. *R. islandica occidentalis*

1. R. curvisiliqua (Hook.) Bessey. Western Yellow Cress. Margins of ponds and small lakes and in boggy ground, San Francisco southward. March–November.

2. R. islandica (Oeder) Borbas var. **occidentalis** (Wats.) Butters and Abbe. Occasional along ponds and streams; Carnadero Creek and Kellogg Lake. July–October. —*R. palustris* (L.) Bess. ssp. *occidentalis* (Wats.) Abrams.

12. Nasturtium R. Br.

1. N. officinale R. Br. Water Cress. Fairly common along the margins of streams and small ponds, San Francisco southward; native of Eurasia. March–November. —*Rorippa nasturtium-aquaticum* (L.) Schinz & Thell.

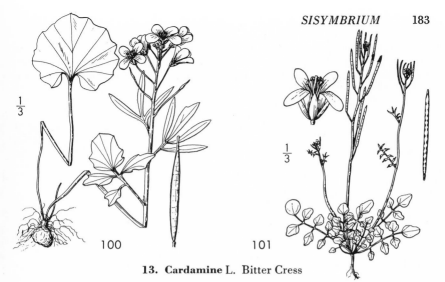

$\frac{1}{3}$

$\frac{1}{3}$

100   101

13. **Cardamine** L. Bitter Cress

1. C. oligosperma Nutt. *Fig. 101.* Few-seeded Bitter Cress. A common species, often forming dense stands, in moist woods and on brush-covered slopes, and becoming a weed in disturbed areas. February–June.

14. **Tropidocarpum** Hook.

1. T. gracile Hook. Slender Tropidocarpum, Dobie Pod. Occasional in grasslands from Stanford southward through the Santa Clara Valley and at Swanton. March–April.

15. **Barbarea** R. Br. Winter Cress

Basal leaves with 2–4 pairs of lateral lobes.   1. *B. orthoceras*
Basal leaves with 4–10 pairs of lateral lobes.   2. *B. verna*

1. B. orthoceras Ledeb. American Winter Cress. Widespread, growing in partial shade in woods, along the edges of chaparral, and in disturbed areas, San Francisco southward. March–July. —*B. americana* Rydb.
2. B. verna (Mill.) Aschers. Early Winter Cress. Occasional as a weed in urban areas; native of Europe. March–July.

16. **Sisymbrium** L.

Pods closely appressed to the rachis, under 2 cm. long, acuminate; pedicels 2 mm.
    long.   1. *S. officinale*
Pods spreading, over 2 cm. long, not acuminate; pedicels 3–10 mm. long.
    Pedicels stout.
        Upper cauline leaves with filiform divisions; pods about 1 mm. in diameter,
            7–10 cm. long.   2. *S. altissimum*
        Upper cauline leaves when present entire or hastate; pods about 2 mm. in
            diameter, 4–6 cm. long.   3. *S. orientale*
    Pedicels very slender.   4. *S. irio*

1. S. officinale (L.) Scop. Hedge Mustard. Weeds of roadsides, disturbed areas, and overgrazed slopes; San Francisco, 10 miles south of San Francisco, Stanford, near Big Basin, and Soda Lake; native of Europe. February–August.

2. S. altissimum L. Tumble Mustard. Occasional as a weed in disturbed areas; San Francisco, summit of Santa Cruz–Los Gatos Highway, and Pajaro River; native of Europe. May–June.

3. S. orientale L. Oriental Sisymbrium. Known definitely only from disturbed areas in San Francisco, but to be expected elsewhere farther south; native of Europe. May–August.

4. S. irio L. London Rocket, Desert Mustard. Occasional as a weed in San Francisco and perhaps elsewhere; native of Europe. March–April.

17. Descurainia Webb & Berthel. Tansy Mustard

Pods 1.5–2.5 cm. long, linear, about 1 mm. or less in diameter.   1. *D. sophia*
Pods 0.8–1.2 cm. long, clavate, about 2 mm. in diameter.

2. *D. pinnata menziesii*

1. D. sophia (L.) Webb ex Prantl. Flixweed, Tansy Mustard. Occasional as a weed in San Francisco; native of Europe. May–August. —*Sisymbrium sophia* L.

2. D. pinnata (Walt.) Britt. ssp. **menziesii** (DC.) Detling. Western Tansy Mustard. Known locally only from fields near Stanford. April.

18. Thelypodium Endl. Thelypodium

Pods reflexed.   1*a. T. lasiophyllum lasiophyllum*
Pods spreading or ascending.   1*b. T. lasiophyllum inalienum*

1a. T. lasiophyllum (H. & A.) Greene var. **lasiophyllum.** California Mustard. Dry stream banks, rock outcroppings, and serpentine ridges; San Francisco, Belmont, Crystal Springs Lake, San Mateo, Woodside, Stanford, near Los Altos, Almaden Canyon, Edenvale, Swanton, Big Basin, and Empire Grade. February–June.

1b. T. lasiophyllum (H. & A.) Greene var. **inalienum** Robins. Serpentine outcroppings, dry open slopes, and on brush-covered areas; San Francisco, San Bruno Mountain, Crystal Springs Lake, Redwood City, Stanford, near San Jose, and Edenvale. February–July.

19. Athysanus Greene

1. A. pusillus (Hook.) Greene. Dwarf Athysanus. Widely distributed on grassy or rocky slopes; San Francisco, Crystal Springs Lake, Portola, Coal Mine Ridge, Lake Lagunita, Saratoga Summit, and Hilton Airport. February–May. One specimen from San Francisco (*Rattan,* May 1881) has glabrous pods; all others from the Santa Cruz Mountains that I have seen have pods with hooked hairs.

20. Thysanocarpus Hook. Lace or Fringe Pod

Pedicels curved uniformly throughout their length.
    Cauline leaves auriculate at the base; basal leaves hirsute, rosulate.
        Pods 3–4 mm. broad, the wings not perforate; petals 1–1.5 mm. long.

1*a. T. curvipes curvipes*

Pods 5–6 mm. broad, the wings usually perforate; petals 2–3 mm. long.
1b. *T. curvipes elegans*
Cauline leaves usually not auriculate; basal leaves usually glabrous, not rosulate. 2. *T. lacinatus crenatus*
Pedicels curved abruptly below the pods, otherwise more or less straight.
3. *T. radians*

1a. **T. curvipes** Hook. var. **curvipes.** *Fig. 102.* Hairy Fringe Pod. Open or
partly shaded slopes, often in gravelly soils, throughout the Santa Cruz Mountains. March–May.
1b. **T. curvipes** Hook. var. **elegans** (F. & M.) Robins. Elegant Fringe Pod.
Usually on somewhat rocky slopes, southern San Mateo County southward,
mainly on the eastern slopes of the Santa Cruz Mountains; Coal Mine Ridge,
Stanford, near Saratoga Summit, Los Gatos Creek, Guadalupe Mines, and Gilroy. March–May. Many plants are intermediate in flower and fruit character
and cannot be placed in either variety.
2. **T. lacinatus** Nutt. var. **crenatus** (Nutt.) Brewer. Crenate Fringe Pod. Dry
rocky slopes; San Francisco, near Saratoga, Guadalupe Canyon, and Waddell
Creek. March–April. —*T. emarginatus* Greene.
3. **T. radians** Benth. Ribbed Fringe Pod. Known locally from near Stanford.
March–April.

21. Iberis L.

1. **I. umbellata** L. Candytuft. Occasional as an escape from cultivation near
Lake Merced in San Francisco and to be expected elsewhere; native of Europe.
May–July.

22. Coronopus Gaertn.

Pods roughened but not tuberculate, about 1.5 mm. long. Common.
1. *C. didymus*
Pods tuberculate, 2.5–3 mm. long. Not common. 2. *C. squamatus*

1. **C. didymus** (L.) Smith. Lesser Wartcress. Weeds of roadsides, disturbed
areas, and coastal bluffs; San Francisco, Pescadero, San Mateo–Santa Cruz
county line, and Bonny Doon Road; native of Europe. March–November.
2. **C. squamatus** (Fors.) Aschers. Occasional as a weed in San Francisco;
native of Europe. —*C. procumbens* Gilib.

23. Capsella Medic.

1. **C. bursa-pastoris** (L.) Medic. Shepherd's Purse. A very common species of
cultivated fields, roadsides, disturbed areas, and on open grassy slopes; native
of Europe. To be found in flower at all times of the year.

24. Hutchinsia R. Br.

1. **H. procumbens** (L.) Desv. Prostrate Hutchinsia. Known locally from saline
areas along San Francisco Bay; Palo Alto and Mayfield. March–April.

25. Cardaria Desv.

1. C. draba (L.) Desv. Hoary Cress. Occasional in disturbed areas and fields from San Francisco southward and especially common in the Pajaro Valley; native of Europe. March–December. —*Lepidium draba* L.

26. Lepidium L. Pepper Grass

Upper cauline leaves cordate-clasping.   1. *L. perfoliatum*
Upper cauline leaves not cordate-clasping.
  Sepals long persistent after anthesis.   2. *L. strictum*
  Sepals not long persistent.
    Pods rounded at the apex, not winged or with acute, divergent teeth. Plants of non-saline or non-alkaline areas.
      Pedicels distinctly flattened, about twice as wide as thick.   3. *L. nitidum*
      Pedicels nearly terete.
        Petals usually longer than the sepals.   4. *L. virginicum pubescens*
        Petals lacking or shorter than the sepals.
          Lower leaves coarsely toothed; pods 2.5–3.5 mm. long.
                                    5. *L. densiflorum*
          Lower leaves pinnatifid, the lobes deeply dentate; pods 1.5 mm. long.   6. *L. pinnatifidum*
    Pods winged at the apex or with acute, divergent teeth. Plants of saline or alkaline areas along San Francisco Bay and in the Santa Clara Valley.
      Pods with divergent teeth at the apex, the teeth less than ⅓ the length of the body; inflorescence not congested.   7. *L. oxycarpum*
      Pods with two wings at the apex, the wings nearly as long as the body; inflorescence congested.   8. *L. latipes*

1. L. perfoliatum L. Shield Grass, Round-leaved Pepper Grass. Disturbed areas, often in saline soils; San Francisco and Redwood City; native of Europe. March–May.
2. L. strictum (Wats.) Rattan. Wayside Pepper Grass. Fairly common as a weed in disturbed ground; San Francisco, Spanishtown, Stanford, Swanton, Saratoga Summit, and Santa Cruz; native of South America. March–May. —*L. pubescens* of authors, *L. bipinnatifidum* of authors.
3. L. nitidum Nutt. Shining Pepper Grass. Common throughout the Santa Cruz Mountains on grassy slopes, rocky areas, and disturbed places. January–May.
4. L. virginicum L. var. **pubescens** (Greene) Hitchc. Wild Pepper Grass. Occasional in the Santa Cruz Mountains, often growing as a weed in disturbed areas; San Francisco, Palo Alto, Stanford, Permanente Creek, Santa Cruz, and Pajaro River. May–November.
5. L. densiflorum Schrader. Common Pepper Grass. Occasional as a weed in San Francisco; introduced locally from western North America.
6. L. pinnatifidum Ledeb. Occasional as a weed in disturbed areas in San Francisco; native of southern Russia.
7. L. oxycarpum T. & G. Sharp-podded Pepper Grass. Saline and alkaline flats along San Francisco Bay and in the Santa Clara Valley; Redwood City, Cooleys Landing, Palo Alto, Mayfield, San Jose, and 3 miles south of San Jose. March–May.

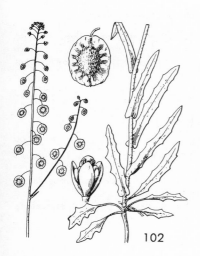

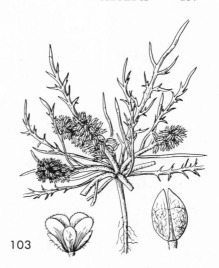

102   103

8. L. latipes Hook. *Fig. 103.* Dwarf Pepper Grass. Alkaline areas bordering the southern portion of San Francisco Bay and in the Santa Clara Valley; Mayfield and 3 miles south of San Jose. March–May.

27. Lunaria L.

1. L. annua L. Moonwort, Honesty. Occasional as an escape from gardens and often persistent; San Francisco, Kings Mountain, Woodside, and Ben Lomond. March–June.

28. Lobularia Desv.

1. L. maritima (L.) Desv. Sweet Alyssum. Escaped from cultivation and established along the coast and in disturbed habitats elsewhere; San Francisco, near Santa Cruz, and near Watsonville. To be found in flower at all seasons of the year. —*Alyssum maritimum* (L.) Lamk., *Koniga maritima* (L.) R. Br.

29. Alyssum L.

1. A. alyssoides L. Yellow Alyssum. Known locally from San Francisco and near Mount Umunhum, but to be expected elsewhere; native of Europe. April–May.

52. RESEDACEAE. Mignonette Family

1. Reseda L. Mignonette

1. R. alba L. White Mignonette. Locally abundant in parts of San Francisco and to be expected elsewhere; native of the Mediterranean region. June–October.

53. CRASSULACEAE. Stone Crop Family

Plants diminutive, annual, aquatic or terrestrial; leaves opposite; flowers under
2 mm. long.   1. *Tillaea*
Plants perennial, terrestrial; leaves alternate or in a basal rosette; flowers over
5 mm. long.
  Petals separate.   2. *Sedum*
  Petals united at the base.   3. *Dudleya*

1. Tillaea L. Pigmyweed

Flowers clustered in the leaf-axils; terrestrial.   1. *T. erecta*
Flowers solitary in the leaf-axils; aquatic or palustrine.   2. *T. aquatica*

1. T. erecta H. & A. *Fig. 104.* Tillaea, Sand Pigmyweed. Very common in sandy
soil throughout the Santa Cruz Mountains. February–May.
2. T. aquatica L. Water Pigmyweed. Vernal pools, drying puddles, and along
the edges of ponds. May–July. —*Tillaeastrum aquaticum* (L.) Britt.

2. Sedum L. Stone Crop, Orpin

Plants shrubby, usually 5 dm. or more tall; petals yellow.   1. *S. dendroideum*
Plants herbaceous, rarely over 2 dm. tall; petals yellow or white.
  Petals white.   2. *S. album*
  Petals yellow.
    Basal leaves lanceolate; follicles widely divergent.
                                                    3. *S. stenopetalum radiatum*
    Basal leaves spatulate; follicles erect or ascending.   4. *S. spathulifolium*

1. S. dendroideum Sesse & Mocino ex DC. Occasional as an escape from cultivation in San Francisco; native of Mexico.
2. S. album L. White Sedum. Becoming fairly common as an escape from cultivation in San Francisco and probably elsewhere; native of Eurasia and North
Africa.
3. S. stenopetalum Pursh ssp. **radiatum** (Wats.) Clausen. Star-fruited Stone
Crop. Rare in the Santa Cruz Mountains; known from San Mateo County and
from Hilton Airport, in the latter locality growing on sandstone outcroppings in
less than 1 cm. of soil. May–July. —*S. radiatum* Wats.
4. S. spathulifolium Hook. Pacific Stone Crop. Cliffs and rocks in exposed
locations; Sutro Hill, Montara Mountain, Crystal Springs Lake, Pescadero
Creek, Stevens Creek, Loma Prieta, Eagle Rock, and near Boulder Creek. April–
July.

3. Dudleya Britt. & Rose. Dudleya, Live-forever

Pedicels stout, usually shorter than the flowers. Plants of coastal bluffs.
                                                              1. *D. farinosa*
Pedicels slender, longer than the flowers. Plants of dry inland areas.
  Petals bright yellow to red; rosette-leaves oblong-lanceolate, 1–4 cm. wide.
  Common.   2a. *D. cymosa cymosa*
  Petals pale yellow; rosette-leaves oblong-lanceolate to triangular-oblong, 0.5–
  2 cm. wide. Restricted to the vicinity of Guadalupe Mine, Madrone, and
  Coyote.   2b. *D. cymosa setchellii*

104

105

1. D. farinosa (Lindl.) Britt. & Rose. *Fig. 105.* Sea Lettuce, Bluff Lettuce, Powdery Dudleya. Rocky slopes and bluffs along the coast. July–August.

2a. D. cymosa (Lemaire) Britt. & Rose ssp. **cymosa.** Lax or Spreading Dudleya. Rocky ridges and slopes away from the ocean from San Bruno Mountain southward. April–July. —*D. laxa* (Lindl.) Britt. & Rose.

2b. D. cymosa (Lemaire) Britt. & Rose ssp. **setchellii** (Jeps.) Moran. Setchell's Dudleya. Local in the Santa Clara Valley on serpentine; Guadalupe Mine, between Coyote and Morgan Hill, and San Juan Bautista Hills. February–June. —*D. setchellii* (Jeps.) Britt. & Rose.

54. SAXIFRAGACEAE. Saxifrage Family

Hypanthium deeply cleft on the lower side; stamens 3.   1. *Tolmiea*
Hypanthium not deeply cleft; stamens 5 or 10.
  Stamens 5.
    Branches of the inflorescence subtended by foliaceous bracts; placentae
      axile.   2. *Boykinia*
    Branches of the inflorescence subtended by reduced linear bracts, these
      sometimes toothed; placentae parietal or basal.   3. *Heuchera*
  Stamens 10.
    Carpels soon unequally 2-valved.   4. *Tiarella*
    Carpels approximately equal.
      Styles normally 3.   5. *Lithophragma*
      Styles normally 2.
        Leaves oblong-ovate to elliptic; petals entire.   6. *Saxifraga*
        Leaves reniform to cordate; petals finely dissected.   7. *Tellima*

1. Tolmiea T. & G.

1. T. menziesii (Pursh) T. & G. Youth-on-Age, Thousand Mothers. Known locally only from near Boulder Creek. May–June.

2. Boykinia Nutt. Boykinia

1. B. elata (Nutt.) Greene. *Fig. 106.* Coast Boykinia, Brook Foam. Moist shaded slopes and stream banks in redwood–Douglas fir forests from central San Mateo County southward; Pescadero, Stevens Creek Canyon, Los Gatos, New Almaden, Big Basin, and Boulder Creek. May–August.

3. Heuchera L. Alum Root, Heuchera

Hypanthium turbinate, thinly pubescent; inflorescence not congested.
1. *H. micrantha*
Hypanthium rounded at the base, densely pilose; inflorescence congested at least
    when young.   2. *H. pilosissima*

1. H. micrantha Dougl. ex Lindl. Alum Root, Small-flowered Heuchera. Common in moist shaded areas and often on stream banks, usually in the redwood–Douglas fir forest, San Francisco southward. May–June. —*H. hartwegii* (Wats. ex Wheelock) Rybd., *H. micrantha* Dougl. ex Lindl. var. *pacifica* Rosend., Butt., & Lakela.
2. H. pilosissima F. & M. Seaside Heuchera. Known locally only from hills near the coast in San Mateo County; Montara Mountain and near Pebble Beach. March–June.

4. Tiarella L.

1. T. unifoliata Hook. Sugar-scoop. Moist stream banks in redwood forests, occurring most commonly on the seaward slopes of the Santa Cruz Mountains; Pescadero, Stevens Creek, Big Basin, Boulder Creek, and Felton. May–July.

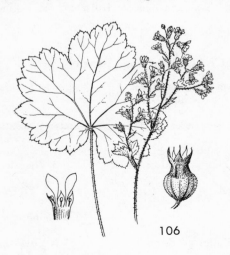

106

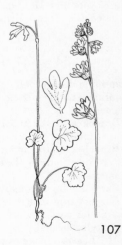

107

5. **Lithophragma** Nutt. Woodland Star

Hypanthium obconic, the base cuneate.   1. *L. affinis*
Hypanthium campanulate, the base truncate or rounded.   2. *L. heterophylla*

1. **L. affinis** Gray. Woodland Star. Moist rocky slopes, meadows, and open oak-madrone forests throughout the Santa Cruz Mountains and foothills. February–May.
2. **L. heterophylla** (H. & A.) T. & G. *Fig. 107.* Hill Star. Moist shaded slopes, widespread; San Francisco, Crystal Springs Lake, Black Mountain, Stanford, Waddell Creek, and Bonny Doon. March–June.

6. **Saxifraga** L. Saxifrage

1. **S. californica** Greene. *Fig. 108.* California Saxifrage. Rocky meadows and oak-madrone woods, usually at lower elevations; Mission Hills, Bayview Hills, Coal Mine Ridge, Alma, Swanton, and Glenwood. February–April.

7. **Tellima** R. Br.

1. **T. grandiflora** (Pursh) Dougl. ex Lindl. Fringe Cups. Moist woods and brush-covered slopes; San Francisco, Pescadero, Coal Mine Ridge, Stevens Creek, and Santa Cruz; often grown in gardens and becoming weedy. March–June.

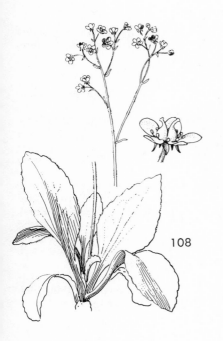

108

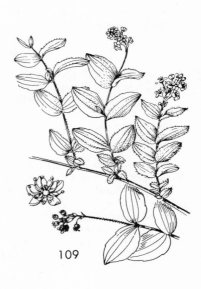

109

55. PARNASSIACEAE. Grass-of-Parnassus Family

1. Parnassia L. Grass-of-Parnassus

1. P. palustris L. var. **californica** Gray. California Grass-of-Parnassus. Known locally only from Loma Prieta and from near Boulder Creek. August–October. —*P. californica* (Gray) Greene.

56. HYDRANGEACEAE. Hydrangea Family

1. Whipplea Torr.

1. W. modesta Torr. *Fig. 109.* Modesty, Yerba de Selva. Common in the drier parts of redwood–Douglas fir forests; Cahill Ridge, Kings Mountain, Searsville, Big Basin, and Zayante. March–June.

57. ESCALLONIACEAE. Escallonia Family

1. Escallonia Mutis ex L. f. Escallonia

1. E. macrantha H. & A. Occasional as escapes from cultivation on coastal slopes in San Francisco; native of Chile.

58. GROSSULARIACEAE. Gooseberry Family

Plants without nodal spines; pedicels with a joint below the ovary.   1. *Ribes*
Plants with nodal spines; pedicels not jointed.   2. *Grossularia*

1. Ribes L. Currant

Hypanthium yellow; plants glabrous.   1. *R. gracillimum*
Hypanthium not yellow; plants variously pubescent.
  Styles glabrous; leaves usually thin, the abaxial surfaces usually lacking tomentum.   2. *R. glutinosum*
  Styles pubescent below; leaves usually thick, the abaxial surfaces with a dense tomentum.   3. *R. malvaceum*

1. R. gracillimum Cov. & Britt. Bugle Currant. Along streams in southern San Mateo, northern Santa Cruz, and northern Santa Clara counties; Portola, Coal Mine Ridge, 3 miles west of Los Altos, and Stanford. April. —*R. aureum* Pursh var. *gracillimum* (Cov. & Britt.) Jeps.
2. R. glutinosum Benth. *Fig. 110.* Winter or Flowering Currant. Usually growing along streams or in moist woods, common from San Francisco southward. March–April. —*R. sanguineum* Pursh var. *glutinosum* (Benth.) Loud.
3. R. malvaceum Smith. Chaparral or California Black Currant. Shaded ravines and chaparral slopes; Masonic Cemetery in San Francisco, Sweeney Ridge, Belmont, Portola, Coal Mine Ridge, Black Mountain, near Los Gatos, Loma Prieta, El Toro, and near Eagle Rock. October–April.

2. Grossularia Mill. Gooseberry

Anthers less than 1 mm. long; berries glabrous.   1. *G. divaricata*
Anthers over 2 mm. long; berries with spines, bristles, or variously pubescent.
  Ovary with long white hairs.   2. *G. senilis*
  Ovary without long white hairs.
    Abaxial leaf-surfaces more or less glabrous.   3. *G. californica*
    Abaxial leaf-surfaces glandular-pubescent.   4. *G. leptosma*

1. G. divaricata (Dougl.) Cov. & Britt. Straggly Gooseberry, Straggle Bush. Occasional from San Francisco southward, usually growing in moist shade and often along or near creeks. February–May. —*Ribes divaricatum* Dougl.
2. G. senilis Cov. Santa Cruz Gooseberry. Endemic in the Santa Cruz Mountains, usually found on wooded slopes; near Saratoga, Loma Prieta, New Almaden, Big Basin, Waddell Creek, Glenwood, Soquel, and Aptos. March–August. —*Ribes menziesii* Pursh var. *senile* (Cov.) Jeps.
3. G. californica (H. & A.) Cov. & Britt. Hill, Hillside, or California Gooseberry. Most common on more or less open or partly wooded hills; San Francisco, Pilarcitos Canyon, Belmont, Portola, Coal Mine Ridge, Stanford, near Los Gatos, Guadalupe River, and El Toro. January–April. —*Ribes californicum* H. & A. Leaf-size and the number of spines per berry varies from plant to plant. Those individuals growing in more or less exposed habitats are gnarled and very woody.
4. G. leptosma Cov. Bay or Canyon Gooseberry. Moist woods and brush-covered slopes, San Francisco south to the vicinity of Loma Prieta. March–June. —*Ribes menziesii* Pursh var. *leptosmum* (Cov.) Jeps.

The *Grossularia menziesii* complex, of which *G. leptosma* is a member, occurs from southern Oregon south to southern California and is a difficult one taxonomically. Numerous local populations have been described as species. *Grossularia senilis* is an example of a fairly distinct local population, whereas *G. leptosma* is much more variable and has a wider geographical distribution. Many specimens of *G. leptosma* tend toward *G. menziesii* (Pursh) Cov. & Britt. in having thicker, more glandular-pubescent leaves.

59. PITTOSPORACEAE. Pittosporum Family

1. Pittosporum Banks

1. P. crassifolium Cunn. Karo, Thick-leaved Pittosporum. Occasional as escapes from cultivation in San Francisco; native of New Zealand. April–July.

60. PLATANACEAE. Sycamore or Plane Tree Family

1. Platanus L. Sycamore

1. P. racemosa Nutt. *Fig. 111.* Sycamore, California Plane Tree. Flood plains and terraces of creeks and rivers in Santa Clara and Santa Cruz counties; near Santa Clara, 4 miles south of San Jose, Carnadero Creek, San Lorenzo River, and Pajaro River. March–April.

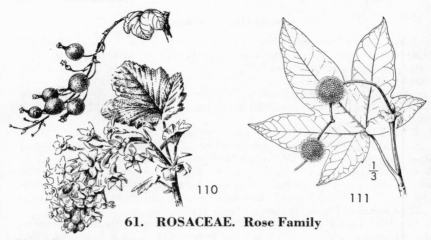

110

111

$\frac{1}{3}$

61. ROSACEAE. Rose Family

Plants herbaceous, not shrubs or vines.
  Petals present, conspicuous, white or yellow; stamens 10 or more; pistils numerous.
    Receptacle enlarging in fruit, becoming red; leaves ternate.   1. *Fragaria*
    Receptacle not enlarging in fruit; leaves pinnate or sometimes ternate.
      Filaments dilated; stamens inserted at a distance from the receptacle.
                                                                    2. *Horkelia*
      Filaments filiform; stamens inserted near the base of the receptacle.
                                                                    3. *Potentilla*
  Petals absent; stamens less than 5; pistils 1 or 2.
    Plants annual, under 1 dm. tall; flowers axillary; stamen 1.   4. *Alchemilla*
    Plants annual or perennial, over 1 dm. tall; flowers in definite inflorescences; stamens 2 or more.
      Hypanthium with barbed prickles.   5. *Acaena*
      Hypanthium without barbed prickles.   6. *Sanguisorba*
Shrubs or vines.
  Fruit of dehiscent follicles; pedicels and hypanthium densely stellate-pubescent.   7. *Physocarpus*
  Fruit of indehiscent achenes or drupelets; pedicels and hypanthium not stellate-pubescent.
    Leaves not divided; plants not prickly; flowers under 5 mm. in diameter, not showy.
      Leaves ovate, over 1.5 cm. long.
        Pistils 5; style not becoming plumose; achenes enclosed within a saucer-shaped hypanthium; leaves usually at least 4 cm. long.
                                                                    8. *Holodiscus*
        Pistil 1; style becoming plumose; hypanthium salverform; leaves usually under 2.5 cm. long.   9. *Cercocarpus*
      Leaves fascicled, linear-subulate, less than 1 cm. long in mature plants.
                                                                    10. *Adenostoma*
    Leaves pinnatifid or palmately lobed; plants usually prickly; flowers showy.
      Fruit of drupelets, not enclosed by a fleshy hypanthium.   11. *Rubus*
      Fruit of achenes, enclosed in a fleshy hypanthium.   12. *Rosa*

1. Fragaria L. Strawberry

Petals white; rhizomes stout.
  Leaflets thin, not leathery. Plants of shaded slopes.   1. *F. californica*
  Leaflets thick, leathery. Plants of the immediate coast.   2. *F. chiloensis*
Petals yellow; rhizomes slender.   3. *F. indica*

1. F. californica C. & S. *Fig. 112.* California Strawberry. Shaded slopes and hillsides at the edge of and in clearings in Douglas fir forests, San Francisco southward. January–June, with mature fruit from May through July.
2. F. chiloensis (L.) Duchesne. Chilean or Beach Strawberry. Coastal sand dunes; San Francisco, San Pedro, Pescadero Beach, Waddell Creek, and Swanton. February–August.
3. F. indica Andr. Mock or Indian Strawberry. Becoming established in lawns and in shaded areas in San Francisco and probably elsewhere; native of India. May–August. —*Duchesnea indica* (Andr.) Focke.

2. Horkelia C. & S.

Pistils more than 50; bractlets ovate, sepaloid.
  Hypanthium cup-shaped, 3–5.5 mm. deep.
    Hypanthium glabrous within; calyx-lobes greenish within.   1. *H. frondosa*
    Hypanthium pubescent within; calyx-lobes purple-flecked within.
                           2. *H. californica*
  Hypanthium saucer-shaped, 1.5–2 mm. deep.
    Leaves glandular-villous to glabrate.   3*a. H. cuneata cuneata*
    Leaves densely silky-pubescent, obscurely glandular.
                3*b. H. cuneata sericea*
Pistils fewer than 40; bractlets lanceolate, smaller than the calyx-lobes.
                4. *H. bolanderi parryi*

1. H. frondosa (Greene) Rydb. Leafy Horkelia. Occasional on the eastern slopes of the Santa Cruz Mountains in southern San Mateo and Santa Clara counties; Woodside, Stanford, and Llagas Creek. April–October. —*Potentilla frondosa* Greene.
2. H. californica C. & S. California Horkelia. Coastal bluffs and grassy slopes, mainly overlooking the ocean; San Francisco, San Bruno Hills, Pescadero Creek, Año Nuevo Point, Stevens Creek, and Aptos. June–October.
3a. H. cuneata Lindl. ssp. **cuneata.** Wedge-leaved Horkelia. Near the ocean and in the inland marine sand deposits of central Santa Cruz County. March–September. —*Potentilla lindleyi* Greene.
3b. H. cuneata Lindl. ssp. **sericea** (Gray) Keck. Along the coast and on the inland sand deposits in Santa Cruz County; Ocean View, Colma, and Mount Hermon. April–June. —*Potentilla lindleyi* Greene ssp. *sericea* (Gray) Howell. Plants with pubescence intermediate between the two subspecies occur occasionally and have been collected near Eccles, Lower Zayante, and from between Scott and Waddell creeks.
4. H. bolanderi Gray ssp. **parryi** (Wats.) Keck. Bolander's Horkelia. Known locally only from San Andreas Valley where it occurs in grasslands. June–July. The collection from San Andreas Valley (*Oberlander 1204*) is typical of *H. bolanderi* ssp. *parryi* with respect to the filament and bractlet shape, however it

varies in the direction of *H. marinensis* (Elmer) Crum ex Keck with respect to the woody rootstock, dense foliage, and fairly short flowering branches.

3. Potentilla L. Five-finger, Cinquefoil

Basal leaves with 3–5 leaflets; styles terminal or subterminal.
  Leaves 5-foliolate at least below.   1. *P. rivalis*
  Leaves 3-foliolate at least below.   2. *P. millegrana*
Basal leaves with 7 to many leaflets; styles terminal to basal.
  Basal leaves pinnately compound.
    Styles terminal or subterminal.   3. *P. hickmanii*
    Styles basal or lateral.
      Leaflets white-silky and tomentose beneath; styles lateral.
                                    4. *P. egedii grandis*
      Leaflets sparingly glandular beneath, not white-silky; styles basal.
                                    5. *P. glandulosa*
  Basal leaves palmately compound.   6. *P. recta*

1. P. rivalis Nutt. ex T. & G. River Cinquefoil. Known from Mountain Lake in San Francisco and from Kelly Lake near Watsonville, but to be expected elsewhere along the margins of lakes and ponds. May–November.
2. P. millegrana Engelm. ex Lehm. Diffuse Cinquefoil. Occasional in marshy and swampy areas; San Francisco, Pilarcitos Canyon, Kelly Lake, and Watsonville. April–October. —*P. rivalis* Nutt. ex T. & G. var. *millegrana* (Engelm.) Wats.
3. P. hickmanii Eastw. Hickman's Cinquefoil. Known locally only from the vicinity of Moss Beach. May.
4. P. egedii Wormskj. var. **grandis** (T. & G.) Howell. Pacific Silverweed. Along the edges of coastal marshes and ponds and occasionally inland as at Crystal Springs Lake. April–June. —*P. pacifica* Howell.
5. P. glandulosa Lindl. *Fig. 113.* Sticky Cinquefoil. Open woods, hillsides, and adventive in disturbed areas, San Francisco southward. March–July.
6. P. recta L. Known locally as a weed in San Francisco and Saratoga; native of Europe. June–July.

4. Alchemilla L.

1. A. occidentalis Nutt. ex T. & G. Dew Cup, Western Lady's Mantle. Common on grassy and sandy slopes throughout the area covered by this flora. February–May. *Alchemilla occidentalis* is an inconspicuous plant and is often overlooked by collectors. Flowering plants may be less than 1 cm. tall.

5. Acaena L. Acaena

Stamens 3–5; spines of the fruiting hypanthium numerous, retrorsely barbed
    throughout their length.   1. *A. californica*
Stamens 2; spines of the fruiting hypanthium few, barbed only at the tip.
                                    2. *A. sanguisorbae*

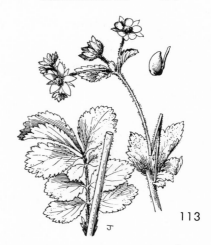

112

113

1. A. californica Bitter. California Acaena. Grassy slopes near the ocean and San Francisco Bay; Twin Peaks, Lake Merced, San Bruno, San Bruno Hills, San Andreas Reservoir, and near Pescadero. March–June.

2. A. sanguisorbae Vahl. Bidi-bidi. Occasional as an escape from cultivation in San Francisco; native of New Zealand. May–July.

6. Sanguisorba L.

1. S. minor (L.) Scop. Burnet. Adventive in disturbed areas and occasionally in grain fields; San Francisco, Corte Madera Creek, near Sunnyvale, and Glenwood; native from the Mediterranean region east to central Russia. May–September.

7. Physocarpus Maxim. Ninebark

1. P. capitatus (Pursh) Kuntze. Pacific Ninebark. Along creek banks, more common in the drainages on the eastern side of the Santa Cruz Mountains; San Andreas Valley, near Redwood City, Coal Mine Ridge, Los Trancos Creek, Permanente Creek, and San Lorenzo River at Glenarbor. April–July.

8. Holodiscus Maxim.

1. H. discolor (Pursh) Maxim. *Fig. 114.* Cream Bush, Ocean Spray. A common shrub in soft chaparral, oak woodland, open Douglas fir forests, and along stream banks, San Francisco southward. April–July. —*H. discolor* (Pursh) Maxim. var. *franciscanus* (Rydb.) Jeps.

9. Cercocarpus HBK. Mountain Mahogany

1. C. betuloides Nutt. ex T. & G. California Mountain Mahogany. Rocky soils from sea level to about 3500 feet elevation, usually in chaparral with various species of *Quercus, Arcostaphylos, Adenostoma fasciculatum, Ceanothus, Garrya fremontii*, and occasionally with *Pinus attenuata* and *P. sabiniana*, southern San Mateo County southward. March–May.

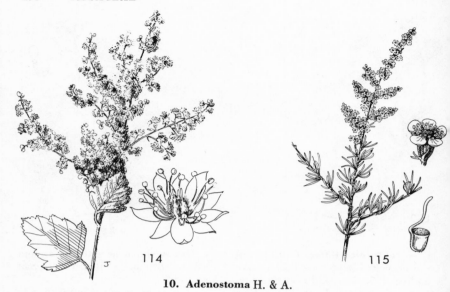

114   115

10. Adenostoma H. & A.

1. A. fasciculatum H. & A. *Fig. 115.* Chamise, Greasewood. Chaparral and exposed slopes from San Francisco southward. May–July. The leaves of seedlings and crown-sprouts are palmately or bipinnately divided, as for example *Abrams*, Oct. 9, 1909, from Black Mountain. Germination during the first rainy season following a fire is rapid and numerous seedlings come up. *Adenostoma* occurs as a common component of chaparral with various species of *Arctostaphylos, Eriodictyon californicum, Photinia arbutifolia, Quercus, Pickeringia montana, Dendromecon rigida,* and *Prunus,* or it occurs in relatively pure stands called chamise.

11. Rubus L. Blackberry, Raspberry

Plants without spines on the stems, petioles, and inflorescences.
    Leaves palmately 5-lobed.   1. *R. parviflorus velutinus*
    Leaves with 3 or 5 leaflets.   2. *R. ulmifolius inermis*
Plants with spines on the stems, petioles, and inflorescences.
    Leaves of the primocanes usually with 3 leaflets or 3-lobed; petals white or
        red; plants erect or prostrate.
        Flowers single or few; sepals ovate; petals red, 1½–2 times as long as the
            sepals.   3. *R. spectabilis franciscanus*
        Flowers clustered; sepals lanceolate, long-acuminate; petals white, about as
            long as the sepals.
            Leaves grayish-white abaxially, 3-foliolate; prickles stout, recurved at the
                apex, prominently enlarged at the base.   4. *R. leucodermis*
            Leaves green abaxially, 3–5-foliolate or 3-lobed; prickles slender, straight,
                little enlarged at the base.   5. *R. ursinus*
    Leaves of the primocanes usually with 5 leaflets; petals pink; plants prostrate
        or trailing.   6. *R. procerus*

1. R. parviflorus Nutt. var. **velutinus** (H. & A.) Greene. Thimble Berry. Streambanks, moist slopes, and protected ravines, throughout the Santa Cruz Mountains. March–August.

2. R. ulmifolius Schott var. **inermis** (Willd.) Focke. Occasional in disturbed areas in San Francisco; native of Europe.

3. R. spectabilis Pursh var. **franciscanus** (Rydb.) Howell. Salmon Berry. Along or near the coast in wooded canyons; San Francisco, Half Moon Bay, Gazos Creek, Swanton, and Liddell Creek. March–May. —*R. spectabilis* Pursh var. *menziesii* (Hook.) Wats.

4. R. leucodermis Dougl. ex T. & G. *Fig 116*. Western or White-stemmed Raspberry. Along streams and in clearings in redwood–Douglas fir forests in Santa Clara and Santa Cruz counties. March–June.

5. R. ursinus C. & S. Pacific or California Blackberry. A common plant, growing along streams, coastal bluffs, road banks, fence rows, and in disturbed areas, San Francisco southward. March–August. —*R. vitifolius* C. & S., *R. ursinus* C. & S. var. *glabratus* Presl, *R. ursinus* C. & S. var. *sirbenus* (Bailey) Howell.

6. R. procerus Muell. ex Boulay. Himalaya Berry. Becoming common as a weed in San Francisco, in central Santa Cruz County, and probably elsewhere; native of Europe. June–August.

12. **Rosa** L. Rose

Abaxial surface of the leaves densely pubescent; plants with stout, usually recurved prickles.   1. *R. californica*

Abaxial surface of the leaves glabrous or with thick, glandular hairs; plants usually with slender, terete, ascending or straight prickles.

Hypanthium densely glandular, the upper part and the sepals persistent.

2. *R. spithamea*

Hypanthium glabrous, the upper portion and the sepals deciduous.

3. *R. gymnocarpa*

1. R. californica C. & S. *Fig. 117*. California Rose, California Wild Rose. Open stream banks, wooded hills, and coastal sand dunes and bluffs, San Francisco southward. April–October. —*R. aldersonii* Greene.

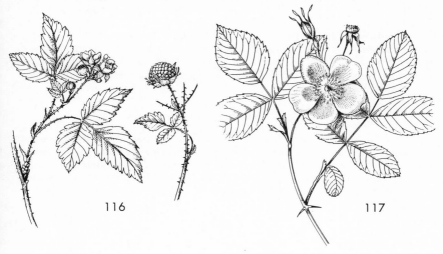

116   117

2. R. spithamea Wats. Ground Rose. Oak-madrone woods, occasionally in chaparral, and in disturbed areas; Mount Umunhum, near San Jose, Saratoga–Big Basin Road, China Grade, and near Felton. May–August. —*R. sonomensis* Greene.

3. R. gymnocarpa Nutt. ex T. & G. Wood Rose. Occasional in open woods and on brushy slopes in partial shade, San Francisco southward. April–September.

62. AMYGDALACEAE. Peach Family

Leaves variously serrate; pistil 1; flowers perfect.   1. *Prunus*
Leaves entire; pistils usually 5; plants dioecious.   2. *Osmaronia*

1. Prunus L. Plum, Cherry, Stone-fruits

Inflorescence corymbose or umbellate.
    Leaves narrowed at the base; drupes red, about 5 mm. long; flowers in corymbose clusters.   1. *P. emarginata*
    Leaves rounded or subcordate at the base; drupes purple, 20–25 mm. long; flowers in umbel-like clusters.   2. *P. subcordata*
Inflorescence racemose.
    Leaves deciduous, thin, not spinose-margined.   3. *P. demissa*
    Leaves persistent, coriaceous, spinose-margined.   4. *P. ilicifolia*

1. P. emarginata (Dougl.) Walp. Bitter Cherry. Streamsides, shaded slopes, and occasionally in chaparral; Pilarcitos Canyon, Cahill Ridge, Searsville, Los Gatos Hills, Loma Prieta Ridge, and near Mount Hermon. April–May. Pedicels and hypanthia vary from glabrous to densely pubescent.

2. P. subcordata Benth. Sierra or Pacific Plum. Southern San Mateo and Santa Clara counties, usually on the eastern slopes of the mountains, growing along creeks and occasionally in chaparral; Portola, Coal Mine Ridge, Page Mill Road, and Permanente Creek. March–April.

3. P. demissa (Nutt.) Walp. Western Choke Cherry. Ravines and woods throughout the Santa Cruz Mountains; San Francisco, Pilarcitos Canyon, Coal Mine Ridge, near summit of Page Mill Road, near Permanente Creek, and Glenarbor. April–June. —*P. virginiana* L. var. *demissa* (Nutt.) Torr.

4. P. ilicifolia (Nutt. ex H. & A.) Walp. *Fig. 118.* Holly-leaved Cherry, Islay. Common in chaparral, on open hills, and often in relatively pure stands in ravines and gullies; San Francisco, San Bruno Mountain, Pilarcitos Dam, Menlo Park, Searsville, between Alma and Soda Springs, Loma Prieta, Saratoga Summit, and Soda Lake. March–June.

2. Osmaronia Greene

1. O. cerasiformis (T. & G. ex H. & A.) Greene. Oso Berry. Shaded slopes and canyons throughout the Santa Cruz Mountains, but more common on the eastern slopes than on the western; San Francisco, San Andreas Dam, Pilarcitos Canyon, Portola Valley, Pescadero Creek, Los Trancos Creek, Black Mountain, Swanton, and Boulder Creek. February–April.

63. MALACEAE. Apple Family

Leaves deciduous; petals 8–10 mm. long.   1. *Amelanchier*
Leaves evergreen; petals 2–4 mm. long.
  Leaves usually 5–10 cm. long, essentially glabrous. Native shrubs.
                                                             2. *Photinia*
  Leaves usually under 3.5 cm. long, densely tomentose abaxially. Introduced
    shrubs.   3. *Cotoneaster*

1. Amelanchier Medic.

1. A. pallida Greene. June or Service Berry. Occasional along streams and on moist slopes, less frequent on the western slopes of the Santa Cruz Mountains than on the eastern; San Francisco, San Mateo Canyon, Searsville, Page Mill Road, Black Mountain, Alma Soda Springs, and 6 miles west of Gilroy. April–May.

2. Photinia Lindl.

1. P. arbutifolia (Ait.) Lindl. *Fig. 119*. Toyon, Christmas Berry. Throughout the Santa Cruz Mountains and surrounding lowlands, common in chaparral, as a member of the understory of open oak-madrone woods, and as isolated shrubs. June–July. —*Heteromeles arbutifolia* (Ait.) Roem.

3. Cotoneaster Medic. Cotoneaster

Petals pink, more or less erect.   1. *C. franchetii*
Petals white, spreading.   2. *C. pannosa*

1. C. franchetii Bois. Occasional as an escape from cultivation in San Francisco; native of China. April–May.
2. C. pannosa Franchet. Escaping from cultivation and becoming established; San Francisco, Stanford, and probably elsewhere; native of China. April–June.

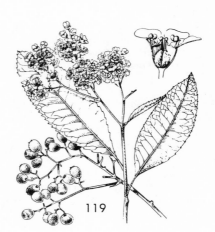

118   119

64. MIMOSACEAE. Mimosa Family

Stamens united at the base, 2–3 cm. long; leaves bipinnately compound, the ultimate divisions 8–12 mm. long.   1. *Albizia*
Stamens distinct, under 5 mm. long; leaves reduced to phyllodia or if bipinnately compound, the ultimate divisions about 4 mm. long.   2. *Acacia*

1. Albizia Dur.

1. A. distachya (Vent.) Macbr. Stink Bean. Becoming established on moist slopes in San Francisco; native of Australia. Flowering at all seasons of the year. —*A. lophantha* (Willd.) Benth.

2. Acacia Mill.

Leaves not reduced to phyllodia.
  Pinnae 8–25 pairs, greenish.   1. *A. decurrens*
  Pinnae 2–5 pairs, grayish.   2. *A. baileyana*
Leaves reduced to phyllodia, occasionally juvenile leaves with a few pinnae.
  Leaves not verticillate, 2.5–15 cm. long.
    Leaves 2–5-veined from the base, usually 10–25 mm. wide.
      Flowers in spikes, often several spikes in the axils of nonsickle-shaped
      phyllodia.   3. *A. longifolia*

      Flowers in glomerulate heads racemosely arranged in the axils of sickle-
      shaped phyllodia.   4. *A. melanoxylon*
    Leaves 1-veined from the base, usually 5–10 mm. wide.   5. *A. retinodes*
  Leaves verticillate, 1–3 cm. long.   6. *A. verticillata*
1. A. decurrens Willd. Green Wattle. Widely planted as an ornamental and becoming naturalized; San Francisco, Palo Alto, Stanford, Big Basin, and Jamison Road; native of Australia. January–March.
2. A. baileyana F. v. M. Bailey Acacia, Cootamundra Wattle. Commonly planted as an ornamental and occasionally producing seedlings; Stanford; native of Australia. Flowering in early spring.
3. A. longifolia Willd. Golden Wattle. Common as an ornamental tree, occasionally becoming established; San Francisco, Santa Cruz, and near Soquel Creek; native of Australia. February–March.
4. A. melanoxylon R. Br. Black or Blackwood Acacia. Commonly planted and becoming naturalized; San Francisco, Palo Alto, and near Sea Cliff Beach State Park; native of Australia. February–March.
5. A. retinodes Schlecht. Water Wattle. Occasionally spontaneous on the Stanford University campus and probably elsewhere; native of Australia.
6. A. verticillata Willd. Whorl-leaf or Star Acacia. Widely planted as ornamentals and occasionally seedlings may be found in the vicinity of mature trees and shrubs; San Francisco and Montalvo; native of Australia.

65. CAESALPINIACEAE. Senna Family

1. Cassia L.

1. C. tomentosa L. f. Senna. Becoming established in Golden Gate Park in San Francisco by means of seeds and underground parts; native of Mexico.

66. FABACEAE. Pea Family

Leaves palmately compound, leaflets usually 5 or more.   1. *Lupinus*
Leaves none, trifoliolate, or variously pinnately compound.
  Shrubs or trees.
    Stems spiny.
      Leaflets 9 or more; flowers white or pale pink; trees.   2. *Robinia*
      Leaflets 1–3 or the leaves reduced to spines; flowers rose-colored or yellowish; shrubs.
        Leaflets 1–3; flowers rose-colored.   3. *Pickeringia*
        Leaves reduced to acicular spines; flowers yellow.   4. *Ulex*
    Stems not spiny.
      Leaflets 9 or more; trees; flowers white or pale pink.   2. *Robinia*
      Leaflets 1–5; shrubs; flowers yellow at least in part.
        Flowers over 1 cm. long; inflorescence not umbellate; pods dehiscent; mature plants usually at least 2 m. tall.
          Calyx 2-lipped; branchlets angular in cross section.   5. *Cytisus*
          Calyx 1-lipped; branchlets round in cross section.   6. *Spartium*
        Flowers under 1 cm. long; inflorescence more or less umbellate; pods indehiscent; mature plants rarely more than 1 m. tall.   7. *Lotus*
  Plants herbaceous, annual, biennial, or perennial.
    Leaves trifoliolate.
      Herbage punctate-dotted.   8. *Psoralea*
      Herbage not punctate-dotted.
        Stamens distinct; flowers about 2 cm. long, bright yellow; inflorescence racemose.   9. *Thermopsis*
        Stamens diadelphous; flowers usually less than 2 cm. long, of various colors; inflorescence various.
          Pods spirally coiled.   10. *Medicago*
          Pods not spirally coiled.
            Flowers in elongate racemes.   11. *Melilotus*
            Flowers solitary, in umbels, or in dense heads.
              Flowers in dense heads, subtended by an involucre or not; leaflets denticulate.   12. *Trifolium*
              Flowers solitary or in few-flowered umbels, the umbels bracteate; leaflets entire.   7. *Lotus*
    Leaves with more than 3 leaflets (if less than 3, the leaf bearing a tendril).
      Pods breaking transversely into 1-seeded segments.   13. *Ornithopus*
      Pods indehiscent or dehiscent, but not breaking into 1-seeded segments.
        Pods indehiscent, bur-like, covered with uncinate bristles; herbage glandular-viscid.   14. *Glycyrrhiza*
        Pods dehiscent, variously pubescent, but lacking uncinate bristles.
          Leaves without tendrils (see also *Lathyrus littoralis*).
            Flowers white, cream, or purple; inflorescence an elongate spike, sometimes somewhat capitate, not bracteate.   15. *Astragalus*
            Flowers yellow, often marked with red; inflorescence umbellate or the flowers solitary, usually bracteate.   7. *Lotus*
          Leaves with well-developed or reduced tendrils.
            Style not flattened, with a tuft or ring of hairs.   16. *Vicia*
            Style more or less flattened, bearded along the adaxial surface.
               Wings of the corolla free from the keel.   17. *Lathyrus*

Wings of the corolla adherent to the keel for about half their
length.
Pods 1–2-seeded; seeds biconvex.   18. *Lens*
Pods many-seeded; seeds globose.   19. *Pisum*

1. **Lupinus** L. Lupine

Plants perennial.
Plants herbaceous, aerial stems annual, arising from a woody caudex.
Keels glabrous.
Leaflets large, 10–20 cm. long; stems usually fistulose; herbage sparsely
hairy.   1. *L. polyphyllus grandifolius*
Leaflets smaller, usually under 5 cm. long; stems not fistulose; herbage
densely pubescent.
Pubescence of the stem appressed.   2a. *L. formosus formosus*
Pubescence of the stem spreading.   2b. *L. formosus bridgesii*
Keels pubescent along the free margins.
Keels pubescent from the claws to the center of the blades; leaflets 4–10
cm. long.
Herbage more or less glabrous.   3a. *L. latifolius latifolius*
Herbage villous.   3b. *L. latifolius dudleyi*
Keels pubescent from the claws to the tip of the blade; leaflets 2–3.5 cm.
long.   4. *L. variicolor*
Plants shrubby, the aerial stems perennial, woody.
Banners glabrous.
Petals usually yellow; herbage sparsely pubescent to glabrate; inflores-
cences over 10 cm. long.   5a. *L. arboreus arboreus*
Wings blue, the banner mainly yellow; herbage spreading-villous; inflo-
rescences under 10 cm. long.   5b. *L. arboreus eximius*
Banners pubescent.
Leaflets about equaling or longer than the petioles, the petioles rarely over
2 cm. long; keels with a few scattered cilia or glabrous. Plants of
coastal and inland sand dunes.   6. *L. chamissonis*
Leaflets shorter than the petioles; lower petioles often several times as
long as the leaflets; keels usually ciliate. Plants of chaparral, canyon
slopes, rocky and open ridges.
Plants usually 1–1.5 m. tall.   7a. *L. albifrons albifrons*
Plants usually under 0.4 m. tall.   7b. *L. albifrons collinus*
Plants annual or rarely biennial.
Flowers not verticillate.
Herbage densely hispid-hairy.   8. *L. hirsutissimus*
Herbage glabrate or with appressed pubescence.   9. *L. truncatus*
Flowers verticillate (at least below).
Seeds 2 per pod, the pods more or less elliptic; cotyledons sessile.
Inflorescences becoming secund; flowers spreading at anthesis.
10. *L. densiflorus*
Inflorescences not becoming secund; flowers erect at anthesis.
11. *L. subvexus*
Seeds more than 2 per pod, the pods elongate; cotyledons petiolate.
Keels ciliate on both margins near the claws.   12. *L. succulentus*

Keels ciliate above near the acumen, or glabrous.
    Pedicels 3–7 mm. long.
        Largest leaflets over 5 mm. wide; pods 7.5–8.5 mm. wide.
            13. *L. affinis*
        Largest leaflets under 5 mm. wide; pods under 6.5 mm. wide.
            14. *L. nanus*
    Pedicels 1–3 mm. long.
        Keels glabrous or rarely with a few cilia.
            Banners 7–8.5 mm. long; pods 7–9 mm. wide.   15. *L. pachylobus*
            Banners 5–6.5 mm. long; pods under 5 mm. wide.   16. *L. bicolor*
        Keels distinctly and copiously pubescent.
            Leaflets glabrous adaxially; reflexed portion of the banners shorter
                than the appressed portion.   17. *L. micranthus*
            Leaflets pubescent on both surfaces; reflexed portion of the ban-
                ners longer than the appressed portion.   16. *L. bicolor*

1. L. polyphyllus Lindl. var. **grandifolius** (Lindl. ex Agardh) T. & G. Large-leaved Lupine. Low moist areas along lagoons and in low marshy ground; Lake Merced, San Pedro, Año Nuevo Point, Camp Evers, Zayante Valley, and near Mount Hermon. April–June.
2a. L. formosus Greene var. **formosus.** Summer or Late Lupine. Open woods, grasslands, and fields; Millbrae, La Honda, Stanford, Black Mountain, Los Altos, Campbell, near Los Gatos, Glenwood, and Felton. December–August.
2b. L. formosus Greene var. **bridgesii** (Wats.) Greene. Bridges' Lupine. Woods and rocky areas; Bayview Hills, San Bruno Mountain, Baden, San Mateo, Coal Mine Ridge, and Black Mountain. May–June.
3a. L. latifolius Agardh var. **latifolius.** Broad-leaved Lupine. Grassy slopes, wooded areas, and on disturbed road embankments; northern San Mateo County southward. December–June.
3b. L. latifolius Agardh var. **dudleyi** Smith. Dudley's Lupine. Endemic, known only from the vicinity of Montara Mountain in northern San Mateo County. March–May.
4. L. variicolor Steud. Lindley's Varied Lupine. Open grasslands and coastal bluffs; San Francisco, South San Francisco, San Bruno, San Andreas Reservoir, Pescadero, Swanton, San Vicente Road, and Santa Cruz. March–June. —*L. franciscanus* Greene. Hybrids between *L. variicolor* and *L. arboreus* occur occasionally; for example from near Montara Mountain (*Smith 47184*) and from Bayview Hills (*Howell 31527*).
5a. L. arboreus Sims var. **arboreus.** Tree, Yellow-flowered, Bush, or Yellow Beach Lupine. Most common in sandy soils near the coast; San Francisco, South San Francisco, Pescadero, Swanton, Scott Creek, Ben Lomond, Santa Cruz, and Robroy. March–September.
5b. L. arboreus Sims var. **eximius** (Davy) Smith. Davy's Bush Lupine. Endemic, known from the vicinity of Montara Mountain and Lake Pilarcitos. March–June.
6. L. chamissonis Eschsch. Chamisso's Bush or Blue Beach Lupine. Coastal sand dunes and inland sand deposits; San Francisco, Mount Hermon, Swanton, Sunset Beach State Park, and Palm Beach. March–August.
7a. L. albifrons Benth. ex Lindl. var. **albifrons.** White-foliaged, Silver, or Ben-

tham's Bush Lupine. Chaparral, brush-covered canyon slopes, and road embankments; Crystal Springs Lake, San Mateo Canyon, Belmont, Campbell, Los Gatos, Loma Prieta, Llagas Post Office, Saratoga Summit, Glenwood, Ben Lomond Ridge, and Santa Cruz. March–June.

7b. **L. albifrons** Benth. ex Lindl. var. **collinus** Greene. Hilly Bush Lupine. Open exposed slopes and rocky summits; San Francisco, San Bruno Hills, and Black Mountain. March–April.

8. L. hirsutissimus Benth. Nettle Annual Lupine. Chaparral, brush-covered slopes, and occasionally on road embankments; Stevens Creek, Swanton, near Big Basin, Bear Creek Canyon, and San Lorenzo Road. March–June.

9. L. truncatus Nutt. ex H. & A. Nuttall's Annual Lupine. Known locally from Santa Cruz. June–July.

10. L. densiflorus Benth. *Fig. 120.* Dense-flowered Platycarpos. Fairly common on moist slopes, grasslands, and occasionally in disturbed areas; Crystal Springs Lake, San Mateo Canyon, Stanford, Los Altos, Guadalupe Reservoir, near Los Gatos, Gilroy, and Pajaro River. April–September. *Lupinus densiflorus* is an exceedingly variable species with respect to the size of the various parts of the plants, the color of the flowers, and the amount, kind, and distribution of pubescence. Twenty-six varieties have been described within this species, but despite its variability *L. densiflorus* is easily recognized by its whorls of 2-seeded secund pods.

11. L. subvexus Smith. Intermediate Platycarpos. Rare locally, known from San Francisco and from near San Jose. April–June.

12. L. succulentus Dougl. ex Koch. Succulent Annual Lupine. Grasslands and areas of clay soil; San Francisco, San Pedro, San Mateo Ravine, La Honda Summit, Portola, Stanford, Black Mountain, Monte Bello Ridge, Los Altos, Saratoga, and Edenvale. April–June.

13. L. affinis Agardh. Grassy slopes and fields; Crystal Springs Lake, San Mateo Ravine, Stanford, Page Mill Road, Wrights Station, and near Los Gatos. April–May. —*L. nanus* Dougl. ex Benth. var. *carnosulus* (Greene) Smith.

14. L. nanus Dougl. ex Benth. Douglas' Annual or Sky Lupine. Common in grasslands throughout the Santa Cruz Mountains and often on coastal sand dunes; San Francisco, San Bruno Mountain, Pescadero, Stanford, Permanente Creek, Campbell, near Los Gatos, Guadalupe Canyon, Eagle Rock, Glenwood, Santa Cruz, and Sunset Beach State Park. March–May. —*L. nanus* Dougl. ex Benth. var. *latifolius* Benth. ex Torr., *L. apricus* Greene.

15. L. pachylobus Greene. Mount Diablo Annual Lupine. Rare in grasslands on the eastern slopes of the Santa Cruz Mountains; Bayview Hills, Woodside, San Antonio Creek, and Los Gatos. March–April.

16. L. bicolor Lindl. Lindley's Annual Lupine. A common spring annual of grasslands, pastures, cultivated fields, open areas in woods and brush-covered slopes, and sometimes forming extensive stands in much disturbed areas. March–May. *Lupinus bicolor* is an exceedingly variable species in which numerous subspecies and varieties have been proposed. The following key will aid in distinguishing the varieties that occur locally:

Keel glabrous, rarely with a few cilia.   16*a.* var. *pipersmithii* (Heller) Smith
Keel distinctly ciliate on the upper margin near the acumen.
  Abaxial calyx-lip tridentate, the lobes usually about 1 mm. long.
                                              16*b.* var. *trifidus* (Torr.) Smith

Abaxial calyx-lip entire to tridentate, the lobes less than 1 mm. long.
Plants much branched from the base, prostrate to decumbent, rarely suberect; verticils commonly 1; banners usually over 8 mm. long.

16c. var. *umbellatus* (Greene) Smith

Plants with a few basal branches, erect to suberect, rarely prostrate; verticils usually several, remote or not; banners usually less than 8 mm.
long.
Banners often not reflexed; acumen blunt; flowers 4–6 mm. long.

16d. var. *microphyllus* (Wats.) Smith

Banners reflexed; acumen slender; flowers 6–8 mm. long.

16e. var. *tridentatus* Eastw.

17. L. micranthus Dougl. ex Lindl. Small-flowered Annual Lupine. Open grassland; San Francisco, San Bruno Hills, Coal Mine Ridge, Stanford, Permanente
Canyon, Los Gatos, Scott Creek, and near Boulder Creek. April–May.

2. **Robinia** L. Locust

1. R. pseudo-acacia L. Black Locust. Occasional as an escape from cultivation;
San Francisco, Boulder Creek, and Watsonville; native of eastern United States.
April–May.

3. **Pickeringia** Nutt.

1. P. montana Nutt. *Fig. 121.* Chaparral Pea, Stingaree Bush. Chaparral
throughout the Santa Cruz Mountains, commonly associated with *Adenostoma
fasciculatum, Arctostaphylos, Quercus durata, Q. dumosa, Prunus ilicifolia,* and
Photinia arbutifolia. May–July.

4. **Ulex** L.

1. U. europaeus L. Gorse, Furse, Prickly Broom. Locally established on overgrazed hills and in disturbed habitats; San Francisco, Visitacion Valley, Año
Nuevo Point, and Scott Valley; native of Europe. February–July.

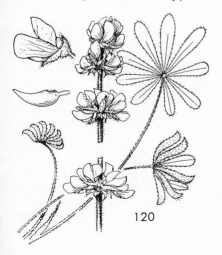

120

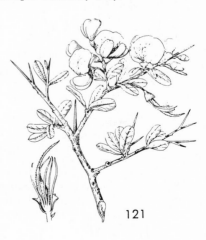

121

5. Cytisus L. Broom

Stems villous, leafy, the leaves not remote; flowers in clusters of 3 or more.
Racemes nearly capitate; petals about 10 mm. long.   1. *C. monspessulanus*
Racemes elongate; petals 13–14 mm. long.   2. *C. maderensis*
Stems more or less glabrous, usually leafless or with the leaves few and remote;
flowers usually solitary or in pairs.   3. *C. scoparius*

1. C. monspessulanus L. French Broom. Roadsides and moist open slopes, often
displacing the native vegetation; San Francisco, near La Honda, Boulder Creek,
Bielawski Lookout, and Soquel Valley; native of southern Europe. March–July.
2. C. maderensis Masf. Occasional, but apparently spreading in oak-woodland
and in disturbed areas; Jasper Ridge, Page Mill Road, and Zayante Creek; native of Madeira. April–May.
3. C. scoparius (L.) Link. Scotch Broom. Escaped from gardens and naturalized along roads, on open slopes, and along the borders of chaparral; Presidio,
Lone Mountain, and near Saratoga Summit; native of Europe. April–June.

6. Spartium L.

1. S. junceum L. Spanish Broom. Occasional as an escape from cultivation;
San Francisco, summit of Page Mill Road, near Saratoga Summit, and Guadalupe Creek; native of Europe. December–June.

7. Lotus L. Bird's Foot Trefoil

Stipules attached by a broad base, not leaflet-like.
Bracts closely subtending the inflorescence.
Plants nearly glabrous; claws of the petals exserted.   1. *L. formosissimus*
Plants somewhat villous; claws of the petals not exserted.
    2. *L. oblongifolius nevadensis*
Bracts at some distance below the inflorescence.
Young stems and branches glabrous.   3. *L. crassifolius*
Young stems and branches hirsute to villous.   4. *L. stipularis*
Stipules reduced to glands.
Plants annual.
Peduncles at least 5 mm. long, bracteate, the bracts occasionally lacking in
L. strigosus.
Flowers yellow, fading to pink or rose; bracts lacking or 1–3-foliolate;
peduncle with 1–5 flowers; rachis of the leaves usually flattened.
Flowers 1–3 per peduncle; bracts lacking or 1–3-foliolate; leaflets 3–9;
styles hairy at the base of the stigmas; seeds rugulose-tuberculate.
    5. *L. strigosus*
Flowers 1–5 per peduncle; bracts 1-foliate; leaflets 4–6; styles glabrous; seeds smooth.   6. *L. salsuginosus*
Flowers white or pinkish, fading to pink or rose; bracts 1- or 3-foliolate;
peduncles with only 1 flower; rachis of the leaves usually terete.
Bracts 1-foliate; calyx-teeth longer than the tube; herbage villous-pubescent to glabrate; leaflets 3.   7. *L. purshianus*
Bracts 3-foliolate; calyx-teeth shorter than the tube; herbage usually
glabrous; leaflets 3–5.   8. *L. micranthus*

Peduncles less than 5 mm. long or the flowers sessile, not bracteate.
Calyx-teeth longer than the tube; pods 0.5–1 cm. long, densely villous.
9. *L. humistratus*
Calyx-teeth about equaling the tube; pods 1–1.5 cm. long, sparsely pubes-
cent. 10. *L. subpinnatus*
Plants perennial.
Pods dehiscent; seeds numerous; peduncles usually at least 5 cm. long.
11. *L. corniculatus*
Pods indehiscent; seeds 2–3; peduncles under 3 cm. long.
Inflorescences bracteate, pedunculate; leaves villous-tomentose to strigose.
Leaves villous-tomentose; peduncles 2–8 mm. long; calyx 4–5 mm.
long; pods pubescent. 12. *L. eriophorus*
Leaves strigose, glabrate in age; peduncles 10–25 mm. long; calyx 6
mm. long; pods glabrous. 13. *L. benthamii*
Inflorescences bractless, sessile or pedunculate; leaves strigose to glab-
rous.
Leaves essentially glabrous; inflorescences sessile, 1–4-flowered.
14. *L. scoparius*
Leaves strigose; inflorescences pedunculate, 2–5-flowered; peduncles
6–20 mm. long. 15. *L. junceus*

1. **L. formosissimus** Greene. Slender or Coast Trefoil. Moist areas along the
coast and inland in boggy ground; Ocean View, near San Bruno, Pilarcitos
Canyon, Pescadero, Pigeon Point, Camp Evers, and Watsonville. March–July.
—*Hosackia gracilis* Benth.
2. **L. oblongifolius** (Benth.) Greene var. **nevadensis** (Gray) Munz. Torrey's
Trefoil. Rare in Santa Cruz County, growing in moist areas; Swanton, Hilton
Airport, Glenarbor, Mount Hermon, and Santa Cruz. June–August. —*L. oblongifolius* (Benth.) Greene var. *torreyi* (Gray) Ottley, *Hosackia torreyi* Gray.
3. **L. crassifolius** (Benth.) Greene. *Fig. 122.* Broad-leaved Trefoil. Dry ridges,
about the edges of chaparral, and as a weed of hilltop vineyards; Black Mountain, Alma Soda Springs, Sierra Azule Road, Mount Umunhum, and near Bielawski Lookout. May–June. —*Hosackia crassifolia* Benth.
4. **L. stipularis** (Benth.) Greene. Stipulate Trefoil. Occasional in moist areas
in coniferous woods and on shaded slopes; between Butano and Pescadero creeks,
Kings Mountain, Loma Prieta Road, near Eagle Rock, and China Grade. April–
July. —*Hosackia stipularis* Benth.
5. **L. strigosus** (Nutt.) Greene. Strigose Trefoil. Common in sandy soils from
San Francisco southward. March–September. —*L. rubellus* (Nutt.) Greene, *Hosackia strigosa* Nutt.
6. **L. salsuginosus** Greene. Coastal Trefoil. Known only from southern Santa
Cruz County near Watsonville, but to be expected elsewhere in loose gravelly
soil. April–May. —*Hosackia maritima* Nutt.
7. **L. purshianus** (Benth.) Clem. and Clem. Spanish Clover, Pursh's Trefoil.
A common annual of grassy slopes, shaded woods, and disturbed areas. April–
October. —*L. americanus* (Nutt.) Bisch., *Hosackia americana* (Nutt.) Piper.
8. **L. micranthus** Benth. Small-flowered Trefoil. Common on open grassy slopes,
edges of woods, and in disturbed areas, San Francisco southward. April–June.
—*Hosackia parviflora* Benth.

9. L. humistratus Greene. Short-podded Trefoil. Common on grassy slopes, in sandy soil, and in moister areas in chaparral. March–July. —*Hosackia brachycarpa* Benth.

10. L. subpinnatus Lag. Chile Trefoil. Common in grassland, on wooded slopes, and in chaparral; San Francisco southward. March–June. —*Hosackia subpinnata* (Lag.) T. & G.

11. L. corniculatus L. Bird's Foot Trefoil. Becoming very common along roads, low adobe fields, and in disturbed areas; native of Europe. June–August.

12. L. eriophorus Greene. Woolly Trefoil. Sandy soils or on open slopes near the coast; San Francisco southward. May–November. —*Hosackia tomentosa* H. & A., *L. heermannii* (D. & H.) Greene var. *eriophorus* (Greene) Ottley.

13. L. benthamii Greene. Bentham's Trefoil. Occasional in rocky soils in western San Mateo County and northern Santa Cruz County; Montara Mountain, San Gregorio, and Swanton. March–April. —*Hosackia cytisoides* Benth.

14. L. scoparius (Nutt.) Ottley. California Broom, Deerweed. A common plant of chaparral and dry open hills. May–October. —*Hosackia glabra* (Vogel) Torr.

15. L. junceus (Benth.) Greene. Rush Trefoil. Occasional in rocky soils and along the edges of chaparral; Kings Mountain, Swanton, and Big Basin. May–July. —*Hosackia juncea* Benth.

8. Psoralea L.

Stems not creeping; petioles usually under 8 cm. long.
    Calyx nearly regular, the lobes triangular.   1. *P. physodes*
    Calyx distinctly irregular, the lobes linear to lanceolate.
        Corolla about 1 cm. long; stems glabrous or nearly so; stipules lanceolate-subulate; inflorescence usually longer than the leaves.
                2. *P. macrostachya*
        Corolla 1.5 cm. long; stems pubescent; stipules lance-ovate; inflorescence usually shorter than the leaves.   3. *P. strobilina*
Stems creeping; petioles usually over 10 cm. long.   4. *P. orbicularis*

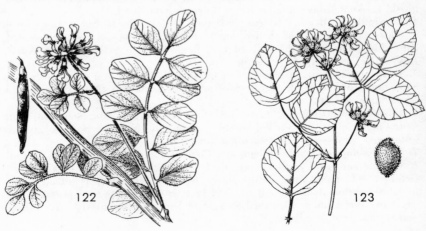

122   123

1. P. physodes Dougl. ex Hook. *Fig. 123.* California Tea. Frequent as a member of the herbaceous understory in oak-madrone woods and less frequent in shaded areas in chaparral. March–July.
2. P. macrostachya DC. California Hemp, Leather Root. Restricted in the Santa Cruz Mountains to the regions from Wrights Station to New Almaden and along Bear Creek Road. June–October. —*P. douglasii* Greene.
3. P. strobilina H. & A. Loma Prieta Psoralea. Usually restricted to serpentine outcroppings, occasional; Guadalupe Valley, near Lexington, Los Gatos, New Almaden, and between Coyote and Madrone. May–September.
4. P. orbicularis Lindl. Round-leaved Psoralea. Occasional in wet ground and along the coast; Sawyer Ridge, Black Mountain Road, San Jose, Eagle Rock, Santa Cruz, and Watsonville. May–July.

9. Thermopsis R. Br.

1. T. macrophylla H. & A. False Lupine, California Thermopsis. Occasional in open areas in chaparral and mixed evergreen forests; Crystal Springs Lake, Ben Lomond Ridge, Warrenella, Zayante Valley, and Santa Cruz. March–July. —*T. velutina* Greene.

10. Medicago L. Medick

Flowers blue; plants perennial. 1. *M. sativa*
Flowers yellow; plants usually annual.
  Pods not spirally coiled except at the tip, black at maturity. 2. *M. lupulina*
  Pods spirally coiled throughout their length, greenish or brownish at maturity.
    Margins of the pods with bristles.
      Leaflets with a small brownish spot at about the center. 3. *M. arabica*
      Leaflets lacking such a spot. 4a. *M. polymorpha vulgaris vulgaris*
    Margins of the pods without bristles.
        4b. *M. polymorpha vulgaris tuberculata*

1. M. sativa L. Alfalfa, Lucerne. A common weed of disturbed habitats and commonly grown as a forage plant; native of Europe. May–September.
2. M. lupulina L. Black Medick. Weeds of disturbed areas, orchards, and cultivated fields; native of Eurasia. April–July. *Medicago lupulina* L. var. *cupaniana* (Guss.) Bois. is a perennial prostrate variety which is sometimes found in lawns. It has also been introduced from Eurasia.
3. M. arabica (L.) Huds. *Fig. 124.* Spotted Bur Clover, Spotted Medick. Occasional on open grassy slopes and in disturbed areas; San Francisco, San Andreas Lake, Stanford, and Santa Cruz; native of Europe and Asia Minor. April–June.
4a. M. polymorpha L. var. **vulgaris** (Benth.) Shinners f. **vulgaris.** Bur Clover. Common in disturbed areas and in open grasslands; native of Europe. March–June. —*M. apiculata* Willd., *M. hispida* Gaertn.
4b. M. polymorpha L. var. **vulgaris** (Benth.) Shinners f. **tuberculata** (Gordon) Shinners. Spineless Bur Clover. Disturbed areas and grassy slopes, San Francisco southward; native of Europe. February–June. —*M. hispida* Gaertn. var. *confinis* (Koch) Burnat.

11. Melilotus Mill. Sweet Clover, Melilot

Petals white.   1. *M. albus*
Petals yellow.
  Flowers about 5 mm. long; pods pubescent.   2. *M. officinalis*
  Flowers about 2.5 mm. long; pods glabrous.   3. *M. indicus*

1. M. albus Desr. ex Lam. White Melilot, White Sweet Clover. Fairly common in disturbed areas; native of Eurasia. February–November.
2. M. officinalis (L.) Lam. Yellow Melilot, Yellow Sweet Clover. Occasional as a weed in disturbed areas; native of Europe. May–August.
3. M. indicus (L.) All. Indian Melilot. A common weed of disturbed areas, fallow fields, and orchards; native of Eurasia. April–October.

12. Trifolium L. Clover.

Inflorescences subtended by an involucre (reduced to a narrow ring in *T. depauperatum*).
  Corollas becoming conspicuously inflated as the pods mature.
    Calyx-teeth glabrous.
      Corollas 1–3 cm. long, greenish or yellowish, occasionally tinged with purple in age or upon drying; stems usually stout.   1. *T. fucatum*
      Corollas under 1 cm. long, yellowish or purplish; stems slender.
        Involucres reduced to a narrow entire ring.   2. *T. depauperatum*
        Involucres divided into lobes, the lobes entire to divided.
                                   3. *T. amplectens*
    Calyx-teeth plumose.
      Inflorescences 1–1.5 cm. in diameter; corollas not white-tipped, about equaling the calyx.   4. *T. barbigerum*
      Inflorescences 1.5–3 cm. in diameter; corollas often white-tipped, much longer than the calyx.   5. *T. grayi*
  Corollas not becoming inflated as the pods mature.
    Involucres cup- or bowl-shaped.
      Calyx-lobes trifid, these lobes often again trifid or bifid, the ultimate segments subulate.   6. *T. cyathiferum*
      Calyx-lobes usually entire, rarely bifid, sometimes erose.
        Involucres glabrous or occasionally with a few hairs abaxially; calyx 3–3.5 mm. long; calyx-lobes 0.5–1.5 mm. long, broadly deltoid, scarious-margined, the margins erose and ciliate; calyx-tube glabrous.   7. *T. microdon*
        Involucres villous; calyx 3.5–4.5 mm. long; calyx-lobes 1.5–2.2 mm. long, deltoid, the margins scarious, more or less entire; calyx-tube usually villous externally.   8. *T. microcephalum*
    Involucres flat.
      Plants perennial; some calyx-lobes cleft into long subulate divisions, the divisions about equal.   9. *T. wormskjoldii*
      Plants annual; calyx-lobes entire or with small lateral teeth.
        Plants glandular-pubescent.   10. *T. obtusiflorum*

Plants not glandular-pubescent.
Calyx-lobes abruptly narrowed at the tip, often with 2 small lateral teeth; corollas at least twice as long as the calyx.
11. *T. tridentatum*
Calyx-lobes gradually narrowed to the tip, without lateral teeth; corollas little exceeding the calyx to twice as long.
Calyx-tube 20-nerved; corolla about twice as long as the calyx, showy.   12. *T. variegatum*
Calyx-tube 10-nerved; corolla little exceeding the calyx, not showy.
13. *T. oliganthum*
Inflorescences not subtended by an involucre, occasionally subtended by the uppermost leaves.
Calyx-tubes becoming corky-thickened, the tubes filled and the pods embedded.
14. *T. angustifolium*
Calyx-tubes not becoming corky-thickened, pods not embedded.
Flowers on short pedicels, reflexed in age.
Corollas yellow.
Flowers 2.5–3.5 mm. long; banners conspicuously diagonally striate; inflorescences 5–15-flowered, 5–7 mm. in diameter. 15. *T. dubium*
Flowers 3.5–4.5 mm. long; banners not conspicuously diagonally striate; heads 20–30-flowered, 8–10 mm. in diameter.
16. *T. procumbens*
Corollas white, pink, or purple.
Plants perennial. Introduced.
Stems creeping and rooting at the nodes; corollas white, rarely pink.
17. *T. repens*
Stems more or less erect, not rooting at the nodes; corollas pink to purple.
Corollas purple; flowers 5–6 mm. long; heads 1–1.4 cm. in diameter.   18. *T. carolinianum*
Corollas pink; flowers 7–10 mm. long; heads 2–3 cm. in diameter.
19. *T. hybridum*
Plants annual. Native or introduced.
Peduncles pilose at least at the top.
Heads with 6–15 fertile flowers, sterile flowers not produced; peduncles erect or ascending in fruit; leaflets longer than broad.
Leaves narrow, deeply notched at the top.
20a. *T. bifidum bifidum*
Leaves broader, shallowly notched at the top.
20b. *T. bifidum decipiens*
Heads with 2–5 fertile flowers, sterile flowers produced at the apex of the peduncle after anthesis of fertile flowers and eventually surrounding them; peduncles recurved and imbedding the developing pods in the ground; leaflets broader than long.
21. *T. subterraneum*
Peduncles glabrous.
Calyx-teeth entire.   22. *T. gracilentum*

Calyx-teeth denticulate and often ciliate.   23. *T. ciliolatum*
Flowers sessile or subsessile, not reflexed in age.
    Corollas red; inflorescences cylindrical, 1.5–2 cm. in diameter; plants
    annual.   24. *T. incarnatum*
    Corollas purple or red (if red, the inflorescence ovoid to globose, 2–3 cm.
        in diameter and the plants perennial; inflorescences ovoid to globose,
        0.5–3 cm. in diameter; plants annual or perennial.
        Plants perennial; inflorescences 2–3 cm. in diameter, closely subtended
          by the upper leaves; corollas red.   25. *T. pratense*
        Plants annual; inflorescences under 2 cm. in diameter, closely sub-
          tended by the upper leaves or not; corollas purple.
          Heads closely subtended by the upper leaves.   26. *T. macraei*
          Heads distant from the upper leaves.
             Corollas at least 1.5 times as long as the calyx.
                              27. *T. dichotomum*
             Corollas about equaling the calyx.   28. *T. albopurpureum*

1. T. fucatum Lindl. Sour Clover. Coastal bluffs, moist grassy areas, and places where water has been plentiful during the winter and early spring; San Francisco southward. March–May. —*T. flavulum* Greene.
2. T. depauperatum Desv. Dwarf Sack Clover. Rare in grasslands; known locally only from Glenwood. March–April.
3. T. amplectens T. & G. Pale Sack Clover. A common, but variable, species of grasslands; San Francisco southward. April–June. —*T. stenophyllum* Nutt., *T. hydrophilum* Greene, *T. truncatum* Greene.
4. T. barbigerum Torr. Bearded Clover. Grasslands from San Francisco southward, most common east of the crests of the Santa Cruz Mountains. March–June.
5. T. grayi Loja. Gray's Clover. Grasslands, mainly near the coast or west of the crests of the Santa Cruz Mountains; San Francisco, Pebble Beach, Big Basin, Jamison Creek, Hilton Airport, and Glenwood. March–June. —*T. barbigerum* Loja. var. *andrewsii* Gray, *T. lilacinum* Greene.

124   125

6. T. cyathiferum Lindl. Bowl Clover. Known locally only from Edenvale in the Santa Clara Valley. May–June.

7. T. microdon H. & A. Valparaiso Clover. Grassy slopes and open fields; San Francisco, Crystal Springs, Belmont, Portola, Stanford, near Los Gatos, 5 miles south of San Jose, and Glenwood. March–July.

8. T. microcephalum Pursh. Small-headed Clover. A common clover of grassy areas. April–June.

9. T. wormskjoldii Lehm. Wormskjold's, Cow, or Coast Clover. Moist coastal bluffs and marshy areas inland; San Francisco, San Bruno, Crystal Springs Reservoir, Arroyo de los Frijoles, Mayfield, Glenwood, and Camp Evers. April–May. —*T. involucratum* Ort., *T. fimbriatum* Lindl., *T. wormskjoldii* Lehm. var. *kennedianum* (McDer.) Jeps.

10. T. obtusiflorum Hook. Clammy or Creek Clover. Santa Clara and Santa Cruz counties, often growing in sandy soils; Llagas Post Office, near Eccles, China Grade, Jamison Creek, Ben Lomond Sand Hills, and Santa Cruz. June–July.

11. T. tridentatum Lindl. *Fig. 125.* Tomcat Clover. A very common clover of grasslands, rocky slopes, and disturbed areas; San Francisco southward. March–May.

12. T. variegatum Nutt. White-tipped Clover. Common in grasslands and grassy areas in woods, San Francisco southward. March–June. Plants with small, few-flowered heads have been called *T. variegatum* Nutt. var. *pauciflorum* (Nutt.) McDer., while plants with large heads and flowers 8–9 mm. long have been called *T. variegatum* Nutt. var. *major* Loja. —*T. variegatum* Nutt. var. *melananthum* (H. & A.) Greene.

13. T. oliganthum Steud. Few-flowered Clover. Open grasslands; Bayview Hills, Stanford, Permanente Canyon, near Los Gatos, Eagle Rock, and Bear Creek Canyon. March–May.

14. T. angustifolium L. Narrow-leaved Clover. Known locally only from Santa Cruz County; Larkin Valley and near Santa Cruz; native of Europe. May–June.

15. T. dubium Sibth. Shamrock. A weed of lawns, gardens, and disturbed areas; San Francisco, Stanford, and probably elsewhere; native of Europe. May–June.

16. T. procumbens L. Low Hop Clover. Occasional as a weed in lawns and open grassland; San Francisco and Stanford; native of Europe. May–June.

17. T. repens L. White Clover. A common weed of lawns, gardens, and moist disturbed areas; native of Europe. April–October.

18. T. carolinianum Michx. Carolina Clover. Known locally from one old collection made in San Francisco; native of the eastern part of the United States. April–May. The specimen upon which this record is based is in the Parish Herbarium of the Dudley Herbarium of Stanford University. As *T. carolinianum* is not otherwise known from California, it is possible that the data on the label is incorrect.

19. T. hybridum L. Alsatian or Alsike Clover. Occasional as a weed in disturbed areas; San Francisco, Stanford, and near Boulder Creek; native of Europe. May–July.

20a. T. bifidum Gray var. **bifidum.** Notch-leaved or Pinole Clover. Grassy slopes and openings in woods, occasionally found in disturbed areas; San Francisco southward but more common to the east of the crests of the Santa Cruz Mountains. April–May.

20b. T. bifidum Gray var. **decipiens** Greene. Deceptive Clover. A common species of grassy slopes, fields, and oak woodlands, San Francisco southward. April–May. The varieties of *T. bifidum* often grow together.
21. T. subterraneum L. Subterranean Clover, Subclover. Known locally only from Waddell Creek, where it has been grown as a cover crop and has escaped; native of Eurasia and North Africa. March–May.
22. T. gracilentum T. & G. Pin-point Clover. Common on grassy slopes, in fields, and as a weed in disturbed areas. March–May. —*T. gracilentum* T. & G. var. *inconspicuum* Fern.
23. T. ciliolatum Benth. Tree Clover. Occasional in grasslands and wooded foothills; Crystal Springs Lake, Portola, Stanford, near Los Gatos, Wrights, Almaden Canyon, Big Basin, Blooms Grade, and Santa Cruz. March–May.
24. T. incarnatum L. Crimson, French, or Italian Clover. Occasional in disturbed areas; Stanford, Pescadero, and Soda Lake; native of Europe. April–June.
25. T. pratense L. Red Clover. Lawns, gardens, and disturbed areas, throughout the Santa Cruz Mountains; native of Eurasia. May–July.
26. T. macraei H. & A. MacRae's Clover. Occasional in grasslands; San Francisco, Baden, Crystal Springs, San Pedro, Coal Mine Ridge, Black Mountain, Saratoga Summit, and Santa Cruz. April–May.
27. T. dichotomum H. & A. Branched Indian Clover. Occasional in grasslands and on serpentine soils; Ocean View, San Mateo Canyon, Palo Alto, Alma, near Los Gatos, 4 miles south of San Jose, Gilroy, and Eagle Rock. April–May. —*T. dichotomum* H. & A. var. *turbinatum* Jeps.
28. T. albopurpureum T. & G. Common Indian or Rancheria Clover. Grassy slopes, northern San Mateo County southward; Crystal Springs Reservoir, Redwood City, Woodside, Stanford, Permanente Canyon, near Los Gatos, San Jose, Edenvale, and Gilroy. April–June. —*T. albopurpureum* T. & G. var. *neolagopus* (Loja.) McDer.

13. Ornithopus L. Bird's Foot

Flowers 1–2 per peduncle; segments of the loments cylindric, circular in cross section.   1. *O. pinnatus*
Flowers 4 or more per peduncle; segments of the loments elliptic, flattened in cross section.   2. *O. roseus*

1. O. pinnatus (Mill.) Druce. Known locally only from Graham Hill between Felton and Santa Cruz; native of Europe. June–July.
2. O. roseus Dufour. Known only from along Graham Hill Road near Felton; native of Europe. June–July.

14. Glycyrrhiza L.

1. G. lepidota Pursh var. **glutinosa** (Nutt.) Wats. Wild or American Licorice. Low ground, grassy flats, and along drainage ditches at low elevations on the eastern side of the Santa Cruz Mountains; Redwood City, San Francisquito Creek, and Stanford. May–August.

15. **Astragalus** L. Milkvetch, Rattleweed, Locoweed

Plants annual.
  Pods about as broad as long, 3–4 mm. long; corollas about 3 mm. long.
  1. *A. gambellianus*
  Pods several times as long as broad, 10–15 mm. long; corollas 7–9 mm. long.
  2. *A. tener*
Plants perennial.
  Pods 7–9 mm. long; flowers in dense racemes.   3. *A. pycnostachyus*
  Pods 3–5 cm. long, inflated and papery in age; flowers in rather loose racemes.
  4. *A. nuttallii virgatus*

1. A. gambellianus Sheldon. *Fig. 126.* Gambell's Dwarf Locoweed. A common species of grasslands and serpentine soils, San Francisco southward, but more common to the east of the crests of the Santa Cruz Mountains. April–May. —*A. nigrescens* Nutt.
2. A. tener Gray. Slender Rattleweed. Known locally only from saline areas along San Francisco Bay; San Francisco and Mayfield. April.
3. A. pycnostachyus Gray. Marsh Locoweed. Salt marshes and salt flats and occasionally inland; Crystal Springs Reservoir, Pescadero, and San Gregorio. June–August.
4. A. nuttallii (T. & G.) Howell var. **virgatus** (Gray) Barneby. San Francisco Rattleweed. Sandy soil of coastal bluffs, San Francisco south to southern San Mateo County. April–October. —*A. franciscanus* Sheldon, *A. menziesii* Gray ssp. *virgatus* (Gray) Abrams.

16. **Vicia** L. Vetch, Tare

Flowers borne on peduncles.
  Flowers small, 3–8 mm. long.
    Pods pubescent, 2-seeded.   1. *V. hirsuta*
    Pods glabrous (rarely with a few scattered hairs), 3–8-seeded.
      Pods under 1.5 cm. long, rounded on each end.   2. *V. tetrasperma*
      Pods over 2.5 cm. long, acute at each end.   3. *V. exigua*
  Flowers larger, at least 10 mm. long.
    Flowers purple-red or yellowish; pods pubescent or glabrous.
      Pods glabrous; herbage sparsely pubescent. Native plants.
      4. *V. gigantea*
      Pods pubescent; herbage densely silky-villous, the pubescence of the stems
        appressed or spreading.   5. *V. benghalensis*
    Flowers blue, purple, or lavender; pods usually glabrous.
      Flowers usually 10 or more per inflorescence.
        Pubescence of the stems spreading; herbage villous.   6. *V. villosa*
        Pubescence of the stems appressed; herbage sparsely pubescent.
        7. *V. dasycarpa*
      Flowers rarely more than 6 per inflorescence.
        Leaflets oblong, ovate, or elliptic.
          Leaflets not truncate.   8a. *V. americana oregana*
          Leaflets truncate.   8b. *V. americana truncata*
        Leaflets linear.   8c. *V. americana minor*

Flowers more or less sessile in the leaf axils.
    Leaflets 4–7 cm. long, tendrils absent or much reduced; flowers 2.5–3 cm. long.
                                                                9. *V. faba*
    Leaflets usually under 3 cm. long, tendrils well developed; flowers usually
        1–2.5 cm. long.
        Flowers 1–2.5 cm. long, lavender-purple; pods glabrous or short-pubescent.
            Flowers 1–1.8 cm. long; leaflets linear to oblanceolate-elliptic; pods
                black at maturity.   10. *V. angustifolia*
            Flowers 2–2.5 cm. long; leaflets obovate; pods brown at maturity.
                                                            11. *V. sativa*
        Flowers 2–2.5 cm. long, white and yellow; pods long-pubescent.
                                                            12. *V. lutea*

1. V. hirsuta (L.) Gray. Hairy Vetch, Tare. Occasional in grassy areas and
along roads; San Francisco and Little Basin; native of Europe, North Africa,
and western Asia. May–July.
2. V. tetrasperma (L.) Moench. Slender Vetch. Occasional as a weed; Pescadero, near Mount Hermon, and Olympia; native of Europe. May–August.
3. V. exigua Nutt. *Fig. 127.* Slender Vetch. Occasional in open wooded areas,
not commonly collected; Mission Hills, Bear Gulch, Stanford, Palo Alto, San
Jose, near Boulder Creek, and Rancho del Oso. March–May. —*V. hassei* Wats.
4. V. gigantea Hook. Giant Vetch. Usually near the coast in moist, partly
shaded habitats, San Francisco southward but rare on the eastern slopes of the
Santa Cruz Mountains. March–July.
5. V. benghalensis L. Occasional in disturbed areas, along roadsides, and on
exposed stream banks, San Francisco southward; native of southern Europe and
North Africa. April–June. —*V. atropurpurea* Desf.
6. V. villosa Roth. Winter or Woolly Vetch. Common as a weed along highways
and along the edges of grasslands, San Francisco southward; native of Europe.
May–August.
7. V. dasycarpa Tenore. Thick-fruited Vetch. Occasional in disturbed areas;
San Francisco, Stanford, and Mount Hermon; native of Europe and North Africa.
April–June. Occasional specimens have pubescent pods (Stanford, *Thomas
7215*, and Edenvale, *Thomas 4948*) and perhaps represent hybrids between *V.
dasycarpa* and *V. benghalensis*.

127

8*a*. V. americana Muhl. ex Willd. var. **oregana** (Nutt.) Nels. American Vetch. Brush-covered areas, open woods, and roadsides, the most common variety of the species in the Santa Cruz Mountains; Twin Peaks, Lake Merced, Seal Cove, Page Mill Road, Coal Mine Ridge, and Stanford. March–May.

8*b*. V. americana Muhl. ex Willd. var. **truncata** (Nutt.) Brewer. Truncate-leaved Vetch. Fairly common on wooded or brush-covered slopes; San Francisco, San Bruno, Cahill Ridge, Spanishtown, Stanford, and Bielawski Lookout. March–May.

8*c*. V. americana Muhl. ex Willd. var. **minor** Hook. Linear-leaved Vetch. Occasional in wooded areas and on grassy slopes; Lake Merced, La Honda, Woodside, Jasper Ridge, Palo Alto, and Swanton. December–May. —*V. linearis* (Nutt.) Greene.

9. V. faba L. Broad or Horse Bean. Occasional as an escape from cultivation; San Francisco, Portola Vally, Stanford, and Loma Prieta Ridge; native of Eurasia. April–July.

10. V. angustifolia L. Smaller Common Vetch. Fairly common in disturbed areas, on grassy slopes, and along the edges of woods, San Francisco southward; native of Europe. April–June. Plants with wide leaves and flowers which are 1.5–1.8 cm. long have been called *V. angustifolia* L. var. *segetalis* (Thuill.) Koch. This variety is often difficult to distinguish from the typical one.

11. V. sativa L. Common or Spring Vetch. Occasional along roadsides, in fields, and in disturbed areas; Bayview Hills, Gazos Creek, Redwood City, Stanford, Hilton Airport, La Selva Beach, and Pajaro River; native of Europe. March–July.

12. V. lutea L. Yellow Vetch. Known locally as roadside weeds near Saratoga Summit and to be expected elsewhere; native of Europe and North Africa. April–June.

17. Lathyrus L. Pea

Leaflets 2, tendrils well developed.
  Petioles with a broad wing; plants perennial; inflorescences 5–15-flowered.
                                                                             1. *L. latifolius*
  Petioles with a narrow wing; plants annual; inflorescences 1–5-flowered.
    Pods pubescent; herbage villous-hirsute; inflorescences 2–5-flowered.
                                                                           2. *L. odoratus*
    Pods glabrous; herbage glabrous; inflorescences 1–3-flowered.
      Flowers solitary, about 1.5 cm. long; pods about 3 cm. long.
                                                                          3. *L. cicera*
      Flowers 1–3, 2.5–3 cm. long; pods 7–10 cm. long.   **4. *L. tingitanus***
Leaflets 4 or more, tendrils well developed or not.
  Tendrils much reduced, not branching or twining.
    Herbage densely silky-pubescent; stipules exceeding the leaflets in size.
    Sand dune plants.   **5. *L. littoralis***
    Herbage sparsely pubescent; stipules smaller than the leaflets. Inland
    plants.   **6. *L. torreyi***
  Tendrils branched and twining.
    Stems winged.   **7. *L. jepsonii californicus***
    Stems angled, but not winged.
      Calyx-tubes and herbage pubescent.
        Plants usually under 3 dm. tall, the internodes usually under 3 cm. long,
          erect rather than scandent.   **8a. *L. vestitus vestitus***
        Plants usually over 4 dm. tall, the internodes usually over 5 cm. long,
          scandent.   **8b. *L. vestitus puberulus***
      Calyx-tubes and herbage glabrous.   **8c. *L. vestitus bolanderi***

1. L. latifolius L. Everlasting Pea. Occasional as an escape from cultivation and established in disturbed areas; San Francisco, near Pescadero, San Jose–Soquel Road, and Felton; native of Europe. June–August.
2. L. odoratus L. Common Sweet Pea. Occasional along roadsides; San Francisco, Big Basin, and Laurel; native of Italy. June–July.
3. L. cicera L. Known only from near Saratoga Summit, where it has been collected as a roadside weed; native of the Mediterranean region. May–June.
4. L. tingitanus L. Tangier Pea. Occasional as a weed; San Francisco, San Gregorio, and Aptos; native of the Mediterranean region. May–September.
5. L. littoralis (Nutt. ex T. & G.) Endl. Silky Beach Pea. Coastal sand dunes from San Francisco south to Watsonville. April–July.
6. L. torreyi Gray. Torrey's or Redwood Pea. Coniferous forests in Santa Cruz County; Big Basin, Ben Lomond, Laurel, San Lorenzo Road, Felton, and Mount Hermon. May–June.
7. L. jepsonii Greene ssp. **californicus** (Wats.) Hitchc. Buff Pea. Rare locally on brush-covered slopes and as an introduced weed in San Francisco; Kings Mountain, Cooleys Landing, Menlo Park, and Stanford. —*L. watsonii* White.
8a. L. vestitus Nutt. ex T. & G. ssp. **vestitus.** *Fig 128.* Common Pacific Pea. Grassy slopes and ridges; Ocean View, San Bruno Hills, and near Saratoga Summit. March–May.

8b. L. vestitus Nutt. ex T. & G. ssp. **puberulus** (White ex Greene) Hitchc. The most common subspecies of this species locally, usually growing in woods, along the edges of chaparral, and along road and highway embankments, San Francisco southward. October–April. —*L. violaceus* Greene.

8c. L. vestitus Nutt. ex T. & G. ssp. **bolanderi** (Wats.) Hitchc. Bolander's Pea. Shaded areas along streams and in open woods; Lake Merced, Crystal Springs Lake, Half Moon Bay, Pescadero Creek, Kings Mountain, and Stanford. January–May.

18. Lens Mill.

1. L. culinaris Medic. Lentil. Rare as a weed in San Francisco; native of Europe. June.

19. Pisum L. Pea

1. P. sativum L. Garden Pea. Rare as an escape from cultivation in San Francisco and probably elsewhere; native of Eurasia. June.

67. GERANIACEAE. Geranium Family

Upper sepal spurred, the spur adnate to the pedicel; petals unequal.
1. *Pelargonium*
Upper sepal not spurred; petals equal.
Anthers 10; beak of the fruit arched when mature; leaves rounded in outline.
2. *Geranium*
Anthers 5; beak of the fruit forming a spiral when mature; leaves ovate to oblong in outline (rounded in *E. macrophyllum*). 3. *Erodium*

1. Pelargonium L'Her. Geranium

Stems not succulent; leaves lobed or divided.
Petals less than 1 cm. long; stems slender, glabrous or nearly so.
1. *P. grossularioides*
Petals more than 1 cm. long; stems stout, soft-villous. 2. *P. vitifolium*
Stems succulent; leaves usually not lobed or divided. 3. *P. zonale*

1. P. grossularioides (L.) Ait. Sparingly introduced on the coastal cliffs at Pebble Beach; native of South Africa. May–June.

2. P. vitifolium (L.) Ait. Grape-leaf Pelargonium. Established locally in San Francisco and along the coast from Pescadero to Pigeon Point; native of South Africa. March–December, but occasional plants can be found in flower at other seasons.

3. P. zonale (L.) Ait. Garden Geranium. Vacant lots and disturbed areas in San Francisco and probably elsewhere; native of South Africa. In flower throughout the year.

2. **Geranium** L. Cranesbill

Sepals not cuspidate or subulate at the tips.
  Carpel-bodies glabrous; stamens 10.   1. *G. molle*
  Carpel-bodies pubescent; stamens 5.   2. *G. pusillum*
Sepals cuspidate or subulate at the tips.
  Carpel-bodies variously pubescent.
    Plants perennial; roots thick.   3. *G. retrorsum*
    Plants annual; roots slender.
      Carpel-bodies short-hirsute, the trichomes of uniform length; seeds with
         rounded pits; inflorescence glandular.   4. *G. dissectum*
      Carpel-bodies villous, the trichomes of different lengths; seeds with elon-
         gated pits; inflorescence rarely glandular.   5. *G. carolinianum*
  Carpel-bodies glabrous.
    Plants perennial; petals rarely over 10 mm. long, pale purple.
                                 6. *G. anemonifolium*
    Plants annual or biennial; petals 12–20 mm. long, crimson.
                          7. *G. robertianum*

1. G. molle L. Cranesbill or Dove's-foot Geranium. Grasslands throughout the Santa Cruz Mountains; native of Europe. February–June.
2. G. pusillum Burm. f. Small-flowered Geranium. Sporadic as a lawn weed in Boulder Creek; native of Europe. May–September.
3. G. retrorsum L'Her. ex DC. New Zealand Geranium. Occasional in disturbed areas and in grasslands in San Francisco and probably elsewhere; native of Australasia. May–September.
4. G. dissectum L. Cut-leaved Geranium. Grassy slopes, pastures, and disturbed habitats, the commonest member of this genus in the Santa Cruz Mountains; native of Europe. April–October.
5. G. carolinianum L. Carolina Geranium. Occasional on wooded slopes and in other shaded areas; known definitely from San Francisco and Stanford. April–October.

128

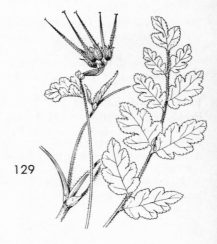

129

6. G. anemonifolium L'Her. Anemone-leaved Geranium. Occasional as an escape from cultivation in Golden Gate Park; native of the Canary Islands and Madeira. July–August.
7. G. robertianum L. Herb Robert, Red Robin. Known from Golden Gate Park as an occasional escape from cultivation; native of eastern North America and Europe. July–August.

3. Erodium L'Her. ex Ait. Storksbill, Filaree, Clocks

Leaves simple, pinnatifid or nearly entire.
  Leaves pinnatifid, adult leaves several times as long as broad, not cordate at base. Common species.
    Concavities at summit of fruit subtended by 1 fold; fruit usually hairy throughout; beak of fruit 5–9 cm. long.   1. *E. obtusiplicatum*
    Concavities at summit of fruit subtended by 2 folds; fruit glabrous above the folds; beak of fruit usually over 9 cm. long.   2. *E. botrys*
  Leaves palmately lobed, adult leaves about as long as broad, cordate at base. Rare.   3. *E. macrophyllum*
Leaves pinnately compound.
  Leaflets pinnately lobed; sepals bearing 1 or 2 bristles at the apex; claws of petals ciliate.   4. *E. cicutarium*
  Leaflets toothed; sepals lacking bristles; claws of petals glabrous.   5. *E. moschatum*

1. E. obtusiplicatum (Maire, Weiller, & Wilczek) Howell. Common in disturbed areas and on grassy slopes throughout the Santa Cruz Mountains; introduced from North Africa. March–May.
2. E. botrys (Cav.) Bertol. Long-beaked or Broad-leaved Filaree. Very common on grassy slopes, disturbed areas, and pastures; native of the Mediterranean region. March–July.
3. E. macrophyllum H. & A. Large-leaved Filaree. Rare in the Santa Cruz Mountains, known definitely only from Pescadero. April.
4. E. cicutarium (L.) L'Her. Red-stemmed Filaree. Common in pastures, in disturbed areas, and on open grassy slopes; native of southern Europe. February–November.
5. E. moschatum (L.) L'Her. ex Ait. *Fig. 129.* White-stemmed or Musk Filaree. A very common weed of disturbed areas and of grassy fields; native of the Mediterranean region. February–October.

68. OXALIDACEAE. Oxalis or Wood Sorrel Family

1. Oxalis L. Wood Sorrel

Petals white or lavender.
  Plants with well developed stems or with stout vertical rootstocks. Introduced species.
    Plants with stout vertical rootstocks; petals 11–12 mm. long; leaflets pubescent.   1. *O. rubra*
    Plants with branched stems; petals 15–17 mm. long; leaflets glabrous.   2. *O. incarnata*

Plants with slender creeping rootstocks. Native species of redwood forest or
rarely found in cultivation. 3. *O. oregana*
Petals yellow.
Leaves all basal; plants somewhat succulent; flowers at least 1.5 cm. long.
4. *O. pes-caprae*
Leaves cauline; plants not succulent; flowers under 1.2 cm. long.
Pedicels sparingly pubescent, the hairs more or less appressed or ascending;
rootstocks slender. Introduced species. 5. *O. corniculata*
Pedicels usually densely pubescent, the hairs spreading; root fusiform,
woody. Native grassland species. 6. *O. pilosa*

1. O. rubra St. Hil. Windowbox Oxalis. Locally established in vacant lots and
in lawns in San Francisco; native of Brazil. June–August.
2. O. incarnata L. Flesh-colored Oxalis. Escaping from cultivation in San
Francisco and probably elsewhere; native of South Africa. April–May.
3. O. oregana Nutt. *Fig. 130.* Redwood Sorrel, Oregon Wood Sorrel. Restricted
to the damp shaded floor of the redwood–Douglas fir forest, commonly associated
with *Asarum caudatum* and *Viola sempervirens.* February–September.
4. O. pes-caprae L. Cape Oxalis, Bermuda Buttercup. Along or near the coast
in moist soil and becoming very common in disturbed areas and lawns; native of
South Africa. February–June. —*O. cernua* Thunb. *Oxalis pes-caprae* was first
noted in the San Francisco Bay region about 40 years ago, and since that time
has become very common, especially along the coast. It is perennial and, once
established, is very difficult to eradicate, since the rootstocks bear bulblets at each
node. The plants reproduce vegetatively as well as sexually.
5: O. corniculata L. Creeping Wood Sorrel. Common as a lawn and garden
weed and in disturbed areas; native of Europe. February–November. A variety
with purplish herbage may be called *O. corniculata* L. var. *atropurpurea* Planch.
—*O. corniculata* L. var. *purpurea* Parlat.
6. O. pilosa Nutt. Hairy Wood Sorrel. Open grassy slopes, mainly on the western slopes of the Santa Cruz Mountains; Lake Merced, Montara Mountain, Peters
Creek, Swanton, near Glenwood, Mount Hermon, Santa Cruz, and near Watsonville. April–November.

69. TROPAEOLACEAE. Tropaeolum Family

1. Tropaeolum L. Nasturtium

1. T. majus L. Garden Nasturtium. Occasionally escaping from cultivation
and persisting; San Francisco, Pebble Beach, Palo Alto, Santa Cruz, and Watsonville; native of South America. April–October.

70. LINACEAE. Flax Family

Petals blue or blue-tinged, drying bluish, 10–15 mm. long; styles 5; mature capsules 5–10 mm. long. Introduced plants. 1. *Linum*
Petals white to pink, drying rose-colored, 2–7 mm. long; styles 3; mature capsules
less than 3 mm. long. Native plants. 2. *Hesperolinon*

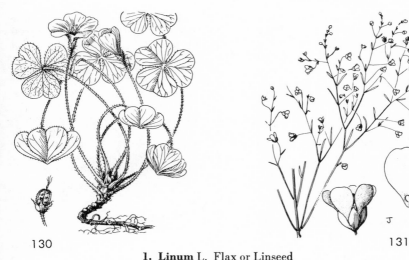

130   131

1. Linum L. Flax or Linseed

Leaves lanceolate; petals usually 1.5 cm. long; mature capsules 7–10 mm. long.
1. *L. usitatissimum*
Leaves linear; petals about 1 cm. long; mature capsules about 5 mm. long.
2. *L. bienne*

1. L. usitatissimum L. Flax, Common Flax. Occasional along roadsides and in disturbed areas at lower elevations; San Francisco, San Bruno Hills, Pescadero, Woodside, Palo Alto, Stanford, and Whitehouse Road; native of Europe. March–July. This species is cultivated extensively in fields along the coast in southern San Mateo County.
2. L. bienne Mill. Narrow-leaved or Small-flowered Flax. Occasionally found along roads, in disturbed areas, and in fallow fields and vineyards; San Francisco, Montara Point, and Glenwood; native of Europe. May–August. —*L. angustifolium* Huds.

2. Hesperolinon (Gray) Small. Dwarf Flax

Sepals glabrous on the abaxial surface; inflorescence not congested.
Petals 2–3.5 mm. long, basal lobes of the petals obsolete.   1. *H. micranthum*
Petals 5–7 mm. long, basal lobes prominent.   2. *H. spergulinum*
Sepals pubescent on the abaxial surface; inflorescence congested, the terminal flower usually nearly sessile.   3. *H. congestum*

1. H. micranthum (Gray) Small. Small-flowered Dwarf Flax. Eastern slopes of the Santa Cruz Mountains, growing in serpentine or in chaparral; San Carlos, Jasper Ridge, Black Mountain, and Los Gatos. May–June. —*Linum micranthum* Gray.
2. H. spergulinum (Gray) Small. *Fig. 131.* Slender Dwarf Flax. Known locally only from Crystal Springs Lake. May–June. —*Linum spergulinum* Gray.

3. **H. congestum** (Gray) Small. Marin or Congested Dwarf Flax. Serpentine slopes in San Francisco and San Mateo counties; Laurel Hill Cemetery, Crystal Springs Lake, Redwood City, and Woodside. May–June. —*Linum congestum* Gray.

71. ZYGOPHYLLACEAE. Caltrop Family

1. Tribulus L. Caltrop

1. **T. terrestris** L. Puncture Vine or Weed, Land Caltrop. Known at present from San Francisco, Palo Alto, Stanford, and from near Gilroy, but to be expected elsewhere along highways and in disturbed areas where the ground is hard-packed; native of Europe. April–September.

72. RUTACEAE. Rue Family

Small trees; fruit a samara.   1. *Ptelea*
Herbs; fruit a capsule.   2. *Ruta*

1. Ptelea L. Hop Tree

1. **P. crenulata** Greene. Western Hop Tree. Known locally only from the foothills west of Gilroy. April–June.

2. Ruta L. Rue

1. **R. chalepensis** L. Occasional as an escape from cultivation in San Francisco; native of the Mediterranean region. April–May.

73. SIMAROUBACEAE. Simarouba or Quassia Family

1. Ailanthus Desf.

1. **A. altissima** (Mill.) Swingle. Tree-of-Heaven. Common as seedlings and small trees near or in the vicinity of trees of this species planted as ornamentals and often becoming a considerable nuisance, especially in cultivated fields; native of China and introduced into California by Chinese laborers more than a century ago. June.

74. POLYGALACEAE. Milkwort Family

1. Polygala L. Milkwort, Polygala

1. **P. californica** Nutt. *Fig. 132.* California Polygala, California Milkwort. Occasional in the Santa Cruz Mountains; near Arroyo de los Fijoles, Butano Ridge, Big Basin, and Soquel Gulch. April–July.

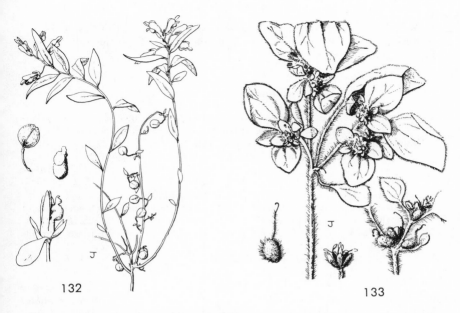

132   133

75. EUPHORBIACEAE. Spurge Family

Leaves entire or at most serrate.	
Plants small shrubs.	1. *Croton*
Plants herbaceous.	
Plants densely stellate-pubescent.	2. *Eremocarpus*
Plants not stellate-pubescent.	
Flowers not subtended by an involucre.	3. *Mercurialis*
Flowers subtended by an involucre.	4. *Euphorbia*
Leaves deeply palmately divided.	5. *Ricinus*

1. Croton L. Croton

1. C. californicus Muell. Arg. California Croton. Known locally only from the sand dunes and sandstone cliffs in the western part of San Francisco. June–November. *Croton californicus* reaches its northern limit of distribution in the Outer Coast Ranges in San Francisco, but it is absent from the rest of the Santa Cruz Mountains.

2. Eremocarpus Benth. Dove Weed

1. E. setigerus (Hook.) Benth. *Fig. 133.* Turkey Mullein. Dry ground of fallow fields, open slopes, stream bottoms, and disturbed areas, most common in the Santa Clara Valley. June–October.

3. Mercurialis L. Mercury

1. **M. annua** L. Herb-Mercury. Occasional as a weed in Golden Gate Park and in cultivated fields in coastal valleys in northern San Mateo County; native of Europe. August–November.

4. Euphorbia L. Spurge

At least the lower cauline leaves alternate (or decussate in *E. lathyris*); plants usually erect; stipules lacking; plants usually branched above.
Glands lacking horns; cauline leaves serrate.
Carpels tuberculate near the summit; umbel-rays usually 3.
    1. *E. spathulata*
Carpels not tuberculate; umbel-rays usually 5.   2. *E. helioscopia*
Glands with horns; cauline leaves entire or crenulate.
Coarse glaucous plants, usually 0.5–1 m. tall; cauline leaves decussate or the lower alternate; carpels over 7 mm. long.   3. *E. lathyris*
Plants slender, usually under 0.4 m. tall; cauline leaves alternate; carpels under 5 mm. long.
    Carpels with slender longitudinal ridges.   4. *E. peplus*
    Carpels lacking ridges.
        Floral leaves narrowly lanceolate.   5. *E. exigua*
        Floral leaves ovate to deltoid.   6. *E. crenulata*
All leaves opposite; plants prostrate or erect; stipules present; plants not umbellately branched above.
Plants prostrate; leaves glabrous or pubescent; capsules glabrous or pubescent.
    Capsules and leaves pubescent.   7. *E. supina*
    Capsules and leaves glabrous.
        Seeds rugose.   8. *E. serpyllifolia*
        Seeds smooth.   9. *E. serpens*
    Plants erect; leaves pubescent; capsules glabrous.   10. *E. maculata*

1. **E. spathulata** Lam. Reticulate-seeded Spurge. Serpentine and grassy slopes, rarely in shaded areas, San Francisco southward. March–June.

2. **E. helioscopia** L. Sun Spurge, Wartweed, Wart Spurge. Occasionally established in fallow fields, deserted gardens, and other disturbed areas; San Francisco, Pescadero, and Santa Cruz; native of Europe. April–September.

3. **E. lathyris** L. Caper Spurge, Compass Plant, Gopher Plant. Widely distributed along the margins of the Santa Cruz Mountains in moist shaded situations, often along streams; San Francisco, Pescadero Creek, Alpine Creek Road, Stanford, Swanton, and near Zayante; native of Europe. February–November.

4. **E. peplus** L. Petty Spurge. A common weed of gardens and other well-watered, disturbed habitats and occasionally in undisturbed areas; native of Europe. To be found in flower at all seasons.

5. **E. exigua** L. Known only from one old collection from Santa Clara; native of Europe.

6. **E. crenulata** Engelm. *Fig. 134.* Chinese Caps. Shaded woods and occasionally in open pastures, San Francisco southward. February–November.
7. **E. supina** Raf. Spotted Spurge. Common as a weed of lawns, gardens, and other moist disturbed habitats; native of eastern North America. May–June.
8. **E. serpyllifolia** Pers. Thyme-leaved Spurge. Low areas which are submerged or very wet during the spring; San Francisco, Crystal Springs Lake, Palo Alto, Stanford, near Alviso, and near Coyote. May–November.
9. **E. serpens** HBK. Occasional as a weed in San Francisco, probably introduced locally from other parts of North America. May–June.
10. **E. maculata** L. Large Spurge. Known locally as a weed from San Francisco and Boulder Creek and to be expected occasionally elsewhere; native of eastern North America and Mexico. April–September.

5. **Ricinus** L.

1. **R. communis** L. Castor Bean. Occasionally spontaneous in disturbed areas, known definitely from Palo Alto; native of Eurasia. June–August.

76. CALLITRICHACEAE. Water Starwort Family

1. **Callitriche** L. Water Starwort

Small inconspicuous aquatic plants such as *Callitriche* are seldom collected. The localities cited below, therefore, reflect inadequate collecting rather than rarity of the species.

Fruit sessile.
  Emersed leaves obovate to spatulate; fruit with 2 bracts.
    Styles shorter than the fruit.   1. *C. verna*
    Styles about twice as long as the fruit.   2. *C. bolanderi*
  All leaves linear, submersed; fruit bractless.   3. *C. hermaphroditica*
Fruit distinctly pedicellate.   4. *C. marginata*

1. **C. verna** L. Vernal Water Starwort. Margins of lakes in northern San Mateo County and to be expected elsewhere. March–August. —*C. palustris* L.
2. **C. bolanderi** Hegelm. Bolander's Water Starwort. Edges of quiet ponds and lakes and on the adjacent muddy banks; San Bruno Creek, Año Nuevo Point, Portola Valley, near Stanford, and Camp Evers. March–May. —*C. heterophylla* Pursh var. *bolanderi* (Hegelm.) Fassett.
3. **C. hermaphroditica** L. Hermaphroditic, Northern, or Autumnal Water Starwort. Fresh-water ponds and quiet streams; known locally only from near Gilroy, but to be expected elsewhere. —*C. autumnalis* L.
4. **C. marginata** Torr. *Fig. 135.* California Water Starwort. Margins of pools, in mud and on moist soil, and on open slopes near springs; San Francisco, Belmont, Hilton Airport, and Camp Evers. February–April. At Hilton Airport *C. marginata* grows on small patches of bare moist soil and is easily overlooked. When growing on moist soil, the pedicels are curved so that the fruit is appressed to the soil surface or the fruit may even be pushed below the surface.

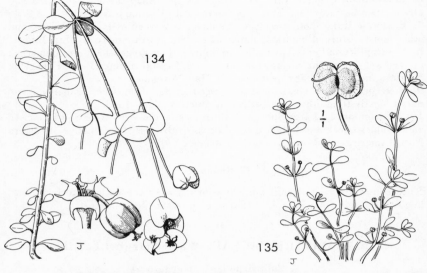

134

135

77. LIMNANTHACEAE. False Mermaid or Meadow Foam Family

1. Limnanthes R. Br. Meadow Foam

Petals of living plants yellow.   1a. *L. douglasii douglasii*
Petals of living plants white.   1b. *L. douglasii nivea*

1a. L. douglasii R. Br. var. **douglasii**. *Fig. 136*. Common Meadow Foam. Throughout most of the Santa Cruz Mountains, but not common; along stream-lets and springs in open meadows; San Bruno, San Francisco Watershed Reserve, Pescadero, Stanford, and 3 miles west of Coyote. March–May. Dried specimens of this variety usually lose their color and are often undistinguishable from var. *nivea*.
1b. L. douglasii R. Br. var. **nivea** Mason. Snow White Meadow Foam. Known locally only from moist meadows between Big Basin and Boulder Creek, but to be expected elsewhere. March–April.

78. ANACARDIACEAE. Sumac Family

Leaves 3-foliolate; berries white to cream-colored.   1. *Rhus*
Leaves pinnate, leaflets numerous; berries red.   2. *Schinus*

1. Rhus L. Sumac, Poison Oak

1. R. diversiloba T. & G. *Fig. 137*. Pacific Poison Oak. Throughout the Santa Cruz Mountains and adjacent coastal areas. March–May. *Rhus diversiloba* is an exceedingly variable species, not only with respect to the size, shape, and amount of pubescence of its leaves but also with respect to its growth form. Within the species are to be found plants which are less than half a meter high, shrubs, small trees, and climbing vines. One vine growing on *Arbutus menziesii* measured twelve meters in length when pulled down, and a portion of the terminal part of the plant still remained in the tree. The stem of the poison oak was about 2 cm. in diameter at the base. Plants of poison oak occur as part of the understory of oak-madrone woods, redwood–Douglas fir forests, and occasionally in stands of redwoods. They are also to be found in disturbed areas such as road embankments, on open slopes, and, as at Sunset Beach, on the low stabilized coastal sand dunes. The autumn foliage lends color to the landscape second to that of no other species in the local flora.

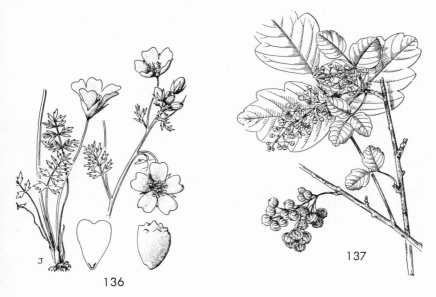

136

137

2. Schinus L.

1. S. molle L. Pepper Tree, Peruvian Pepper Tree. Seedlings are commonly found in the vicinity of ornamentals and if not disturbed commonly grow to maturity; native of the American tropics.

79. CELASTRACEAE. Burning Bush or Staff-Tree Family

1. Euonymus L. Burning Bush, Western Wahoo

1. E. occidentalis Nutt. ex Torr. *Fig. 138.* Pawnbroker or Western Burning Bush. Moist shaded areas in redwood–Douglas fir forests, usually along or near streams and lake banks; Pilarcitos Creek, San Andreas Lake, Butano Creek, Searsville Lake, Big Basin, San Vicente Creek, and Felton. April–June.

opposite leaves
samara fruit

80.   ACERACEAE. Maple Family

1.   Acer L. Maple

Leaves palmately lobed; flowers polygamous; petals present.

1. *A. macrophyllum*

Leaves trifoliolate; flowers dioecious; petals absent.   2. *A. negundo californica*

1. A. macrophyllum Pursh. *Fig. 139.* Canyon or Big-leaved Maple. Along streams and on moist wooded slopes often well up a hillside from the stream; occasionally, as along Skyline Drive near Saratoga Summit, *A. macrophyllum* occurs in fairly pure stands far removed from streams. March–May.

2. A. negundo L. var. **californicum** (T. & G.) Sarg. California Box Elder. Along the lower portions of major streams where the gradient is not steep, usually below about 500 feet elevation; San Francisco southward. March–April. —*A. negundo* L. ssp. *californicum* (T. & G.) Wesmael.

81.   HIPPOCASTANACEAE. Horse Chestnut or Buckeye Family

1.   Aesculus L. Buckeye, Horse Chestnut

1. A. californica (Spach) Nutt. *Fig. 140.* California Buckeye. A characteristic tree of dry slopes throughout the Santa Cruz Mountains and its foothills from sea level to about 3,000 feet, San Francisco southward. April–June. *Aesculus californica* changes markedly as the growing season progresses. Leaves appear as early as the beginning of February, transforming bare trees into lush green. By the beginning of June the inflorescences stand out like white or flesh-colored candles. By the end of July or the first part of August the leaves usually have

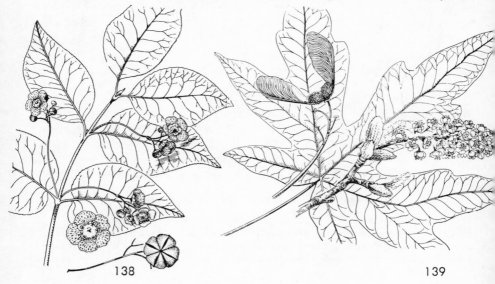

138   139

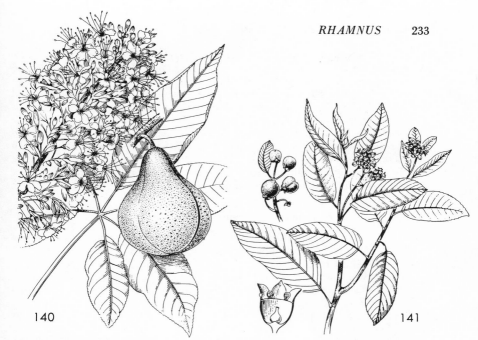

140   141

fallen, and in place of each inflorescence there are one to several chestnut-like fruits. Germination is good, and it is not uncommon to find several dozen seedlings under the parent tree by the next spring. Of these, however, only a very few, often none, survive. The shape of the mature tree is variable. On more exposed slopes, the branches depart close to the ground; no main trunk is formed. In more favorable localities a small tree with a rounded crown results. Common associates of the Buckeye are *Quercus agrifolia, Diplacus aurantiacus, Artemisia californica,* and *Rhus diversiloba.*

82.   RHAMNACEAE.   Buckthorn Family

Petals absent or if present not clawed, greenish; fruit berry-like, not lobed.
                                                                    1. *Rhamnus*
Petals present, clawed, white, blue, or purplish; fruit a capsule, usually some-
      what 3-lobed.   2. *Ceanothus*

1. Rhamnus L.   Buckthorn, Cascara

Leaves 3–7 cm. long; petals present; berries black.
    Leaves essentially glabrous abaxially.   1*a. R. californica californica*
    Leaves distinctly tomentulose abaxially.   1*b. R. californica tomentella*
Leaves usually under 3 cm. long; petals none or minute; berries red.
    Small branches divaricate, somewhat spinescent; leaves 10–17 mm. long. Com-
        mon.   2*a. R. crocea crocea*
    Small branches not divaricate, not spinescent; leaves 18–30 mm. long. Rare.
                                                      2*b. R. crocea ilicifolia*

1a. R. californica Esch. ssp. **californica.** *Fig. 141.* California Coffee Berry. A common shrub from San Francisco southward, growing in chaparral, wooded areas, and on brush-covered slopes. May–July.

1b. R. californica Esch. ssp. **tomentella** (Benth.) Wolf. Open hillsides and chaparral; Kings Mountain, Stevens Creek Road, Los Gatos Creek, near Los Gatos, Saratoga Summit, Hilton Airport, and Jamison Creek Road. May–June.

2a. R. crocea Nutt. ssp. **crocea.** Red Berry. A common shrub of chaparral and oak-madrone woods, San Francisco southward, but more common on the eastern slopes of the Santa Cruz Mountains. February–May.

2b. R. crocea Nutt. ssp. **ilicifolia** (Kell.) Wolf. Holly-leaved Buckthorn. Known locally only from near Llagas Post Office. March–June.

2. **Ceanothus** L. California Lilac, Ceanothus

Leaves alternate; stipules thin, deciduous; stomata of the leaves not sunken into pits; fruits lacking horns but often crested or ridged.
Leaves 3-veined from the base.
Plants with spinose branchlets, the young branches glaucous. 1. *C. incanus*
Plants lacking spinose branchlets, but the branchlets often divergent and stiff.
Leaves entire, usually deciduous. 2. *C. integerrimus*
Leaves with serrate or glandular-denticulate margins.
Flowers white; leaves shiny and varnished on the adaxial surface. Rare.
3. *C. velutinus laevigatus*
Flowers blue to purple, rarely whitish. Common.
Leaves pale on the abaxial surface, the adaxial surface plane; branchlets rigid and divaricate, not angled. 4. *C. sorediatus*
Leaves green on the abaxial surface, the adaxial surface depressed above the major veins; branchlets usually flexible, ascending, angled. 5. *C. thyrsiflorus*
Leaves 1-veined from the base, occasionally the first pair of lateral veins more prominent than the others.
Leaves entire. 2. *C. integerrimus*
Leaves variously denticulate.
Leaves glandular-papillate adaxially, 2–5 cm. long. Common.
6. *C. papillosus*
Leaves not glandular-papillate adaxially, usually under 2 cm. long. Rare.
Leaf-margins conspicuously revolute. 7. *C. dentatus*
Leaf-margins not revolute, often undulate. 8. *C. foliosus*
Leaves usually opposite; stipules more or less conical, corky, persistent; stomata of the leaves sunken into pits; fruit with horns.
Flowers white.
Mature leaves entire. Common. 9. *C. cuneatus dubius*
Mature leaves with at least a few teeth. Rare. 10. *C. ferrisae*
Flowers blue. 11. *C. ramulosus*

1. C. incanus T. & G. *Fig. 142.* Coast Whitethorn. Open ridges and moist non-shaded areas; San Francisco, Swanton, Big Basin, Glenwood, near Ben Lomond, Felton, and Lower Zayante. May–June.

2. C. integerrimus H. & A. Deer Brush. A variable chaparral species; known locally from near Portola, Loma Prieta, and Ben Lomond. May–June. —*C. andersonii* Parry. Hybrids between *C. integerrimus* and *C. papillosus* are sometimes found as, for example, at Boulder Creek and Ben Lomond.

3. C. velutinus Dougl. ex Hook. var. **laevigatus** (Hook.) T. & G. Sticky Laurel, Tobacco Bush, Varnishleaf Ceanothus. Known locally only from Swanton. March–April.

4. C. sorediatus H. & A. Jim Bush. Common in chaparral and along the edges of woods from northern San Mateo County southward. April–June.

5. C. thyrsiflorus Esch. Blue Brush, Blue Blossom. A common shrub on moist wooded hills from San Francisco southward. March–May. Hybrids between *C. thyrsiflorus* and *C. sorediatus* are occasionally found.

6. C. papillosus T. & G. Warty-leaved Ceanothus. Chaparral and along the margins of woods; Kings Mountain, Alma Soda Springs, Wrights, Loma Prieta Peak, Bielawski Lookout, Big Basin, Empire Grade, Eagle Rock, Boulder Creek, Ben Lomond, Santa Cruz, and Aptos Hills. April–May. Hybrids between *C. papillosus* and *C. thyrsiflorus* are frequently found and have been called *C.* × *regius* (Jeps.) McMinn. The leaves tend to have 3 veins from the base and few papillae. *Ceanothus papillosus* also forms hybrids with *C. sorediatus*. The branches tend to be rigid and divaricate; the leaves are 3-veined from the base and with or without papillae on the adaxial surface.

7. C. dentatus T. & G. Dwarf or Cropleaf Ceanothus. Rare in the Santa Cruz Mountains, known only from the vicinity of Larkin Valley southward. March–June.

8. C. foliosus Parry. Wavyleaf Ceanothus, Indigo Brush. Chaparral, known from Loma Prieta and Mount Madonna. April–May.

9. C. cuneatus (Hook.) Nutt. var. **dubius** Howell. Common Buck Brush, Santa Cruz Buck Brush. Endemic in the Santa Cruz Mountains, the most common species of *Ceanothus* in chaparral, frequently forming dense, almost impenetrable thickets, northern San Mateo County southward. February–August.

142

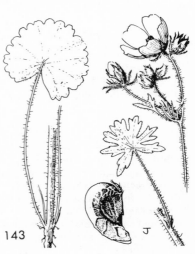

143

10. C. ferrisae McMinn. Coyote or Ferris' Ceanothus. Occasional in Santa Cruz and Santa Clara counties; Los Gatos Creek and near Mount Hermon. January–March.

11. C. ramulosus (Greene) McMinn. Coast Ceanothus, Blue Buck Brush. Occasional, known from near Ben Lomond, Big Basin, and Scott Valley. March–April.

83. MALVACEAE. Mallow Family

Carpels with horns. 1. *Modiola*
Carpels lacking horns.
   Bracts lacking. 2. *Sidalcea*
   Bracts present.
      Bracts united basally, the lobes ovate to orbicular; anthers scattered on the tube.
         Petals 1–4 cm. long; axis of the fruit projecting beyond the carpels.
            3. *Lavatera*
         Petals 5–10 cm. long; axis of the fruit not projecting beyond the carpels.
            4. *Althaea*
      Bracts not united but sometimes adnate to the calyx, linear to ovate; anthers near the top of the tube.
      Plants not usually stellate-pubescent. Introduced plants. 5. *Malva*
      Plants densely stellate-pubescent. Native plants.
         Petals glabrous in bud except toward the base, white to pink or rose.
            6. *Malacothamnus*
         Petals stellate-pubescent in bud, usually yellowish. 7. *Sida*

1. Modiola Moench

1. M. caroliniana (L.) G. Don. Wheel Mallow. A weed of disturbed areas, but not common; native in the southeastern portion of the United States, Central America, and South America. April–August.

2. Sidalcea Gray ex Benth. Checker

Plants annual; stipules divided into narrow segments. 1. *S. diploscypha*
Plants perennial; stipules entire.
   Leaves thin, vitiform; plants 1–2 m. tall. 2. *S. malachroides*
   Leaves thick, round in outline or deeply parted; plants usually under 5 dm. tall. 3. *S. malvaeflora*

1. S. diploscypha (T. & G.) Gray ex Benth. *Fig. 143.* Fringed Sidalcea. Occasional on serpentine outcroppings or in shallow clay soils, mainly in the eastern foothills of the Santa Cruz Mountains; near Redwood City, Woodside, Stanford, near San Jose, and near Gilroy. April–May.

2. S. malachroides (H. & A.) Gray. Maple-leaved Sidalcea. Rare locally; known only from near the coast in the vicinity of Santa Cruz. May–June.

3. S. malvaeflora (DC.) Gray ex Benth. Wild Hollyhock, Checker Bloom. A very common perennial herb of open grassy slopes and coastal bluffs. February–June. Two subspecies may be recognized in the Santa Cruz Mountains and may be distinguished by the following key:

Cauline leaves usually dissected into linear segments.

3*a*. *S. malvaeflora* ssp. *laciniata* Hitchc.

Cauline leaves orbicular to reniform, not dissected into linear segments, shallowly 7–9-lobed, often divided ¾–⅘ of their length.

3*b*. *S. malvaeflora* ssp. *malvaeflora*

Numerous intergrades exist between these two subspecies. The former is more common inland, while the latter is more common along the coast.

3. Lavatera L.

Bracts longer than the calyx, broadly ovate to orbicular; leaves densely pubescent on both sides.

1. *L. arborea*

Bracts shorter than the calyx, deltoid to broadly ovate; leaves less pubescent abaxially than adaxially.

Plants annual herbs; petals under 3 cm. long; leaves 3–5-lobed, the margins crenate.

2. *L. cretica*

Plants perennial shrubs; petals over 3 cm. long; leaves 5–7-lobed, the margins coarsely dentate.

3. *L. assurgentiflora*

1. **L. arborea** L. Tree Mallow. Coastal bluffs and sand dunes along the coast and along the margins of San Francisco Bay; San Francisco, Half Moon Bay, Pebble Beach, near Pescadero, and near Palo Alto; native of Europe. May–July.

2. **L. cretica** L. Cretan Lavatera. Becoming common as a weed in disturbed areas, in fallow fields, and along roadsides; San Francisco, between Princeton and Miramar, San Bruno Mountain, Palo Alto, and Watsonville; native of Europe. March–November, but to be expected in flower at all times of the year.

3. **L. assurgentiflora** Kell. Malva Rose. Originally planted along the coast as a windbreak and now becoming established; San Francisco and Palo Alto; native of the California Channel Islands. To be found in flower at all times of the year.

4. Althaea L. Hollyhock

1. **A. rosea** Cav. Hollyhock. Occasional as a weed in disturbed areas and spontaneous in gardens; San Francisco, Stanford, and Palo Alto; native of China. Summer months.

5. Malva L. Mallow, Cheeses

Bracts linear, acute.

Petals barely longer than the calyx; carpels rugose-reticulate.

1. *M. parviflora*

Petals about 2 times as long as the calyx; carpels not reticulate.

2. *M. neglecta*

Bracts lanceolate to broadly ovate.

Petals about 2–3 times as long as the calyx; bracts ovate.

3. *M. nicaeensis*

Petals about 4–5 times as long as the calyx; bracts oblong-lanceolate.

4. *M. sylvestris*

1. **M. parviflora** L. Cheese-weed. A common weed of fields, gardens, orchards, and in disturbed areas; native of Europe. March–October.

238   *HYPERICACEAE*

2. M. neglecta Wallr. Occasional as a weed in disturbed areas in San Francisco; native of Eurasia. April–October.
3. M. nicaeensis All. *Fig. 144.* Bull Mallow. A common weed of disturbed areas, along roads and highways, in fields, and in river and stream beds; native of Europe. April–November.
4. M. sylvestris L. High Mallow. Occasional in vacant lots in San Francisco and perhaps elsewhere; native of Europe. April–October.

6. Malacothamnus Greene

Abaxial leaf-surface densely covered with shaggy-tomentose stellate scales, the
    epidermis usually not visible in mature leaves.   1. *M. arcuatus*
Abaxial leaf-surface covered with stellate scales, the epidermis clearly visible.
                                                        2. *M. hallii*

1. M. arcuatus (Greene) Greene. Northern Malacothamnus. Most common in chaparral or in oak-madrone woods on the eastern slopes of the Santa Cruz Mountains from northern San Mateo County southward; Sawyer Ridge, Crystal Springs Lake, near La Honda, Alviso, near Stanford, Black Mountain, Santa Clara, Loma Prieta, and China Grade. May–September. —*Malvastrum arcuatum* (Greene) Robinson.
2. M. hallii (Eastw.) Kearney. Hall's Malacothamnus. Known locally only from the vicinity of Uvas and Llagas creeks in southern Santa Clara County, often growing with *Salvia mellifera* and *Adenostoma fasciculatum*. June–August. —*Malvastrum hallii* Eastw.

7. Sida L.

1. S. leprosa (Ort.) K. Schum. var. **hederacea** (Dougl. ex Hook.) K. Schum. Alkali Mallow. Occasional in heavy alkaline and adobe soils in the Santa Clara Valley and as a weed in San Francisco. July–October. —*S. hederacea* (Dougl. ex Hook.) Torr.

84. STERCULIACEAE. Cacao or Sterculia Family

1. Fremontodendron Cov. Flannel Bush, Fremontia

1. F. californicum (Torr.) Cov. ssp. **crassifolium** (Eastw.) Thomas. California Fremontia. Known only from three localities in the Santa Cruz Mountains, where it grows on rocky chaparral slopes; Butano Ridge, Big Basin, and Loma Prieta. May–August. —*Fremontia californica* Torr. ssp. *crassifolia* (Eastw.) Abrams.

85. HYPERICACEAE. St. John's Wort Family

1. Hypericum L. St. John's Wort

Petals usually under 1 cm. long; leaves under 3 cm. long.
    Petals usually twice as long as the sepals; stems erect, plants not prostrate
        mats.

Leaves oval, 2–3 cm. long, usually with clasping base; sepals ovate, 3–4 mm.
    long.   1. *H. formosum scouleri*
Leaves oblong, 1–2 cm. long, base not clasping; sepals lanceolate, 5–6 mm.
    long.   2. *H. perforatum*
Petals about equaling the sepals or shorter; stems prostrate or ascending,
    plants often forming dense mats.   3. *H. anagalloides*
Petals 2.5–3.5 cm. long; leaves 4–6 cm. long.   4. *H. calycinum*

1. H. formosum HBK. var. **scouleri** (Hook.) Coult. Scouler's St. John's Wort.
Known only from San Francisco, Loma Prieta, and Boulder Creek, but to be
expected elsewhere in moist soil. June–October. —*H. scouleri* Hook.
2. H. perforatum L. *Fig. 145.* Klamath Weed, Common St. John's Wort. Occasional in disturbed habitats throughout the Santa Cruz Mountains; San Francisco, Crystal Springs Lake, Saratoga, and Ben Lomond Mountain; native of
Europe, now almost cosmopolitan as a weed. June–September.
3. H. anagalloides C. & S. Tinker's Penny, Creeping St. John's Wort. Low, moist
or swampy ground along the coast or occasionally inland; San Francisco, Pebble
Beach, Swanton, Ben Lomond Mountain, Camp Evers, and Santa Cruz. March–
July.
4. H. calycinum L. Aaronsbeard St. John's Wort. Widely planted as an ornamental and becoming established in San Francisco and perhaps elsewhere; native
of the Mediterranean region and Asia Minor.

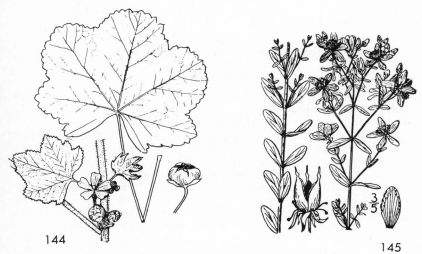

144

145

86.   ELATINACEAE.   Waterwort Family

1. Elatine L. Waterwort

1. E. triandra Schkuhr var. **brachysperma** (Gray) Fassett. Short-seeded Waterwort. Lake margins and vernal pools; Mountain Lake in San Francisco, Pilarcitos Lake, and near Eagle Rock. April–June. —*E. brachysperma* Gray.

240   *VIOLACEAE*

87. FRANKENIACEAE. Frankenia Family

1. Frankenia L.

1. F. grandifolia C. & S. *Fig. 146.* Yerba Reuma, Alkali Heath. Upper limits of salt marshes, on dikes and levees, and on coastal bluffs; San Francisco southward. May–November. Where plants are subjected to the action of waves at high tide, they become somewhat woody, gnarled, and shrubby.

88. CISTACEAE. Rock-Rose Family

1. Helianthemum Mill.

Flowers few; petals 5–7 mm. long; inflorescence very leafy; low and spreading plants, somewhat matted.   1a. *H. scoparium scoparium*
Flowers numerous; petals 4–6 mm. long; inflorescence almost leafless; tall, erect, rush-like plants.   1b. *H. scoparium vulgare*

1a. H. scoparium Nutt. var. **scoparium.** Common Rush Rose, Broom Rose. Mainly on the western slopes of the Santa Cruz Mountains on dry rocky slopes, sandy areas, and sometimes in chaparral; Crystal Springs Lake,.Glenwood, Ben Lomond Sand Hills, and Santa Cruz. April–August. This variety is often a rapid colonizer of dry disturbed areas such as road banks and cuts.
1b. H. scoparium Nutt. var. **vulgare** Jeps. Usually on the eastern slopes of the Santa Cruz Mountains; Jasper Ridge, Black Mountain, Loma Prieta, and Big Basin. April–September.

89. VIOLACEAE. Violet Family

1. Viola L. Violet

Stipules laciniate to lyrate-pinnatifid.   1. *V. tricolor*
Stipules entire or toothed.
    Inner surface of petals white, blue, violet, or purple.
        Inner surface of petals violet-purple; lateral petals with a tuft of elongated hairs at base; spur conspicuous, 4–5 mm. long. Growing on grassy slopes or as an escape from cultivation.
            Stolons lacking; stipules linear-lanceolate. Native plants.   2. *V. adunca*
            Stolons rooting at the nodes; stipules ovate-lanceolate. Introduced plants.
                3. *V. odorata*
        Inner surface of lateral petals purple or white with a small purple spot at the base, the others white; lateral petals with a tuft of clavate hairs at base; spur shallow, scarcely 2–3 mm. long. Growing on wooded slopes.
                4. *V. ocellata*
    Inner surface of petals yellow except for veining.
        Leaves narrowed or truncate at base, rarely subcordate.
            Rootstalks deep-seated; leaves broadly ovate, the bases sometimes cordate; flowers 1–2 cm. long. Growing on open or shaded grassy slopes.
                5. *V. pedunculata*

Rootstalks shallow; upper leaves elongate or ovate-lanceolate, lower ones
    broader; flowers usually 1 cm. or less long. Associated with *Quercus*
    and *Pinus* where these species occur with chaparral.
                                          6. *V. quercetorum*
Leaves distinctly cordate at base.
  Stems creeping, stoloniferous, evergreen. Growing in redwood forests.
                                         7. *V. sempervirens*
  Stems erect from stout horizontal rootstalks, not evergreen. Growing along
    shaded stream banks.   8. *V. glabella*

1. V. tricolor L. Wild Pansy. Occasionally persisting in gardens in San Francisco; native of Europe.
2. V. adunca Smith. Western Dog or Blue Violet. Open fields and hillsides on
the western slopes of the Santa Cruz Mountains from sea level to about 2,000 feet
elevation; San Francisco, Montara Mountain, Pescadero, and Bonny Doon. February–April.
3. V. odorata L. English or Sweet Violet. Common in gardens and occasionally
becoming established; reported from San Francisco and Palo Alto; native of
Eurasia. December–March.
4. V. ocellata T. & G. Two-eyed Violet, Western Heart's Ease. Wooded slopes
throughout the Santa Cruz Mountains; Cahill Ridge, Pescadero Creek, Saratoga
Hills, Los Gatos Canyon, Uvas Creek, Boulder Creek, Felton, and Aptos. March–
June.
5. V. pedunculata T. & G. *Fig. 147.* Johnny-Jump-Up, Yellow Pansy, California
Golden or Yellow Violet. Open grassy hillsides; San Francisco, South San Fran-
cisco, Stanford, Saratoga Hills, San Martin, Swanton, and Santa Cruz. March–
April.

146   147

6. V. quercetorum Baker & Clausen. Mountain Violet. Known locally from only a few localities where it is associated in chaparral with various species of *Quercus*, *Pinus attenuata*, and *P. sabiniana*; Sierra Azule Ridge and Ben Lomond Mountain. March–May. —*V. purpurea* of authors, in part.

7. V. sempervirens Greene. Evergreen or Redwood Violet. Redwoods or redwood–Douglas fir forests. January–October. —*V. sarmentosa* Dougl. ex Hook.

8. V. glabella Nutt. ex T. & G. Stream or Smooth Yellow Violet. Moist stream banks in redwood–Douglas fir forests. March–May.

90. LOASACEAE. Loasa Family

1. **Mentzelia** L. Stick-leaf, Blazing Star

Petals 20–35 mm. long.   1. *M. lindleyi*
Petals less than 5 mm. long.
    Bracts ovate, concealing flowers and lower portions of the capsules; outer filaments dilated; seeds not grooved at the angles.   2. *M. micrantha*
    Bracts linear-lanceolate, not concealing flowers and lower parts of capsules; filaments not dilated; seeds grooved at the angles.   3. *M. dispersa*

1. M. lindleyi T. & G. *Fig. 148.* Lindley's Blazing Star. Occasional in the Santa Clara Valley and adjoining foothills, often in serpentine soil; near Almaden Canyon, 4 miles south of San Jose, Edenvale, and Uvas Creek. April–August.

2. M. micrantha (H. & A.) T. & G. Small-flowered Stick-leaf. The most common species of Mentzelia locally, on dry hillsides and in disturbed areas; Black Mountain, Permanente Canyon, Loma Prieta, Saratoga Summit, and Quail Hollow. April–August.

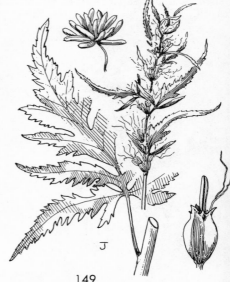

148   149

3. M. dispersa Wats. Nada Stick-leaf. Known locally only from Alma Soda Springs and Loma Prieta Ridge. May–August.

91. DATISCACEAE. Datisca Family

1. Datisca L.

1. D. glomerata (Presl) Baill. *Fig. 149.* Durango Root. Creek banks and stream beds at lower elevations in the Santa Cruz Mountains; San Francisquito Creek, near Stevens Creek, Jamison Creek, and Santa Cruz. May–August.

92. THYMELAEACEAE. Daphne or Mezereum Family

1. Dirca L. Leatherwood

1. D. occidentalis Gray. *Fig. 150.* Western Leatherwood. Moist wooded hillsides, often with *Arbutus menziesii* and *Quercus agrifolia*, sometimes forming thickets; Montara Mountain, San Andreas Lake, Searsville, and foothills near Stanford University. January–March.

93. LYTHRACEAE. Loosestrife Family

Leaves mainly alternate; floral tube cylindric.   1. *Lythrum*
Leaves opposite; floral tube campanulate to globular.   2. *Ammannia*

150

151

244   *MYRTACEAE*

1. Lythrum L. Loosestrife

Petals 4–6 mm. long; flowers pedicellate; seeds linear-lanceolate.
1. *L. californicum*
Petals 1–2.5 mm. long; flowers sessile to subsessile; seeds ovoid.
2. *L. hyssopifolia*

1. L. californicum T. & G. *Fig. 151.* California Loosestrife. Occasional in the Santa Clara Valley from southern San Mateo County southward; Menlo Park, Sunnyvale, and Santa Clara. June–September.
2. L. hyssopifolia L. Grass Poly, Hyssop Loosestrife. Moist places, where water has stood during the winter and early spring; common throughout the Santa Cruz Mountains. April–October.

2. Ammannia L.

1. A. coccinea Rottb. Long-leaved Ammannia. Known definitely only from a pond about 4 miles south of Pescadero, but to be expected elsewhere in similar habitats. May–October.

94. MYRTACEAE. Myrtle Family

Petals and calyx fused to form a calyptra.   1. *Eucalyptus*
Petals and calyx-lobes free.   2. *Leptospermum*

1. Eucalyptus L'Her.

1. E. globulus Labill. Blue Gum. This species has been extensively planted all through the Santa Cruz Mountains to form wind breaks and as an ornamental, and produces seedlings quite commonly; San Francisco, near Burlingame, Stanford, and elsewhere; native of Australia. December–May.

Eucalpytus viminalis Labill., also a native of Australia, is spontaneous in Marin County and is to be expected in the Santa Cruz Mountains. *Eucalyptus viminalis* has smooth capsules in clusters of 3, while those of *E. globulus* are single and wrinkled. The hybid between *E. globulus* and *E. viminalis, E.* × *mortoniana* Kinney, is occasionally spontaneous in Golden Gate Park in San Francisco.

2. Leptospermum Forst.

1. L. laevigatum F. Muell. Australian Tea Tree. Becoming established at several localities in San Francisco, and to 'be expected occasionally elsewhere as an escape from cultivation; native of Australia.

Eugenia apiculata DC. is occasionally spontaneous in Golden Gate Park in San Francisco. It is a native of Chile and may be distinguished from *Eucalyptus* and *Leptospermum* by the presence of berries instead of woody or leathery capsules.

95. ONAGRACEAE. Evening Primrose Family

Capsules indehiscent.
> Petals small, less than 8 mm. long, pink or red; leaves lanceolate, sinuate;
> calyx-lobes not persistent. Plants of dry land.   1. *Gaura*
> Petals large, 10 mm. or more long, yellow; leaves orbicular to ovate; calyx-
> lobes persistent. Aquatic plants.   2. *Jussiaea*

Capsules dehiscent.
> Sepals deciduous after flowering; capsules several times as long as broad.
>> Seeds comose.
>>> Flowers large; petals red; calyx-tube at least 1 cm. long.   3. *Zauschneria*
>>> Flowers small; petals white to lavender; calyx-tubes lacking or very short.
>>>   4. *Epilobium*
>> Seeds not comose.
>>> Anthers basifixed or nearly so; petals white, lavender, or red.
>>>> Calyx-lobes not reflexed; petals not clawed.   5. *Boisduvalia*
>>>> Calyx-lobes reflexed with the tips usually united by 2's or 4's; petals
>>>> usually clawed.   6. *Clarkia*
>>> Anthers versatile; petals usually yellow.   7. *Oenothera*
> Sepals persistent after flowering; capsules to twice as long as broad.
>   8. *Ludwigia*

1. Gaura L.

Plants glabrate; calyx-tube 2.5–3 mm. long; filaments 8–10 mm. long.
                                                            1. *G. sinuata*
Plants appressed-pubescent; calyx-tube 5–8 mm. long; filaments 5–6 mm. long.
                                                            2. *G. odorata*

1. **G. sinuata** Nutt. ex Seringe. Occasional in fields; southern San Mateo County and near Santa Cruz; introduced from Oklahoma, Arkansas, Texas, and northern Mexico. May–June.

2. **G. odorata** Sesse ex Lag. Reported as a pasture weed near Capitola; native of Texas and Mexico.

2. Jussiaea L.

1. **J. repens** L. var. **peploides** (HBK.) Griseb. Yellow Water Weed. Lakes and marshes in San Francisco and southern Santa Cruz counties, but to be expected elsewhere. June–November. —*J. californica* (Wats.) Jeps.

3. Zauschneria Presl. California Fuchsia

1. **Z. californica** Presl. *Fig. 152.* California Fuschia. Dry foothills and mountains throughout the Santa Cruz Mountains from northern San Mateo County southward, usually in poor or rocky soil in chaparral and on serpentine. July–November. The plants of this species in the Santa Cruz Mountains are variable with respect to pubescence and flower size. The pubescence varies from tomentose to slightly glandular-villous and the flowers range from 2 to 4 cm. in length. *Zauschneria californica* is able to invade the rocky soil of some steep highway embankments. Artificial plantings of this attractive, late-flowering perennial might be tried in order to cover some of the ugly scars left on so many of our hills.

152

153

4. Epilobium L. Willow Herb

Plants annual, usually growing in dry ground; epidermis exfoliating readily.
   Plants usually under 2.5 dm. tall, puberulent throughout; calyx-tube about 1
      mm. long.   1. *E. minutum*
   Plants usually over 3 dm. tall, pubescent above; calyx-tube over 2 mm. long.
                                                                      2. *E. paniculatum*
Plants perennial, usually growing in moist ground or in low areas where water
      has stood; epidermis not exfoliating.
   Plants producing turions, under 4 dm. tall, unbranched.   3. *E. halleanum*
   Plants not producing turions, often over 4 dm. tall, usually branched.
      Petals 6–10 mm. long, red to purple; leaves usually opposite.
         Herbage densely pubescent.   4. *E. watsonii*
         Herbage sparsely pubescent.   5. *E. franciscanum*
      Petals 2–6 mm. long, white to pink or rose; upper leaves usually alternate.
         Upper portion of the stem and inflorescence glandular with spreading
            hairs; inflorescence usually dense, with numerous flowers.
                                                          6. *E. adenocaulon occidentale*
         Upper portion of the stem and inflorescence not glandular (occasionally
            slightly so), pubescence of incurved hairs; inflorescence usually open,
            few-flowered.   7. *E. californicum*

1. E. minutum Lindl. ex Hook. Minute Willow Herb. Occasional in chaparral
and in areas where *Pinus attenuata* is common; San Francisco, west of San Mateo, near Woodside, Stevens Creek, west of Los Gatos, Alma, Eagle Rock, Ben
Lomond San Hills, and Loma Prieta. April–May.
2. E. paniculatum Nutt. Panicled Willow Herb. Common throughout the Santa
Cruz Mountains and adjacent lowlands, in fields, rocky ridges, serpentine outcroppings, and in disturbed ground. May–October. *Epilobium paniculatum* is
a variable species, in which several varieties and forms have been described.
The following key will serve as a means of distinguishing the varieties and forms
that occur locally:

Calyx-tube in mature flowers over 6 mm. long.
>2a. *E. paniculatum* var. *jucundum* (Gray) Trel.
Calyx-tube in mature flowers under 6 mm. long.
>Calyx-tube 4 mm. long.   2b. *E. paniculatum* f. *laevicaule* (Rydb.) St. John
>Calyx-tube 2–3 mm. long.
>>Pedicels and capsules somewhat glandular-pubescent.
>>>2c. *E. paniculatum* var. *paniculatum*
>>Pedicels and capsules densely glandular-pubescent.
>>>2d. *E. paniculatum* f. *adenocladon* Hausskn.

Epilobium paniculatum var. *paniculatum* and *E. paniculatum* f. *adenocladon* are the two most common subspecific entities locally. Although this is an annual species, the stems often become very woody toward the base and often between 1 and 1.5 cm. in diameter near the base.
3. E. halleanum Hausskn. Hall's Willow Herb. Known from the bog at Camp Evers. April–May.
4. E. watsonii Barbey. Watson's Willow Herb. Occasional in marshy ground along the coast and at Camp Evers. June–July. Specimens intermediate in character between *E. watsonii* and *E. franciscanum* are occasionally found.
5. E. franciscanum Barbey. *Fig. 153.* San Francisco Willow Herb. Usually along or near the coast in bogs, marshes, and low moist areas; San Francisco, Colma, Pebble Beach, Pigeon Point, San Jose, Swanton, Camp Evers, and La Selva Beach. April–June. —*E. watsonii* Barbey var. *franciscanum* (Barbey) Jeps.
6. E. adenocaulon Hausskn. var. **occidentale** Trel. Northern Willow Herb. Common in low moist areas, where water has stood during the spring, along streams, and in boggy areas; San Francisco, Crystal Springs Reservoir, Searsville, Portola Valley, Stevens Creek, Black Mountain, Loma Prieta, Glenwood, and Santa Cruz. June–October. —*E. californicum* Hausskn. var. *occidentale* (Trel.) Jeps.
7. E. californicum Hausskn. California Willow Herb. Along streams, springs, and in moist shaded ground; San Francisco, Stevens Creek Road, foothills west of Los Gatos, Loma Prieta, Big Basin, and Bracken Brae. May–September. —*E. adenocaulon* Hausskn. var. *parishii* (Trel.) Munz.

5. Boisduvalia Spach

Capsules septicidal, septa separating from the valves of the capsule.
>1. *B. densiflora*
Capsules loculicidal, septa remaining attached to the valves of the capsule.
>Capsules straight, 5–8 mm. long; bracts ovate-oblong; petals 2–4 mm. long.
>>2. *B. glabella*
>Capsules with the tips curved away from the stem, 6–10 mm. long; bracts
>linear; petals 1.5–2 mm. long.   3. *B. stricta*

1. B. densiflora (Lindl.) Wats. Dense-flowered Boisduvalia. Low moist areas in which water stands during the early part of the growing season; Bayview Hills, San Andreas Lake, Crystal Springs Lake, Alviso, Hilton Airport, and Aptos Creek. June–September. A minor variant, *B. densiflora* var. *imbricata* Greene, has the bracts of the inflorescence conspicuously imbricate and concealing the flowers and capsules. In *B. densiflora* var. *densiflora*, the flowers and fruits are not hidden by the bracts.

2. B. glabella (Nutt.) Walp. Smooth Boisduvalia. Low wet meadows and margins of pools in the Santa Clara Valley and bordering foothills; Coal Mine Ridge, Stanford University, and between Gilroy and Morgan Hill. June–September.
3. B. stricta (Gray) Greene. Narrow-leaved Boisduvalia. Known from Page Mill Road, New Almaden, and from near Boulder Creek. May–August.

6. Clarkia Pursh. Clarkia, Godetia

Stamens 4; hypanthium over 10 mm. long.
  Hypanthium about 15 mm. long; petals about twice as long as broad; plants usually over 2 dm. tall.   **1. C. concinna**
  Hypanthium about 25 mm. long; petals about as long as broad; plants usually under 2 dm. tall.   **2. C. breweri**
Stamens 8; hypanthium under 10 mm. long.
  Petals distinctly clawed, the claw equaling the limb.
    Petals 6–12 mm. long, the claw with a pair of teeth near the base, the limb lanceolate to rhomboid, often obscurely 3-lobed; rachis of the inflorescence reflexed. Rare.   **3. C. rhomboidea**
    Petals 10–20 mm. long, the claw lacking teeth, the limb deltoid to rhomboid; rachis of the inflorescence straight. Common.   **4. C. unguiculata**
  Petals not clawed, or the claw very short.
    Rachis of the inflorescence reflexed; corollas 5–12 mm. long.
      Petals white to cream-colored, unmarked, under 1 cm. long; hypanthium 0.5–1.5 mm. long.   **5. C. epilobioides**
      Petals colored, or if pale then flecked with purple, over 1 cm. long; hypanthium 1–3 mm. long.   **6. C. modesta**
    Rachis of the inflorescence straight; petals 5–30 mm. long.
      Leaves ovate, obovate, elliptic, or oblanceolate, usually 2–3 times as long as broad; plants often prostrate to decumbent. Coastal plants.
        Petals 25–30 mm. long; leaves elliptic to broadly lanceolate; capsules usually pedicellate. Common.   **7b. C. rubicunda blasdalei**
        Petals 5–11 mm. long; leaves elliptic, ovate, or oblanceolate; capsules sessile. Rare.   **8. C. davyi**
      Leaves linear to lanceolate, usually 5 or more times as long as broad; plants usually erect. Plants usually inland.
        Capsules always sessile, mature ones 1–2 cm. long, distinctly quadrangular in cross section; immature ones 8-ribbed, less than 8 times as long as broad.
          Petals 15–25 mm. long; stamens shorter than the mature stigma, free from the stigma.   **9a. C. purpurea viminea**
          Petals under 15 mm. long; stamens about equaling the mature stigma, connivent to it.   **9b. C. purpurea quadrivulnera**
        Some capsules almost always pedicellate, the pedicels to 4 cm. long, mature capsules 1.5–4 cm. long, rounded to quadrangular in cross section; immature capsules 10 or more times as long as broad.
          Mature stigmas elevated above the stamens.
            Sepals 12–20 mm. long; hypanthium over 4 mm. long. Common.   **7a. C. rubicunda rubicunda**
            Sepals 4–5 mm. long; hypanthium 1–3 mm. long. Rare, known only from the Presidio in San Francisco.   **10. C. franciscana**
          Mature stigmas not elevated above the stamens.   **11. C. affinis**

1. C. concinna (F. & M.) Greene. Lovely Clarkia. Usually in open woods on the eastern slopes of the Santa Cruz Mountains; San Francisco, Stevens Creek, near Saratoga, Congress Springs, and Loma Prieta. May–June.

2. C. breweri (Gray) Greene. Brewer's Clarkia. Known locally only from Loma Prieta, growing in rocky ground. May–June.

3. C. rhomboidea Dougl. in Hook. Rhomboid Clarkia. Known locally only from near Santa Cruz. May–June.

4. C. unguiculata Lindl. *Fig. 154.* Elegant Clarkia. One of the more common species of Clarkia in the Santa Cruz Mountains from southern San Mateo County southward, usually in open woods or in grasslands; Alpine School, Coal Mine Ridge, near Stanford, Black Mountain, Campbell Creek, Los Gatos Canyon, near San Jose, near Bielawski Lookout, Camp Evers, near Felton, and Santa Cruz. April–August. —*C. elegans* Dougl.

5. C. epilobioides (Nutt.) Nels. & Macbr. Willow Herb Godetia. Rare locally in open woods and along the borders of chaparral; San Francisco and Redwood City. April–May. —*Godetia epilobioides* (Nutt.) Wats.

6. C. modesta Jeps. Modest Clarkia. Rare, known locally only from the vicinity of Los Gatos, associated with oaks and *Pinus sabiniana.* April–July.

7a. C. rubicunda (Lindl.) Lewis & Lewis ssp. **rubicunda.** Farewell-to-Spring. Grasslands and grassy areas in oak woodlands, common from San Francisco southward. April–August.

7b. C. rubicunda (Lindl.) Lewis & Lewis ssp. **blasdalei** (Jeps.) Lewis & Lewis. Blasdale's Godetia. Coastal bluffs; San Francisco, Mussel Rock, near San Gregorio, Pebble Beach, Pigeon Point, Santa Cruz, and Seabright. June–October. —*Godetia blasdalei* Jeps.

8. C. davyi (Jeps.) Lewis & Lewis. Davy's Godetia. Sandy soils of coastal bluffs; San Francisco, Pigeon Point, and near Santa Cruz. May–June. —*Godetia quadrivulnera* (Dougl.) Spach var. *davyi* Jeps.

9a. C. purpurea (Curtis) Nels. & Macbr. ssp. **viminea** (Dougl.) Lewis & Lewis. Large Godetia. Dry open grasslands, San Francisco southward; San Francisco, Pescadero–La Honda Road, Peters Creek, Stanford, near Los Gatos, near Big Basin, and Locatelli Ranch. May–July. —*Godetia viminea* (Dougl.) Spach.

154

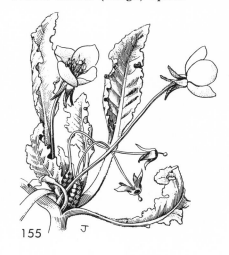

155

9*b*. C. purpurea (Curtis) Nels. & Macbr. ssp. **quadrivulnera** (Dougl.) Lewis & Lewis. Four-spotted Godetia. Grasslands and open wooded areas, common from San Francisco southward. April–July. —*Godetia quadrivulnera* (Dougl.) Spach. Intermediates between the two subspecies of *C. purpurea* are often found.

10. C. franciscana Lewis & Raven. San Francisco Clarkia. Endemic in San Francisco, known only from a serpentine slope in the Presidio. June–July.

11. C. affinis Lewis & Lewis. Open woods, known from the vicinity of Stanford southward, mainly to the east of the crests of the Santa Cruz Mountains; near Stanford, near Cupertino, Stevens Creek, Uvas-Almaden Road, and Mount Charlie Road. April–June.

7. Oenothera L. Evening Primrose

Petals yellow.
  Stigma linearly 4-lobed; petals over 2.5 cm. long.
    Petals 3.5–5 cm. long; cauline leaf-blades at least ⅓ as wide as long and strongly crinkled.   1. *O.* × *erythrosepala*
    Petals 2.5–4 cm. long; cauline leaf-blades usually less than ¼ as wide as long, tending to be plane.
      Tips of the sepals 3–6 mm. long; buds attenuate; sepals 3–3.5 cm. long.
                          2*a*. *O. hookeri hookeri*
      Tips of the sepals 1–2.5 mm. long; buds blunt; sepals mostly 2–2.5 cm. long.   2*b*. *O. hookeri montereyensis*
  Stigma spherical or capitate; petals 2 cm. long or less.
    Ovary partly sterile above, sterile portion 1–12 cm. long; plants caespitose to very short-stemmed.
      Capsule winged, under 1 cm. long; annuals; leaves less than 1 cm. wide, sessile.   3. *O. graciliflora*
      Capsule not winged, over 1 cm. long; perennials; leaves 1–4 cm. wide, distinctly petiolate.   4. *O. ovata*
    Ovary fertile above; plants not caespitose.
      Capsules terete; leaves 1–4 mm. wide, generally linear-oblong.
        Capsules 15–25 mm. long, not beaked; stems densely pubescent with short, appressed or incurved hairs. Plants of coastal sand dunes or inland sand deposits.   5*a*. *O. contorta strigulosa*
        Capsules usually over 25 mm. long; stems glabrous or with spreading pubescence. Inland plants, usually growing in nonsandy soils.
                          5*b*. *O. contorta epilobioides*
      Capsules quadrangular in cross section; leaves 5–15 mm. wide.
        Flowers small, petals 2–4 mm. long.
          Stems semiprostrate; cauline leaves oblong-lanceolate, obtuse, sessile, but not clasping. Plants of sand dunes and sandy soils.
                      6*a*. *O. micrantha micrantha*
         Stems erect or ascending; cauline leaves oblong-ovate to broadly ovate, acute, with subcordate clasping base. Inland plants, usually in nonsandy soils.   6*b*. *O. micrantha jonesii*
        Flowers larger, petals 5–9 mm. long.   7. *O. cheiranthifolia*
Petals rose to purple.   8. *O. rosea*

1. O. × erythrosepala Borb. Lamarck's Evening Primrose. Occasional in the Santa Cruz Mountains, known definitely only from San Francisco; of hybrid origin. June–July.

2a. O. hookeri T. & G. ssp. **hookeri.** Hooker's Evening Primrose. Occasional in moist areas, usually away from the coast; Lake Merced, Colma, Cahill Ridge, Redwood City, Sargents, Alma, and near Big Basin. May–August.

2b. O. hookeri T. & G. ssp. **montereyensis** Munz. Monterey Evening Primrose. Along or near the coast on bluffs, in moist pastures, and in meadows. June–October.

3. O. graciliflora H. & A. Slender-flowered Primrose. Occasional on serpentine outcroppings and on open grassy slopes, eastern foothills of the Santa Cruz Mountains; Spring Valley Lake, near Woodside, Kings Mountain, Stanford, near Los Gatos, San Jose, and San Juan Bautista Hills. February–April.

4. O. ovata Nutt. *Fig. 155.* Suncup, Golden Eggs. A common plant of open grassy slopes and meadows, San Francisco southward. February–June. Young leaves of *O. ovata* make a passable substitute for lettuce, but, according to one Italian resident of Boulder Creek, have a mild physic action.

5a. O. contorta Dougl. ex Hook. var. **strigulosa** (F. & M.) Munz. Contorted Primrose. Common on well-stabilized coastal sand dunes and on inland marine sand deposits; Presidio, Lake Merced, Bonny Doon, Ben Lomond Sand Hills, near Watsonville, and Sunset Beach State Park. March–June.

5b. O. contorta Dougl. ex Hook. var. **epilobioides** (Greene) Munz. Occasional in areas of bare soil, but usually absent along the coast; Stanford, Swanton, Bielawski Lookout, Jamison Creek Road, and Mount Hermon. April–June.

6a. O. micrantha Hornem. ex Spreng. var. **micrantha.** Small Primrose. Mainly on coastal, well-stabilized sand dunes and on inland sand deposits, San Francisco southward. May–June.

6b. O. micrantha Hornem. ex Spreng. var. **jonesii** (Levl.) Munz. Jones' Evening Primrose. More or less bare ground along the borders of chaparral and occasionally in deserted orchards and vineyards; San Jose, Los Gatos, Loma Prieta, Big Basin, Empire Grade, Glenwood, Boulder Creek, and Santa Cruz. April–July.

7. O. cheiranthifolia Hornem. ex Spreng. Beach Primrose. A common plant of coastal sand dunes from San Francisco southward, often growing with *Convolvulus soldanella, Cakile,* and *Abronia.* April–July, with occasional plants to be found with flowers at most times of the year.

8. O. rosea Ait. Rose Sundrops. Occasional as a weed in gardens and as an escape from cultivation, known from San Francisco and Saratoga; native in the southern part of the United States southward into northern South America.

8. Ludwigia L. Marsh Purslane

1. L. palustris (L.) Ell. var. **pacifica** Fern. & Griscom. Pacific Marsh Purslane. Occasional along the edges of ponds and puddles. June–September.

96. HALORAGIDACEAE. Water Milfoil Family

Leaves finely dissected; aerial leaves sometimes entire; stamens 8.
1. *Myriophyllum*
Leaves entire; stamen 1.
2. *Hippuris*

1. Myriophyllum L. Water Milfoil

Flowers borne in the axils of submerged leaves.   1. *M. brasiliense*
Flowers borne above the water in the axils of bracts.
  Bracts entire or toothed.   2. *M. exalbescens*
  Bracts pinnately dissected into linear lobes.   3. *M. verticillatum*

1. M. brasiliense Camb. Parrot Feather. Growing along the edges of lakes in San Francisco and perhaps elsewhere; native of South America. June–August.
2. M. exalbescens Fern. American Milfoil. Lakes, ponds, and reservoirs, often locally abundant; San Francisco southward. July–September. —*M. spicatum* L. ssp. *exalbescens* (Fern.) Hult.
3. M. verticillatum L. Whorl-leaved Milfoil. San Andreas and Upper Crystal Springs Lake and to be expected in other lakes and reservoirs. Probably to be found with flowers or young inflorescences at all times of the year. In 1955, *Myriophyllum verticillatum* was very abundant in San Andreas Lake in a band 10–20 feet wide in water between about 5 and 10 feet deep. Until several years ago, the lake level fluctuated considerably, but during the past several years it has remained fairly constant. It has been during the period of lake level stability that *M. verticillatum* has become abundant. During 1955, a crew of about 10 men kept busy constantly cutting out the plants. Since then, the lake level has been lowered and the infestation has been reduced, at least temporarily.

2. Hippuris L.

1. H. vulgaris L. Mare's Tail. Rare in ponds and pools along creeks in San Francisco and northern San Mateo counties. July–September.

97. ARALIACEAE. Aralia or Ginseng Family

Erect perennial herbs; leaves pinnate.   1. *Aralia*
Vines; leaves simple, palmately lobed.   2. *Hedera*

1. Aralia L. Spikenard

1. A. californica Wats. *Fig. 156.* California Spikenard, Elk Clover. Deeply shaded slopes, along streams, and in dense redwood forests; San Francisco Watershed Reserve, Stevens Creek, Loma Prieta, New Almaden, Alba Road, and near Felton. July–August.

2. Hedera L. Ivy

1. H. helix L. English Ivy. Becoming established in disturbed areas; San Francisco, San Francisco Watershed Reserve, and probably elsewhere; native of Europe. Rarely flowering locally.

98. UMBELLIFERAE. Carrot or Parsley Family

Fruits variously spiny, bristly, or with scales or papillae.
  Bracts of the involucre spine-tipped; fruit scaly.   1. *Eryngium*
  Bracts of the involucre not spine-tipped, or the bracts lacking; fruit not scaly.

Mature fruits linear, at least 4 times as long as broad including the beak.
    Plants annual; beak long.   2. *Scandix*
    Plants perennial; beak short.   3. *Osmorhiza*
Mature fruits not linear, not more than twice as long as broad.
    Umbels sessile or nearly so; plants annual.
        Herbage scabrous; fruit with uncinate bristles and tubercles.
                                              4. *Torilis*
        Herbage glabrous; fruit papillate.   5. *Apiastrum*
    Umbels pedunculate; plants annual or perennial.
        Flowers yellow or purple; plants perennial.   6. *Sanicula*
        Flowers white or pinkish; plants annual (except one species of *Daucus*
           and one species of *Torilis*).
           Involucre lacking, or occasionally of one bract.
                Fruit with a prominent beak.   7. *Anthriscus*
                Fruit not beaked.   4. *Torilis*
           Involucre present; fruits lacking a prominent beak.
                Rays unequal, longer than the bracts.   8. *Caucalis*
                Rays equal, shorter than the bracts.   9. *Daucus*
Fruits not spiny, bristly, or with scales or papillae, often pubescent.
    Fruits not winged, more or less circular in cross section or flattened laterally.
        Leaves simple.
           Plants glabrous or if pubescent, not stellate; perennial, stems creeping.
                                10. *Hydrocotyle*
           Plants stellate-pubescent; annual, stems erect or ascending.
                                11. *Bowlesia*
        Leaves compound.
           Divisions of the leaves linear to filiform.
               Flowers yellow; plants usually over 1.5 m. tall; involucre and involucels
                lacking; leaves with a licorice taste.   12. *Foeniculum*
               Flowers white to pink; plants usually under 1.5 m. tall; involucre and
                involucels present; leaves without licorice taste.   13. *Perideridia*
           Divisions of the leaves lanceolate to ovate.
               Flowers yellow.
                  Stylopodium absent; plants caulescent, perennial. Introduced.
                                14. *Petroselinum*
                  Stylopodium low, conical; plants usually acaulescent. Native.
                                15. *Tauschia*
               Flowers white to pink.
                  Umbels nearly sessile; rays usually longer than the peduncles; plants
                  annual.   16. *Apium*
                Umbels not sessile; peduncles longer than the rays; plants biennial or
                  perennial.
                  Plants of wet habitats; some part of the fruit corky-thickened.
                      Leaves once-pinnate; ribs slender.   17. *Berula*
                      Leaves 2–3-pinnate; ribs thick.
                          Rays over 3 cm. long; leaflets lanceolate.   18. *Cicuta*
                          Rays under 3 cm. long; leaflets ovate.   19. *Oenanthe*
                  Plants of dry habitats; fruits not corky-thickened.
                      Bracts conspicuously divided into filiform segments.   20. *Ammi*

Bracts inconspicuous or lacking, not divided into filiform segments.

Plants 2–3 m. tall; stems with red spots.	21. *Conium*

Plants under 1 m. tall; stems not with red spots.

The ultimate divisions of the upper leaves not markedly narrower than those of the lower leaves; pedicels 2–5 mm. long. Native plants.	22. *Ligusticum*
The ultimate divisions of the upper leaves much narrower than those of the lower leaves; pedicels 5–10 mm. long. Introduced plants.	23. *Coriandrum*

Fruits winged, circular in cross section, more or less flattened dorsally.

Leaves once-pinnate.

Leaflets 3; flowers white.	24. *Heracleum*
Leaflets more than 5; flowers yellow red.	25. *Pastinaca*

Leaves 2–3-pinnate.

Flowers yellow; leaf-divisions usually fine.	26. *Lomatium*

Flowers white; leaf-divisions coarse.

Leaves pubescent.	27. *Angelica*
Leaves glabrous.	28. *Conioselinum*

1. **Eryngium** L. Button Snakeroot

Involucral bracts entire, lacking marginal teeth. Plants mainly of coastal bluffs.	1. *E. armatum*
Involucral bracts not entire, with 2 or more spine-like teeth. Inland plants.	2. *E. aristulatum*

1. E. armatum (Wats.) C. & R. *Fig. 157.* Coast or Prickly Eryngo, Coyote Thistle. Along the coast on coastal bluffs; San Francisco, Pebble Beach, Pigeon Point, Camp Evers, and Santa Cruz. April–June.

2. E. aristulatum Jeps. Jepson's or Ground Eryngo, Coyote Thistle. Areas where water has stood during the spring, most common in the Santa Clara Valley and adjacent foothills; Jasper Ridge, Stanford, Palo Alto, Stevens Creek, and between Gilroy and Morgan Hill. May–October.

156

157

2. Scandix L.

1. S. pectin-veneris L. Venus' or Lady's Comb, Shepherd's Needle. A common weed of gardens, moist disturbed areas, meadows, and fields; native of Eurasia. March–June.

3. Osmorhiza Raf. Sweet Cicely

Involucel lacking.   1. *O. chilensis*

Involucel conspicuous.   2. *O. brachypoda*

1. O. chilensis H. & A. *Fig. 158.* Wood or Mountain Sweet Cicely. Moist woods, usually in deep shade, commonly growing with *Arbutus menziesii*, *Quercus kelloggii*, *Q. wislizenii*, and *Pseudotsuga menziesii*; near La Honda, Kings Mountain, Coal Mine Ridge, Black Mountain Road, near Saratoga, near Wrights, near Los Gatos, Waddell Creek, Swanton, and Glenwood. April–September. —*O. nuda* Torr.

2. O. brachypoda Torr. ex Durand. California Sweet Cicely. Known locally only from Castle Rock Ridge in northern Santa Cruz County, growing in shaded woods, but to be expected elsewhere in similar habitats. May–June.

4. Torilis Adans.

Umbels sessile or nearly so.   1. *T. nodosa*

Umbels long-pedunculate.   2. *T. arvensis*

1. T. nodosa (L.) Gaertn. Knotted Hedge Parsley. Usually in shaded areas, along stream banks, in grassy clearings, and at the edges of woods, most common in the lowlands and foothills on the eastern side of the Santa Cruz Mountains; San Francisco, near Millbrae, Woodside to Crystal Springs, Stanford, San Francisquito Creek, Saratoga, and Saratoga Summit; native of Europe. May–June.

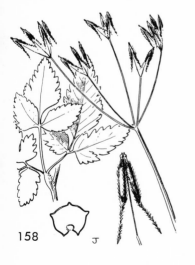

158   159

2. T. arvensis (Huds.) Link. Occasional as a weed in central Santa Cruz County; Jamison Creek; native of Europe. June–July.

5. Apiastrum Nutt.

1. A. angustifolium Nutt. Mock Parsley, Wild Celery. Sandy and rocky soils, on serpentine, open hills, stabilized coastal sand dunes, and in chaparral; San Francisco, Crystal Springs Lake, Woodside, San Francisquito Creek, near Los Gatos, Swanton, near Davenport, and Sunset Beach. February–May.

6. Sanicula L. Snake Root, Sanicle

Basal leaves subentire, lobed or divided or trifoliolate.
  Herbage and bracts yellowish-green; plants more or less prostrate; bracts much longer than the umbels.   1. *S. arctopoides*
  Herbage and bracts bright green; plants erect; bracts shorter than the umbels.
    Leaves subentire to 3-parted, rarely divided further, somewhat succulent; fruit with bristles only above. Plants of coastal sea bluffs in San Francisco.   2. *S. maritima*
    Leaves deeply 3–5-parted, often laciniate, not succulent; fruit with bristles all over. Plants of inland habitats.
      Plants usually over 3 dm. tall; leaf-margins serrate to lobed, not laciniate; plants usually branched above.   3. *S. crassicaulis*
      Plants usually under 3 dm. tall; leaf-margins laciniate; plants usually branched below.   4. *S. laciniata*
Basal leaves pinnatifid or compound.
  Rachis of the leaves toothed.   5. *S. bipinnatifida*
  Rachis of the leaves not toothed.
    Fruit with some uncinate prickles; plants with taproots.   6. *S. bipinnata*
    Fruit lacking uncinate prickles; plants with a globose tuberous root.
                                                                7. *S. tuberosa*

1. S. arctopoides H. & A. Yellow Mats, Footsteps-of-Spring, Snake Root, Bear's Foot Sanicle. Open grassy slopes, windswept summits of coastal hills, and coastal bluffs, often growing with *Ranunculus californicus, Lomatium caruifolium,* and *Sidalcea malvaeflora*; Twin Peaks, Colma, Ocean View, San Bruno Mountain, Montara Mountain, near Pescadero, Sweeney Ridge, Swanton, and Bonny Doon–Davenport Road. March–April.
2. S. maritima Kell. ex Wats. Adobe, Saltmarsh, or Dobie Sanicle. Known only from near the coast in San Francisco, probably now extinct. March–May.
3. S. crassicaulis Poepp. ex DC. *Fig. 159.* Pacific Sanicle, Gamble Weed. Fairly common throughout the Santa Cruz Mountains, usually away from the coast on shaded slopes, commonly under *Quercus agrifolia, Arbutus menziesii, Lithocarpus densiflora,* and *Pseudotsuga menziesii.* February–June. —*S. menziesii* H. & A.
4. S. laciniata H. & A. Coast Sanicle. Occasional in San Mateo and northern Santa Clara counties in oak-madrone woods and along the edges of chaparral; Cahill Ridge, San Francisquito ridges, and probably elsewhere. April–May. —*S. laciniata* H. & A. var. *serpentina* (Elmer) Jeps.

5. S. bipinnatifida Dougl. ex Hook. Purple Sanicle. Common in oak-grasslands, open slopes, serpentine areas, and inland sand deposits, San Francisco southward. February–May.

6. S. bipinnata H. & A. Poison Sanicle. Foothills of the Santa Cruz Mountains along their eastern slope in Santa Clara County, growing in oak-grasslands; Stanford, near Stanford, near Los Gatos, and slope of Loma Prieta. March–April.

7. S. tuberosa Torr. Turkey Pea, Tuberous Sanicle. Rare, known definitely only from San Francisco and the Woodside serpentine. March–May.

7. Anthriscus Hoffm.

1. A. scandicina (Weber) Mansf. Bur Chervil. Moist areas along stream banks, grasslands, in oak-madrone woods, on sand hills, and in disturbed areas; San Francisco, Crystal Springs Lake, Sawyer Ridge, Kings Mountain, Coal Mine Ridge, Stanford, near Saratoga, near Los Gatos, near Llagas Post Office, Swanton, and Ben Lomond Sand Hills; native of Europe. April–June. —*A. neglecta* Boiss. & Reuter var. *scandix* (Scop.) Hyl., *A. vulgaris* Pers.

8. Caucalis L.

1. C. microcarpa H. & A. California Hedge Parsely. Borders of chaparral, open slopes, and in oak-madrone woods; San Francisco, Sawyer Ridge, near San Mateo, Coal Mine Ridge, Black Mountain, near Los Gatos, and 7 miles north of Davenport. April–June.

9. Daucus L.

Plants lacking a fleshy taproot; annuals; divisions of the bracts of the involucre oblong; plants usually under 0.5 m. tall.   1. *D. pusillus*
Plants with a fleshy taproot; biennials; divisions of the bracts of the involucre linear; plants usually over 0.5 dm. tall.   2. *D. carota*

1. D. pusillus Michx. Rattlesnake Weed. Fairly common on grassy slopes, rocky and serpentine areas, and on coastal bluffs, San Francisco southward; San Francisco, Sawyer Ridge, San Mateo Creek, Woodside, Stanford, near Almaden Canyon, Mount Charlie Road, Santa Cruz, and Sunset Beach. April–June. Plants from along the coast, from coastal bluffs, and inland marine sand deposits are usually stouter and have more compact inflorescences.

2. D. carota L. Carrot, Queen Anne's Lace. Occasional as a plant of disturbed areas, roadsides, and coastal bluffs; San Francisco, near San Antonio Creek, Alviso, Wrights–Loma Prieta Road, and Santa Cruz Point; native of Europe. May–November.

10. Hydrocotyle L. Marsh Pennywort

Leaves peltate.   1. *H. verticillata*
Leaves not peltate, but reniform.
  Terrestrial; fruit sessile; adaxial leaf-surface with long trichomes on the veins.
     2. *H. sibthorpioides*
  Aquatic; fruit pedicellate; leaves glabrous.   3. *H. ranunculoides*

1. H. verticillata Thunb. *Fig. 160.* Spike or Whorled Marsh Pennywort. Muddy ground along streams, ponds, and marshes; San Francisco, Crystal Springs Lake, Boulder Creek, Camp Evers, Santa Cruz, and Watsonville. June–September. Plants with peduncles equaling the leaves and with long pedicels have been called *H. verticillata* Thunb. var. *triradiata* (Rich.) Fern. *—H. prolifera* Kell.

2. H. sibthorpioides Lam. Growing as a lawn weed in San Mateo and to be expected occasionally elsewhere; native of Asia and Africa. April–May.

3. H. ranunculoides L. f. Floating Marsh Pennywort. Growing in soggy humus soil at the edges of ponds and lakes and in fresh-water marshes; San Francisco, Coal Mine Ridge, Corte Madera, near Alviso, Camp Evers, Santa Cruz, and Watsonville Slough. February–October.

11. Bowlesia R. & P.

1. B. incana R. & P. Occasional in moist, open, grassy areas, in oak-madrone woods, and on brush-covered slopes, commonly growing with *Nemophila* and *Montia*; San Francisco, San Francisco Watershed Reserve, La Honda, Purisima Creek, Stanford, and Swanton. March–May. *—B. lobata* R. & P.

12. Foeniculum Adans.

1. F. vulgare Mill. Sweet Fennel. A common weed of disturbed areas, along fences and roads, and occasionally as a weed in gardens; native of the Mediterranean region. May–October.

13. Perideridia Reichenb.

Plants with a fleshy tuberous root; styles long, slender.   1. *P. gairdneri*
Plants with a number of woody roots arranged in a cluster; styles short, the base
     thick and conical.   2. *P. kelloggii*

1. P. gairdneri (H. & A.) Mathias. Gairdner's Yampah, Squaw Potato. Occasional in open grassy areas and in moist flats along the coast; near Pebble Beach, Año Nuevo Point, Palo Alto, Big Basin, and Santa Cruz. June–September.

2. P. kelloggii (Gray) Mathias. Kellogg's Yampah, Dobie Spindleroot. Occasional in grasslands and on wooded or brush-covered slopes; San Francisco, Belmont, near Crystal Springs Lake, near Los Gatos, Empire Grade, and near Aptos Creek. June–August. *—Carum kelloggii* (Gray) C. & R.

14. Petroselinum Hoffm.

1. P. crispum (Mill.) Mansfield. Parsley. Occasional as an escape from cultivation; San Francisco and near Boulder Creek; native of Europe. June–August.

15. Tauschia Schlecht.

Bracts of the umbellets conspicuous; fruit about 4–7 mm. long.   1. *T. hartwegii*
Bracts of the umbellets inconspicuous; fruit about 3–5 mm. long.

               2. *T. kelloggii*

1. T. hartwegii (Gray) Macbr. *Fig. 161.* Hartweg's Tauschia. Occasional in northern Santa Clara County, usually on wooded slopes and to be expected farther south. April–May. —*Velaea hartwegii* (Gray) C. & R.

2. T. kelloggii (Gray) Macbr. Kellogg's Tauschia. Known from San Francisco and San Mateo counties, commonly in grassland, along the edges of chaparral, and on wooded slopes; San Francisco, Pilarcitos Canyon, Crystal Springs Lake, San Mateo Ravine, and Calero Valley. March–May. —*Velaea kelloggii* (Gray) C. & R.

16. Apium L.

1. A. graveolens L. Celery, Smallage. Usually in moist heavy soils and in salt marshes; San Francisco, Palo Alto, Mountain View–Alviso Road, San Antonio Creek, and Santa Cruz; native of Europe. July–October.

17. Berula Hoffm. ex Bess.

1. B. erecta (Huds.) Cov. Cut-leaved Water Parsnip. Along the edges of lakes and ponds; San Francisco, San Francisco Watershed Reserve, and Stanford. July–August.

18. Cicuta L. Water Hemlock

1. C. douglasii (DC.) C. & R. Douglas or Western Water Hemlock. Occasional along sloughs, creeks, and marshes; San Francisco, near Mountain View, Glenarbor, Kings Creek, Gazos Creek, and Watsonville. June–August.

Cicuta bolanderi Wats., which occurs in salt marshes from central to southern California, is to be expected locally. It differs from *C. douglasii* in having very oily fruits and intercostal intervals equaling or wider than the ribs, whereas *C. douglasii* has fairly non-oily fruits and intercostal spaces narrower than the ribs.

19. Oenanthe L.

1. O. sarmentosa Presl. Pacific or American Oenanthe. Fairly common along or near the coast, along San Francisco Bay, and occasionally inland in marshes; San Francisco, near San Bruno, Baden Brook, Alpine School, Searsville Lake, Alviso, near Mountain View, Swanton, Davenport Landing, Camp Evers, Santa Cruz, and Watsonville. May–October.

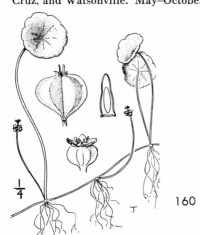

$\frac{1}{4}$ 160

$\frac{1}{3}$ 161

20. Ammi L.

1. A. visnaga (L.) Lam. Toothpick Ammi, Bishop's Weed. Known locally only from the vicinity of Saratoga, but probably to be expected elsewhere in the Santa Clara Valley; native of Eurasia. July–September.

21. Conium L.

1. C. maculatum L. Poison Hemlock. A common weed of fields, disturbed areas, and along creeks; native of Eurasia and North Africa. April–August.

22. Ligusticum L. Lovage

1. L. apiifolium (Nutt.) Gray. Wood, Celery-leaved, or Pacific Lovage. Wind-swept grassy summits and among shrubby vegetation in exposed areas; Mission Hills, Twin Peaks, near Ocean View, San Bruno Mountain, and San Mateo. April–June.

23. Coriandrum L.

1. C. sativum L. Coriander. Occasional as an escape from cultivation; San Francisco, Stanford, and probably elsewhere; native of southern Europe.

24. Heracleum L.

1. H. maximum Bartr. *Fig. 162.* Cow Parsnip. Common on moist slopes, wooded slopes, at the edges of brush, and on coastal bluffs; San Francisco, Pilarcitos Creek Dam, Coal Mine Ridge, San Francisquito Creek, Page Mill Road, San Antonio Creek, near Saratoga Summit, and near Olympia. April–July. —*H. lanatum* Michx.

25. Pastinaca L.

1. P. sativa L. Parsnip. Occasional in bottom lands as an escape from cultivation; San Francisco, near Mountain View, and Boulder Creek; native of Europe. April–June.

26. Lomatium Raf. Hog Fennel

Plants usually over 3 dm. tall; leaf-divisions coarse; fruits glabrous.
  Fruits elliptic, not notched at the base and the apex, wings narrower than the body of the fruit. 1. *L. californicum*
  Fruits nearly orbicular, notched at the base and the apex, wings about twice as broad as the body of the fruit. 2. *L. parvifolium*
Plants usually under 3 dm. tall; leaf-divisions fine; fruits glabrous or pubescent.
  Bracts of the involucel obovate.
    Petiolules of the primary leaflets shorter than the compound blade; cauline leaves usually several; wings usually broader than the body of the fruit. 3. *L. utriculatum*
    Petiolules of the primary leaflets longer than or as long as the compound blade; cauline leaves usually 1 (occasionally more); wings usually narrower than the body of the fruit. 4. *L. caruifolium*
  Bracts of the involucel not obovate, usually oblong.
    Petals glabrous; fruit oblong. 5. *L. macrocarpum*
    Petals pubescent; fruit nearly orbicular. 6. *L. dasycarpum*

1. L. californicum (Nutt.) Math. & Const. California Lomatium, Chu-Chu-Pate. Occasional, usually on shaded slopes; Crystal Springs Reservoir and Permanente Creek. April–May. —*Leptotaenia californica* Nutt.

2. L. parvifolium (H. & A.) Jeps. Coast Parsnip. Known only from southern Santa Clara County near Gilroy. April–May.

3. L. utriculatum (Nutt.) C. & R. Common Lomatium, Bladder Parsnip. Open grassy slopes and ridges, occasionally on serpentine; San Francisco, Crystal Springs Lake, Alpine Grade, Coal Mine Ridge, Stanford, Almaden Canyon, and near Saratoga Summit. March–April.

4. L. caruifolium (H. & A.) C. & R. *Fig. 163.* Caraway-leaved Lomatium, Alkali Parsnip. Grassy slopes and fields, often on windswept summits, less frequent in low heavy soil in the Santa Clara Valley; San Francisco, San Bruno Hills, near Brisbane, Crystal Springs Lake, near San Mateo, Black Mountain, and near San Jose. March–April.

5. L. macrocarpum (H. & A.) C. & R. Large-fruited Lomatium, Sheep Parsnip. Commonly growing on open serpentine outcroppings, northern San Mateo County southward along the eastern slopes of the Santa Cruz Mountains; Crystal Springs Lake, San Mateo Ravine, near Woodside, Stanford, Black Mountain, and Guadalupe Hills. April–May.

6. L. dasycarpum (T. & G.) C. & R. Woolly-fruited Lomatium, Lace Parsnip. Common in rocky soils of ridges, serpentine outcroppings, and chaparral, San Francisco southward. February–May.

27. Angelica L.

Leaves densely tomentose on the abaxial surface. Plants of coastal bluffs.
                                                        1. *A. hendersonii*
Leaves villous to scabrous on the abaxial surface. Inland plants. 2. *A. tomentosa*

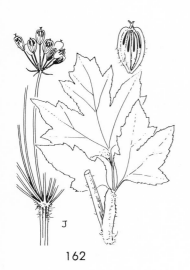

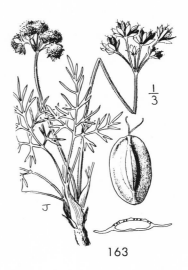

162   163

1. A. hendersonii C. & R. Coast or Henderson's Angelica. Coastal bluffs in San Francisco and along the coast in San Mateo County; San Francisco, Mussel Rock, and near Pescadero. April–June.

2. A. tomentosa Wats. Wood or California Angelica. Brush-covered slopes, occasionally to be found from northern San Mateo County southward; Crystal Springs Lake, Stanford, Black Mountain, Stevens Creek, near Raymonds, and Santa Cruz. July–October.

28. Conioselinum Hoffm.

1. C. chinense (L.) BSP. Hemlock Parsley. Reported from Mission Hills in San Francisco by E. L. Greene.

99. CORNACEAE. Dogwood Family

*opposite leaves
red stems*

1. Cornus L. Dogwood.

Inflorescence cymose; large, conspicuous white bracts absent; shrubs.
　　Branches of the inflorescence glabrous or with a few appressed hairs; petals strap-shaped; leaves with few appressed hairs on the abaxial surface, veins not prominent; drupes white. 　　　　　　　　　1. *C. glabrata*
　　Branches of the inflorescence pubescent, either with appressed or spreading hairs; petals ovate-oblong; leaves with spreading hairs on the abaxial surface, veins prominent; drupes bluish. 　　　　　　2. *C. californica*
Inflorescence capitate; large, conspicuous white bracts present; trees.
　　　　　　　　　　　　　　　　　　　　　　　　　3. *C. nuttallii*

1. C. glabrata Benth. Brown or Smooth Dogwood. Along or near streams on the eastern slopes of the Santa Cruz Mountains at low elevations; Crystal Springs, Los Trancos Creek, Black Mountain, Page Mill Road, and Los Altos. March–June.

2. C. californica C. A. Mey. *Fig. 164.* Creek or Western Red Dogwood. Along streams and moist shaded slopes throughout the Santa Cruz Mountains from near sea level to about 2500 feet; San Francisco, Pescadero Creek, San Mateo Creek, Loma Prieta, New Almaden, Waddell Creek, and Zayante Creek. April–June.

3. C. nuttallii Audubon ex T. & G. Mountain Dogwood. Known definitely from one locality between Los Gatos and Santa Cruz and reported from San Francisco. April–June.

100. GARRYACEAE. Silk Tassel Family

1. Garrya Dougl.

Leaves densely tomentose beneath, margins undulate. 　　　　1. *G. elliptica*
Leaves glabrous or somewhat sparingly pubescent, blades flat. 　　2. *G. fremontii*

164

165

1. G. elliptica Dougl. *Fig. 165.* Coast Silk Tassel, Silk Tassel Bush, Quinine Bush. Occasional in chaparral; San Francisco, Cahill Ridge, Portola, Coal Mine Ridge, Black Mountain, Los Altos, near Watsonville, and Eagle Rock. December– April.

2. G. fremontii Torr. Bear Brush, Fremont's Silk Tassel. Chaparral on the eastern slope of Loma Prieta, usually above 3000 feet elevation; Loma Prieta, Loma Prieta Ridge, and Mount Umunhum. November–April.

101. PYROLACEAE. Wintergreen Family

1. Pyrola L. Wintergreen

1. P. picta Smith f. **aphylla** (Smith) Camp. *Fig. 166.* Leafless Pyrola, Red Canker. Occasional in San Mateo and Santa Cruz counties, usually growing with *Arbutus menziesii, Pseudotsuga menziesii,* and *Lithocarpus densiflora*; Pescadero, Butano Ridge, Big Basin, and Peavine Creek Canyon. June–July. Plants of this species are usually leafless, but individuals are occasionally found with small leaves.

102. MONOTROPACEAE. Indian Pipe Family

Corolla sympetalous; flower-parts pubescent.   1. *Hemitomes*
Petals united only at the base; flower-parts glabrous.   2. *Pleuricospora*

1. Hemitomes Gray

1. H. congestum Gray. Gnome Plant, Hemitomes. Occasional in redwood– Douglas fir forests; Butano Ridge, Big Basin, Boulder Creek, and Brookdale. May–June. —*Newberrya congesta* (Gray) Torr.

2. Pleuricospora Gray

1. P. fimbriolata Gray. Sierra Sap, Fringed Pine-sap. Rare in the redwoods in Santa Cruz Mountains; between Pescadero Creek and Butano Creek, Big Basin, Waddell Creek, and near Felton. June.

103. ERICACEAE. Heath Family

Corolla of separate petals; leaves resinous-dotted abaxially.   1. *Ledum*
Corolla of united petals; leaves not resinous-dotted.
    Corolla funnelform to campanulate, over 3 cm. long.   2. *Rhododendron*
    Corolla urceolate, 1 cm. long or less.
        Fruit a capsule; calyx becoming fleshy; small, weak-stemmed shrubs, usually
            in moist forests.   3. *Gaultheria*
        Fruit a berry or drupe-like; calyx not becoming fleshy; trees and shrubs with
            hard stems and branches.
            Cells of the ovary many-seeded; surface of ovary granular; leaves usually
                over 6 cm. long; large trees.   4. *Arbutus*
            Cells of the ovary with a single seed; surface of the ovary glabrous or
                glandular, not granular; leaves usually less than 6 cm. long; shrubs,
                rarely small trees.   5. *Arctostaphylos*

1. Ledum L.

1. L. glandulosum Nutt. ssp. **columbianum** (Piper) Hitchc. var. **australe** Hitchc.
Coastal Labrador Tea. Known only from the vicinity of Boulder Creek. July.

2. Rhododendron L.

Leaves thick, coriaceous, evergreen, brown abaxially; stamens 10, not exserted.
                                             1. *R. macrophyllum*
Leaves thin, deciduous, green abaxially; stamens 5, exserted.   2. *R. occidentale*

1. R. macrophyllum D. Don ex G. Don. California Rose Bay, California Rhododendron. Occasional in southern San Mateo and Santa Cruz counties, usually
near the coast or occasionally inland; Swanton, Big Basin, Pine Mountain, Bonny
Doon, and Scott Valley. April–July. —*R. californicum* Hook.
2. R. occidentale (T. & G.) Gray. *Fig. 167.* Western Azalea. Along creeks and
about the edges of moist meadows; Pescadero Creek, Butano Creek, Swanton,
Big Basin, Hilton Airport, and Santa Cruz. May–July. —*Azalea californica* T.
& G.

166   167

3. Gaultheria L.

1. G. shallon Pursh. *Fig. 168.* Salal. Occasional in the understory of redwood–Douglas fir forests and less frequent in broad-leaved forests; San Francisco. Kings Mountain, Pescadero Creek, Swanton, San Vicente Creek and Big Trees. April–July.

4. Arbutus L.

1. A. menziesii Pursh. *Fig. 169.* Madroño, Madrone. Wooded slopes throughout the Santa Cruz Mountains, most commonly associated with *Lithocarpus densiflora, Pseudotsuga menziesii,* and *Quercus* spp., but occasionally forming pure but small stands; San Francisco, Woodside, Page Mill Road, Morgan Hill, near Glenwood, Big Basin, near Hilton Airport, and near Felton. March–May. *Arbutus menziesii* is one of the handsomest trees in the Coast Ranges. The white flowers begin to appear in March. By August they have been replaced by clusters of bright red berries. In early summer the old layers of bark peel, leaving the trunk a pale-green color that soon changes through successive shades to a rich reddish-brown. Only on old trunks and branches do the layers of bark remain for more than a year or two.

5. Arctostaphylos Adans. Manzanita

Arctostaphylos is a difficult genus taxonomically and one that has evolved rapidly during the Pliocene and Pleistocene; it is probably still undergoing rapid evolution, which has resulted in many small isolated populations that are morphologically distinct, for example *A. silvicola* and *A. glutinosa*, and in widespread variable populations, as for example *A. crustacea* and *A. glandulosa*. In order to determine a specimen of *Arctostaphylos* satisfactorily, it is essential that adequate field notes be taken. These should include the following: presence or absence of a burl; height of the shrub and general habit, whether prostrate, procumbent, or erect; pubescence, bloom, and/or viscidness of the leaves, twigs, young ovaries, and mature fruits; and the position of the nascent inflorescences, whether erect or drooping.

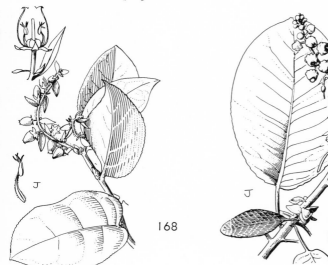

168

169

Flowers 4-merous; leaves 12–20 mm. long, glossy on both surfaces; erect shrubs, 0.5–1.5 m. tall.   1. *A. sensitiva*
Flowers 5-merous.
  Leaves, petioles, and young branches glaucous; nutlets united into one stone; shrubs or small trees, 2–3 m. tall. Eastern slopes of the Santa Cruz Mountains from the vicinity of Los Gatos south to Loma Prieta.   2. *A. glauca*
  Leaves, petioles, and young branches not glaucous; nutlets separable.
    Low, procumbent or prostrate shrubs. San Francisco, San Bruno Mountain, and near Watsonville.
      Leaves essentially elliptic, 1–2.5 cm. long, cuneate or rounded at the base, petiolate.
        Racemes few. San Francisco.   3. *A. franciscana*
        Racemes numerous. Near Watsonville.   4. *A. hookeri*
      Leaves oblong-elliptic, 2.5–3 cm. long, deeply cordate at the base, sessile and clasping. San Bruno Hills.   5*b. A. andersonii imbricata*
    Erect shrubs.
      Leaves auriculate or deeply cordate at the base, sessile, clasping; plants not stump-sprouting, killed by fire.
        Leaves nearly glabrous adaxially; nascent inflorescences more or less erect.
          Leaves 4–6 cm. long. Crystal Springs southward.
                5*a. A. andersonii andersonii*
          Leaves 2.5–3 cm. long. San Bruno Hills.   5*b. A. andersonii imbricata*
        Leaves canescent adaxially; nascent inflorescences drooping. Northwestern Santa Cruz County.   6. *A. glutinosa*
      Leaves usually acute, rounded, truncate, or subcordate at the base (if auriculate or deeply cordate then not sessile and clasping).
        Plants not stump-sprouting, killed by fire.
          Young twigs with long or short, bristly hairs. Northern Santa Cruz County, growing on Monterey shale.   6. *A. glutinosa*
          Young twigs canescent, lacking bristly hairs.
            Leaves oblong, ovate to oval, rounded to auriculate at the base.
              Leaves rounded at the base. Eastern slopes of the Santa Cruz Mountains near Loma Prieta.   7. *A. canescens*
              Leaves auriculate to deeply cordate at the base. Northern Santa Cruz County, growing on Monterey shale.   6. *A. glutinosa*
            Leaves lanceolate to ovate-lanceolate, usually cuneate at the base. Marine sand deposits, Santa Cruz County.   8. *A. silvicola*
        Plants sprouting from an enlarged burl, not killed by fire.
          Stomata present on both leaf-surfaces in approximately equal numbers; leaf-bases rounded to truncate, rarely subcordate; young twigs usually with long glandular hairs.   9. *A. glandulosa*
          Stomata restricted to the abaxial leaf-surfaces; leaf-bases rounded, truncate, to subcordate; young twigs not glandular.
            Young twigs almost always densely setose-bristly, the individual trichomes 1–3 times as long as the diameter of the twig; shrubs 2–3 m. tall. San Mateo County southward.
                10*a. A. crustacea crustacea*
            Young twigs not setose-bristly; shrubs usually under 1 m. tall. Lake Merced, San Francisco.   10*b. A. crustacea rosei*

1. A. sensitiva Jeps. Sensitive Manzanita. Sparse chaparral, dry open ridges, and inland marine sand deposits, commonly growing with *Pinus attenuata, Adenostoma fasciculatum, Dendromecon rigida,* and other species of *Arctostaphylos*; Cahill Ridge, near Crystal Springs Reservoir, Butano Ridge, Big Basin, Pine Mountain, Bonny Doon, near Felton, and Mount Hermon. January–May. —*A. nummularia* Gray var. *sensitiva* (Jeps.) McMinn.

2. A. glauca Lindl. Big-berried Manzanita. Chaparral on the eastern slopes of the Santa Cruz Mountains from Los Gatos southward, often forming dense, nearly pure stands; Los Gatos, Mount Umunhum, Loma Prieta, Uvas Creek, and near New Almaden. January–February.

3. A. franciscana Eastw. San Francisco Manzanita. Serpentine and rocky outcroppings, endemic in San Francisco County, and probably now extinct. December–April. —*A. hookeri* G. Don ssp. *franciscana* (Eastw.) Munz.

4. A. hookeri G. Don. Monterey or Hooker's Manzanita. Known from the sand hills along the coast near Watsonville. March–April.

5a. A. andersonii Gray var. **andersonii.** *Fig. 170.* Heart-leaved or Santa Cruz Mountains Manzanita. A common endemic species of *Arctostaphylos* from Crystal Springs Lakes southward along the crests of the Santa Cruz Mountains, usually above 1,000 feet in elevation; Crystal Springs Lakes, Pescadero Creek, Butano Ridge, Woodside, Kings Mountain, Empire Grade, Alba Road, Big Basin, Aptos, Mount Madonna Road, and near Mount Madonna. November–March.

5b. A. andersonii Gray var. **imbricata** (Eastw.) Adams ex McMinn. A local endemic variety found only on the windswept summits of the San Bruno Hills. February–March.

6. A. glutinosa Schreiber. Glutinous Manzanita. Known only from a small area in Santa Cruz County on the ridge between Scott and Mill creeks, southwest of Eagle Rock, growing in chaparral on Monterey shale with *Pinus attenuata, Adenostoma fasciculatum, Dendromecon rigida, Quercus chrysolepis,* and *A. glandulosa*; endemic. March–April.

7. A. canescens Eastw. Hoary Manzanita. Chaparral along the crests of the Santa Cruz Mountains from Mount Umunhum to Loma Prieta. February–April.

8. A. silvicola Jeps. & Wiesl. Silver-leaved Manzanita. Marine sand deposits in Santa Cruz County, occasionally elsewhere, growing with *Pinus ponderosa, P. attenuata, Adenostoma fasciculatum, Ceanothus* ssp., *Eriodictyon californicum, Dendromecon rigida,* and *Quercus agrifolia*; endemic. November–March.

9. A. glandulosa Eastw. Eastwood's Manzanita. Occasional in chaparral; Mount Umunhum, Loma Prieta, and near Eagle Rock. February–April.

170

171

10a. A. crustacea Eastw. var. **crustacea.** Brittle-leaved Manzanita. Common in chaparral from San Francisco southward; San Francisco, near San Andreas Lake, Kings Mountain, near Searsville, Black Mountain, Castle Rock, near Los Gatos, Loma Prieta, Big Basin, Bonny Doon, Graham Hill Road, Aptos, Larkin Valley, and Watsonville. February–May. Plants with densely white-tomentose abaxial leaf surfaces have been called *A. crustacea* Eastw. var. *tomentosiformis* Adams. Numerous intergrades occur between this variety and the typical variety, which has glabrous to pubescent abaxial leaf surfaces.

10b. A. crustacea Eastw. var. **rosei** (Eastw.) McMinn. Rose's Manzanita. Known only from the vicinity of Lake Merced in San Francisco, where it is endemic. February–April.

104. VACCINIACEAE. Huckleberry Family

1. Vaccinium L.

Leaves thick, leathery, margins dentate; filaments hairy. Common.   1. *V. ovatum*
Leaves thin, margins entire; filaments glabrous. Rare.   2. *V. parvifolium*

1. V. ovatum Pursh. *Fig. 171.* Evergreen or Short Huckleberry. Throughout the mountainous area of the Santa Cruz Peninsula; San Francisco, San Bruno Hills, Pescadero Creek, Summit Springs, Chalks, Big Basin, and Santa Cruz. February–June. *Vaccinium ovatum* occupies a wide range of habitats in the Santa Cruz Mountains. Most commonly it is the conspicuous shrubby species in the understory of the redwood–Douglas fir forest. In some localities, as at Pine Mountain and the Chalks, it is a part of the chaparral, but the individual plants are often no more than one-half meter tall, compared with a height of two meters or more in the forests. Plants from the chaparral also have smaller leaves. The fruit is edible and heavy crops are not unusual. A minor variant of *V. ovatum*, which has glaucous berries, has been called *V. ovatum* Pursh var. *saporosum* Jeps. and ranges with the typical variety, *V. ovatum* var. *ovatum*.

2. V. parvifolium Smith. Red Bilberry or Red Huckleberry. Redwood–Douglas fir forests in southern San Mateo and northern Santa Cruz counties; Pescadero Creek, Butano Creek, Gazos Creek, and Big Basin. March–June.

105. PRIMULACEAE. Primrose Family

Ovary superior, free.
  Stems elongate; leaves not in a basal rosette.
    Leaves borne in a whorl at the summit of the stem, leaves 4 cm. or more long;
      plants with tuberous rootstocks.   1. *Trientalis*
    Leaves not whorled, under 2 cm. long; plants lacking tuberous rootstocks.
      Corollas lacking; capsules valvate; leaves fleshy.   2. *Glaux*
      Corollas present; capsules circumscissile; leaves not fleshy.
        Leaves opposite; corollas conspicuous.   3. *Anagallis*
        Leaves alternate; corollas minute.   4. *Centunculus*
  Stems short; leaves in a basal rosette.
    Corolla-lobes not reflexed at anthesis, shorter than the calyx; scapes under
      4 cm. tall, slender.   5. *Androsace*
    Corolla-lobes reflexed at anthesis, longer than the calyx; scapes over 5 cm.
      long, stout.   6. *Dodecatheon*
Ovary half inferior.   7. *Samolus*

1. Trientalis L. Starflower

1. T. latifolia Hook. *Fig. 172.* Pacific Starflower. Shaded slopes, moist woods, and occasionally in the open; Crystal Springs, Pescadero Creek, Coal Mine Ridge, Swanton, Blooms Grade, Loma Prieta School, and Soquel Gulch. March–June. —*T. europaea* L. var. *latifolia* (Hook.) Torr.

2. Glaux L.

1. G. maritima L. Sea Milkwort or Black Saltwort. Known from specimens collected in San Francisco and Palo Alto, but to be expected elsewhere in salt marshes. May–June.

3. Anagallis L. Pimpernel

1. A. arvensis L. Scarlet Pimpernel. Common in disturbed areas, overgrazed pastures, gardens, and coastal sand dunes from San Francisco southward; native of Europe. Plants with flowers can be found at most times of the year. A blue-flowered form, *Anagallis arvensis* L. f. *azurea* Hylander, has been collected in San Francisco and is to be expected occasionally elsewhere. —*A. arvensis* L. f. *caerulea* (Schreb.) Baumg.

4. Centunculus L.

1. C. minimus L. Chaffweed or False Pimpernel. Rare locally, but to be expected occasionally in wet grassy flats and along the margins of ponds; San Francisco, Año Nuevo Point, and Basin Way. April–July. —*Anagallis minima* (L.) Krause.

5. Androsace L.

1. A. acuta Greene. California Androsace. Occasional in grasslands and on serpentine in the Santa Clara Valley and adjacent foothills; near Stanford, 4 miles south of San Jose, and Edenvale. March–May. —*A. elongata* L. ssp. *acuta* (Greene) Robbins.

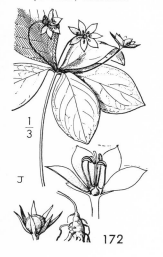

172

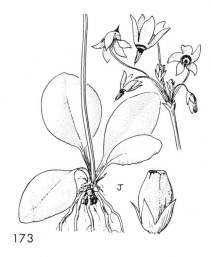

173

6. Dodecatheon L. Shooting Star

Rice-grain bulblets present at flowering time; filament-tube lacking a yellow spot
below each anther.

Flowers 5-merous.   1*a*. *D. hendersonii hendersonii*
Flowers 4-merous.   1*b*. *D. hendersonii cruciatum*

Rice-grain bulblets lacking; filament-tube with a yellow spot below each anther.
Anthers 3.5–5 mm. long, the tips acute to obtuse; pollen-sacs usually yellow.

     2*a*. *D. clevelandii sanctarum*

Anthers 1.5–4 mm. long, the tips retuse to obtuse; pollen-sacs dark.

     2*b*. *D. clevelandii patulum*

1*a*. D. hendersonii Gray ssp. **hendersonii.** *Fig. 173.* Henderson's Shooting Star.
Wooded slopes and on serpentine, southern San Mateo County southward at low
elevations on the eastern slopes of the Santa Cruz Mountains; Coal Mine Ridge,
Guadalupe Reservoir, and Edenvale. February–April.

1*b*. D. hendersonii Gray ssp. **cruciatum** (Greene) Thompson. Wooded slopes
and serpentine outcroppings; San Francisco, Montara Mountain, San Francisco
Watershed Reserve, Woodside, Stanford, Page Mill Road, Bean Creek Drainage,
Bielawski Lookout, Glenwood, and Jamison Creek Road. February–April.

2*a*. D. clevelandii Greene ssp. **sanctarum** (Greene) Abrams. Padres' Shooting
Star. Southern San Mateo County southward; Año Nuevo Pines, Swanton, Hilton
Airport, and near Gilroy. January–April.

2*b*. D. clevelandii Greene ssp. **patulum** (Greene) Thompson. Lowland Shooting
Star. Open grasslands from San Francisco southward, mainly on the eastern
slopes of the Santa Cruz Mountains; San Francisco, Woodside, and near Stanford. January–March. —*D. patulum* Greene var. *bernalinum* Greene.

7. Samolus L.

1. S. floribundus HBK. Known from one collection from a marsh in San Francisco, but to be expected occasionally elsewhere, especially in the Pajaro River
Valley. —*S. parviflorus* Raf.

106. PLUMBAGINACEAE. Thrift, Leadwort, or Plumbago Family

Inflorescence capitate, terminating a naked scape; leaves linear.   1. *Armeria*
Inflorescence cymose-paniculate; leaves oblong-obovate.   2. *Limonium*

1. Armeria Willd.

1. A. maritima (Mill.) Willd. var. **californica** (Boiss.) Lawr. *Fig. 174.* Sea
Pink, California Thrift. Low coastal bluffs along the ocean and occasionally
inland; San Francisco, Granada, Pescadero, and Mount Hermon. March–August.
—*Statice arctica* (Cham.) Blake var. *californica* (Boiss.) Blake, *A. arctica*
(Cham.) Wallr. ssp. *californica* (Boiss.) Abrams.

2. Limonium Mill.

Leaves obovate to oblong-obovate, 5–16 cm. long, gradually narrowed into a petiole. Native plants.   1. *L. californicum*
Leaves oblong-spatulate, usually less than 5 cm. long, abruptly narrowed at the
base. Introduced plants.   2. *L. perfoliatum*

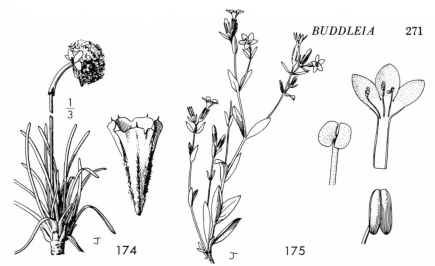

1
3

174   175

1. L. californicum (Boiss.) Heller. Sea Lavender, California Marsh Rosemary. Coastal salt marshes and marshes along San Francisco Bay; San Francisco, near Dumbarton Bridge, Cooleys Landing, Palo Alto Yacht Harbor, and Alviso. July–November.

2. L. perfoliatum (Karel ex Boiss.) Kuntze. Locally established in San Francisco as an escape from cultivation; native of the Caspian Sea region.

107. OLEACEAE. Olive Family

1. Fraxinus L. Ash

Shrubs or small trees, 2–7 m. tall; leaves glabrous; leaflets usually 7; petals 2; flowers perfect.   1. *F. dipetala*

Trees, 10–25 m. tall; leaves pubescent; leaflets usually 3–5; petals none; flowers dioecious.   2. *F. latifolia*

1. F. dipetala H. & A. Foothill or Flowering Ash. Dry slopes and hillsides in the vicinity of Uvas and Llagas Creeks. April–May.

2. F. latifolia Benth. Oregon Ash. Occasional in the stream bottoms on the eastern side of the Santa Cruz Mountains; below San Andreas Dam, San Mateo, and near Palo Alto. March–April.

Ligustrum ovalifolium Hassk., California Privet, a native of eastern Asia, has been found to spread by means of suckers and to produce seedlings in San Francisco. It may be distinguished from *Fraxinus* by its simple leaves.

108. LOGANIACEAE. Logania Family

1. Buddleia L.

1. B. davidii Franchet. Summer Lilac. Becoming common along roads and in disturbed areas; San Francisco, near Woodside, Redwood City, and Palo Alto; native of China. June–September.

109.   GENTIANACEAE.   Gentian Family

Calyx campanulate, lobes reduced to teeth less than 1 mm. long.   1. *Microcala*
Calyx not campanulate, divided nearly to the base into linear-lanceolate lobes.
2. *Centaurium*

1.   Microcala Hoffm. & Link

1. M. quadrangularis (Lam.) Griseb. Timwort, American Microcala. Grassy
slopes, mainly on the eastern side of the Santa Cruz Mountains; San Francisco,
Stanford, Jasper Ridge, and Santa Cruz. May–June.

2.   Centaurium Hill. Centaury

Corolla-lobes under 6 mm. long; anthers under 2 mm. long.
  Rudimentary flowers present in the axils of at least the terminal flowers; flowers
    sessile or nearly so.   1. *C. floribundum*
  Rudimentary flowers lacking; flowers sessile to pedicellate.
    Flowers sessile or subsessile; pedicels under 0.5 mm. long; corolla-lobes
      lanceolate, about 3.5 mm. long.   2. *C. muhlenbergii*
    At least some flowers long-pedicellate; corolla-lobes ovate, 4–6 mm. long.
                                                                    3. *C. davyi*
Corolla-lobes 8 mm. long or longer; anthers over 2 mm. long.   4. *C. trichanthum*

1. C. floribundum (Benth.) Robinson. June Centaury. Rare locally; near
Stanford and near Boulder Creek. July–September.
2. C. muhlenbergii (Griseb.) Wight. Monterey Centaury. Occasional, usually
in fields and pastures; Stanford, Palo Alto, and Graham Hill Road. June–August.
3. C. davyi (Jeps.) Abrams. *Fig. 175.* Davy's Centaury. Widely distributed in
grassy or disturbed areas, but not usually abundant; San Francisco, near Pescadero, Año Nuevo Point, Sand Hill Road, Clarita Vineyard, near Los Gatos, Loma
Prieta, Swanton, San Vicente Creek, and Santa Cruz. April–July.
4. C. trichanthum (Griseb.) Robinson. Alkali Centaury. Occasional in the
Santa Cruz Mountains on fairly open slopes; Belmont, near San Jose, and near
Boulder Creek. May–June.

110.   MENYANTHACEAE.   Buckbean Family

1.   Menyanthes Tourn.

1. M. trifoliata L. Buckbean. This species was reported from a marsh in San
Francisco by Dr. H. H. Behr and is now extinct locally.

111.   APOCYNACEAE.   Dogbane Family

Flowers large, over 2 cm. long, blue; seeds lacking a coma. Introduced plants.
                                                                    1. *Vinca*
Flowers small, under 1 cm. long, white to flesh-colored; seeds with a coma. Native plants.   2. *Apocynum*

1. Vinca L.

1. V. major L. *Fig. 176.* Periwinkle. Commonly escaped from cultivation along roads, creek banks, railroad embankments and in other areas; San Francisco, Cahill Ridge, Rancho del Oso, and near Sea Cliff Beach; native of the Mediterranean region. To be found in flower at all times of the year.

2. Apocynum L. Dogbane, Indian Hemp

Corollas at least twice as long as the calyces.
  Corollas campanulate, about 3 times as long as the calyx-lobes; leaves ovate, not mucronate, glabrous or pubescent.   1. *A. pumilum*
  Corollas cylindric, about twice as long as the calyx-lobes; leaves elliptic to broadly lanceolate, often mucronate, glabrous.
    2. *A. medium floribundum*
Corollas less than twice as long as the calyces.   3. *A. cannabinum*

1. A. pumilum (Gray) Greene. Mountain Hemp. Rare locally, growing on exposed slopes and in vineyards; La Honda Road and Ben Lomond Mountain. July–August. —*A. pumilum* (Gray) Greene var. *rhomboideum* (Greene) Beg. & Bel.

2. A. medium Greene var. **floribundum** (Greene) Woodson. Western Dogbane. Known locally only from near Laurel in Santa Cruz County. July–August. —*A. cannabinum* L. var. *floribundum* (Greene) Jeps.

3. A. cannabinum L. Common Dogbane, Indian Hemp. Occasional in the central part of the Santa Cruz Mountains; near Los Gatos, near Felton, and Soquel Road. June–August. —*A. cannabinum* L. var. *pubescens* (Mitchell) A. DC.

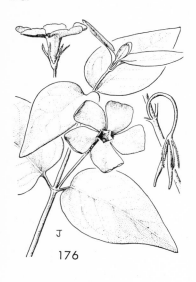

176

177

112. ASCLEPIADACEAE. Milkweed Family

1. Asclepias L. Milkweed

1. A. fascicularis Decne. *Fig. 177.* Narrow-leaved Milkweed. Dry slopes and valleys, commonly in disturbed ground; Redwood City, Stanford, Mountain View, Black Mountain, Uvas Creek Road, near San Jose, and Felton. July–September. —*A. mexicana* of California authors.

113. CONVOLVULACEAE. Morning-Glory Family

Plants parasitic; stems orange, lacking chlorophyll.   1. *Cuscuta*
Plants not parasitic; stems green, chlorophyll present.
  Corollas large, 2–6 cm. long; flowers subtended by a pair of bracts.
    Stigmas filiform to more or less cylindric.   2. *Convolvulus*
    Stigmas capitate.   3. *Ipomoea*
  Corollas small, under 6 mm. long; flowers not subtended by a pair of bracts.
    Leaves oblong-ovate to broadly lanceolate; stems tufted, not creeping or rooting at the nodes.   4. *Cressa*
    Leaves suborbicular to reniform; stems creeping, rooting at the nodes.
        5. *Dichondra*

1. Cuscuta L. Dodder

Corollas lacking scales on the inner surface of the corolla-tube; capsules globose or depressed globose.
  Flowers sessile; anthers about 0.5 mm. long, oval; corolla-lobes spreading.
    1. *C. occidentalis*
  Flowers pedicellate; anthers 0.75–1 mm. long, elongate; corolla-lobes reflexed.
    2. *C. californica*
Corollas with scales on the inner surface of the corolla-tube; capsules depressed globose to elongate.
  Capsules globose to depressed globose. Occasional.
    Corolla-tubes about as broad as long, corolla-lobes about equaling the tube.
      3. *C. campestris*
    Corolla-tubes longer than broad, the corolla-lobes shorter than the tube.
      4. *C. suaveolens*
  Capsule elongate. Common.
    Flowers 3–4.5 mm. long; corolla-tubes more or less campanulate. Plants of salt marshes.   5. *C. salina major*
    Flowers 5–6 mm. long; corolla-tubes cylindric. Plants of inland habitats.
      6. *C. subinclusa*

1. C. occidentalis Millsp. Western Dodder. Occasional in the Santa Clara Valley, growing on various native and introduced herbs and shrubs; 4 miles south of San Jose and San Juan Bautista Hills. May–August.
2. C. californica H. & A. Chaparral, California, or Common Dodder. Occasional in the Santa Cruz Mountains, growing on various native shrubs and herbs; San Francisco, San Francisco Watershed Reserve, near Crystal Springs Lake, and near Boulder Creek. May–August. —*C. californica* Choisy.

3. C. campestris Yuncker. Western Field Dodder. Common on *Xanthium strumarium* in Lake Lagunita, Stanford, and occasionally on such cultivated plants as Pelargonium and to be expected on *Baccharis, Salix, Franseria,* and *Medicago.* August–October.

4. C. suaveolens Ser. Fringe Dodder. Occasional, growing on alfalfa; near Stanford, and Santa Cruz Mountains; native of South America. August–October.

5. C. salina Engelm. var. **major** Yuncker. Salt Marsh or Alkali Dodder. Salt marshes along the ocean and San Francisco Bay, commonly parasitic on various species of *Salicornia*; San Francisco, near Menlo Park, Cooleys Landing, near Mountain View, near Alviso, Santa Cruz, and near mouth of Pajaro River. June–October.

6. C. subinclusa Dur. & Hilg. *Fig. 178.* Canyon Dodder. The most common non-salt marsh species of *Cuscuta* in the Santa Cruz Mountains, parasitic on a large number of plants including *Pickeringia, Baccharis, Aesculus, Rhus, Photinia, Eriogonum, Lupinus, Stachys, Ceanothus, Eriodictyon, Populus, Artemisia,* and various grasses; San Francisco southward. June–November. —*C. ceanothi* Behr.

2. Convolvulus L. Morning-Glory, Bindweed

Bracts sepal-like, broadly ovate or broader, closely subtending the flowers.
 Stems and leaves glabrous or nearly so; stems long, twining or trailing, definitely not acaulescent.   1. *C. soldanella*
 Stems and leaves variously pubescent; stems very short or trailing, some plants definitely acaulescent.
  Stems and leaves densely white-pubescent.   2. *C. malacophyllus collinus*
  Stems and leaves villous-pubescent, often sparsely so.   3. *C. subacaulis*
Bracts not sepal-like, linear, not closely subtending the flowers.
 Stems and leaves glabrous; leaves usually acute; corollas over 3 cm. long. Native plants.   4. *C. occidentalis*
 Stems and leaves pubescent, rarely glabrous; leaves usually obtuse; corollas less than 3 cm. long. Introduced weeds.   5. *C. arvensis*

1. C. soldanella L. *Fig. 179.* Beach Morning-Glory. Coastal sand dunes and beaches; San Francisco, Año Nuevo Point, Swanton, Waddell Creek, Santa Cruz, and near Watsonville. May–July.

2. C. malacophyllus Greene var. **collinus** (Greene) Abrams. Woolly Morning-Glory. Rare in serpentine or in rocky soil on the eastern slopes of the Santa Cruz Mountains in Santa Clara County; Alamitos, New Almaden, and Madrone. May.

3. C. subacaulis (H. & A.) Greene. Hill Morning-Glory. Open grassy slopes, mainly to the east of the crests of the Santa Cruz Mountains; San Francisco, Crystal Springs Lake, Half Moon Bay Road, near Burlingame, Woodside, Coal Mine Ridge, Stanford, Permanente Creek, near Los Gatos, Almaden Ridge, and Hilton Airport. April–June.

4. C. occidentalis Gray. Chaparral, Western, or Bush Morning-Glory. Moist wooded slopes and on brush-covered hills, San Francisco southward; San Francisco, Sawyer Ridge, Crystal Springs Lake, Pescadero, San Mateo Creek, Stanford, Permanente Creek, Los Gatos Creek, Loma Prieta, Swanton, Santa Cruz, and near Burrell. April–July. —*C. occidentalis* Gray var. *purpuratus* (Greene) Howell, *C. occidentalis* Gray var. *solanensis* (Jeps.) Howell. The corolla color of this species varies from purple through pink to white.

5. C. arvensis L. Field Bindweed, Orchard Morning-Glory. A very common weed of orchards, fallow fields, grasslands, and disturbed areas; native of Europe. May–October.

3. Ipomoea L. Morning-Glory

Leaves sparsely pubescent abaxially; calyx-lobes acute, not long attenuate.
                                                                1. *I. purpurea*
Leaves densely canescent abaxially; calyx-lobes long attenuate.   2. *I. mutabilis*

1. I. purpurea (L.) Roth. Common Morning-Glory. Occasional as an escape from cultivation; Palo Alto and Boulder Creek; native of Mexico. August–September.

2. I. mutabilis Lindl. Well established in disturbed ground at the Presidio in San Francisco; native of Mexico.

4. Cressa L. Alkali Weed

1. C. truxillensis HBK. var. **vallicola** (Heller) Munz. Cressa. Occasional in salt marshes along San Francisco Bay and in low alkaline areas in the Santa Clara Valley; Cooleys Landing, near Alviso, and San Jose. June–October —*C. cretica* L. var. *truxillensis* (HBK.) Choisy.

5. Dichondra Forst. & Forst. f.

1. D. repens Forst. & Forst. f. Dichondra. Occasional as an escape from cultivation in San Francisco, northern San Mateo County, Palo Alto, and probably elsewhere; native of both the New- and the Old-World tropics. February–June.

114. POLEMONIACEAE. Phlox or Gilia Family

Leaves opposite, those of the inflorescence sometimes alternate.
    Leaves entire.   1. *Phlox*
    Leaves palmately divided.   2. *Linanthus*
Leaves all alternate.
    Corollas funnelform, 1.5–2.5 cm. in diameter; plants perennial.
                                                                3. *Polemonium*
    Corollas tubular to salverform, 1.5 cm. in diameter or less; plants annual or
        perennial.
        Calyx-lobes spiny.
            Inflorescence densely white-woolly.   4. *Eriastrum*
            Inflorescence not densely white-woolly.   5. *Navarretia*
        Calyx-lobes not spiny.
            Fruiting calyx with a small reflexed lobe in each sinus, calyx growing with
                the capsule, chartaceous in age.   6. *Collomia*
            Fruiting calyx lacking reflexed appendages, calyx not growing with the
                capsule, usually breaking..
                Seeds 10–35 per capsule; stamens equally inserted in the throat or tube
                    of the corolla.   7. *Gilia*
                Seeds 3 per capsule; stamens unequally inserted in the throat of the
                    corolla.   8. *Allophyllum*

1. Phlox L.

1. P. gracilis (Hook.) Greene. Slender Phlox. Open grassy slopes throughout the Santa Cruz Mountains; San Francisco, San Bruno Hills, Kings Mountain, Alpine Grade, Coal Mine Ridge, Stanford, near Guadalupe Creek, Edenvale, and near Saratoga Summit. February–April. —*Gilia gracilis* Hook., *Microsteris gracilis* (Hook.) Greene.

2. Linanthus Benth.

Flowers not in dense heads.
  Mature corollas scarcely longer than the calyces.   1. *L. pygmaeus*
  Mature corollas at least 2 or more times as long as the calyces.
    Corollas 2–3 cm. long; calyces 9–15 mm. long; plants usually not branched.
                                              2. *L. dichotomus*
    Corollas under 2 cm. long; calyx 4–8 mm. long; plants usually well branched.
      Plants usually branched above; corolla-tubes of mature flowers shorter than the calyces; corolla-lobes over ½ the length of the corolla; filaments hairy at the base.   3. *L. liniflorus*
      Plants usually branched below; corolla-tubes of mature flowers longer than the calyces; corolla-lobes ⅓ or less the length of the corolla; filaments glabrous.   4. *L. ambiguus*
Flowers in dense heads.
  Corolla-tubes about equaling the calyces; corolla-lobes usually about 12 mm. long.   5. *L. grandiflorus*
  Corolla-tubes distinctly longer than the calyces; corolla-lobes under 8 mm. long.
    Corolla-lobes 5–8 mm. long.
      Calyx-tubes glabrous.
        Segments of the bracts linear to awl-shaped; corollas usually multicolored. Inland plants.   6. *L. androsaceus*
        Segments of the bracts spatulate to obovate; corollas yellow. Coastal plants.   7. *L. parviflorus*
      Calyx-tubes puberulent, often glandular.   7. *L. parviflorus*
    Corolla-lobes 3–5 mm. long.
      Plants usually over 2 dm. tall; calyx with a conspicuous hyaline membrane in the sinuses.   8. *L. ciliatus*
      Plants commonly under 1.5 dm. tall; calyx with an inconspicuous hyaline membrane.
        Corollas completely yellow.   9. *L. acicularis*
        Corollas white, pink, or red, only the throat yellow.   10. *L. bicolor*

1. L. pygmaeus (Brand) Howell. Pigmy Linanthus. Rare in grasslands and on serpentine soils; Woodside, near Stanford, and Ben Lomond Mountain. April–May.
2. L. dichotomus Benth. *Fig. 180.* Evening Snow. Dry hill slopes, serpentine, and sandy soils, southern San Mateo County southward, mainly in the foothills on the eastern slopes of the Santa Cruz Mountains; Jasper Ridge, near Searsville Lake, Stanford, Page Mill Road, Black Mountain, San Jose, Edenvale, and Felton. February–May.

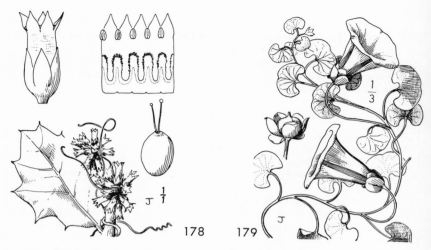

178   179

3. L. liniflorus (Benth.) Greene. Flax-flowered Linanthus. Dry slopes, serpentine, and borders of chaparral and grasslands, San Francisco southward along the eastern slopes of the Santa Cruz Mountains; San Francisco, Crystal Springs Lake, near Millbrae, near San Carlos, Woodside, Searsville, Stevens Creek, Alamitos Creek, and Loma Prieta. April–July. *Linanthus liniflorus* is a variable species. The branching in *L. liniflorus* ssp. *liniflorus* is usually alternate, while in *L. liniflorus* (Benth.) Greene ssp. *pharnaceoides* (Benth.) Mason, it is commonly opposite. In the typical subspecies the throat of the corolla is 3–4 times as long as the corolla-tube; in ssp. *pharnaceoides* it is only 1–2 times as long. The characters used to separate the two subspecies may not be significant in that numerous intermediates occur in the Santa Cruz Mountains and some plants, for example *Oberlander 84* from Crystal Springs, approach ssp. *pharnaceoides*. Specimens from San Francisco tend to be more compact, to have shorter internodes, and to branch more freely than those from other areas.

4. L. ambiguus (Rattan) Greene. Serpentine Linanthus. Most common on serpentine outcroppings, less frequent on sandstone, Crystal Springs Lake southward, usually on the eastern slopes of the Santa Cruz Mountains; Crystal Springs Lake, Woodside, Stevens Creek, near San Jose, Edenvale, Uvas Creek, Gilroy-Watsonville Road, and Jamison Creek. March–June. —*L. rattanii* Greene var. *ambiguus* (Rattan) Jeps.

5. L. grandiflorus (Benth.) Greene. Large-flowered Linanthus. Dry open ridges, mainly along the crests of the Santa Cruz Mountains; San Francisco, near La Honda, Loma Azule, Uvas Road, near Gilroy, Aptos Creek, and Soquel Canyon. May–July.

6. L. androsaceus (Benth.) Greene. Shower or Common Linanthus. A common species of open grasslands and grassy slopes, usually away from the sea; Crystal Springs Lake, near San Mateo, Pescadero–La Honda Road, Woodside, Coal Mine Ridge, near Stanford, Black Mountain, Permanente Creek, Los Gatos Hills, near Los Gatos, Gilroy Valley, Swanton, Big Basin, near Eagle Rock, Santa Cruz, and Mount Charlie Road. April–July.

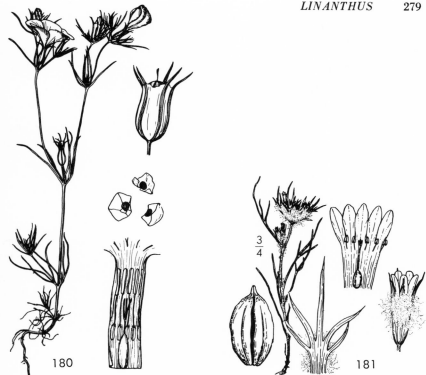

180 181

7. L. parviflorus (Benth.) Greene. Small-flowered Linanthus. A common species of coastal bluffs, inland marine sand deposits, and grassy slopes; San Francisco, San Bruno, Moss Beach, Pescadero Beach, Coal Mine Ridge, near Stanford, near Los Gatos, New Almaden, Guadalupe Mines, Hilton Airport, Bonny Doon, Felton, and Glenwood. April–June. —*L. androsaceus* (Benth.) Greene ssp. *croceus* (Milliken) Mason, *L. androsaceus* (Benth.) Greene ssp. *luteus* (Benth.) Mason, *L. rosaceus* (Hook. f.) Greene, *L. longituba* Heller. *Linanthus parviflorus* is a variable species that can be and has been divided *ad infinitum* by various botanists.

8. L. ciliatus (Benth.) Greene. Whisker Brush, Bristly-leaved Linanthus. Occasional in northern Santa Cruz County on grassy slopes; near Bielawski Lookout and near Saratoga Summit. May–June.

9. L. acicularis Greene. Bristly Linanthus. Rare in grasslands and on serpentine soils, foothills along the eastern slope of the Santa Cruz Mountains; Woodside and Coal Mine Ridge. April–May.

10. L. bicolor (Nutt.) Greene. Bicolored Linanthus. Grasslands, mainly in the foothills on the eastern slopes of the Santa Cruz Mountains; Woodside, Coal Mine Ridge, Jasper Ridge, Stanford, Black Mountain, near Los Gatos, and near Saratoga Summit. March–May.

3. Polemonium L.

1. P. carneum Gray. Great Polemonium. Known locally only from the vicinity of Pilarcitos Reservoir and Pilarcitos Creek in northern San Mateo County. May–September.

4. Eriastrum Woot. & Standl.

1. E. abramsii (Elmer) Mason. *Fig. 181.* Abrams' Eriastrum. Occasional in chaparral; Emerald Lake, Jasper Ridge, and Black Mountain. June–July. —*Navarretia abramsii* Elmer, *Hugelia abramsii* (Elmer) Jeps. & Bailey.

5. Navarretia R. & P.

Basal portion of the bracts broad, ovate.
    Distal segment of the bracts 3-toothed, bracts not digitately divided, distinctly longer than broad; stamens included.   1. *N. atractyloides*
    Distal segment of the bracts not 3-toothed; bracts digitately divided, about as broad as long; stamens exserted.   2. *N. heterodoxa*
Basal portion of the bracts narrow, linear to lanceolate.
    Corollas 10–17 mm. long; corolla-lobes 4–6 mm. long; middle cauline leaves with more or less equally spaced teeth, rarely bipinnate.   3. *N. viscidula*
    Corollas 6–12 mm. long; corolla-lobes 1.5–3 mm. long; middle cauline leaves with irregularly spaced divisions, usually bipinnate.
        Corollas 9–12 mm. long; stamens 1–4 mm. long; filaments inserted near the base of the corolla-tube; plants usually over 2 dm. tall.
                                               4. *N. squarrosa*
        Corollas 6–7 mm. long; stamens under 1.5 mm. long; filaments inserted near the middle of the corolla-tube; plants usually under 2 dm. tall.
                                                 5. *N. mellita*

1. N. atractyloides (Benth.) H. & A. Holly-leaved Navarretia. Rocky and sandy soils in Santa Clara and Santa Cruz counties; Wrights, Big Basin, near Jamison Creek, Glenwood, Mount Hermon, Ben Lomond Sand Hills, and near Gilroy. May–August.

2. N. heterodoxa (Greene) Greene. *Fig. 182.* Calistoga Navarretia. Serpentine and rocky chaparral slopes, Crystal Springs Lake southward; Crystal Springs Lake, near Redwood City, San Carlos, Searsville Lake, Black Mountain, near Los Gatos, and New Almaden. June–July.

3. N. viscidula Benth. Sticky Navarretia. Rare in southern San Mateo and northern Santa Clara counties, usually on grassy slopes in heavy soil; Searsville, Portola, and near Searsville Lake. May–June.

4. N. squarrosa (Esch.) H. & A. Skunkweed. The most common species of *Navarretia* locally, growing in dry, hard-packed soils, on grassy slopes, vernal pools, and as a weed in disturbed areas, San Francisco southward. June–October.

5. N. mellita (Greene) Greene. Honey-scented Navarretia. Occasional in rocky or sandy soils, usually in chaparral or along its borders; Belmont, Black Mountain, near Eagle Rock, and Big Basin. June–July.

6. Collomia Nutt.

Leaves pinnately parted or lobed, if entire then the blades wide; corollas under
1.2 cm. long, pink to red-purple. 1. *C. heterophylla*
Leaves entire, lanceolate to linear; corollas over 2 cm. long, pinkish-yellow.
2. *C. grandiflora*

1. C. heterophylla Hook. Varied-leaved Collomia. Fairly common in sandy
soils of inland sand deposits, along open creek banks, and on brush-covered
slopes; San Francisco (as a weed in a flower bed), Pilarcitos Canyon, near
Pescadero Creek, near Palo Alto, Los Gatos Canyon, Congress Springs, Saratoga–
Big Basin Road, Big Basin, Swanton, near Felton, Loma Prieta, and Soquel
Valley. April–July.
2. C. grandiflora Dougl. ex Hook. Large-flowered Collomia. Occasional in the
Santa Cruz Mountains, usually growing in the mixed evergreen forest; San Francisco, near Saratoga, Bielawski Lookout, and near Boulder Creek. May–July.

7. Gilia R. & P.

Stems conspicuously leafy, the cauline leaves about as large as the basal ones, but
reduced gradually upwards.
Inflorescences capitate, usually with 50–100 flowers.
Corolla-lobes linear, 1–2 mm. wide; heads very sparingly floccose at the base
or glabrous. 1*a. G. capitata capitata*
Corolla-lobes oval, 2.2–3.3 mm. wide; heads densely floccose at the base.
Corollas deep blue-violet; basal leaves bipinnately dissected. Sand dune
plants. 1*b. G. capitata chamissonis*
Corollas pale blue-violet; basal leaves pinnately dissected. Inland plants.
1*c. G. capitata staminea*
Inflorescences capitate or glomerulate, usually with less than 25 flowers, or the
flowers solitary.
Corollas with purple spots in the throat.
Corollas 11–16 mm. long, 9–14 mm. across. 2. *G. tricolor*
Corollas 6–8 mm. long, 3–5 mm. across. 3. *G. clivorum*
Corollas lacking purple spots in the throat.
Green part of the calyx-lobes 0.8–1 mm. wide; corollas 6–8 mm. long;
capsules with over 24 seeds; plants usually under 3 dm. tall.
3. *G. clivorum*
Green part of the calyx-lobes 0.5–0.8 mm. wide; corollas 5–20 mm. long;
capsules with 10–18 seeds; plants 1–7 dm. tall. 4. *G. achilleaefolia*
Stems not conspicuously leafy, the cauline leaves much smaller than the basal
ones. 5. *G. tenuiflora*

1*a*. G. capitata Sims ssp. **capitata.** Globe or Blue Field Gilia. Occasional in
the Santa Cruz Mountains; Black Mountain and Alba Grade. June–July.
1*b*. G. capitata Sims ssp. **chamissonis** (Greene) Grant. Dune Gilia. Sand dunes
in San Francisco. April–June. —*G. achilleaefolia* Benth. ssp. *chamissonis*
(Greene) Brand.

POLEMONIACEAE282.

1c. G. capitata Sims ssp. **staminea** (Greene) Grant. Range Gilia. Rather rare in the Santa Cruz Mountains; San Francisco, Cupertino, and Boulder Creek. April–May. —*G. achilleaefolia* Benth. ssp. *staminea* (Greene) Mason & Grant.

2. G. tricolor Benth. Birds Eyes, Tricolor Gilia. Open grasslands, often on serpentine soil, foothills on the eastern side of the Santa Cruz Mountains from the vicinity of Stanford southward; Stanford, Edenvale, and near Gilroy. February–April.

3. G. clivorum (Jeps.) Grant. Open fields and grassy or rocky slopes; San Francisco, Crystal Springs Lake, near Redwood City, Searsville, Coal Mine Ridge, Black Mountain, Campbell, Waddell Creek, near Saratoga Summit, and near Watsonville. March–June. —*G. multicaulis* Benth. var. *clivorum* Jeps.

4. G. achilleaefolia Benth. California Gilia. A common species of open slopes, brush-covered areas, chaparral, and shaded oak-woodland; Woodside, Portola, Coal Mine Ridge, near Stanford, Permanente Creek, Black Mountain, near Los Gatos, Loma Prieta, New Almaden, near San Jose, Madrone Station, Swanton, near Saratoga Summit, and Felton. February–July. —*G. peduncularis* Eastw. ex Milliken. *Gilia achilleaefolia* is exceedingly variable. Dr. Verne Grant, a student of the Polemoniaceae, has written (Aliso **3**:7, 1954) that "Almost every local population in *Gilia achilleaefolia* constitutes a separate race with distinctive characters. These races do not group themselves into broad geographical assemblages, as in most species including the related *Gilia capitata*, but the broader subdivision in the species is rather along ecological lines. The large-flowered races with dense heads occupy sunny hillsides in grassland and oak savannah; the small-flowered races with loose cymes occur in the shade of oak woodland or redwood forest; and there are numerous transitional forms in the semi-shade of open oak woods."

5. G. tenuiflora Benth. *Fig. 183.* Slender-flowered Gilia. Known locally from the sand hills near Felton and Ben Lomond. April–May.

8. Allophyllum (Nutt.) A. & V. Grant

1. A. divaricatum (Nutt.) A. & V. Grant. Straggling Gilia. Occasional in Santa Clara and Santa Cruz counties, usually above 2000 feet elevation in mixed evergreen and Douglas fir forests and along the edges of chaparral; near Saratoga, Loma Prieta, near Eagle Rock, and Empire Grade. May–September. —*Gilia divaricata* Nutt., *G. gilioides* (Benth.) Greene ssp. *volcanica* (Brand) Mason & Grant.

115. HYDROPHYLLACEAE. Waterleaf or Phacelia Family

Plants herbaceous.
Ovary 1-celled.
Herbage commonly prickly, not viscid or scented; calyces often with appendages in the sinuses.
Herbage succulent; stems prickly; flowers in terminal cymes.
1. *Pholistoma*
Herbage not succulent; stems usually not prickly; flowers solitary.
2. *Nemophila*
Herbage viscid, scented; appendages lacking.
3. *Eucrypta*
Ovary partially or wholly divided by the placenta.
Plants lacking tomentose tubers at the base; style not entire.
Corollas blue, violet, or white, not persistent.
4. *Phacelia*
Corollas yellow, persistent.
5. *Emmenanthe*
Plants with tomentose tubers at the base; style entire.
6. *Romanzoffia*
Plants shrubby.
7. *Eriodictyon*

1. Pholistoma Lilja ex Lindb.

Calyx with auricles, enveloping the capsule; mature capsules 5–10 mm. in diameter. Common.
1. *P. auritum*
Calyx lacking auricles, not enveloping the capsule; mature capsules 2–4 mm. in diameter. Rare in the Santa Clara Valley.
2. *P. membranaceum*

1. P. auritum (Lindl.) Lilja ex Lindb. Common Fiesta Flower, Climbing Nemophila. Most common on the eastern slopes of the Santa Cruz Mountains, commonly growing among grasses in partial shade of *Quercus agrifolia* and *Aesculus californica*; San Francisco, South San Francisco, Pescadero Creek, Stanford, Stevens Creek, and near Saratoga. March–June. —*Nemophila aurita* Lindl., *Ellisia aurita* (Lindl.) Jeps.
2. P. membranaceum (Benth.) Constance. White Fiesta Flower. Occasional in grasslands on open hills in the Santa Clara Valley; 4 miles south of San Jose, Evergreen, and Edenvale. January–May. —*Nemophila membranacea* (Benth.) Greene, *Ellisia membranacea* Benth.

2. Nemophila Nutt. ex Barton

Corollas usually 1 cm. or more in diameter.
Corollas conspicuously blue-veined, usually blue at least toward the periphery, if black-punctate the dots toward the center.
1a. *N. menziesii menziesii*

Corollas not usually conspicuously blue-veined, usually white, black-punctate toward the periphery.   1*b. N. menziesii atomaria*
Corollas less than 1 cm. in diameter.
Auricles in the sinuses of the calyx usually ⅓ as long as the calyx-lobes or longer; corollas marked with blue, purple, or black. Usually growing in unshaded areas.   2. *N. pedunculata*
Auricles less than ⅓ as long as the calyx-lobes; corollas not marked. Usually growing in shaded areas.
Basal leaves with 5–7 distinct divisions, these petiolulate, orbicular; style 2–3 mm. long. Eastern slopes of the Santa Cruz Mountains.
                    3. *N. heterophylla*
Basal leaves with 5 divisions, these not petiolulate nor orbicular; style to 1.5 mm. long. Both slopes of the Santa Cruz Mountains.
                    4. *N. parviflora*

1*a. N. menziesii* H. & A. var. **menziesii.** Baby Blue Eyes. Open grassy slopes or growing in the shade of *Quercus agrifolia, Q. lobata, Arbutus menziesii,* and *Rhus diversiloba*; San Francisco, La Honda, Searsville, Coal Mine Ridge, Black Mountain, near Los Gatos, New Almaden, Morgan Hill, Swanton, Bielawski Lookout, Deer Ridge Farm, and Chittenden Pass. February–April.
1*b. N. menziesii* H. & A. var. **atomaria** (F. & M.) Chandler. Northern San Mateo County southward, commonly on grassy slopes; Colma, San Andreas Valley, Montara Mountain, Half Moon Bay, Pebble Beach, Kings Mountain, Portola, Black Mountain, and Loma Prieta. March–April.
2. **M. pedunculata** Dougl. ex Benth. *Fig. 184.* Meadow Nemophila. San Francisco southward, growing in moist open grassy areas; San Francisco, Kings Mountain, Black Mountain, Stevens Creek, Morgan Hill, Jamison Road, Hilton Airport, Boulder Creek, and Felton. February–April.
3. **N. heterophylla** F. & M. Canon or Variable-leaved Nemophila. Mainly on the eastern slopes of the Santa Cruz Mountains, commonly growing in the shade of *Quercus agrifolia, Sambucus, Aesculus californica,* and *Rhus diversiloba* with such herbs as *Dentaria californica* and *Montia perfoliata*; San Francisco, San Bruno Hills, Pilarcitos Canyon, Black Mountain, Woodside, Stanford, near Los Gatos, New Almaden, and Morgan Hill. January–April.
4. **N. parviflora** Dougl. ex Benth. Woodland or Small-flowered Nemophila. Shaded and wooded slopes; San Francisco, Pilarcitos Canyon, Crystal Springs Lake, Pescadero Creek, Stanford, Wrights, Loma Prieta, Rancho del Oso, Swanton, Big Basin, Jamison Creek, Boulder Creek, and Glenwood. March–June.

3. Eucrypta Nutt.

1. **E. chrysanthemifolia** (Benth.) Greene. Common Eucrypta. Moist ravines on the eastern slopes of the Mountains from northern Santa Clara County southward; Hidden Villa and Permanente Ravine. March–June.

4. Phacelia Juss.

Plants perennial.
Divisions of the leaves toothed to crenate.   1. *P. ramosissima*
Leaves or their divisions entire.

Corollas 4–5 mm. long; inflorescences glandular; stems usually solitary; cauline leaves or the terminal lobe of the cauline leaves broadly lanceolate to elliptic to broadly ovate.   2. *P. nemoralis*

Corollas 5–7 mm. long; inflorescences not glandular; stems usually several; cauline leaves or the terminal lobe of the cauline leaves lanceolate to linear-lanceolate.

Corollas usually blue to purplish; lobes of the corollas spreading after anthesis; calyx-lobes 7–8 mm. long in fruit, 2–2.5 mm. wide, oblong-lanceolate, not imbricate; basal leaves usually with 1–2 pairs of lateral lobes.   3. *P. californica*

Corollas whitish; corolla-lobes somewhat incurved after anthesis; calyx-lobes 5–10 mm. long in fruit, 1.5–4 mm. wide, somewhat imbricate, lanceolate to ovate; basal leaves with 3 or more pairs of lateral lobes.
                                                           4. *P. imbricata*

Plants annual.

Stems and leaves distinctly bristly-hispid, the individual trichomes 2–4 mm. long.

Stamens exserted; corollas 5–7 mm. long.   5. *P. malvaefolia*

Stamens included; corollas 2–4 mm. long.   6. *P. rattanii*

Stems and leaves variously pubescent or glabrous, not bristly-hispid.

Pedicels 1–5 cm. long in fruit; inflorescences not congested, the flowers very remote from each other in fruit.   7. *P. douglasii*

Pedicels under 0.3 cm. long in fruit or the flowers sessile; inflorescences congested, the flowers close together in fruit.

Leaves entire, toothed, or rarely with a pair of small lobes at the base.

Leaves coarsely toothed; corollas tubular-funnelform, usually under 10 mm. long, 5–10 mm. across; corolla-lobes 1.5–2 mm. long.
                                                           8. *P. suaveolens*

Leaves entire, occasionally with a pair of basal lobes; corollas open-campanulate, 10–18 mm. long, 15–24 mm. across; corolla-lobes 4–5 mm. long.   9. *P. divaricata*

Leaves pinnatifid, leaflets again pinnatifid or toothed.

Calyx-lobes conspicuously enlarged in fruit, ovate, the veins distinct, glabrous to subglabrous, ciliate on the margins, 8–10 mm. long, 2–5 mm. broad; leaf-divisions coarse. Plants of heavy clay soils, mainly in lowlands.   10. *P. ciliata*

Calyx-lobes little enlarged in fruit, lanceolate to oblanceolate or obovate, the veins indistinct, densely pubescent, 4–7 mm. long, 0.5–3 mm. broad; leaf-divisions fine. Plants of rocky or sandy soils, foothills.   11. *P. distans*

1. P. ramosissima Dougl. ex Lehm. Branching Phacelia. Coastal sand dunes from San Francisco southward; San Francisco, Año Nuevo Point, Santa Cruz, Sunset Beach State Park, and near mouth of Pajaro River. May–September.

2. P. nemoralis Greene. Shade or Stinging Phacelia. Wooded slopes from San Francisco southward to the vicinity of Loma Prieta. April–August.

3. P. californica Cham. California Phacelia. Common, usually in rocky soil of open hillsides, coastal bluffs, and canyon slopes; San Francisco, Colma, San Bruno Mountains, near Daly City, Crystal Springs Lake, near Half Moon Bay, and near Santa Cruz. April–September.

4. P. imbricata Greene. Imbricate Phacelia. Rocky slopes and road embankments, usually away from the coast, northern San Mateo County southward; San Mateo Reservoir, Coal Mine Ridge, Stanford, near Los Gatos, near San Jose, San Juan Bautista Hills, San Vicente Creek, Boulder Creek, and near Glenwood. February–June. —*P. magellanica* Cov. var. *calycosa* (Gray) Jeps. & Bailey.

5. P. malvaefolia Cham. Stinging Phacelia. Usually near the coast on partly shaded slopes; San Francisco, near Colma, San Bruno Hills, San Andreas Lake, near San Mateo, Kings Mountain, Waddell Creek, Swanton, Bonny Doon, and Soda Lake. April–July.

6. P. rattanii Gray. Rattan's Phacelia. Occasional in loose sandy soil, forests, woods, and chaparral, Santa Clara County from the vicinity of Stanford southward; Stanford, Page Mill Road, San Antonio Creek, Stevens Creek, Uvas Road, near Camp Evers, San Vicente Creek, and near Felton. May–June.

7. P. douglasii (Benth.) Torr. Douglas' Phacelia. Known locally only from sandy soil at Lake Merced in San Francisco, from the inland sand deposits in Santa Cruz County, and occasionally from well-stabilized coastal sand dunes; San Francisco, near Ben Lomond, and near Felton. April–May.

8. P. suaveolens Greene. Sweet-scented Phacelia. Known locally only from the vicinity of Mount Umunhum and Loma Prieta in Santa Clara County and from Eagle Rock in Santa Cruz County, but to be expected in any recently burned or cleared chaparral; Mount Umunhum, near Loma Prieta, Loma Prieta Summit, and Eagle Rock. April–July.

9. P. divaricata (Benth.) Gray. Divaricate Phacelia. Rocky soils, serpentine outcroppings, and grasslands, Crystal Springs Lake southward, mainly east of the crest of the Santa Cruz Mountains; Crystal Springs Lake, Peters Creek, La Honda, Emerald Lake, Woodside serpentine, between Saratoga Summit and Page Mill Road, Black Mountain, and near Los Gatos. February–June.

10. P. ciliata Benth. Great Valley Phacelia. Heavy clay and adobe soils, San Francisco southward, but most common in the Santa Clara Valley; San Francisco, San Miguel Hills, Burlingame, Palo Alto, between Mountain View and Los Altos, near San Jose, and south of San Jose. March–May, occasionally plants to be found with flowers as late as November. *Phacelia ciliata* thrives in fields that have been allowed to lie fallow and in orchards that have not been plowed. Albino plants are occasionally encountered.

11. P. distans Benth. *Fig. 185.* Common Phacelia. Serpentine outcroppings, sand dunes and deposits, disturbed rocky areas, and occasionally in grasslands, common from San Francisco southward. March–June.

5. Emmenanthe Benth.

1. E. penduliflora Benth. Whispering Bells. Chaparral slopes, mainly those that have been disturbed either by fire or by chaparral clearance; near Stevens Creek Reservoir, Black Mountain Road, near Los Gatos, Mount Umunhum, Loma Prieta, Quail Hollow, and Castle Rock. April–July.

6. Romanzoffia Cham. ex Nees

1. R. suksdorfii Greene. Suksdorf's Romanzoffia. San Francisco south to the vicinity of Big Basin, usually in shade on moist rocks, bluffs, and ledges with mosses and *Sedum*; San Francisco, San Bruno Mountains, Pilarcitos Canyon, Crystal Springs Lake, San Mateo Canyon, and Big Basin. March–April. —*R. californica* Greene.

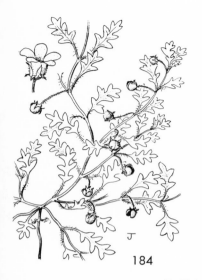

184   185

7. Eriodictyon Benth.

1. E. californicum (H. & A.) Torr. *Fig. 186.* California Mountain Balm, Yerba Santa. A common chaparral species, often growing with *Castanopsis chryso- phylla, Ceanothus papillosus, Arctostaphylos, Rhus diversiloba*, and *Adenostoma fasciculatum*, and invading road embankments and cleared areas; San Andreas Lake, Sweeney Ridge, Belmont, Kings Mountain, Stanford, Black Mountain, Permanente Creek, near Los Gatos, Big Basin, Alba Grade, San Lorenzo River, Glenwood, and near Felton. April–July. *Eriodictyon californicum* sprouts from the roots after fires as well as seeding itself. The leaves and young stems are often infected with a sooty fungus, a member of the genus *Heterosporium*, which turns them black.

116. BORAGINACEAE. Borage Family

Corollas yellow, orange, or blue.
    Corollas blue (rarely white in *Cynoglossum*).
        Corollas nearly rotate; stamens with an abaxial appendage.   1. *Borago*
        Corollas with a distinct tube; stamens unappendaged.
            Corollas somewhat zygomorphic; stamens long exserted; plants to 2 m.
                tall.   2. *Echium*
            Corollas actinomorphic; stamens included; plants to 1 m. tall.
                Nutlets reticulate or covered with barbed spines; corollas about 1 cm.
                    in diameter.
                    Nutlets covered with barbed spines; leaves mainly basal.
                                      3. *Cynoglossum*
                  Nutlets reticulate; stems leafy throughout.   4. *Anchusa*
                Nutlets smooth; corollas about 0.5 cm. in diameter.   5. *Myosotis*
    Corollas yellow to orange.   6. *Amsinckia*

Corollas white, often yellow at the center.
Herbage conspicuously succulent, glaucous, glabrous; ovary not lobed; plants
    perennial.   7. *Heliotropium*
Herbage not succulent (somewhat so in some species of *Allocarya*), not glau-
    cous, usually conspicuously pubescent; ovary deeply 4-lobed; plants
    annual.
    Nutlets widely divergent at maturity, with uncinate bristles at least at the
        tips.   8. *Pectocarya*
    Nutlets not divergent at maturity, lacking uncinate bristles.
        Nutlets with a longitudinal groove on the adaxial surface which is forked
            at the base.   9. *Cryptantha*
        Nutlets lacking such a groove, but with an evident attachment scar below
            the middle and often with a distinct adaxial keel.
            Lower leaves opposite.   10. *Allocarya*
            Lower leaves alternate.
                Receptacle conical; nutlets attached more or less laterally.
                    Basal rosette usually not persistent; nutlets with a distinct baso-
                        lateral stipe-like projection.   11. *Echidiocarya*
                    Basal rosette usually present; nutlets lacking a projection.
                                                              12. *Plagiobothrys*
                Receptacle flat; nutlets attached by their bases.   13. *Lithospermum*

1. Borago L.

1. B. officinalis L. Borago. Occasional as an escape from gardens; San Francisco, Colma, Stanford, and Los Altos; native of Europe. Probably flowering
throughout the year.

2. Echium L. Viper's Bugloss

1. E. fatuosum Ait. Locally established in San Francisco, often on coastal
bluffs; native of the Canary Islands.

3. Cynoglossum L.

1. C. grande Dougl. ex Lehm. *Fig. 187.* Grand or Western Hound's Tongue.
Oak-madrone woods, open Douglas fir forests, and shrubby slopes; San Fran-
cisco, La Honda, Butano, Coal Mine Ridge, near Los Gatos, Swanton, near Big
Basin, Bear Creek, and near Santa Cruz. February–April.

4. Anchusa L.

1. A. sempervirens L. Everlasting Alkanet. Becoming well established in
Golden Gate Park in San Francisco; native of Europe.

5. Myosotis L.

1. M. latifolia Poir. Wood Forget-Me-Not. Becoming increasingly common along
roadsides and streams in the redwood–Douglas fir forests and occasionally elsewhere; San Francisco, Tunitas Creek, Pescadero Creek, near Holy City, Glenwood, and near Santa Cruz; native of Europe and North Africa. March–May.
—*M. sylvatica* Ehrh. ex Hoffm.

6. Amsinckia Lehm. Fiddleneck

At least 2 of the calyx-lobes usually united; leaves erose-dentate.

                                        1. *A. spectabilis*
Calyx-lobes free; leaves entire.
  Flowers essentially radially symmetrical; the corolla-tube not bent; plants homostylic.
    Corollas pale yellow, barely exserted, 4–6 mm. long, the limb 2–3 mm. across; plants appearing grayish-green.   2. *A. menziesii*
    Corollas orange to orange-yellow, conspicuously exserted, 7–20 mm. long, the limb 5–10 mm. across; plants appearing green.
      Corolla-throat open.   3. *A. intermedia*
      Corolla-throat closed by hairy crests.   4. *A. lycopsoides*
  Flowers somewhat irregular; the corolla-tube bent; plants heterostylic.
                                        5. *A. lunaris*

1. A. spectabilis F. & M. Seaside Amsinckia. Coastal sand dunes and beaches; San Francisco, Pigeon Point, Año Nuevo Point, Waddell Creek, Sunset Beach, and Watsonville Slough. March–July.

2. A. menziesii (Lehm.) Nels. & Macbr. Menzies' or Rigid Fiddleneck. Grassy fields and open hillsides; near Palo Alto, Stanford, Permanente Creek, near Los Gatos, Swanton, and Big Basin. March–July. —*A. retrorsa* Suksd.

3. A. intermedia F. & M. *Fig. 188.* Common Fiddleneck. Common in fields, grassy slopes, and disturbed areas; San Francisco, San Mateo Ravine, Portola, Alviso, Stanford, near Los Gatos, near New Almaden, near Swanton, Santa Cruz, near Big Basin, and Scott Valley. March–June.

4. A. lycopsoides Lehm. Known locally from the Presidio in San Francisco. April–June.

5. A. lunaris Macbr. Macbride's Fiddleneck. Occasional in shade on moist wooded slopes; Hillsborough and San Mateo–Half Moon Bay Road. May–June.

7. Heliotropium L.

1. H. curassavicum L. var. **oculatum** (Heller) Johnston ex Tidestrom. Seaside Heliotrope, Chinese Pusley. Along sandy river banks, sea beaches, and becoming increasingly common in disturbed habitats along roads and highways. March–October.

8. Pectocarya DC. ex Meisn.

Nutlets divergent in pairs; calyx with straight hairs at the apex of the lobes.
                                                                                    1. *P. penicillata*
Nutlets not divergent in pairs; calyx with uncinate hairs at the apex of the lobes.
                                                                                    2. *P. pusilla*

1. P. penicillata (H. & A.) A. DC. Winged Pectocarya. Sandy soil of inland sand deposits and probably in chaparral and in other dry habitats; San Francisco, Stanford, and Ben Lomond Sand Hills. April–May.
2. P. pusilla (A. DC.) Gray. Little Pectocarya. Chaparral, open woods, and on serpentine; near San Mateo, Black Mountain, Page Mill Road, Jasper Ridge, Permanente Creek, and Morgan Hill. April–May.

9. Cryptantha Lehm. ex F. & M. Nievitas, White Forget-Me-Not

At least some nutlets in each flower papillate or muricate.
   Mature calyx about 1 mm. long; nutlets papillate, under 1 mm. long, one often
      larger than the others and often smooth.   1. *C. micromeres*
   Mature calyx 2–4 mm. long; nutlets muricate, 1.5–2.5 mm. long, all equal in
      size and equally muricate.   2. *C. muricata jonesii*
Surface of the nutlets smooth.
   Nutlets usually 1 per flower.
      Hairs on the calyx, especially toward the base, with pallid encrustations,
         these usually curved; fruiting calyx 2–5 mm. long.   3. *C. flaccida*
      Hairs on the calyx lacking encrustations, straight; fruiting calyx 1–2 mm.
         long.   4. *C. microstachys*
   Nutlets usually 4 per flower.
      Nutlets broadly ovoid.   5. *C. torreyana pumila*
      Nutlets lanceolate to ovate-lanceolate.
         Spikelets usually bractless or with a few bracts near the base, elongate,
            not dense.   6. *C. hispidissima*
         Spikelets bracteate, short, dense.   7. *C. leiocarpa*

1. C. micromeres (Gray) Greene. Minute-flowered Cryptantha. Chaparral burns, margins of chaparral, and loose sandy soil of dry ridges; near Searsville, Stanford, Permanente Canyon, near Los Gatos, near Big Basin, and Soquel Valley. April–June.
2. C. muricata (H. & A.) Nels. & Macbr. var. **jonesii** (Gray) Johnston. *Fig. 189.* Prickly Cryptantha. Dry chaparral, usually above 1,500 feet elevation; Permanente Creek, Alma Soda Spring, Loma Prieta, Loma Prieta Ridge, Big Basin, Glenwood, and Powder Mill Canyon. May–August.

3. C. flaccida (Dougl.) Greene. Flaccid Cryptantha, Nievitas. Common on serpentine outcroppings, rarely in other rocky areas, mainly east of the crest of the Santa Cruz Mountains; San Francisco, San Andreas Lake, Crystal Springs Lake, Redwood City, Woodside, near Menlo Park, Jasper Ridge, Alamitos Creek, near San Jose, near Los Gatos, Loma Prieta, Edenvale, and Bear Creek Canyon. April–June.

4. C. microstachys (Gray) Greene. Tejon Cryptantha. Chaparral, usually in poor or rocky soil; Coal Mine Ridge, near Stanford, Black Mountain, Stevens Creek, Eagle Rock, near Felton, and Soquel Canyon. April–July.

5. C. torreyana (Gray) Greene var. **pumila** (Heller) Johnston. Torrey's Cryptantha. Occasional in grasslands; San Andreas Lake, Coal Mine Ridge, Stanford, Stevens Creek, near Los Gatos, Swanton, and near Boulder Creek.

6. C. hispidissima Greene. Cleveland's Cryptantha. Occasional in sandy soils and in the grassy margins of chaparral; San Francisco, La Honda Grade, near Eagle Rock, near Big Basin, near Glenwood, near Mount Hermon, and Ben Lomond Sand Hills. May–July.

7. C. leiocarpa (F. & M.) Greene. Coast Cryptantha. Stabilized coastal sand dunes and occasionally on inland sand dunes; San Francisco, near Scott Creek, Sunset Beach, and Pajaro River. April–May.

10. Allocarya Greene

Nutlet attachment basal or basolateral, the scar not broad, not linear; pedicels commonly shorter than the calyx, or the flowers sessile.

    Stems somewhat fistulose or succulent; nutlet attachment basal, often substipitate; calyx with well-developed costae; the lower $\frac{1}{2}$–$\frac{2}{3}$ of the mature nutlets hidden by the calyx-tube.   1. *A. glabra*

    Stems not fistulose or succulent; nutlet attachment basolateral, not substipitate; calyx lacking well-developed costae; mature nutlets not hidden by the calyx-tube.

        Adaxial keel of the nutlets not in a groove.   2. *A. bracteata*

        Adaxial keel of the nutlets in a distinct groove.

            Pericarp thick, tuberculate and somewhat granulate. Endemic in San Francisco.   3. *A. diffusa*

            Pericarp thin, granulate. Santa Cruz County.   4. *A. californica*

Nutlet attachment lateral, the scar linear; pedicels equaling the calyx to several times as long, or the flowers sessile.

    Corollas 1.5–2.5 mm. across; keel on the adaxial surface of the nutlet not in a groove. Rare.   5. *A. undulata*

    Corollas 4–10 mm. across; keel in a definite groove.

        Stems trailing, usually not branched below the 4th–6th internode; lower internodes elongate; pedicels 2–30 mm. long, usually longer than the calyx, often recurved; corolla 6–10 mm. broad.

                 6a. *A. chorisiana chorisiana*

        Stems branched from the base; lower internodes short; pedicels commonly shorter than the calyx; corolla 4–6 mm. broad.

                 6b. *A. chorisiana hickmanii*

1. A. glabra (Gray) Macbr. Glabrous Allocarya. Low alkaline soils in the Santa Clara Valley and rarely in the adjacent foothills; Santa Clara, 3 miles south of San Jose, and Los Gatos. April–May. —*Plagiobothrys glaber* (Gray) Johnston.

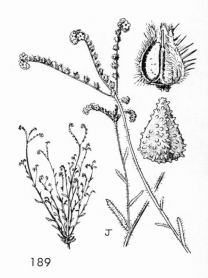

188   189

2. A. bracteata Howell. Bracted Allocarya. Common in moist grassy areas, ditches, and low wet places, at low elevations on the eastern side of the Santa Cruz Mountains; San Francisco, Crystal Springs Lake, near Woodside, Coal Mine Ridge, near Stanford, Permanente Canyon, near Campbell, Santa Clara, and near Wrights. March–May. —*A. cusickii* Greene var. *vallicola* Jeps., *Plagiobothrys bracteatus* (Howell) Johnston.

3. A. diffusa Greene. Diffuse Allocarya. Occasional in clay soils in San Francisco, where it is endemic. April–June. —*Plagiobothrys diffusus* (Greene) Johnston.

4. A. californica (F. & M.) Greene. California Allocarya. Known only from Santa Cruz County; about 4 miles north of Santa Cruz. April–May. —*Plagiobothrys reticulatus* (Piper) Johnston var. *rossianorum* Johnston.

5. A. undulata Piper. Coast Allocarya. Occasional in the Santa Cruz Mountains in marshy areas and in wet clay soils; Pilarcitos Lake, San Andreas Lake, and Madrone Station. April–July. —*Plagiobothrys undulatus* (Piper) Johnston.

6a. A. chorisiana (Cham.) Greene var. **chorisiana.** *Fig. 190.* Artist's Allocarya. Endemic; low boggy ground, San Francisco southward to central Santa Cruz County, generally absent east of the crest of the Santa Cruz Mountains; San Francisco, San Bruno Mountain, Sweeney Ridge, Crystal Springs Lake, Pebble Beach, Peters Creek, near La Honda, Belmont, Año Nuevo Point, Corte Madera, Hilton Airport, and Camp Evers. March–July. —*Plagiobothrys chorisianus* (Cham.) Johnston var. *chorisianus.*

6b. A. chorisiana (Cham.) Greene var. **hickmanii** (Greene) Jeps. Hickman's Allocarya. Low moist ground, coastal bluffs, and inland sand deposits; Pescadero, near Woodside, near Stanford, Swanton, Jamison Creek, Big Basin, near Olympia, and Camp Evers. March–August. —*A. chorisiana* (Cham.) Greene var. *myriantha* (Greene) Jeps., *Plagiobothrys chorisianus* (Cham.) Johnston var. *hickmanii* (Greene) Johnston.

11. Echidiocarya Gray

1. E. californica Gray. California Echidiocarya. Known locally only from Eagle Rock in Santa Cruz County, but to be expected elsewhere in rocky soils along the margins of chaparral. April–May. —*Plagiobothrys californicus* (Gray) Greene.

12. Plagiobothrys F. & M. Popcorn Flower

Calyx circumscissile; corolla-limb 6–8 mm. broad; margins and midribs of the
    leaves staining purple; nutlets 2–2.5 mm. long.   1. *P. nothofulvus*
Calyx not circumscissile; corolla-limb 1.5–4 mm. broad; leaves not staining
    purple; nutlets 1.5–3 mm. long.
    Nutlets 2–3 mm. long, triangular-ovoid to orbicular-ovoid; stems 1 to several
        from the base; plants usually 3–6 dm. tall; calyx about 4–6 mm. long.
        Stems 1–2 from the base; nutlets triangular-ovoid, 2.5–3 mm. long, with an
            annular caruncle.   2. *P. campestris*
        Stems several from the base; nutlets orbicular-ovoid, about 2 mm. long, with
            a solid caruncle.   3. *P. canescens*
    Nutlets 1.5–2 mm. long, thick-cruciform; stems usually many from the base;
        plants usually 0.5–3 dm. tall; calyx about 3 mm. long.   4. *P. tenellus*

1. P. nothofulvus (Gray) Gray. *Fig. 191.* Foothill Snowdrops, Rusty Popcorn Flower. A common annual of grassy fields and slopes, San Francisco southward. February–June. —*Plagiobothrys nothofulvus* is readily identified by its circumscissile calyx and purple-staining leaves. The superficial cells of the margins and the midribs of the leaves contain an anthocyanin pigment that stains the hands when the leaves are crushed or the newsprint when the plants are pressed.
2. P. campestris Greene. Fulvous Popcorn Flower. Occasional in grasslands in the Santa Clara Valley; Stanford and near San Jose. March–June. —*P. fulvus* (H. & A.) Johnston var. *campestris* (Greene) Johnston.
3. P. canescens Benth. Valley Popcorn Flower. Occasional in oak-woodland; near Felton, near Gilroy, and Sargents. April–May.
4. P. tenellus (Nutt.) Gray. Slender Popcorn Flower. Open grassland, rocky soils, and sandy soils of inland marine sand deposits; Menlo Park, Jasper Ridge, Stanford, Palo Alto, Black Mountain, near Los Gatos, Castle Rock, Jamison Creek, near Felton, and Ben Lomond Sand Hills. March–May.

190   191

13. Lithospermum L.

1. L. arvense L. Puccoon, Corn Gromwell. Reported from San Francisco and known from other parts of the San Francisco Bay area; native of Europe. May–June.

117. VERBENACEAE. Verbena or Vervain Family

Calyx 5-lobed; flowers in spikes; corollas blue to purplish.   1. *Verbena*
Calyx 2-lipped; flowers in dense heads; corollas white to pink or rose-colored.
                                                                      2. *Phyla*

1. Verbena L. Vervain

Adaxial leaf-surfaces not scabrous; spikes loosely flowered in fruit.
                                                          1. *V. lasiostachys*
Adaxial leaf-surfaces scabrous; spikes densely flowered in fruit.   2. *V. robusta*

1. V. lasiostachys Link. *Fig. 192.* California Vervain, Western Verbena. Dry ground of disturbed areas, creek bottoms, roadsides, and along the edges of brushy vegetation; San Francisco, Pilarcitos Canyon, Pescadero, Coal Mine Ridge, near Stanford, Gilroy, Glenwood, China Grade, Swanton, and Watsonville. May–September.
2. V. robusta Greene. Robust Vervain. Margins of ponds, river beds, and low moist ground, where water has stood during the winter, mainly on the eastern side of the Santa Cruz Mountains; Crystal Springs Lake, near Woodside, Stanford, near Los Altos, and Chittenden. June–October.

2. Phyla Lour. Mat Grass

1. P. nodiflora (L.) Greene var. **rosea** (D. Don) Moldenke. Garden Lippia. Occasional in marshes, in lawns, and in disturbed areas; San Francisco, Crystal Springs Lake, Stanford, near Gilroy, and near Felton. May–October. —*Lippia nodiflora* Michx. var. *rosea* (D. Don) Munz.

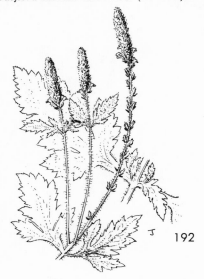

192   193

118. LABIATAE. Mint Family

Calyx 2-lipped.   *Laminaceae (new name)*
  Fertile stamens 2.
    Anther-cells 2, close together; bracts with 7–9 radiating spines; herbs.
                                                      1. *Acanthomintha*
    Anther-cells 1 or 2, if 2 the anther-cells separated by an elongated connective; bracts lacking spines or with only a terminal spine; herbs or shrubs.   2. *Salvia*
  Fertile stamens 4.
    Calyx-lips entire, the adaxial hood-like.   3. *Scutellaria*
    Calyx-lips toothed, not hood-like.
      Corollas white or slightly tinged with lavender; flowers in axial clusters; bracts lacking.   4. *Melissa*
      Corollas violet; flowers in dense terminal inflorescences; bracts present.   5. *Prunella*
Calyx regular or nearly so.
  Stamens long-exserted, at least twice as long as the corolla, 4; nutlets united.   6. *Trichostema*
  Stamens included or exserted, less than twice as long as the corolla, 2 or 4; nutlets distinct.
    Calyx-teeth 10, with hooked spines at the tips.   7. *Marrubium*
    Calyx-teeth 5, lacking hooked spines.
      Corollas regular or nearly so.
        Flowers in terminal heads; stamens 4. Dry habitats.   8. *Monardella*
        Flowers in axillary clusters or in verticils; stamens 2 or 4. Wet habitats.
          Stamens 2.   9. *Lycopus*
          Stamens 4.   10. *Mentha*
      Corollas strongly 2-lipped.
        Adaxial pairs of stamens longer than the abaxial.
          Leaves trifoliate.   11. *Cedronella*
          Leaves simple, variously toothed.
            Stems erect; leaves ovate to oblong.   12. *Nepeta*
            Stems creeping with ascending branches; leaves suborbicular.   13. *Glecoma*
        Adaxial pair of stamens shorter than or equaling the abaxial.
          Upper lip of the corollas concave.
            Corollas lacking a ring of hairs within.   14. *Lamium*
            Corollas with a ring of hairs within.   15. *Stachys*
          Upper lip of the corollas plane.
            Flowers solitary in the axils of the leaves or leaf-like bracts; plants perennial.
              Corollas 25–30 mm. long; shrubby plants, 1–1.5 m. tall, not trailing.   16. *Lepechinia*
              Corollas 6–8 mm. long; stems trailing.   17. *Satureja*
            Flowers in dense clusters; plants annual or perennial.
              Plants annual; bracts not colored; corollas 3–5 mm. long; calyx glabrous within.   18. *Pogogyne*
              Plants perennial; bracts colored; corollas 5–6 mm. long; calyx pubescent within.   19. *Origanum*

1. Acanthomintha Gray ex Benth. & Hook.

1. A. obovata Jeps. ssp. **duttonii** Abrams. Dutton's Thornmint. A local endemic of serpentine areas in southern San Mateo County; Crystal Springs Lake, Redwood City, near Menlo Golf Club, Emerald Lake, and Woodside serpentine. April–July.

2. Salvia L. Sage

Annual herbs; stamens with 2 anther-cells; leaves pinnatifid.   1. *S. columbariae*
Perennial herbs or shrubs; stamens with 1 anther-cell; leaves not pinnatifid.
    Perennial herbs; leaves hastate to truncate at the base; corollas about 3 cm.
      long.   2. *S. spathacea*
    Shrubs; leaves gradually narrowed at the base, not hastate or truncate; corollas
      about 1.2 cm. long.   3. *S. mellifera*

1. S. columbariae Benth. Chia. Dry open grassy slopes, edges of chaparral, and bare rocky slopes; Crystal Springs Lake, Coal Mine Ridge, Black Mountain, Alma Soda Springs, 4 miles south of San Jose, and Loma Prieta. March–July.
2. S. spathacea Greene. Canon, Pitcher, or Crimson Sage. Occasional on open slopes in San Francisco and northern San Mateo County; San Francisco and San Bruno Hills. March–May.
3. S. mellifera Greene. *Fig. 193.* California Black Sage. Dry slopes and chaparral, Santa Clara and Santa Cruz counties; Los Gatos Hills, near Los Gatos, El Toro, near Gilroy, Glenwood, Ben Lomond Mountain, Santa Cruz, and Pajaro River. March–August.

3. Scutellaria L. Skull Cap

Flowers blue; rootstocks terminating in tubers. Common.   1. *S. tuberosa*
Flowers white; rootstocks slender, lacking tubers. Rare.   2. *S. bolanderi*

1. S. tuberosa Benth. Dannie's Skull Cap. Oak-madrone woods, borders of shrubby vegetation, and in central Santa Cruz County on the inland marine sand deposits; San Francisco, Cahill Ridge, Crystal Springs Lake, Coal Mine Ridge, Stanford, Black Mountain, near Los Gatos, Loma Prieta, Swanton, near Glenwood, and near Felton. March–June.
2. S. bolanderi Gray. Sierra or Bolander's Skull Cap. Known in the Santa Cruz Mountains only from Uvas Creek in southern Santa Clara County. May.

4. Melissa L.

1. M. officinalis L. Garden or Lemon Balm. Roadsides and shaded canyons, most common in the central portion of the Santa Cruz Mountains; La Honda, Palo Alto, Wrights, Loma Prieta Ridge, Glenwood, Mount Hermon, and Boulder Creek; native of Europe. July–October.

5. Prunella L.

Leaves ovate to oblong-ovate, rounded at the base, ½ as broad as long. Lawn
    weeds.   1*a. P. vulgaris vulgaris*
Leaves lanceolate, gradually narrowed at the base, about ⅓ as broad as long.
    Native plants.   1*b. P. vulgaris lanceolata*

1a. P. vulgaris L. ssp. **vulgaris.** Self-heal, Heal-all. Occasional as a weed in lawns; San Francisco, Stanford, and probably elsewhere; native of Europe. June–July. —*P. vulgaris* L. var. *parviflora* (Poir.) DC.
1b. P. vulgaris L. ssp. **lanceolata** (Barton) Hulten. Along the coast on bluffs and inland in shaded woods; San Francisco, Cahill Ridge, Crystal Springs Lake, Pescadero, Pigeon Point, Kings Mountain, Boulder Creek, and Powder Mill Canyon. April–December. An indistinct coastal variety, *P. vulgaris* L. var. *atropurpurea* Fernald, has larger flowers and bracts that are less ciliate on the margins.

6. Trichostema L. Blue Curls

1. T. lanceolatum Benth. *Fig. 194.* Turpentine, Vinegar, or Camphor Weed. Dry open fields and slopes, southern San Mateo County southward; near Stanford, Black Mountain, Saratoga Summit, near Los Gatos, Waddell Creek, Hilton Airport, Felton, and near Watsonville. June–October.

7. Marrubium L.

1. M. vulgare L. Common or White Hoarhound. Widely distributed in disturbed areas, especially common along roadsides and in hard-packed soils; native of Eurasia. February–July.

8. Monardella Benth. Monardella

Plants annual, not woody at the base; leaves oblong, oblanceolate, to linear, undulate-margined or not.
  Leaves undulate-margined; bracts with the lateral veins more or less parallel to the midvein.   1. *M. undulata*
  Leaves not undulate-margined; bracts with the lateral veins not parallel to the midvein.   2. *M. douglasii*
Plants perennial, usually woody at the base; leaves ovate to lanceolate, or orbicular, not undulate-margined.
  Leaves thin, ovate to lance-ovate, not orbicular, nearly glabrous to somewhat villous.   3a. *M. villosa villosa*
  Leaves thick, broadly ovate to orbicular, densely villous to tomentose.
    3b. *M. villosa franciscana*

1. M. undulata Benth. *Fig. 195.* Curly-leaved Monardella. Coastal sand dunes in San Francisco and on the inland marine sand deposits in Santa Cruz County, at the latter locality growing in chaparral with *Adenostoma fasciculatum, Arctostaphylos silvicola, A. crustacea, Salvia mellifera,* and with occasional trees of *Pinus ponderosa* and *P. attenuata;* San Francisco, near Zayante, near Ben Lomond, Quail Hollow Road, and near Olympia. May–August.
2. M. douglasii Benth. Fenestra Monardella. Known locally only from serpentine outcroppings near Almaden Reservoir. June–July.
3a. M. villosa Benth. var. **villosa.** Pennyroyal, Coyote Mint. Fairly common in the Santa Cruz Mountains, often on rocky ridges and brushy canyon sides, San Francisco southward. May–September. —*M. villosa* Benth. var. *subglabra* Hoover.

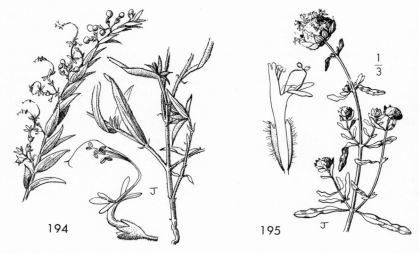

194   195

3b. M. villosa Benth. var. **franciscana** (Elmer) Jeps. Franciscan Coyote Mint. Mainly along or near the coast; San Francisco, San Bruno Hills, Edgemar, Montara Mountain, and San Pedro. May–July. *Monardella villosa* var. *franciscana*, a well-marked subspecific taxon of *M. villosa*, has thick, nearly orbicular, densely villous leaves and tomentose stems above. Intermediates between var. *villosa* and var. *franciscana* exist and are usually more inland than var. *franciscana*. Intermediates have been collected at Ocean View, Big Basin, Eagle Rock, and Swanton.

9. Lycopus L.

1. L. americanus Muhl. ex Barton. Cut-leaved Water Hoarhound. Known definitely only from Mountain Lake in San Francisco, but perhaps to be expected elsewhere. June–September.

10. Mentha L. Mint

Verticils remote, subtended by ordinary foliage leaves.
  Leaves subtending the verticils about equaling the cauline leaves, the leaves
    usually with 5–6 pairs of lateral veins.   1. *M. arvensis lanata*
  Leaves subtending the verticils conspicuously smaller than the cauline leaves,
    the leaves usually with 2–3 pairs of lateral veins.   2. *M. pulegium*
Verticils usually close together, subtended by bracts; inflorescence spike-like.
  Calyx under 3 mm. long; verticils numerous; inflorescence spike-like.
    Herbage glabrous or nearly so.   3. *M. spicata*
    Herbage tomentose, especially so abaxially.   4. *M. rotundifolia*
  Calyx 3–4 mm. long; verticils few to many.
    Inflorescence of 1–3 verticils, these usually head-like; calyx lavender-purple,
      densely beset with yellowish sessile glands; petioles of the main leaves
      usually at least 5 mm. long.   5. *M. citrata*
    Inflorescence of several to many verticils; calyx greenish-purple, glabrous to
      sparingly beset with yellowish sessile glands; petioles usually under 5
      mm. long or none.   6. *M. piperita*

1. M. arvensis L. var. **lanata** Piper. Field or Marsh Mint. Occasional along coastal marshes and the borders of ponds and lakes; San Francisco, Granada, Mountain View, and near Watsonville. August–October.

2. M. pulegium L. Pennyroyal. Weeds in disturbed areas, commonly of roadside ditches, low areas where water has stood, and in dry lake beds, occasionally abundant locally; Pilarcitos Canyon, near Crystal Springs Reservoir, near Woodside, Coal Mine Ridge, and Palo Alto; native of Europe. June–October.

3. M. spicata L. Spearmint. Occasional in disturbed areas; San Francisco, Stanford, near Raymonds, and Loma Prieta; native of Europe. July–October.

4. M. rotundifolia L. Apple Mint. Locally established near Boulder Creek; native of Europe. July–September.

5. M. citrata Ehrh. Bergamot Mint. Established in moist areas along creeks and ponds; San Francisco, Palo Alto, Campbell Creek, and Santa Cruz; native of Europe. July–October.

6. M. piperita L. Peppermint. Low moist places; near Alviso, between Alviso and Coyote Creek, and Aptos; native of Europe. August–November.

11. Cedronella Moench

1. C. canariensis (L.) Willd. ex Webb & Berth. Canary Balm, Herb-of-Gilead, Balm-of-Gilead. Occasional as an escape from cultivation in San Francisco; native of the Canary Islands. Summer months.

12. Nepeta L.

1. N. cataria L. Catnip, Catmint. Rare as an escape from cultivation; Raymonds; native of Europe. July–September.

13. Glecoma L.

1. G. hederacea L. Gill-over-the-ground, Ground Ivy. Reported from Little Arthur Creek, growing in moist shade; native of Europe. March–May. —*Nepeta hederacea* (L.) Trev.

14. Lamium L. Henbit

Upper leaves sessile to clasping.	1. *L. amplexicaule*
Upper leaves petiolate.	2. *L. purpureum*

1. L. amplexicaule L. Common Henbit, Dead Nettle, Giraffe Head. Common as a garden weed and as a weed of cultivated fields; native of Europe. January–August.

2. L. purpureum L. Red Henbit, Dead Nettle. Known locally only as a weed from San Mateo; native of Europe. April–September.

15. Stachys L. Hedge Nettle

Ring of hairs in the corolla-tube not producing a constriction visible on the surface.
  Ring of hairs horizontal, near the base of the tube; corolla-tube 8–10 mm. long; calyx 6–7 mm. long.   1. *S. bullata*

Ring of hairs oblique or horizontal, near the middle of the tube; corolla-tube
    15–25 mm. long; calyx 11–15 mm. long.   2. *S. chamissonis*
Ring of hairs producing a visible constriction.
    Leaves rounded to subcordate at base, usually ovate.
        Flower-verticils forming an interrupted spike; corolla rose-colored or
            purple; herbage not conspicuously glandular-villous.
                                                        3. *S. rigida quercetorum*
        Flower-verticils forming a short dense spike; corolla whitish; herbage
            glandular-villous.   4. *S. pycnantha*
    Leaves gradually narrowed at the base, not rounded or subcordate, oblong,
        silky-canescent.   5. *S. ajugoides*

1. S. bullata Benth. *Fig. 196.* California Hedge Nettle. Common throughout
the Santa Cruz Mountains, usually in moist areas in canyons, ravines, and woods.
April–September.
2. S. chamissonis Benth. Swamp Stachys, Coast Hedge Nettle. Along or near
the coast from San Francisco southward to Año Nuevo Point; San Francisco,
Pilarcitos Stone Dam, Arroyo de Los Frijoles, and Año Nuevo Point. May–
October.
3. S. rigida Nutt. ex Benth. ssp. **quercetorum** (Heller) Epling. Rigid Hedge
Nettle. Common from San Francisco southward, usually on bushy slopes and on
open grassy slopes, more common on the eastern slopes of the Santa Cruz Mountains. March–July. —*S. ajugoides* Benth. var. *quercetorum* (Heller) Jeps.
Stachys rigida ssp. *quercetorum* is commonly confused with *S. bullata*. In addition to the characters mentioned in the key, the following may be helpful in
distinguishing between the two species:

Ring of hairs at about the middle of the corolla-tube; corolla-tube usually at least
    somewhat saccate abaxially; calyx-lobes widely divergent at maturity; leaves
    3–7 cm. long, generally sparsely pubescent on the adaxial surface.
                                                        3. *S. rigida quercetorum*
Ring of hairs near the base of the corolla-tube; corolla-tube not saccate; calyx-
    lobes not widely divergent; leaves 3–15 cm. long, villous-tomentose on the
    adaxial surface.   1. *S. bullata*

4. S. pycnantha Benth. Coast Stachys, Short-spiked Hedge Nettle. Moist ground
near springs, ponds, and the borders of marshes, often on serpentine; Crystal
Springs Lake, Redwood City, Palo Alto, Black Mountain, Wrights Station, and
Goertz Ravine. July–October.
5. S. ajugoides Benth. Bugle Hedge Nettle. Usually in low, often marshy areas;
San Francisco, Crystal Springs Lake, Belmont, Cahill Ridge, Granada, Redwood
City, Stanford, Moffett Field, between Gilroy and Morgan Hill, Camp Evers, and
Santa Cruz. May–October.

16. Lepechinia Willd.

1. L. calycina (Benth.) Epling. *Fig. 197.* Pitcher Sage. Chaparral and brush-
covered slopes throughout the Santa Cruz Mountains, often growing with *Adenostoma fasciculatum, Pickeringia montana, Eriodictyon californica, Diplacus
aurantiacus,* and *Photinia arbutifolia,* San Francisco southward. April–July.
—*Sphacele calycina* Benth.

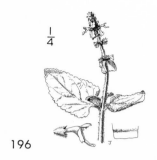

196   197

17. Satureja L.

1. S. douglasii (Benth.) Briq. Yerba Buena. Common on open slopes, in brush, and in redwood–Douglas fir forests, San Francisco southward. May–August. —*Micromeria douglasii* Benth.

18. Pogogyne Benth.

1. P. serpylloides (Torr.) Gray. Thyme-leaved Pogogyne. Occasional in grasslands and as a weed in paths and walks that have had puddles of water standing on them in the winter and early spring; San Francisco, near Crystal Springs Lake, Stanford, Permanente Creek, Los Gatos, and Santa Cruz. April–June.

19. Origanum L.

1. O. vulgare L. Wild Marjoram. Known from central Santa Cruz County, where it is thoroughly naturalized; near Glenwood and Felton; native of Europe. August–September.

119.   SOLANACEAE.   Nightshade Family

Plants spiny shrubs.   1. *Lycium*
Plants herbaceous or suffrutescent, if spiny then not shrubs.
  Corollas rotate or nearly so.
    Corollas blue or white; anthers lacking terminal sterile portions; berries
      small, yellow, reddish, or black.   2. *Solanum*
    Corollas yellow; anthers with terminal sterile portions; berries very large,
      red.   3. *Lycopersicon*
  Corollas not rotate, rather from campanulate to tubular.
    Calyx conspicuously enlarged and inflated in fruit.
      Corollas yellow.   4. *Physalis*
      Corollas purplish or reddish.   5. *Atropa*
    Calyx not conspicuously enlarged in fruit.
      Corollas tubular or prismatic, usually yellow or white.
        Calyx 3.5–10 cm. long; corollas white; fruit spiny.   6. *Datura*
        Calyx under 2.5 cm. long; corollas white to yellow (occasionally
          purplish); fruit not spiny.   7. *Nicotiana*
      Corollas urceolate to campanulate, purple, blue, pink, or white.
        Corollas funnelform, purple, blue, pink, or white.   8. *Petunia*
        Corollas urceolate, white.   9. *Salpichroa*

1. Lycium L.

1. L. halimifolium Mill. Matrimony Vine. Levees and borders of salt marshes and occasionally along streams, lower part of the Santa Clara Valley; Alviso Slough, San Jose, and Saratoga; native of southeastern Europe and western Asia. April–September.

2. Solanum L. Nightshade

Plants with spines; pubescence commonly stellate.
  Plants perennial; leaves lobed or sinuate.
    Leaves ovate, densely tomentose abaxially, glabrate adaxially.
                                            1. *S. marginatum*
    Leaves linear to lanceolate, silvery-canescent on both surfaces.
                                    2. *S. elaeagnifolium*
  Plants annual; leaves bipinnatifid.   3. *S. rostratum*
Plants without spines; pubescence various.
  Leaves simple, entire to variously lobed, not hastate at the base.
    Corollas white, rarely tinged with blue, deeply lobed; plants herbaceous, occasionally subsuffrutescent.
      Flowers large, the corolla-lobes 6–11 mm. long; anthers 2.6–4 mm. long; ripe berries black.
        Peduncles deflexed at maturity; berries dehiscent immediately when ripe.   4. *S. furcatum*
        Peduncles erect at maturity; berries not dehiscent immediately when ripe.   5. *S. douglasii*
      Flowers small, the corolla-lobes about 2–4 mm. long; anthers 1.2–2.5 mm. long; ripe berries green, yellowish, reddish, or black.
        Herbage conspicuously villous or hirsute, often glandular; anthers 2–2.5 mm. long; berries green, yellowish, or reddish; seeds over 1.8 mm. long.   6. *S. sarrachoides*
        Herbage subglabrous; anthers 1.2–1.5 mm. long; berries black; seeds under 1.8 mm. long.   7. *S. nodiflorum*
    Corollas blue, shallowly lobed; plants suffrutescent.
      Stems glabrous.   8. *S. aviculare*
      Stems pubescent.
        Leaves under 7 cm. long; more or less erect shrubs, under 1 m. tall. Native plants.
          Pubescence of the stem of forked or branched trichomes. Common.   9. *S. umbelliferum*
          Pubescence of the stem of nonbranched trichomes, often glandular. Infrequent.   10. *S. xanti intermedium*
        Leaves to 25 cm. long; climbing shrubs, 2–3 m. tall. Introduced plants.   11. *S. gayanum*
  Leaves pinnately compound or hastate at the base.
    Leaves pinnate; stems not climbing; plants with tubers.   12. *S. tuberosum*
    Leaves hastate at the base; stems climbing; plants lacking tubers.
                              13. *S. dulcamara*

1. S. marginatum L. f. White-margined Nightshade. Occasional as a weed in San Francisco, but well established; native of Africa. May–August.

2. S. elaeagnifolium Cav. Bull-Nettle, Silver-leaved Nettle. Occasional as a spontaneous weed in San Francisco; native of the central portion of the United States.

3. S. rostratum Dunal. Buffalo Bur. Occasional as a weed; San Francisco, Santa Cruz, and probably elsewhere; native from South Dakota southward into Mexico. May–September.

4. S. furcatum Dunal. Forked Nightshade. Occasional in disturbed areas near the coast from San Francisco to central San Mateo County; San Francisco, Colma, and Pedro Point; native of Chile. April–October.

5. S. douglasii Dunal. Douglas' Nightshade. Along or near the coast; San Francisco, Sharp Park, San Bruno Mountain, Half Moon Bay, Año Nuevo Point, San Mateo–Santa Cruz County line, Swanton, between Scott Valley and Santa Cruz, Aptos, and Sunset Beach State Park. March–December.

6. S. sarrachoides Sendt. ex Mart. Hairy Nightshade. A weed of cultivated fields, gardens, and vineyards; San Francisco, Pedro Valley, Moss Beach, Palo Alto, and 6 miles south of San Jose; native of South America. August–November.

7. S. nodiflorum Jacq. Small-flowered Nightshade. A common weed of disturbed areas and cultivated fields; native of the New World tropics. March–December. —*S. nigrum* of authors.

8. S. aviculare Forst. f. Poporo. Occasional as a weed in urban areas; San Francisco and probably elsewhere; native of New Zealand. June–September.

9. S. umbelliferum Esch. *Fig. 198.* Blue Witch. A common, widely distributed plant occurring on wooded and brush-covered slopes and along the margins of chaparral, often growing with *Arbutus menziesii, Baccharis pilularis, Ceanothus*, and *Quercus*; San Francisco, San Bruno Hills, Peters Creek, Woodside, Kings Mountain, near Stanford, near Los Gatos, Loma Prieta, near Gilroy, Rancho del Oso, near Big Basin, Glenwood, Santa Cruz, New Brighton, and near Watsonville. January–September. White-flowered individuals are occasionally encountered. Plants with densely white-tomentose stems may be referred to *S. umbelliferum* Esch. var. *incanum* Torr., and are so far known only from southern Santa Cruz County.

10. S. xanti Gray var. **intermedium** Parish. Purple Nightshade. Rare in the Santa Cruz Mountains, known locally only from near Saratoga. February–September.

11. S. gayanum (Remy) Phil. f. Occasionally spontaneous in Golden Gate Park in San Francisco; native of Chile.

12. S. tuberosum L. Potato. Occasional in disturbed ground in urban areas; San Francisco and probably elsewhere; native of South America. April–September.

13. S. dulcamara L. Climbing Nightshade, Bitter Sweet. Rare as an escape from cultivation in San Francisco; native of Europe. May–September.

3. Lycopersicon Mill.

1. L. esculentum Mill. Tomato. Occasional as an escape from cultivation; San Francisco, near Alviso, Los Gatos Canyon, and Pajaro River; native of South America. March–September.

4. Physalis L. Ground Cherry

Plants essentially glabrous.	1. *P. ixocarpa*
Plants distinctly pubescent.	2. *P. pubescens*

1. P. ixocarpa Brot. ex Hornem. Tomatillo. Known locally from southern Santa Cruz and Santa Clara counties, but to be expected occasionally elsewhere; near Watsonville and Gilroy; native of Mexico. June–October.
2. P. pubescens L. Low Hairy Ground Cherry. Locally established in Boulder Creek; native of tropical America. June–December.

5. Atropa L.

1. A. belladonna L. Belladonna, Deadly Nightshade. Occasional as an escape from cultivation in Golden Gate Park; native of Eurasia.

6. Datura L.

Corollas under 10 cm. long; plants annual.   1. *D. stramonium*
Corollas over 15 cm. long; plants perennial.   2. *D. inoxia*

1. D. stramonium L. Jimson Weed. Occasional in disturbed habitats; San Francisco, Redwood City, Palo Alto, Stanford, near Alviso, Los Altos, and San Jose; native of Asia. June–September. *Datura stramonium* L. var. *tatula* (L.) Torr., Purple Thorn Apple, has lavender corollas and has been found at Saratoga and Freedom.
2. D. inoxia Mill. Tolguacha. Occasional in disturbed areas; San Francisco, Stanford, and Coyote; native of other parts of the United States and Mexico. April–September. —*D. meteloides* DC. ex Dunal.

7. Nicotiana L. Tobacco

Small trees or large shrubs; corollas yellow.   1. *N. glauca*
Herbs; corollas greenish to white, sometimes tinged with red.
  Corollas 3–7 cm. long, gradually widening from the base to the limb.
    Corolla-limb usually under 1.5 cm. in diameter; stamens inserted in the lower half of the corolla-tube; cauline leaves petiolate.
        2. *N. acuminata multiflora*
    Corolla-limb 2.5–5 cm. in diameter; stamens inserted in the upper half of the corolla-tube; cauline leaves more or less sessile.   3. *N. bigelovii*
  Corollas 6.5–8.5 cm. long, not gradually widening from the base to the limb, but rather with a gradual or an abrupt constriction below the limb.
    Constriction below the limb gradual; corollas white, about 6 times as long as the calyx; panicles short, somewhat congested.   4. *N. sylvestris*
    Constriction below the limb abrupt; corollas about 3–4 times as long as the calyx; panicles elongate.
      Corolla-tube yellowish-green, tinged with red, the limb reddish.
        5. *N.* × *sanderae*
      Corolla-tube greenish-white, the limb white.   6. *N. alata*

1. N. glauca Graham. *Fig. 199.* Tree or Mexican Tobacco. Stream beds and disturbed areas; San Francisco, Menlo Park, Stanford, Mountain View, and Pajaro River; native of South America. March–September.
2. N. acuminata (Graham) Hook. var. **multiflora** (Philippi) Reiche. Many-flowered Tobacco. Occasional in sandy soils and in disturbed areas; Stanford, near Coyote, Boulder Creek, near Freedom, and Watsonville; native of Chile. May–September.

3. N. bigelovii (Torr.) Wats. Indian Tobacco. Occasional on the eastern slopes of the Santa Cruz Mountains; San Francisco, Black Mountain, and Soda Springs. May–July.

4. N. sylvestris Spegazzini & Comes. Occasional as a spontaneous plant on the Stanford University campus; native of Argentina. June–September.

5. N. × sanderae Wats. Cultivated in Golden Gate Park in San Francisco and well established; a hybrid garden ornamental, derived from *N. alata* Link & Otto and *N. forgetiana* Hemsley, both native of South America.

6. N. alata Link & Otto. Commonly cultivated in gardens and occasionally escaping and becoming established; Stanford; native of South America. May–August.

8. Petunia Juss.

Corollas under 1 cm. long. Native plants.	1. *P. parviflora*
Corollas over 5 cm. long. Garden escapes.	2. *P. hybrida*

1. P. parviflora Juss. Wild Petunia. Sandy soils along rivers, ponds, and lakes, and becoming a weed in Golden Gate Park; San Francisco, Spring Valley Lakes, Crystal Springs Lake, Searsville Lake, Santa Cruz, Chittenden Pond, and Pajaro River. June–September.

2. P. hybrida Vilm. Common Garden Petunia. Widely grown as an ornamental and occasionally becoming established locally; Stanford; of hybrid origin of South American species of *Petunia*. Flowering throughout most of the year.

    Nicandra physalodes (L.) Gaertn. has been collected as a weed in a Brussels sprouts field near Moss Landing in northern Monterey County and grows as an ornamental in Golden Gate Park in San Francisco. It is to be expected in the Santa Cruz Mountains. *Petunia parviflora* has entire leaves, while *Nicandra* has coarsely toothed leaves.

198

199

9. Salipchroa Meirs

1. S. rhomboidea (Gill. & Hook.) Miers. Lily-of-the-Valley Vine. Occasional in disturbed areas; San Francisco, Santa Cruz, and probably elsewhere; native of South America. June–October.

120. SCROPHULARIACEAE. Figwort Family

Stamens 5; corollas yellow. 1. *Verbascum*
Stamens 2 or 4, a fifth sterile filament sometimes present.
  Corollas spurred or saccate at the base on the lower side of the corolla.
    Corollas spurred at the base.
      Plants erect, glabrous; leaves linear. 2. *Linaria*
      Plants prostrate, glabrous or pubescent; leaves ovate to orbicular, entire or lobed.
        Leaves entire, pinnately veined; plants pubescent. 3. *Kickxia*
        Leaves lobed, palmately veined; plants glabrous. 4. *Cymbalaria*
    Corollas saccate at the base.
      Capsules symmetrical, not opening by pores. 5. *Asarina*
      Capsules asymmetrical, opening by pores. 6. *Antirrhinum*
  Corollas not spurred or saccate at the base, but sometimes gibbous at the base on the upper side.
    Upper lip of the corolla not galeate.
      Stamens with anthers 4.
        Corollas essentially regular; prostrate aquatic plants. 7. *Limosella*
        Corollas distinctly irregular; plants, if aquatic, not prostrate.
          Median lobe of the lower lip of the corolla keeled. 8. *Collinsia*
          Median lobe of the lower lip of the corolla not keeled.
            Flowers small, the corollas less than 5 mm. long. 9. *Tonella*
            Flowers larger, the corollas over 5 mm. long.
              Sterile filament present.
                Corollas 8–15 mm. long. 10. *Scrophularia*
                Corollas 25–35 mm. long. 11. *Penstemon*
              Sterile filament lacking.
                Corollas campanulate, the lobes erect; flowers pendent; stems usually over 1 m. tall, herbaceous. 12. *Digitalis*
                Corollas not campanulate, the lobes not erect; flowers pendent to erect; stems usually under 1 m. tall or, if taller, then the plants woody.
                  Plants woody. 13. *Diplacus*
                  Plants herbaceous. 14. *Mimulus*
      Stamens with anthers 2.
        Corollas tubular.
          Leaves linear-lanceolate; stems thick, fleshy; sepals exceeding the mature capsules. 15. *Gratiola*
          Leaves ovate; stems capillary, not fleshy; sepals not exceeding the mature capsules. 16. *Lindernia*
        Corollas rotate. 17. *Veronica*

Upper lip of the corollas galeate.

Anther-cells equal in size and borne side by side; seed-coat not conspicuously reticulate; leaves opposite or alternate.

Corollas reddish-purple; plants perennial; leaves pinnatifid, alternate.
18. *Pedicularis*

Corollas pale purple to white; plants annual; leaves simple, opposite.
19. *Bellardia*

Anther-cells often unequal in size, not borne side by side, or 1-celled; seed-coat obviously reticulate; leaves alternate.

Plants annual.

Plants over 4 dm. tall, very much branched; bracts not colored.
20. *Cordylanthus*

Plants under 3 dm. tall, simple or with a few branches; bracts commonly colored.
21. *Orthocarpus*

Plants perennial (see also *Castilleja stenantha*).

Bracts not colored; calyx spathe-like.
20. *Cordylanthus*

Bracts commonly colored; calyx tubular.
22. *Castilleja*

1. Verbascum L. Mullein

Herbage densely stellate-pubescent; calyx densely stellate-pubescent, not glandular.

Spike dense, the flowers completely obscuring the axis of the inflorescence; capsules 8–10 mm. long.
1. *V. thapsus*

Raceme not dense, the axis visible; capsules 4–6 mm. long.
2. *V. speciosum*

Herbage glabrous to thinly pubescent; calyx with at least some glandular hairs.

At least some nodes with more than one flower; pedicels 3–5 mm. long; capsules with some stellate or branched hairs; leaves pubescent.
3. *V. virgatum*

Only one flower per node; pedicels 10–15 mm. long; capsules lacking stellate or branched hairs; leaves glabrous.
4. *V. blattaria*

1. V. thapsus L. Common or Woolly Mullein. Occasional in pastures, along roadsides, and in other disturbed habitats; San Francisco, near Woodside, Saratoga, near Coyote, and Highland Way; native of Eurasia. June–November.

2. V. speciosum Schrad. Showy Mullein. Known locally only from near the summit of the Santa Cruz–Los Gatos Highway; native of Eurasia. July–September.

3. V. virgatum Stokes ex With. Wand Mullein. Occasional as a weed in fallow fields; China Grade and Santa Cruz; native of Eurasia. May–October.

4. V. blattaria L. Moth Mullein. Occasional in dry creek beds and in disturbed areas; San Francisco, near Stanford, and near Gilroy; native of Eurasia. March–October.

2. Linaria Mill. Toadflax

Corollas purple; spurs curved.

Spur stout; palate well developed; corollas violet-purple. Introduced plants.
1. *L. bipartita*

Spur filiform; palate not well developed; corollas violet. Native plants.
2. *L. texana*
Corollas yellow; spurs straight.
3. *L. vulgaris*

1. L. bipartita Willd. Known as an occasional escape from cultivation in San Francisco; Twin Peaks and Lake Merced; native of the Mediterranean region. February–June.
2. L. texana Scheele. Blue Toadflax. Sandy or loose soil throughout the Santa Cruz Mountains; San Francisco, Crystal Springs Lake, Tunitas Creek, Stanford, Ben Lomond Mountain, near Ben Lomond, and Sunset Beach State Park. February–May. —*L. canadensis* (L.) Dumort. var. *texana* (Scheele) Pennell.
3. L. vulgaris Hill. Butter-and-Eggs. Occasional as a garden escape; Stanford; native of Eurasia. Summer months.

3. **Kickxia** Dumort. Fluellin

1. K. spuria (L.) Dumort. Round-leaved Fluellin. Occasional as a weed in hard-packed soil; Menlo Park, Stanford, and Palo Alto; native of Eurasia and Africa. June–November.

4. **Cymbalaria** Hill

1. C. muralis Gaertn., Mey. & Scherb. Kenilworth Ivy. Becoming established as an escape from cultivation in San Francisco and at Stanford; native of Europe. April–September.

5. **Asarina** Mill.

1. A. stricta (H. & A.) Pennell. Lax Snapdragon. Occasional in clearings in chaparral; La Honda Road and near Los Gatos. April–May. —*Antirrhinum kelloggii* Greene.

6. **Antirrhinum** L. Snapdragon

Sepals all equal; corollas 3–5 cm. long. Introduced plants.   1. *A. majus*
Sepals not equal, the adaxial one usually longer than the others; corollas under 2 cm. long.
 Seeds not cup-shaped; calyx-lobes usually less than 1 cm. long. Native plants.
  Inflorescence usually congested; corollas 1.5–2 cm. long; lower leaf-blades 4–7 cm. long.   2. *A. multiflorum*
  Inflorescence usually lax; corollas under 1.5 cm. long; lower leaf-blades 1.5–2.5 cm. long.   3. *A. vexillo-calyculatum*
 Seeds cup-shaped; calyx-lobes usually 1.2–2 cm. long. Introduced plants.
    4. *A. orontium*

1. A. majus L. Garden Snapdragon. Rare as a garden escape, possibly not persisting; Bayview Hills and Stevens Creek; native of Europe. March–June.
2. A. multiflorum Pennell. Withered Snapdragon. Sandy soil in chaparral; Los Gatos Canyon, Mount Umunhum, Loma Prieta, Big Basin, Glenwood, Ben Lomond Sand Hills, and Santa Cruz. June–September.

3. A. vexillo-calyculatum Kell. *Fig. 200.* Wiry Snapdragon. Chaparral, serpentine, and open rocky areas, mainly on the eastern slopes of the Santa Cruz Mountains; Woodside, Coal Mine Ridge, Los Gatos, Oakhill Cemetery, Loma Prieta, Big Basin Road, and below Saratoga Summit. April–October. —*A. elmeri* Rothm.

4. A. orontium L. Corn Snapdragon. Occasional as a weed and persisting in gardens; Stanford and 10 miles north of Santa Cruz; native of the Mediterranean region and Eurasia. April–June.

7. Limosella L.

1. L. aquatica L. Northern Mudwort. Mud of margins of lakes and ponds; Lone Mountain, Crystal Springs, Año Nuevo Point, and Felt Lake. May–October.

8. Collinsia Nutt. Chinese Houses

Upper pair of stamens each with an upward-projecting, awn-like process about
   2 mm. long at the base of the filaments.   1. *C. heterophylla*
Upper pair of stamens lacking an upward-projecting, awn-like process or this
   less than 1 mm. long.
  Pedicels shorter than the calyces.
    Lateral corolla-lobes not bearded; leaves essentially glabrous abaxially;
      plants not glandular-pubescent.
      Upper corolla-lobes less than ½ as long as the lower; inflorescences very
        congested, usually of only 1 terminal verticil.   2. *C. corymbosa*
      Upper corolla-lobes more than ½ as long as the lower; inflorescences less
        congested, usually of several verticils.   3. *C. bartsiaefolia hirsuta*
    Lateral corolla-lobes bearded; leaves pubescent abaxially; plants densely
      glandular-pubescent.   4. *C. tinctoria*
  At least some of the lower pedicels longer than the calyces.
    Only the lower pedicels longer than the calyces; corollas 15 mm. long or
      longer; inflorescences with gland-tipped hairs.   5. *C. franciscana*
    All pedicels longer than the calyces; corollas under 10 mm. long; inflores-
      cences lacking gland-tipped hairs.   6. *C. sparsiflora collina*

1. C. heterophylla Buist ex Grah. *Fig. 201.* Innocence, Purple-and-white Chinese Houses. Shaded grassy slopes and fields, commonly associated with *Quercus agrifolia, Q. kelloggii,* and *Arbutus menziesii*; Lake Merced, La Honda Road, Stanford, near Los Gatos, Guadalupe Hills, Bear Creek Road, Mount Charley Road, Swanton, and Santa Cruz. April–July. —*C. bicolor* Benth., *C. multicolor* Lindl. & Paxt. *Collinsia heterophylla* is the most common species of this genus in the Santa Cruz Mountains.

2. C. corymbosa Herder. Round-headed Chinese Houses. Sand dunes along the coast in San Francisco; Lake Merced and Mountain Lake. May–June.

3. C. bartsiaefolia Benth. var. **hirsuta** (Kell.) Pennell. White Chinese Houses. Occasionally in sandy soil; Lake Merced, Tuxedo, and Quail Hollow Road. April–May.

4. C. tinctoria Hartw. Iodine Collinsia. Known locally from a single specimen collected at Lake Merced. May–August.

200

201

5. C. franciscana Bioletti. Franciscan Blue-eyed Mary. Open or shaded slopes throughout the Santa Cruz Mountains, occasionally on serpentine; San Francisco, San Bruno Hills, San Mateo Creek, Stanford, Edenvale, and Waddell Creek. March–May. —*C. sparsiflora* F. & M. var. *franciscana* (Bioletti) Jeps.

6. C. sparsiflora F. & M. var. **collina** (Jeps.) Newsom. Open grassy fields and slopes from San Francisco southward, on the eastern margins of the Santa Cruz Mountains; Mission Hills, San Andreas Lake, Crystal Springs Lake, Stanford, and Guadalupe Creek. March–May. —*C. solitaria* Kell.

9. Tonella Nutt. ex Gray

1. T. tenella (Benth.) Heller. Small-flowered Tonella. Occasional in the oak belt on the eastern slopes of the Santa Cruz Mountains; Stanford, Coal Mine Ridge, Saratoga Springs, Los Gatos Creek, and slope of Loma Prieta. March–May.

10. Scrophularia L. Figwort

1. S. californica C. & S. *Fig. 202.* Coast Figwort, California Bee-Plant. Woods, borders of thickets, occasionally in chaparral, and in coastal scrub; San Francisco, San Bruno Hills, Half Moon Bay Road, Coal Mine Ridge, near Los Gatos, San Jose, Edenvale, Swanton, and near Aptos. February–July. —*S. multiflora* Pennell, *S. californica* C. & S. var. *floribunda* Greene.

11. Penstemon Mitch. Beard-Tongue

Plants perennial herbs; leaves over 5 cm. long. 1. *P. rattanii kleei*
Plants low woody shrubs; leaves under 5 cm. long.
    Corollas red, buds red; leaves elliptic, usually serrate; inflorescence puberulent. 2. *P. corymbosus*
    Corollas rose-violet, buds yellowish; leaves linear, entire; inflorescence usually glabrous. 3. *P. heterophyllus*

1. P. rattanii Gray ssp. **kleei** (Greene) Keck. Rattan's Penstemon. Endemic in the Santa Cruz Mountains, usually in areas of somewhat sandy soil in chaparral or burned chaparral; Ben Lomond Mountain, Eagle Rock, Loma Prieta, and Aptos Creek. April–July.
2. P. corymbosus Benth. Redwood Penstemon. Occasional in crevices of dry rocky outcroppings and ledges; Stevens Creek Canyon, Mount Umunhum, Loma Prieta, and Eagle Rock. July–August.
3. P. heterophyllus Lindl. Foothill Penstemon. Rare locally, known only from Los Gatos, Mount Umunhum, and Madrone. June–July.

12. Digitalis L.

1. D. purpurea L. Purple Foxglove. Occasional as an escape from gardens; San Francisco, San Francisco Watershed Reserve, China Grade, and San Vicente Creek; native of Europe. May–September. Flower color usually varies within a local colony from nearly pure white to pink to purple.

13. Diplacus Nutt.

1. D. aurantiacus (Curtis) Jeps. *Fig. 203.* Sticky or Bush Monkey Flower. One of the commonest shrubs of dry hillsides, chaparral, rock outcroppings, and coastal scrub; commonly associated with *Eriodictyon californicum, Adenostoma fasciculatum, Quercus agrifolia, Q. chrysolepis, Rhus diversiloba*, and various species of *Arctostaphylos*; San Francisco southward. March–July. —*Mimulus aurantiacus* Curtis.

202   203

14. **Mimulus** L. Monkey Flower

Flowers yellow.
  Plants densely shaggy-villous.   1. *M. pilosus*
  Plants not shaggy-villous.
    Calyx-teeth conspicuously unequal, the upper much larger than the others;
      calyx much inflated; corolla-throat partly closed.   2. *M. guttatus*
    Calyx-teeth all nearly equal, the upper not conspicuously larger; calyx not
      conspicuously inflated; throat of the corolla open.
      Calyx-teeth broadly deltoid, rounded at the apex; plants annual, not
        rhizomatous.   3. *M. floribundus*
      Calyx-teeth narrowly deltoid, acute at the apex; plants perennial, rhi-
        zomatous.   4. *M. moschatus*
Flowers red, purple, or violet.
  Plants perennial, over 2.5 dm. tall; anthers ciliate; calyx 25–30 mm. long.
    Plants of wet areas.   5. *M. cardinalis*
  Plants annual, under 1.5 dm. tall; anthers glabrous; calyx 9–12 mm. long.
    Plants of dry areas.
    Pedicels at least 3 times as long as the calyx.   6. *M. androsaceus*
    Pedicels shorter than the calyx.
      Corolla 15–40 mm. long; lobes of the lower lip of the corolla smaller than
        those of the upper lip or nearly lacking; capsule asymmetrical.
        Corolla 30–40 mm. long; lower lip of the corolla almost lacking.
                                         7. *M. douglasii*
        Corolla 15–25 mm. long; lower lip of the corolla well developed.
                                         8. *M. congdonii*
      Corolla 8–10 mm. long; lobes of the corolla all nearly equal; capsule
        symmetrical.   9. *M. rattanii*

1. M. pilosus (Benth.) Wats. Downy Mimetanthe. Known definitely only from near Eagle Rock, but to be expected elsewhere in the nearly bare soil of disturbed areas and in sandy soil along streams. June–July. —*Mimetanthe pilosa* (Benth.) Greene.

2. M. guttatus DC. Common Large Monkey Flower. A very common species which can be found at most any location that has springs, seeps, ponds, or swales, San Francisco southward. January–October. *Mimulus guttatus* is a common but variable species in which many subspecies and varieties have been described and which has been divided into a number of segregate species. The following key, together with the more important synonyms, indicates some of the modes of variation in *M. guttatus*. Of the five varieties that occur in the Santa Cruz Mountains, none shows any definite geographical or ecological distribution patterns except perhaps *M. guttatus* var. *grandis*, which tends to be more coastal.

Corolla less than 1.5 times the length of the calyx, under 1 cm. long.
                        2a. var. *micranthus* (Heller) Campbell
Corolla at least twice as long as the calyx, over 1.5 cm. long.
  Calyx more or less truncate at both the top and bottom, sinuses glabrous.
                        2b. var. *arvensis* (Greene) Grant
  Calyx not truncate, sinuses pubescent.

Calyx 6–22 mm. long, glabrous to sparsely pubescent; capsule 7–8 mm. long.
Inland plants.
Stems 2–13 mm. in diameter, not geniculate; inflorescence never scorpi-
oid; upper calyx-tooth usually 2 times as long as the others.
2c. var. *guttatus*
(—*M. langsdorfii* Donn, *M. scouleri* Hook., *M. guttatus* var. *depauper-
atus* (Gray) Grant, *M. platycalyx* Pennell)
Stems 5–7 mm. in diameter, often geniculate; inflorescence often scorpioid
when young; upper calyx-tooth usually 3 times as long as the others.
2d. var. *gracilis* (Gray) Campbell
(—*M. nasutus* Greene, *M. glareosus* Greene)
Calyx 17–30 mm. long, pubescent to tomentose; capsule 8–12 mm. long.
Coastal plants.   2e. var. *grandis*
(—*M. guttatus* ssp. *litoralis* Pennell, *M. guttatus* ssp. *arenicola* Pennell)

3. M. floribundus Dougl. ex Lindl. Floriferous Monkey Flower. Rare in the
Santa Cruz Mountains; Swanton, near Boulder Creek, Felton, and Santa Cruz.
April–September.
4. M. moschatus Dougl. ex Lindl. Musk Flower. Occasional in stream beds in
Santa Cruz County and as an escape from cultivation in San Francisco. May–
November. —*M. moschatus* Dougl. ex Lindl. var. *longiflorus* Gray, *M. moschatus*
Dougl. ex Lindl. var. *sessifolius* Gray.
5. M. cardinalis Dougl. ex Benth. *Fig. 204.* Scarlet Monkey Flower. Occasional,
from San Francisco southward, along springs and streams; San Francisco, Sawyer Ridge, Lake San Andreas, near Stanford, near Black Mountain, Wrights,
Loma Prieta, and Swanton. June–July.
6. M. androsaceus Curran ex Greene. Androsace Monkey Flower. Known
locally only from Ben Lomond Sand Hills. April.
7. M. douglasii (Benth.) Gray. Purple Mouse-Ears. Occasional on bare sand,
serpentine, grassy slopes, and in clearings on brush-covered slopes; near Woodside, Coal Mine Ridge, Stevens Creek, Black Mountain, near Los Gatos, and near
Boulder Creek. February–April. Plants that produce cleistogamous flowers have
been called *M. cleistogamus* Howell. The formation of cleistogamous flowers
appears to be correlated with a water shortage.
8. M. congdonii Robins. Congdon's Monkey Flower. Known locally only from
the vicinity of Boulder Creek. March–May.
9. M. rattanii Gray. Rattan's Monkey Flower. Rocky and sandy soils; Mount
Umunhum, Loma Prieta, and Ben Lomond Sand Hills. April–June. —*M. rattanii* Gray var. *decurtatus* (Grant) Pennell.

15. Gratiola L.

1. G. ebracteata Benth. Bractless Hedge-hyssop. Occasional along the muddy
shores of ponds near the southern end of Crystal Springs Lake. April–May.

16. Lindernia All.

1. L. anagallidea (Michx.) Pennell. Long-stalked Lindernia. Known locally
from Pinto Lake in southern Santa Cruz County, but to be expected elsewhere in
shallow waters along lakes and ponds. July–November.

17. Veronica L. Speedwell

Herbs.
All leaves opposite; the main stem not terminating in an inflorescence.
Plants essentially glabrous. Growing in moist or wet habitats.
    Leaves sessile and clasping.   1. *V. connata glaberrima*
    Leaves short-petiolate, not clasping.   2. *V. americana*
Plants pubescent. Growing in dry habitats.   3. *V. chamaedrys*
Upper leaves often bract-like, alternate; main stem terminating in an inflorescence.
Pedicels longer than the subtending leaf-like bracts; corollas over 5 mm. in diameter.
    Plants annual; stems stout; leaves usually longer than broad.
        4. *V. persica*
    Plants perennial; stems filiform; leaves usually broader than long.
        5. *V. filiformis*
Pedicels shorter than the subtending leaf-like bracts; corollas less than 5 mm. in diameter.
    Plants perennial; stems creeping, rooting at the nodes; style about as long as the mature capsule.   6. *V. serpyllifolia*
    Plants annual, not creeping, the lower nodes occasionally with adventitious roots; style less than ½ the length of the mature capsule.
        Leaves ovate to suborbicular, crenate; capsules deeply obcordate.
        7. *V. arvensis*
        Leaves oblong to lanceolate, obscurely crenate; capsules shallowly notched.   8. *V. peregrina xalapensis*
Shrubs.
    Leaves obovate to obovate-oblong, 5–10 cm. long; corollas reddish-purple.
        9. *V.* × *franciscana*
    Leaves elliptic, about 3 cm. long; corollas lilac-purple.   10. *V. speciosa*

1. V. connata Raf. ssp. **glaberrima** Pennell. Broad-fruited Water Speedwell. Known locally only from southern Santa Cruz County, but to be expected occasionally elsewhere in moist sand of river beds; Corralitos and Pajaro River. May–October. —*V. comosa* Richt.

2. V. americana (Raf.) Schwein. ex Benth. *Fig. 205.* American Brooklime. Along the margins of ponds, streams, and ditches; San Francisco, Pilarcitos Canyon, Crystal Springs Lake, Moss Beach, Felt Lake, Mountain View–Alviso Road, Swanton, and Watsonville Slough. May–October.

3. V. chamaedrys L. Germander Speedwell. Occasional as a lawn weed in San Francisco; native of Europe.

4. V. persica Poir. Persian Speedwell. Fairly common along stream beds, as a weed in gardens and lawns, and in disturbed areas, San Francisco southward; native of Eurasia. March–September. —*V. buxbaumii* Ten.

5. V. filiformis Sm. Occasional as a weed in San Francisco; native of the Caucasus and Asia Minor.

6. V. serpyllifolia L. Thyme-leaved Speedwell. Occasional as a lawn and garden weed; San Francisco, Palo Alto, and Montalvo; native of Europe. April–August.

7. V. arvensis L. Corn Speedwell. Weeds of gardens and disturbed areas; San Francisco, Stanford, Montalvo, and Swanton; native of Europe. April–June.

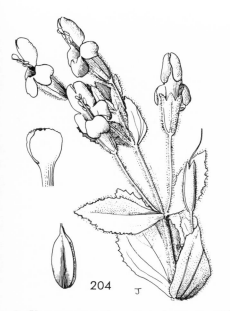

204   J   205   J

8. V. peregrina L. ssp. **xalapensis** (HBK.) Pennell. Purselane Speedwell. Fairly common in marshy ground and in soils where water has stood during the spring; San Francisco, San Andreas Lake, Crystal Springs Lake, Stanford, Uvas Creek, and Swanton. April–June.

9. V. × franciscana Eastw. Commonly cultivated as a hedge plant in San Francisco and naturalized on ocean bluffs. Flowering in the summer and fall. —*Hebe × franciscana* (Eastw.) Souster. *Veronica × franciscana* is a hybrid, the result of a cross between *V. speciosa* from New Zealand and *V. decussata* from the Falkland Islands, first made about 1862 by Isaac Anderson-Henry of Edinburgh. The hybrids were grown under the common name of "Blue Gem" in England, by 1868 they were being grown in Melbourne, and by 1889 they were grown in Golden Gate Park in San Francisco. In addition to the common name of Blue Gem, the hybrids passed botanically as *V. decussata*. Miss Alice Eastwood recognized that the plants growing in Golden Gate Park were not typical *V. decussata* and applied the name *V. franciscana*.

10. V. speciosa R. Cunn. ex Hook. Occasional as an escape from cultivation in San Francisco, where it is grown as a hedge; native of New Zealand. Probably flowering all year. —*Hebe speciosa* (R. Cunn. ex Hook.) Cockayne & Allan.

18. Pedicularis L. Lousewort

Pinnae oblong-lanceolate; corollas usually at least 2.5 cm. long, the lower lip
    about 2 mm. long, erect and usually appressed.   **1.** *P. densiflora*
Pinnae elliptic-oblong to ovate; corollas usually under 2 cm. long, the lower lip
    4–6 mm. long, divergent from the galea.   **2.** *P. dudleyi*

1. P. densiflora Benth. ex Hook. *Fig. 206.* Indian Warrior. Oak woodland and along the edges of chaparral from northern San Mateo County southward, but much more common on the eastern slopes of the Santa Cruz Mountains; Sawyer Ridge, near Belmont, Woodside, Searsville, Black Mountain, near Los Gatos, Mount Umunhum, El Toro, Swanton, and near Eagle Rock. January–July.

2. P. dudleyi Elmer. Dudley's Lousewort. Endemic in the Santa Cruz Mountains, known from only a few localities in the redwoods; Pescadero Creek, San Lorenzo River, Aptos, and Old Page Mill. April–June.

19. Bellardia All.

1. B. trixago (L.) All. Bellardia. Weeds of disturbed roadsides and overgrazed pastures; San Francisco, San Bruno Mountain, San Andreas Valley, and Stanford; introduced from the Mediterranean region. April–May. This species has probably spread considerably in the past fifteen years. The oldest specimen from California in the Dudley Herbarium was collected in Sonoma County in 1902. The majority of the specimens were collected after about 1940, and the first from the Santa Cruz Mountains (*Wiggins 11162*) was collected in 1945. The label bears the note, "apparently introduced in 1944."

20. Cordylanthus Nutt. ex Benth. Bird's-beak

Leaves oblong-lanceolate or wider; inflorescence spicate. Plants of salt marshes.
    Stamens 4; herbage finely pubescent.   1. *C. maritimus*
    Stamens 2; herbage hirsute-hispid.   2. *C. mollis*
Leaves or their divisions linear; inflorescence of capitate clusters of 2 or more
    flowers. Plants of dry inland areas.
    Capitate clusters each with 3–15 flowers, dense; bracts several; leaves entire
        or with a pair of lobes; calyx 12–15 mm. long.   3. *C. rigidus*
    Capitate clusters each of 2–3 flowers, not dense; bracts few; leaves entire;
        calyx 17–19 mm. long.   4. *C. pilosus*

1. C. maritimus Nutt. ex Benth. Salt Marsh Bird's-beak. Salt marshes along the borders of San Francisco Bay; San Francisco, Redwood City, Palo Alto, and near Alviso. July–October.
2. C. mollis Gray. Soft Bird's-beak. Known from one specimen collected in a salt marsh in San Francisco. July.
3. C. rigidus (Benth.) Jeps. *Fig. 207.* Stiffly-branched Bird's-beak. Central San Mateo County southward on both slopes of the Santa Cruz Mountains, often on serpentine, road embankments, and in openings in chaparral; San Andreas Valley, near Crystal Springs Lake, Page Mill Road, Permanente Creek, Saratoga, Los Gatos Creek, near Loma Prieta, Big Basin, Blooms Grade, and Aptos Creek. August–October.
4. C. pilosus Gray. Hairy Bird's-beak. Eastern slopes of the Santa Cruz Mountains from northern San Mateo County southward to the vicinity of Los Gatos, usually on dry, open hill slopes; Crystal Springs Lake, Woodside, near Portola, Coal Mine Ridge, near Stanford, Palo Alto, Black Mountain, Los Gatos Canyon, and Raymonds Ranch. July–October.

21. Orthocarpus Nutt. Owl's Clover, Orthocarpus

Anthers 2-celled.
    Bracts green throughout; galea about equaling the lower lip; teeth of the lower
        lip inconspicuous; corollas yellow to white.
        Bracts entire; anthers glabrous.   1. *O. campestris*

Bracts cleft; anthers pubescent. 2. *O. lithospermoides*

Bracts tipped with yellow or purple; galea exceeding the lower lip; the teeth conspicuous; corollas yellow, white, to purple.

Spikes conspicuous, wider than the leaves; corollas not linear, broader above; lower lip more than 2 mm. deep.

Galea hooked at the tip, bearded.

Divisions of the floral bracts 1 mm. or less wide.

3a. *O. purpurascens purpurascens*

Divisions of the floral bracts 1–2 mm. wide.

3b. *O. purpurascens latifolius*

Galea straight, pubescent.

Leaves oblong, not attenuate, 3 or more mm. wide, entire or with rounded teeth. Coastal. 4. *O. castillejoides*

Leaves lanceolate, attenuate, usually under 3 mm. wide, entire or more usually with lanceolate divisions. Usually inland plants.

5. *O. densiflorus*

Spikes inconspicuous, not usually wider than the leaves; corollas linear; lower lip less than 2 mm. deep. 6. *O. attenuatus*

Anthers 1-celled.

Stamens not exceeding the galea.

Flowers conspicuous; main axis distinct; galea straight or slightly curved.

Herbage pubescent; galea purple.

Corollas yellow; lower lip 3–4 mm. deep. 7a. *O. erianthus erianthus*

Corollas white, aging pink; lower lip 4–5 mm. deep.

7b. *O. erianthus roseus*

Herbage glabrous, the inflorescence often puberulent; galea yellowish.

8. *O. faucibarbatus albidus*

Flowers inconspicuous; many branches from near the base, obscuring the main axis; galea sharply curved. 9. *O. pusillus*

Stamens exceeding the galea. 10. *O. floribundus*

206

207

1. O. campestris Benth. Field Orthocarpus. Known from one specimen collected near Stanford; native of Oregon and other parts of California. April.

2. O. lithospermoides Benth. Cream Sacs. Grasslands and serpentine outcroppings from northern San Mateo County southward, commonly on the eastern slopes of the Santa Cruz Mountains; San Andreas Lake, Half Moon Bay Road, near Woodside, Searsville Ridge, near Morgan Hill, and Gilroy–Almaden Road. April–June.

3a. O. purpurascens Benth. var. **purpurascens.** *Fig. 208.* Escobita, Common Owl's Clover. A common spring annual of grasslands, cultivated fields, and oak grasslands, San Francisco southward. March–May.

3b. O. purpurascens Benth. var. **latifolius** Wats. Occasional along or near the coast on dry bluffs, commonly growing with *Fragaria chiloensis, Castilleja foliolosa, Hemizonia corymbosa,* and various grasses; Lake Merced, near Ocean View, Montara Point, and Sunset Beach State Park. May–June.

4. O. castillejoides Benth. Paint Brush Orthocarpus. Along or near the coast, growing in moist areas with grasses and sedges, often subject to ocean spray; San Bruno Hills, San Pedro, Montara Point, and Pescadero Creek. April–June, occasionally through September.

5. O. densiflorus Benth. Owl's Clover. A very common annual of grasslands, open fields, and serpentine soils, San Francisco southward. March–May.

6. O. attenuatus Gray. Narrow-leaved Orthocarpus. Grassy slopes, meadows, and sandy flats; San Francisco, Menlo Heights, Stanford, Jasper Ridge, Permanente Canyon, and Boulder Creek. April–May.

7a. O. erianthus Benth. var. **erianthus.** Butter-and-Eggs, Johnny-Tuck. Open fields and grassy slopes; San Francisco, near Palo Alto, near San Martin, Gilroy, between Scott and Waddell creeks, and Santa Cruz. March–April.

7b. O. erianthus Benth. var. **roseus** Gray. Pelican or Popcorn Flower. Along or near the coast from San Francisco southward to Santa Cruz County, usually growing on open grassy or sandy slopes and fields; Lake Merced, Ocean View, Daly City, Colma, San Bruno Hills, and near Santa Cruz. April–May.

8. O. faucibarbatus Gray var. **albidus** (Keck) Howell. Smooth Orthocarpus. Along the coast in sandy soils of coastal bluffs and grassy flats, also inland at Camp Evers in Santa Cruz County; San Francisco, near Pebble Beach, Pigeon Point, Año Nuevo Point, Davenport Landing, Camp Evers, Santa Cruz, and near Watsonville. April–June.

9. O. pusillus Benth. Dwarf Orthocarpus. A common annual of shaded or open grasslands; San Francisco, San Bruno Hills, San Andreas Valley, Crystal Springs Reservoir, Stanford, Permanente Canyon, near Eagle Rock, and Santa Cruz. March–May.

10. O. floribundus Benth. San Francisco Orthocarpus. Moist grassland, fields, and marshy areas, San Francisco southward into San Mateo County; San Francisco, San Bruno Mountain, Colma, Burlingame, San Mateo, Seal Cove, San Andreas Lake, and Belmont. March–May.

22. Castilleja Mutis ex L. f. Indian Paint Brush

Plants annual; leaves and bracts entire; bracts green at anthesis.
1. *C. stenantha*
Plants perennial; leaves and bracts entire to lobed; bracts usually colored at anthesis.

Stems and leaves densely white-tomentose. 2. *C. foliolosa*
Stems and leaves variously pubescent, but not densely white-tomentose.
 Lower lip of the corolla exserted, evident; leaves linear-lanceolate to linear,
  usually entire.
  Corollas 3.5–4.5 cm. long; leaves occasionally with a small pair of lobes.
   Common. 3. *C. franciscana*
  Corollas 2–3.5 cm. long; leaves usually entire. Rare. 4. *C. miniata*
 Lower lip of the corolla rarely exserted; leaves various.
  Leaves oblong-lanceolate to orbicular, entire or occasionally lobed, thick;
   stems diffuse to erect.
   Leaves oblong to orbicular, rounded, usually not more than twice as long
    as broad; bracts entire; plants hirsute to villous-hirsute, the tri-
    chomes glandless. 5a. *C. latifolia latifolia*
   Leaves oblong-lanceolate, usually at least 3 times as long as broad;
    bracts trifid; plants more or less hirsute, usually with glandular
    trichomes. 5b. *C. latifolia wightii*
  Leaves lanceolate to linear, thin, usually lobed; stems erect. 6. *C. affinis*

1. C. stenantha Gray. Large-flowered Paint Brush. Known locally from Harkins
Slough near Watsonville. June–July.
2. C. foliolosa H. & A. *Fig. 209.* Woolly Paint Brush. A common species of
Castilleja, growing along the edges of chaparral, dry rocky slopes, or in sandy
soils of inland marine sand deposits; San Francisco, Sawyer Ridge, near Crystal
Springs Lakes, near Stanford, Black Mountain, Guadalupe Hills, near Los Gatos,
Swanton, Eagle Rock, Glenwood, and near Tuxedo. March–August.
3. C. franciscana Pennell. Franciscan Paint Brush. Brush-covered hills and
along the margins of chaparral; San Francisco, San Bruno Mountain, Cahill
Ridge, Pilarcitos Lake, San Mateo Creek, Crystal Springs Lake, near La Honda,
and Butano Creek. March–September.
4. C. miniata Dougl. ex Hook. Great Red Paint Brush. Known only from one
old specimen from a now extinct marsh in San Francisco.
5a. C. latifolia H. & A. var. **latifolia.** Monterey or Seaside Paint Brush. Stabilized coastal sand dunes; San Francisco and mouth of Pajaro River. March–
September.

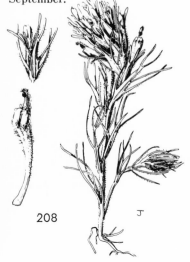

208

209

5b. C. latifolia H. & A. var. **wightii** (Elmer) Zeile. Wight's Paint Brush. A common member of this genus, occurring on coastal bluffs and sand dunes, moist fields and pastures, and in coniferous forests; San Francisco, Cahill Ridge, San Pedro, Mussel Rock, Pescadero, San Mateo–Santa Cruz county line, near Swanton, Davenport, and near Santa Cruz. March–October. Plants with less crowded inflorescences, usually rounded calyx-lobes, and bracts and calyces which are bright red distally have been called *C. wightii* Elmer ssp. *rubra* Pennell. *Castilleja wightii*, sensu Pennell, has relatively crowded inflorescences, acute to rounded calyx-lobes, and yellowish to red bracts and calyces, and tends to be more coastal than ssp. *rubra*.

6. C. affinis H. & A. Indian Paint Brush. Open slopes, borders of chaparral, and wooded areas, usually away from the immediate coast; San Francisco, Crystal Springs Lake, Kings Mountain, Portola State Park, Coal Mine Ridge, Stanford, near Saratoga, near Los Gatos, Loma Azule, New Almaden, San Jose, near Swanton, Empire Grade, and near Boulder Creek. March–August.

121. LENTIBULARIACEAE. Bladderwort Family

1. Utricularia L.

1. U. vulgaris L. ssp. **macrorhiza** (Le Conte) Clausen. Common Bladderwort. Known locally only from the Mud Lakes on Coal Mine Ridge in southern San Mateo County.

122. OROBANCHACEAE. Broomrape Family

Stamens not pubescent at base; capsules 2-valved.   1. *Orobanche*
Stamens pubescent at base; capsules 4-valved.   2. *Boschniakia*

1. Orobanche L. Broomrape

Plants lacking floral bracts; pedicels slender, elongate.
    Stems stout, 3–10 mm. thick; pedicels 3–12, equaling or shorter than the stem.
        Calyx-lobes broadly triangular, equaling or shorter than the tube.
            1a. *O. fasciculata fasciculata*
        Calyx-lobes lanceolate, longer than the tube.   1b. *O. fasciculata franciscana*
    Stems slender, 2–4 mm. thick, short; pedicels 1–3, exceeding the stem in length.
            2. *O. uniflora*
Plants with two subulate floral bracts; flowers pedicellate or not; inflorescence spicate, corymbose, or paniculate.
    Palatal folds of the corolla lacking or poorly developed; inflorescence compact; flowers sessile; base of the stem thickened.   3. *O. bulbosa*
    Palatal folds well developed; inflorescence less compact; flowers sessile or pedicellate; base of the stem not conspicuously thickened.
        Calyx 10–16 mm. long, the lobes longer than the tube. Common.
            Inflorescence corymbose; anthers densely woolly.
                Corollas over 4 cm. long, purple; calyx 2–2.5 cm. long. Coastal bluffs and sand dunes.   4a. *O. grayana violacea*
                Corollas under 3.5 cm. long, pinkish; calyx 1.5–2 cm. long. Inland plants.   4b. *O. grayana jepsonii*
            Inflorescence subracemose to paniculate; anthers glabrous or sparsely hairy.   5. *O. californica*
        Calyx under 10 mm. long, the lobes equaling the tube. Rare.   6. *O. pinorum*

1*a*. **O. fasciculata** Nutt. var. **fasciculata.** Clustered Broomrape. Occasional in chaparral and forested areas; Stanford, San Jose, Loma Prieta, and between Ben Lomond and Zayante. April–August.

1*b*. **O. fasciculata** Nutt. var. **franciscana** Achey. Franciscan Broomrape. Occasional in chaparral or on wooded slopes; San Francisco, near Colma, Black Mountain, Los Gatos Hills, Loma Prieta, near Bielawski Lookout, and Redwood Park. May–July. Both varieties are parasitic on plants of a number of genera; among them are *Eriodictyon, Eriogonum, Artemisia,* and *Phacelia.*

2. **O. uniflora** L. Naked Broomrape. Occasional in the Santa Cruz Mountains, usually parasitic on various members of the Compositae, Crassulaceae, and Saxifragaceae; San Mateo Creek, San Mateo Canyon, and Big Basin. March–April. —*O. uniflora* L. ssp. *occidentalis* (Greene) Abrams ex Ferris, *O. uniflora* L. ssp. *purpurea* (Heller) Achey, *O. uniflora* L. var. *minuta* (Suksd.) Beck, *O. uniflora* L. var. *sedi* (Suksd.) Achey.

3. **O. bulbosa** Beck. Chaparral Broomrape. Chaparral slopes on the eastern slopes of the Santa Cruz Mountains, commonly parasitic on *Adenostoma fasciculata*; Coal Mine Ridge, Searsville Lake, Loma Prieta, and Eagle Rock. May–October.

4*a*. **O. grayana** Beck var. **violacea** (Eastw.) Munz. Gray's Broomrape. Along the coast on bluffs and sand dunes, usually parasitic on various species of *Grindelia*; Montara Point, Half Moon Bay, and Pescadero. June–October.

4*b*. **O. grayana** Beck var. **jepsonii** Munz. Jepson's Broomrape. Occasional in the foothills of the Santa Cruz Mountains, parasitic on *Baccharis, Grindelia,* and perhaps other members of the Compositae; Palo Alto, near Palo Alto, and Santa Cruz. July–September.

5. **O. californica** C. & S. California Broomrape. Occasional in the Santa Cruz Mountains, most common as a parasite on *Eriophyllum staechadifolium* and probably on other members of the Compositae; San Francisco, between Cupertino and Saratoga, and Coyote. July–September.

6. **O. pinorum** Geyer ex Hook. Pine Broomrape. Known locally only from the Redwood forests north of Boulder Creek. July–August.

2. Boschniakia Meyer ex Bong.

1. **B. strobilacea** Gray. *Fig. 210.* California Ground Cone. Occasional in chaparral and on wooded slopes, parasitic on various species of *Arctostaphylos* and *Arbutus menziesii*; near Año Nuevo Point, Wrights, Boulder Creek, Brookdale, Felton, and Sierra Azule Ridge. April–May.

123. PLANTAGINACEAE. Plantain Family

1. Plantago L.

Plants with a well-developed stem.   1. *P. indica*
Plants acaulescent.
  Leaves laciniately divided; inflorescence nodding prior to anthesis.
          2. *P. coronopus*
  Leaves entire or occasionally irregularly dentate; inflorescence erect prior to anthesis.
    Plants biennial or perennial; leaves ovate to lanceolate; if linear, then fleshy.

Leaves conspicuously fleshy. Plants of maritime or salt marsh habitats.
    Leaves about equaling the scapes, both erect or nearly so; seeds usually
      2 mm. or longer; leaves thin on drying.   3*a. P. juncoides juncoides*
    Leaves shorter than the scapes, both spreading; seeds usually under
      2 mm. long; leaves thick on drying.   3*b. P. juncoides californica*
Leaves not fleshy. Plants of various habitats.
    Leaves ovate; inflorescence usually glabrous; seeds 4.   4. *P. major*
    Leaves oblong to lanceolate; inflorescence pubescent; seeds 2 or 3.
      Mature spikes 9–30 cm. long; flowers polygamous; seeds 3, not ex-
        cavated on the face.   5. *P. hirtella galeottiana*
      Mature spikes usually under 9 cm. long; flowers perfect; seeds 2,
        deeply excavated on the face.   6. *P. lanceolata*
Plants annual; leaves linear, not fleshy.
    Corolla-lobes over 2 mm. long; stamens 4. Grassy hillsides.   7. *P. erecta*
    Corolla-lobes under 1 mm. long; stamens 2. Maritime habitats or of low-
      lying flats.   8. *P. bigelovii*

1. P. indica L. Sand Plantain. Known from sand dunes in San Francisco; native of Europe. July–August.

2. P. coronopus L. Cut-leaved Plantain. Sandy soil in disturbed habitats, mainly along the coast, but occasionally inland along railroad tracks; San Francisco, Pescadero, Santa Cruz, Laurel, and Robroy; native of Eurasia. March–August.

3a. P. juncoides Lamk. var. **juncoides.** Pacific Seaside Plantain. Occasional in salt marshes and on coastal bluffs as far south as San Mateo County; San Francisco, Redwood City, and Mayfield. June–November. —*P. maritima* L. ssp. *juncoides* (Lamk.) Hulten.

3b. P. juncoides Lamk. var. **californica** Fern. California Seaside Plantain. Common on dry exposed bluffs along the coast; San Francisco, San Francisco–San Mateo county line, Pescadero, and Santa Cruz. April–November. —*P. maritima* L. var. *californica* (Fern.) Pilger.

4. P. major L. Common Plantain, White Man's Foot. Common in gardens, lawns, and moist disturbed habitats throughout our area; native of Europe. May–November.

5. P. hirtella HBK. var. **galeottiana** (Decne.) Pilger. Mexican Plantain. Mainly near the coast and the margins of San Francisco Bay, occasionally inland in swampy habitats; San Francisco, near Pescadero, Portola Road, Black Mountain, Swanton, Camp Evers, and Watsonville. May–November. —*P. subnuda* Pilger.

6. P. lanceolata L. Ribwort, English Plantain, Buckhorn. Very common along roadsides, disturbed areas, and as a troublesome lawn weed; native of Europe. April–October.

7. P. erecta Morris. *Fig. 211.* California Plantain. Common throughout the Santa Cruz Mountains especially on the eastern slopes in grasslands and serpentine soils; San Francisco, San Bruno Hills, Pebble Beach, Woodside, near Stanford, Permanente Creek, near Los Gatos, near San Jose, Swanton, Glenwood, and Santa Cruz. January–June. —*P. hookeriana* F. & M. var. *californica* (Greene) Poe. *Plantago erecta* is extremely variable with respect to size, number of leaves, and size of inflorescence. The variation is due largely to environmental factors.

8. P. bigelovii Gray. Annual Coast Plantago. Ocean cliffs, edges of salt marshes, and low pastures; San Francisco, Año Nuevo Point, Mayfield, Alviso, and near San Jose. March–June.

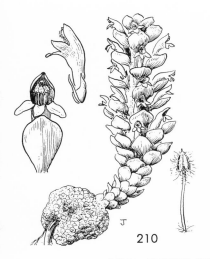

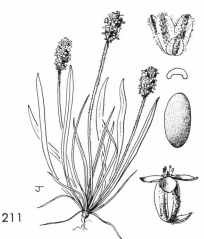

210   211

124. RUBIACEAE. Madder Family

Plants shrubs; leaves opposite, shiny above.   1. *Coprosma*
Plants herbaceous; leaves whorled, dull.
    Corollas funnelform; flowers blue, subtended by an involucre.   2. *Sherardia*
    Corollas rotate; flowers white or yellow, not subtended by an involucre.
                                                    3. *Galium*

1. Coprosma Forst.

1. C. repens Hook. f. Mirrorbush. Known only from San Francisco, where it has been found as an occasional escape from cultivation and where it forms extensive stands on coastal bluffs; native of New Zealand. Flowering in the late summer and fall.

2. Sherardia L.

1. S. arvensis L. Blue Field Madder. Common as a lawn weed, occasional along roadsides, and on grassy slopes; San Francisco, Cahill Ridge, near Half Moon Bay, Hillsborough, Stanford, Jasper Ridge, San Jose, Glenwood, and near Santa Cruz; native of Europe. March–September.

3. Galium L. Bedstraw

Flowers yellow.   1. *G. verum*
Flowers white.
  Leaves 4 or occasionally 5 per node.
    Plants annual, stems very slender, often rooting at the nodes; fruit longer
        than broad, with hooked hairs terminally and sometimes on the lateral
        surfaces.   2. *G. murale*
    Plants perennial, usually with stout stems, often woody, not rooting at the
        nodes; fruit broader than long, glabrous, or if pubescent the hairs more
        or less evenly distributed.
      Leaves distinctly linear-subulate, acute; plants usually matted, under 10
        cm. tall.   3. *G. andrewsii*

Leaves ovate to lanceolate, obtuse; plants not matted, usually over 10 cm.
tall (some specimens of *G. californicum* under 5 cm. tall).
Stems weak; herbaceous throughout; leaves 4 or 5 per node; plants of
marshy or very moist habitats.   4. *G. trifidum subbiflorum*
Stems somewhat woody at least below; leaves 4 per node; plants of dry
habitats.
Stems and leaves scabrous, the trichomes usually retrorse; plants
climbing, often 1.5 m. tall; fruit glabrous.   5. *G. nuttallii*
Stems and leaves pilose-hispidulose, the trichomes usually spreading;
plants usually under 2–3 dm. tall, more or less tufted; fruit
pubescent or glabrous.   6. *G. californicum*
Leaves 6 or 8 per node, occasionally some plants with 2 or 4 per node.
Fruit smooth or tuberculate, never with bristles.
Stems glabrous or with spreading hairs, not scabrous; plants perennial;
fruit smooth, on slender erect pedicels.   7. *G. mollugo*
Stems scabrous; plants annual; fruit tuberculate or rough, on stout re-
curved pedicels.   8. *G. tricornutum*
Fruit with hooked bristles.
Plants perennial; leaves elliptic to ovate, 6 per node.   9. *G. triflorum*
Plants annual; leaves oblong-linear to linear, 6 or 8 per node, occasion-
ally 2 or 4.
Fruit under 1 mm. long; plants slender, diffusely branched.
10. *G. parisiense*
Fruit over 1.5 mm. long; plants coarser, usually branched only below.
11. *G. aparine*

1. G. verum L. Yellow Bedstraw. Very common as a lawn weed on the Stanford
University campus, also reported from near San Mateo; native of Eurasia. July–
September.
2. G. murale (L.) All. Wall Bedstraw. Known from only two localities, but to
be expected elsewhere in lawns and on open grassy slopes; Redwood City and
near Stanford; native of Europe. March–May.
3. G. andrewsii Gray. Phlox-leaved Bedstraw. Rocky areas near the summit of
Loma Prieta. March–August.
4. G. trifidum L. var. **subbiflorum** Wieg. Trifid Bedstraw. Occasional in very
moist or marshy areas; San Francisco, San Andreas Lake, Santa Cruz, and near
Pajaro River. June–November. —*G. trifidum* L. var. *pacificum* Wieg., *G. tinctorium* L. var. *subbiflorum* (Wieg.) Fern.
5. G. nuttallii Gray. Climbing Bedstraw. Common on shrubby slopes and coastal
bluffs; San Francisco, Montara Mountain, San Mateo, near Swanton, near Felton,
and Soquel Valley. March–August.
6. G. californicum H. & A. California Bedstraw. A rather variable group of
plants growing on partially shaded slopes and brushy hills; San Francisco, San
Bruno Hills, Cahill Ridge, Pescadero Creek, La Honda, Kings Mountain, Stevens
Creek, Big Basin, Empire Grade, and Santa Cruz. May–July.
7. G. mollugo L. Wild Madder. Occasionally encountered as orchard weeds;
Menlo Park and Los Altos; native of Europe. May–September.
8. G. tricornutum Dandy. Rough-fruited Bedstraw. Occasional as a weed in
fallow fields and gardens; near Palo Alto and Llagas Post Office; native of
Europe. April–July. —*G. tricorne* Stokes.

9. G. triflorum Michx. *Fig. 212.* Sweet-scented or Fragrant Bedstraw. Throughout the wooded and forested parts of the Santa Cruz Mountains; Cahill Ridge, San Gregorio, Stevens Creek, Swanton, and Glencarry Road. April–August.

10. G. parisiense L. Wall Bedstraw. Occasional as a weed in shaded canyons and on shaded slopes; Boulder Creek and probably elsewhere; native of Europe. May–September.

11. G. aparine L. Goose Grass, Cleavers, Bedstraw. Very widely distributed, especially in thickets and open woodland, the most common species of *Galium* locally; San Francisco, Cahill Ridge, Crystal Springs, San Mateo, Stanford, Edenvale, near Waddell Creek, near Glenwood, and near Watsonville. March–August. —*G. vaillantii* DC., *G. spurium* L. var. *echinospermum* (Wallr.) Hayak.

125. CAPRIFOLIACEAE. Honeysuckle Family

Leaves compound; corollas rotate. 1. *Sambucus*
Leaves simple; corollas tubular.
  Berries white; corollas not 2-lipped. 2. *Symphoricarpos*
  Berries red or black; corollas 2-lipped. 3. *Lonicera*

1. Sambucus L. Elderberry

Inflorescence flat-topped; fruit blue, usually glaucous. Inland plants.
1. *S. mexicana*
Inflorescence subglobose to pyramidal; fruit red. Coastal plants.
2. *S. callicarpa*

1. S. mexicana Presl ex DC. *Fig. 213.* Blue Elderberry. Common on brushy slopes and along the borders of chaparral, usually away from the immediate coast; San Francisco, Sawyer Ridge, Coal Mine Ridge, Stanford, Black Mountain, San Antonio Creek, near Los Gatos, Carnadero Creek, near Saratoga Summit, and near Soda Lake. April–June. —*S. glauca* and *S. coerulea* of California authors.

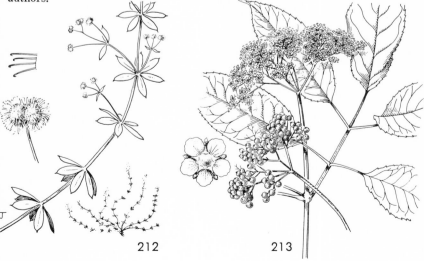

212   213

2. S. callicarpa Greene. Coast Red Elderberry. Fairly common along or near the coast in shaded and wooded canyons; San Francisco, Pilarcitos Canyon, Butano Creek, Waddell Creek, Liddell Creek, and near Santa Cruz. March–April. —*S. pubens* Michx. var. *aborescens* T. & G.

2. Symphoricarpos Duhamel. Snowberry

Leaves more or less glabrous adaxially, often subcoriaceous; young twigs glabrous to minutely puberulent; erect shrubs, usually 1–2 m. tall; fruit 10–15 mm. in diameter.   1. *S. albus laevigatus*
Leaves pubescent adaxially, thin; young twigs pubescent; erect or trailing shrubs, commonly under 5 dm. tall; fruit 4–6 mm. in diameter.   2. *S. mollis*

1. S. albus (L.) Blake var. **laevigatus** (Fern.) Blake. Common Snowberry. Common on brush-covered slopes and open, wooded hillsides, often growing with *Baccharis pilularis* and *Rhus diversiloba*; San Francisco, San Mateo Creek, Crystal Springs, Woodside, Coal Mine Ridge, Stanford, Black Mountain, near Los Gatos, near Glenwood, Blooms Grade, and Santa Cruz. May–July. —*S. rivularis* Suksd.
2. S. mollis Nutt. Creeping or Trailing Snowberry. Usually growing in partial shade in brushy and woody areas; near Redwood City, Kings Mountain, Black Mountain, Permanente Creek, near Los Gatos, near New Almaden, Big Basin, Blooms Grade, and Watsonville. April–June.

3. Lonicera L. Honeysuckle.

Flowers borne in pairs in the axils of 2 bracts; leaves not connate; shrubs.
    Flowers 1–2 cm. long; stems erect. Native plants.   1. *L. involucrata*
    Flowers 3–4 cm. long; stems twining. Introduced plants.   2. *L. japonica*
Flowers borne in whorls; upper leaves usually connate; vines.
    Inflorescence glandular-pubescent; corollas glandular-pubescent externally. Common.   3. *L. hispidula*
    Inflorescence glabrous; corolla not glandular-pubescent. Southern Santa Clara County.   4. *L. interrupta*

1. L. involucrata (Richards.) Banks ex Spreng. *Fig. 214.* Twinberry. Along streams in moist, shaded areas; San Francisco, San Andreas Lake, near Pescadero, San Gregorio, Lobitos Creek, Stanford, Zayante Creek, and near Watsonville. March–June. —*L. ledebourii* Esch.
2. L. japonica Thunb. Japanese or Garden Honeysuckle. Escaping from cultivation in San Francisco and to be expected elsewhere; native of Japan.
3. L. hispidula Dougl. ex Lindl. Hairy Honeysuckle. Common in evergreen forests, wooded areas, openings in chaparral, usually away from the immediate coast; San Francisco, Sawyer Ridge, Pescadero Creek, Belmont, Kings Mountain, Portola, Black Mountain, near Los Gatos, near Saratoga Summit, Swanton, and Blooms Grade. May–June. —*L. hispidula* Dougl. ex Lindl. var. *vacillans* Gray.
4. L. interrupta Benth. Chaparral Honeysuckle. Known locally only from the southern part of Santa Clara County in the vicinity of Uvas and Llagas creeks. May–June.

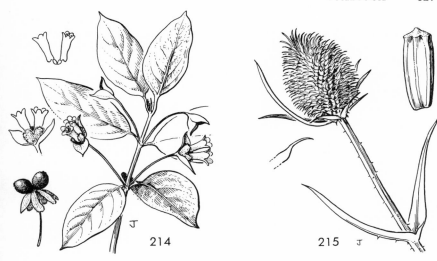

214   215

Two additional members of the Honeysuckle family, *Leycesteria formosa* Wallich, a native of the Himalayas, and *Viburnum tinus* L., a native of the Mediterranean region, have become at least locally established in San Francisco. *Leycesteria* has simple leaves, a 5-lobed corolla, and dark red berries. *Viburnum* has simple leaves and a rotate corolla.

126. DIPSACACEAE. Teasel Family

Stems and leaves prickly; receptacular bracts longer than the flowers.
1. Dipsacus

Stems and leaves not prickly; receptacular bracts not longer than the flowers.
2. Scabiosa

1. Dipsacus L. Teasel

Receptacular bracts flexible, straight.   1. *D. sylvestris*
Receptacular bracts stout, the tips recurved.   2. *D. fullonum*

1. D. sylvestris Huds. Wild Teasel. Occasional in disturbed areas in San Francisco and probably elsewhere; native of Europe. July–September.
2. D. fullonum L. *Fig. 215.* Fuller's Teasel. Common along roads, in fields, pastures, and disturbed areas in general, San Francisco southward; native of Europe. March–October. —*D. sativus* Honck.

2. Scabiosa L.

1. S. atropurpurea L. Mourning Bride, Pincushion. Becoming increasingly common along roads and highways and in disturbed areas; San Francisco, Spanish Town, Stanford, near Alviso, Seabright, and Camp Evers; native of Europe. February–November, and probably flowering the year around. The corolla color of *S. atropurpurea* varies from pure white through various shades of pink, lavender, and purple to deep purple-black. Usually the total range of color variation is present within each population, but occasionally the extremes are lacking. Some plants become distinctly woody below, especially if they have been mutilated or cut during the growing season.

127. VALERIANACEAE. Valerian Family

Plants perennial, 0.5 m. or more tall; corollas over 1 cm. long.

1. *Centranthus*

Plants annual, less than 0.5 m. tall; corollas less than 1 cm. long.

Plants not branched; inflorescence spike-like. Native plants.   2. *Plectritis*

Plants dichotomously branched; inflorescence flat-topped. Introduced plants.

3. *Valerianella*

1. Centranthus DC.

1. C. ruber (L.) DC. Jupiter's Beard, Red Valerian. Escaped from cultivation and well established, especially near cities and towns; San Francisco southward; native of the Mediterranean region. To be found in flower at all times of the year.

2. Plectritis DC.

Cotyledons and the first few pairs of leaves bright green, usually only the midvein conspicuous; fruit-body triangular in cross section, keel rounded or grooved, wings present or absent, if present the wings thin, their margins not revolute; spur of the corolla usually less than ⅓ the length of the corolla.

Fruit with an acute keel, keel sometimes with a longitudinal groove.

1*a. P. congesta congesta*

Fruit rounded on the back, not with a groove.   1*b. P. congesta brachystemon*

Cotyledons and the first few pairs of leaves silvery, the lateral veins as well as the midvein conspicuous; fruit-body obcompressed, the keel usually with a groove, wings present or absent, if present the wings thick, the margins revolute; spur of the corolla very short to over ½ the length of the corolla.

Corollas slender, the spur usually longer than the ovary, with purple spots in the sinuses between the upper and lower lips; fruits winged or not, usually with a dense band of hairs on either side of the keel.

Corollas usually over 5 mm. long; trichomes of the fruits cylindric.

2*a. P. ciliosa ciliosa*

Corollas under 4 mm. long; trichomes of the fruits distinctly clavate.

2*b. P. ciliosa insignis*

Corollas stout, the spur usually shorter than the ovary, lacking purple spots; fruits winged or not, lacking dense bands of hairs on either side of the keel.   3. *P. macrocera*

1*a.* **P. congesta** (Lindl.) DC. ssp. **congesta.** *Fig. 216.* Pink Plectritis. Wooded slopes, serpentine, and grasslands; San Francisco, San Bruno Hills, Año Nuevo Point, Coal Mine Ridge, Stanford, near Eagle Rock, and Hilton Airport. February–May.

1*b.* **P. congesta** (Lindl.) DC. ssp. **brachystemon** (F. & M.) Morey. The most common species of this genus locally, borders of oak woods, wooded slopes, edges of chaparral and in grasslands; San Francisco, San Bruno Hills, Crystal Springs, Stevens Creek, near Madrone Springs, and Swanton. March–May. —*P. anomala* (Gray) Suksd., *P. aphanoptera* (Gray) Suksd., *P. congesta* (Lindl.) DC. ssp. *nitida* (Heller) Morey, *P. gibbosa* Suksd., *P. magna* (Greene) Suksd., *P. samolifolia* Hoeck.

2a. P. ciliosa (Greene) Jeps. ssp. **ciliosa.** Long-spurred Plectritis. Rare in the Santa Cruz Mountains, growing on grassy slopes near Guadalupe Creek. March–April. —*P. californica* (Suksd.) Dyal, *P. macroptera* (Suksd.) Rydb.

2b. P. ciliosa (Greene) Jeps. ssp. **insignis** (Suksd.) Morey. Occasional on exposed grassy slopes; Coal Mine Ridge and near Stanford. April–May.

3. P. macrocera T. & G. White Plectritis. Wooded slopes, occasionally on serpentine, central San Mateo County southward; Crystal Springs, near San Mateo, Woodside, Stanford, and Black Mountain. April–May.

3. Valerianella Mill.

1. V. locusta (L.) Betcke. Lamb's Lettuce, Corn Salad. Reported from San Francisco and now known from several localities in California; native of Europe. —*V. olitoria* (L.) Pollard.

128. CUCURBITACEAE. Gourd Family

Fruits large, dry, green, spiny. Native plants.   1. *Marah*
Fruits small, fleshy, red, not spiny. Introduced plants.   2. *Bryonia*

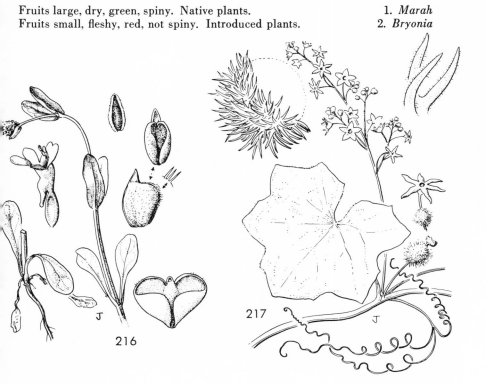

216

217

1. Marah Kell.

Corollas of the mature flowers campanulate; carpels 2–3, 1–2- (rarely 3–4) seeded; fruit attenuate at the ends, spines weak, few, to 6 mm. long.

1. *M. oreganus*

Corollas of mature flowers rotate; carpels 4, commonly 1-seeded; fruit globose, spines many, rigid, 12 mm. long.   2. *M. fabaceus*

1. M. oreganus (T. & G.) Howell. Western Wild Cucumber, Coast Manroot. Fairly common along the borders of grasslands and in brushy areas; San Francisco, Cahill Ridge, Crystal Springs Reservoir, Kings Mountain, Black Mountain, near Saratoga Summit, and Blooms Grade. March–May. —*Echinocystis oreganus* (T. & G.) Congdon.

2. M. fabaceus (Naud.) Greene. *Fig. 217.* Valley Manroot, Wild Cucumber. Common on coastal sand dunes, brushy slopes, and along roadsides, scrambling over other vegetation; San Francisco, Sweeney Ridge, San Mateo, Pigeon Point, Coal Mine Ridge, Stanford, near Los Gatos, Swanton, Saratoga Summit, Ben Lomond Mountain, Santa Cruz, Sunset Beach State Park, and near Watsonville. March–April. —*Echinocystis fabaceus* Naud.

2. Bryonia L.

1. B. dioica Jacq. Bryony. Well established in Golden Gate Park in San Francisco; native of Eurasia.

In addition to the above, various species of *Cucurbita*, particularly *C. pepo* L., the Field Pumpkin, and *Citrullus vulgaris* Schrad., the Watermelon, are to be expected as escapes from cultivation in disturbed areas.

129.   CAMPANULACEAE.   Bellflower Family

Plants perennial.
    Corollas divided nearly to the base.   1. *Asyneuma*
    Corolla-lobes about ½ the length of the tube.   2. *Campanula*
Plants annual.
    Flowers pedicellate.   2. *Campanula*
    Flowers sessile.
        Calyx-lobes 1 cm. or more long in fruit; capsules dehiscent at the apex; leaves narrow.   3. *Githopsis*
        Calyx-lobes less than 0.5 cm. long in fruit; capsules not dehiscent at the apex; leaves usually broad.
        Capsules longer than broad; calyx-lobes narrowly deltoid, entire.
            4. *Triodanus*
        Capsules about as long as broad; calyx-lobes ovate, toothed.
            5. *Heterocodon*

1. Asyneuma Griseb. & Schenk

1. A. prenanthoides (Durand) McVaugh. *Fig. 218* (p. 340). California Harebell. Usually in partial shade in wooded areas, southern San Mateo County southward; near La Honda, Kings Mountain, Stevens Creek, Wrights, Los Gatos, Bielawski Lookout, Empire Grade, Glenwood, Big Basin, and Santa Cruz. June–September. —*Campanula prenanthoides* Durand.

2. Campanula L.

Plants annual, with a tap root.   1. *C. angustiflora*
Plants perennial, with creeping rhizomes.   2. *C. californica*

1. C. angustiflora Eastw. Eastwood's Campanula. Known from near Boulder Creek, where it grows in rocky soil in chaparral. May–June.
2. C. californica (Kell.) Heller. Swamp Harebell. Known locally only from swamps at Camp Evers. June–July. —*C. linnaeifolia* Gray.

3. Githopsis Nutt.

1. G. specularioides Nutt. Common Bluecup. Occasional on shaded slopes in woodsy and brushy areas; San Francisco, Coal Mine Ridge, near Los Gatos, and near Big Basin. April–June.

4. Triodanus Raf.

1. T. biflora (R. & P.) Greene. Small Venus' Looking-glass. Occasional in brushy areas, chaparral, or grassy flats, often overlooked by collectors; San Francisco, Jasper Ridge, Stanford, Los Gatos, Swanton, Big Basin, and Santa Cruz. April–June. —*Specularia biflora* (R. & P.) F. & M.

5. Heterocodon Nutt.

1. H. rariflorum Nutt. Heterocodon. Occasional in moist grassy areas, usually in partial shade; San Francisco, Jamison Creek Road, and doubtless elsewhere. April–July.

130. LOBELIACEAE. Lobelia Family

Corolla 2-lipped.   1. *Downingia*
Corolla-lobes nearly equal.   2. *Legenere*

1. Downingia Torr.

Corolla-tube 3–4 mm. long; upper corolla-lobes 3.5–5 mm. long, shorter than the lower 3 lobes; lower lobes with 1 red-purple spot near the base.
                                            1. *D. concolor tricolor*
Corolla-tube 2–3 mm. long; upper corolla-lobes 6–8 mm. long, about as long as the lower 3 lobes; lower lobes with 3 dark purple spots near the base.
                                            2. *D. pulchella*

1. D. concolor Greene var. **tricolor** (Greene) Jeps. Maroon-spotted Downingia. Spring pools and mud flats in the Santa Clara Valley, probably near extinction at present; near Mayfield, between Morgan Hill and San Martin, and between Morgan Hill and Gilroy. May–June.
2. D. pulchella (Lindl.) Torr. Flat-faced Downingia. Santa Clara Valley from San Jose southward in alkaline areas, low-lying fields, and vernal pools, probably now extinct; near San Jose and near Gilroy. April–June.

2. Legenere McVaugh

1. L. limosa (Greene) McVaugh. Legenere. Known locally only from ponds on Coal Mine Ridge in southern San Mateo County. May. —*Howellia limosa* Greene.

131. COMPOSITAE. Sunflower Family

KEY TO THE TRIBES

Corollas within one head all ligulate; plants with milky juice.   1. ·*Cichorieae*
Corollas within one head never all ligulate; plants without milky juice.
  Ray-flowers none; anthers caudate at the base.
    Plants never spiny or prickly, the herbage usually at least somewhat white-tomentose when young; involucral bracts when present white to tawny, scarious and often hyaline.   4. *Inuleae*
    Plants usually with spiny herbage, involucral bracts, or both; herbage variously pubescent, rarely white-tomentose; involucral bracts always present, the margins occasionally scarious.   12. *Cynareae*
  Ray-flowers present or none; anthers not caudate at the base.
    Receptacle without bristles or chaffy bracts.
      Flowers never yellow; heads discoid; branches of the styles thickened above; stigmatic surface confined to the lower half of the style-branch.
          2. *Eupatorieae*
      Flowers commonly yellow; heads discoid or radiate; branches of the styles not thickened above; style-branch with a stigmatic line from the base to the top.
        Involucral bracts usually conspicuously imbricate (poorly so in *Bellis*, *Erigeron*, and *Conyza*).
          Pappus of bristles or awns, rarely lacking; involucral bracts usually herbaceous at least in part.   3. *Astereae*
          Pappus none or represented by a low ring or a crown at the top of the achene; involucral bracts dry, scarious.   10. *Anthemideae*
        Involucral bracts usually not conspicuously imbricate.
          Pappus of paleae, awns, stout bristles, or none; involucral bracts in 1–3 series.
            Disk-flowers fertile; achenes not curved.   8. *Helenieae*
            Disk-flowers sterile; achenes curved to form a nearly complete circle.   9. *Calenduleae*
          Pappus of soft capillary bristles; involucral bracts mainly in 1 series (sometimes with very short outer bracts in addition to the main series).   11. *Senecioneae*
    Receptacle with chaffy bracts or bristles, but not every disk-flower necessarily subtended by a receptacular bract.
      Ray-flowers absent; pappus none; heads often unisexual; fruit usually a bur.   7. *Ambrosieae*
      Ray-flowers present, the ligules sometimes inconspicuous; pappus of paleae or awns or none.
        Involucral bracts enclosing or closely enveloping the achenes of the ray-flowers; plants often with conspicuous, tack-shaped glands.
          6. *Madieae*
        Involucral bracts not enclosing or closely enveloping the achenes of the ray-flowers; plants rarely conspicuously glandular.
          Ray-flowers yellow; involucral bracts herbaceous.   5. *Heliantheae*
          Ray-flowers white to rose-colored; involucral bracts dry, scarious.
              10. *Anthemideae*

1. Cichorieae. Chicory Tribe

Achenes lacking pappus.   1. *Lapsana*
Achenes with pappus.
  Pappus of paleae or at least paleaceous at the base.
    Margins of leaves spinose; involucral bracts spine-tipped.   2. *Scolymus*
    Margins of leaves not spinose; involucral bracts not spine-tipped.
      Heads sessile; corollas white, pink, or blue; paleae lacking awns.
        3. *Cichorium*
      Heads scapose; corollas yellow; some paleae in each head with awns.
        Involucral bracts in 1 series, enfolding the outer achenes, becoming
          corky.   4. *Hedypnois*
        Involucral bracts in more than 1 series, not enfolding the outer achenes,
          not becoming corky.
          All achenes with awns, these at most subplumose.   5. *Microseris*
          Only the inner achenes with awns, these conspicuously plumose.
            6. *Leontodon*
  Pappus of awns or bristles, not paleaceous.
    Pappus conspicuously plumose.
      Achenes beakless.   7. *Stephanomeria*
      Achenes, at least the inner ones, beaked.
        Receptacles chaffy; leaves basal.   8. *Hypochaeris*
        Receptacles lacking chaff; stems at least somewhat leafy.
          Herbage prickly; corollas yellow.   9. *Picris*
          Herbage not prickly; corollas rarely yellow.
            Achenes including the beak over 3 cm. long; plants perennial;
              corollas lavender or yellow.   10. *Tragopogon*
            Achenes including the beak under 1.5 cm. long; plants annual;
              corollas white.   11. *Rafinesquia*
  Pappus not plumose, merely barbed or barbellate.
    Stems scapose; leaves all basal, or nearly so.
      Achenes spinose or at least tuberculate above, 4–5-ribbed.
        12. *Taraxacum*
      Achenes not spinose or tuberculate above, 10-ribbed.   13. *Agoseris*
    Stems leafy.
      Achenes flattened.
        Achenes beaked; involucre cylindrical in flower.   14. *Lactuca*
        Achenes not beaked; involucre campanulate in flower. 15. *Sonchus*
      Achenes terete.
        Heads nodding in bud; leaves, especially when young, with conspicu-
          ous tufts of wool along the margins.   16. *Malacothrix*
        Heads erect in bud; leaves lacking tufts of wool.
          Leaves and stem with long yellowish bristly hairs; pappus usually
            stiff.   17. *Hieracium*
          Leaves and stem lacking such hairs; pappus soft.   18. *Crepis*

2. Eupatorieae. Eupatory Tribe

Leaves alternate.   19. *Brickellia*
Leaves opposite.   20. *Eupatorium*

3. Astereae. Aster Tribe

Plants dioecious; heads discoid.   21. *Baccharis*
Plants monoecious; heads radiate or discoid.
  Heads radiate, the ray-flowers sometimes rather inconspicuous.
    Ray-flowers yellow.
      Heads 1 cm. or more in diameter, usually few per stem.
        Heads gummy; pappus of 2–8 awns.   22. *Grindelia*
        Heads not gummy; pappus of numerous bristles.
          Leaves glabrous.   23. *Haplopappus*
          Leaves variously pubescent.
            Leaves toothed or coarsely serrate.   24. *Heterotheca*
            Leaves entire.   25. *Chrysopsis*
      Heads less than 1 cm. in diameter, usually numerous.
        Leaves linear or broader, over 3 cm. long, not fascicled.   26. *Solidago*
        Leaves linear-terete, under 1 cm. long, fascicled.   23. *Haplopappus*
    Ray-flowers white, pink, lavender, or blue.
      Leaves basal; heads solitary, pedunculate; pappus none; ray-flowers white
        to pink.   27. *Bellis*
      Some cauline leaves present; heads rarely solitary or long-pedunculate;
        pappus usually present; ray-flowers usually not pure white.
        Pappus-bristles usually 3–5, occasionally none; heads usually solitary.
                                                                          28. *Chaetopappa*
        Pappus-bristles more than 5; heads usually several, rarely only one.
          Plants densely white-tomentose, sometimes becoming glabrate in age.
                                                                          29. *Corethrogyne*
          Plants variously pubescent but not white-tomentose.
            Ligules of the ray-flowers inconspicuous, scarcely exceeding the
              involucre.
              Heads usually about 10 mm. in diameter.   30. *Aster*
              Heads usually about 6 mm. or less in diameter.   31. *Conyza*
            Ligules of the ray-flowers conspicuous, much exceeding the invo-
              lucre.
              Involucral bracts in 1 series, under 0.8 mm. wide, gradually nar-
                rowed toward the tip, the tips not herbaceous.   32. *Erigeron*
              Involucral bracts in 3 series, over 1.5 mm. wide (rarely only 1.2
                mm. wide), abruptly narrowed toward the tip, the tips her-
                baceous.   30. *Aster*
  Heads discoid (in *Lessingia* the marginal flowers enlarged, the lobes spread-
    ing).
    Plants annual.
      Pappus-bristles 3–5, rarely none.   28. *Chaetopappa*
      Pappus-bristles numerous.
        Plants usually white-tomentose, at least when young; marginal flowers
          with expanded corollas, pink to purplish (yellow in *Lessingia ger-
          manorum* var. *germanorum*).   33. *Lessingia*
        Plants not white-tomentose; marginal flowers with a very inconspicuous
          ligule, yellowish.   31. *Conyza*
    Plants perennial.
      Involucres 2–3 times as long as broad.   34. *Chrysothamnus*

Involucres about as long as broad.
  Leaves punctate-glandular, narrowly linear.   23. *Haplopappus*
  Leaves not punctate-glandular, linear or broader.
    Involucres 5–6 mm. high; lateral heads on widely divergent branches.
                                                     32. *Erigeron*
    Involucres 8–10 mm. high; lateral heads on ascending to erect
      branches.
      Involucral bracts gradually narrowed to the tip.   25. *Chrysopsis*
      Involucral bracts abruptly narrowed at the tip.   23. *Haplopappus*

4. Inuleae. Everlasting Tribe

Leaves deltoid, the blades 8–12 cm. long, cordate at the base, green adaxially,
    densely white-tomentose abaxially; plants perennial; achenes with stalked
    glands.   35. *Adenocaulon*
Leaves various, not deltoid; plants annual or perennial; achenes lacking stalked
    glands.
  Pistillate marginal flowers not enclosed or closely subtended by receptacular
      bracts; involucral bracts numerous; pappus of capillary bristles; plants
      annual or perennial.
    Involucral bracts pearly white; plants polygamodioecious, perennial.
                                                         36. *Anaphalis*
    Involucral bracts shiny, not pearly white; plants monoecious, annual or
        perennial.   37. *Gnaphalium*
  Pistillate flowers enclosed or closely subtended by receptacular bracts; invo-
      lucral bracts lacking or few; pappus usually absent (present in *Filago*
      and some species of *Stylocline*); plants annual.
    Leaves opposite; heads globose; plants usually diffusely branched and form-
        ing mats.   38. *Psilocarphus*
    Leaves alternate; heads globose to cylindrical; plants erect or variously
        branched.
      Pappus present, conspicuous, the tips of the bristles usually exserted.
                                                         39. *Filago*
      Pappus present or lacking, never conspicuous, the tips never exserted.
        Receptacular bracts closely subtending but not enclosing the pistillate
            flowers, the bracts persistent, not falling away with the achenes;
            leaves spatulate, the ones toward the top of the stem much exceed-
            ing the stem-apex and the terminal flowers.   40. *Evax*
        Receptacular bracts enclosing the pistillate flowers, the bracts decidu-
            ous, falling away with the achenes; leaves linear, oblanceolate, to
            oblong, the ones toward the top of the stem not usually exceeding
            the stem-apex or terminal flowers.
          Receptacle broader than long; central (staminate) flowers lacking
              pappus; margins of the marginal receptacular bracts lacking
              hyaline margins; involucral bracts present.   41. *Micropus*
          Receptacle longer than broad; central flowers with an inconspicuous
              pappus; margins of the marginal receptacular bracts with hya-
              line margins; involucral bracts present or absent.
                                                         42. *Stylocline*

5. Heliantheae. Sunflower Tribe

Heads solitary on long peduncles.
Involucral bracts in 2 or more series, similar.
  Lower cauline leaves alternate; achenes not flattened.   43. *Wyethia*
  Lower cauline leaves opposite; achenes conspicuously flattened.
                                                                  44. *Helianthella*
Involucral bracts in 2 series, dissimilar.   45. *Coreopsis*
Heads not solitary, usually several, clustered at the ends of the stems and
  branches.
  Leaves opposite, sessile. Plants of wet areas or garden escapes.
    Achenes with pappus.
      Ray-flowers yellow; involucral bracts in 2 series; pappus of the disk-
        flowers of 2–4 retrorsely barbed awns.   46. *Bidens*
      Ray-flowers white; involucral bracts all more or less alike; pappus of
        the disk-flowers of 8–14 paleae.   47. *Galinsoga*
    Achenes lacking pappus.   48. *Guizotia*
  Leaves alternate, petiolate. Plants of dry areas.   49. *Helianthus*

6. Madieae. Tarweed Tribe

Achenes of the ray-flowers compressed dorso-ventrally, each hidden within the
  involucral bracts, the margins of which more or less overlap.
  Disk-flowers 6 or more.
    Achenes 10-ribbed and tuberculate-scabrous; pappus of paleae, oblong,
      obtuse.   50. *Achyrachaena*
    Achenes not ribbed or scabrous; pappus present or absent, not obtuse.
      Achenes of the disk-flowers sterile, poorly developed; heads closing during
        midday; herbage villous at least when young.   51. *Lagophylla*
      Achenes of the disk-flowers fertile; heads open all day; plants variously
        pubescent.   52. *Layia*
  Disk-flower 1.   53. *Madia*
Achenes of the ray-flowers usually compressed laterally, each not hidden by the
  enveloping lateral margins of the involucral bracts.
  Ligules of the ray-flowers 3-toothed or 3-lobed, the lobes more or less parallel;
    herbage lacking tack-shaped glands.
    Achenes of the ray-flowers laterally compressed and longitudinally striate,
      completely enfolded (but not hidden) by the involucral bracts; basal
      leaves subentire.   53. *Madia*
    Achenes of the ray-flowers not laterally compressed or striate, only partially
      enfolded by the involucral bracts; basal leaves pinnatifid or toothed,
      rarely subentire.
      Upper leaves and involucral bracts without terminal pit-like glands; pap-
        pus present or lacking.   54. *Hemizonia*
      Upper leaves and involucral bracts with terminal pit-like glands; pappus
        none.   55. *Holocarpha*
  Ligules of the ray-flowers 3-cleft or -parted into lobes which are palmately
    spreading; tack-shaped glands present.   56. *Calycadenia*

7. Ambrosieae. Ragweed Tribe

Leaves entire; within one head, the marginal flowers pistillate, the disk-flowers
perfect; mature achenes not enclosed within the involucre.   57. *Iva*
Leaves serrate, lobed, or dissected; heads unisexual; mature achenes enclosed
within the involucre; fruiting involucre often spiny.
    Fruiting involucre not spiny.   58. *Ambrosia*
    Fruiting involucre spiny.
        Fruiting involucre 1–2.5 cm. long; spines of the involucre hooked; plants
annual; stems usually erect.   59. *Xanthium*
        Fruiting involucre under 1 cm. long; spines of the involucre not hooked;
plants perennial; stems usually spreading or prostrate.   60. *Franseria*

8. Helenieae. Sneezeweed Tribe

Leaves alternate at least above.
    Leaves lobed or variously divided.
        Plants perennial, woody at least at the base, usually over 3 dm. tall.
                61. *Eriophyllum*
        Plants annual, usually under 1.5 dm. tall.
            Herbage somewhat tomentose; pappus present.   62. *Chaenactis*
            Herbage glabrous; pappus none.   63. *Blennosperma*
    Leaves entire or toothed.
        Ligules of the ray-flowers reflexed at maturity; receptacle globose to hemi-
spherical; leaves decurrent; plants perennial.   64. *Helenium*
        Ligules of the ray-flowers not reflexed at maturity; receptacle flat or conical;
leaves not decurrent; plants annual.
        Involucral bracts rarely over 10.
            Plants white-tomentose; pappus none; rays conspicuous; receptacle
conical.   65. *Monolopia*
            Plants commonly glabrous, if somewhat pubescent never tomentose;
pappus of 3–5 subulate awns; rays inconspicuous; receptacle flat.
                66. *Rigiopappus*
        Involucral bracts 25 or more.   67. *Hulsea*
Leaves opposite.
    Involucral bracts in several series; herbage fleshy; pappus none.
                68. *Jaumea*
    Involucral bracts in 1 series; herbage usually not fleshy; pappus usually
present.
        Involucral bracts not united into a cup.   69. *Baeria*
        Involucral bracts united into a shallow cup.   70. *Lasthenia*

9. Calenduleae. Calendula Tribe

The only genus.   71. *Calendula*

10. Anthemideae. Mayweed Tribe

Heads radiate.
    Inflorescence corymbose; heads numerous, small; ray-flowers usually 5, the
ligules about as long as broad.   72. *Achillea*

Inflorescence not corymbose; heads few, larger; ray-flowers 15 or more, the ligules 2–4 times as long as broad.

Receptacle conical, bracts present toward the top; ray-flowers white.
73. *Anthemis*

Receptacle more or less flat, naked; ray-flowers yellow or white.
74. *Chrysanthemum*

Heads discoid.

Receptacle chaffy.
75. *Santolina*

Receptacle not chaffy.

Heads sessile, solitary, inconspicuous in the axils; plants annual.   76. *Soliva*

Heads not sessile, rarely solitary, conspicuous; plants annual, biennial, or perennial.

Receptacle conical.
77. *Matricaria*

Receptacle flat to convex.

Plants herbaceous, rarely over 1.5 dm. tall; marginal flowers pedicellate.
78. *Cotula*

Plants herbaceous to woody, usually at least 3 dm. tall, often to 1.5 m. tall; marginal flowers sessile.

Inflorescence elongate; heads small, less than 8 mm. in diameter.
79. *Artemisia*

Inflorescence more or less flat-topped, not elongate; heads large, about 12 mm. in diameter.
80. *Tanacetum*

11. Senecioneae. Groundsel Tribe

Leaves opposite.
81. *Arnica*

Leaves alternate or basal.

Basal leaves palmately lobed; plants not climbing.
82. *Petasites*

Leaves not palmately lobed, or if so plants climbing.
83. *Senecio*

12. Cynareae. Thistle Tribe

Leaves not spiny, or if slightly so the mature heads 8–15 cm. in diameter.

Basal leaves usually at least 1 m. long; mature heads 8–15 cm. in diameter; pappus plumose.
84. *Cynara*

Basal leaves under 0.5 m. long; mature heads under 3 cm. in diameter; pappus not plumose.

Involucral bracts hooked at the tip.
85. *Arctium*

Involucral bracts not hooked at the tip.
86. *Centaurea*

Leaves spiny.

Leaves marked with white along the veins.
87. *Silybum*

Leaves not marked with white along the veins.

Attachment of the achenes lateral; involucral bracts with lateral spines.

Uppermost leaves nearly obscuring and surpassing the involucre; achenes terete.
88. *Cnicus*

Uppermost leaves not obscuring the involucre; achenes quadrangular.
89. *Carthamus*

Attachment of the achenes basal; involucral bracts usually entire, the outer sometimes with lateral spines.

Pappus plumose.
90. *Cirsium*

Pappus not plumose.
91. *Carduus*

1. Cichorieae. Chicory Tribe

1. Lapsana L.

1. L. communis L. Nipplewort. Occasional in disturbed areas and in weedy gardens; San Francisco and near La Honda; native of Europe, North Africa, and western and central Asia. June–August.

2. Scolymus L.

1. S. hispanicus L. Golden Thistle. Known only from Los Gatos; native of the Mediterranean region. Summer months.

3. Cichorium L.

Leaves rough-pubescent; pappus about ⅛ the length of the achene; peduncles of the terminal heads not conspicuously thickened.   1. *C. intybus*
Leaves essentially glabrous; pappus about ¼ the length of the achene; peduncles of the terminal heads conspicuously thickened.   2. *C. endivia*

1. C. intybus L. Chicory. Very common along roadsides and in disturbed areas in the Santa Clara Valley; native of Eurasia. Flowering all year.
2. C. endivia L. Endive. Occasional in disturbed areas in San Francisco; native of Eurasia.

4. Hedypnois Mill.

1. H. cretica (L.) Willd. Crete Hedypnois. Occasional in disturbed areas; San Francisco and Guadalupe Canyon; native of the Mediterranean region. March–May.

5. Microseris D. Don. Microseris

Plants perennial; taproot fleshy.   1. *M. paludosa*
Plants annual; taproot slender.
　Paleae lanceolate or linear-lanceolate, bifid at the apex; awns usually equal to or shorter than the paleae.
　　Paleae silvery-scarious, deciduous; awns filiform; heads nodding at least in bud.   2. *M. linearifolia*
　　Paleae silvery to dull, persistent; awns stouter; heads always erect.
　　　Paleae usually dull, shorter than to longer than the achenes; tips of the ribs thick, flaring.   3. *M. heterocarpa*
　　　Paleae usually silvery, much shorter than the achenes; tips of the ribs not thickened nor flaring.   4. *M. decipiens*
　Paleae ovate to ovate-lanceolate, irregularly margined but not bifid; awns usually at least 1.5 times as long as the paleae.
　　Paleae thin, smooth to minutely scabrous, flat at maturity at least in the upper half; awns minutely spiculate; achenes smooth to minutely scabrous on the ribs.
　　　Achenes never dark-spotted, the apex of greater diameter than the body; paleae under 2 mm. long. Plants of the eastern side of the Santa Cruz Mountains.   5. *M. elegans*
　　　Achenes often dark-spotted, the apex of lesser diameter than the body; paleae 1–4 mm. long. Coastal plants.   6. *M. bigelovii*

Paleae usually thick, smooth to more commonly scabrous or villous, curved at maturity or almost lacking; awns barbellulate; achenes scabrous. Paleae 1–6 mm. long; ribs of the achenes flaring at the apex.
7a. *M. douglasii douglasii*
Paleae usually under 1 mm. long; ribs of the achenes not flaring at the apex.   7b. *M. douglasii tenella*

1. **M. paludosa** (Greene) Howell. Marsh Scorzonella. Occasional on wet grassy slopes near the coast; San Francisco, Pebble Beach, and Santa Cruz. April–June. —*Scorzonella paludosa* Greene.
2. **M. linearifolia** (Nutt.) Sch.-Bip. Uropappus. Fairly common in rocky soils, serpentine, chaparral, and on open grassy slopes; San Francisco, San Francisco Watershed Reserve, Redwood City, Coal Mine Ridge, Stanford, near Los Gatos, Loma Prieta Ridge, Edenvale, near Boulder Creek, and Waddell Creek. April–May. —*Uropappus linearifolius* Nutt., *Calais lindleyi* DC.
3. **M. heterocarpa** (Nutt.) Chambers. Derived Microseris. Grassy slopes and fields, mainly on the eastern slopes of the Santa Cruz Mountains; San Francisco, near Woodside, Stanford, near Los Gatos, and near Boulder Creek. March–May. —*Uropappus heterocarpus* Nutt.
4. **M. decipiens** Chambers. *Fig. 219.* Thomas' Microseris. Known so far only from two localities in northwestern Santa Cruz County; near Eagle Rock and Waddell Creek. April–May. *Microseris decipiens* is an allotetraploid species, its parents presumably being *M. linearifolia* and *M. bigelovii*. Additional populations of *M. decipiens* are to be expected in areas where the two parent species are sympatric.
5. **M. elegans** Greene ex Gray. Elegant Microseris. Known locally only from Stevens Creek. April–May.

218

219

6. M. bigelovii (Gray) Sch.-Bip. Coast or Bigelow's Microseris. Along or near the coast and on coastal bluffs and areas of sandy soil; San Francisco, Pebble Beach, Swanton, near Eagle Rock, and Sunset Beach. May–June.

7a. M. douglasii (DC.) Sch.-Bip. ssp. **douglasii.** Douglas' Microseris. Grassy slopes and flats on the eastern side of the Santa Cruz Mountains, often on serpentine soils; San Francisco, Crystal Springs Lake, Woodside, Mayfield, Jasper Ridge, near Los Gatos, Uvas and Llagas roads, near Coyote, Almaden Canyon, and Edenvale. March–May.

7b. M. douglasii (DC.) Sch.-Bip. ssp. **tenella** (Gray) Chambers. Grassy slopes on the eastern side of the Santa Cruz Mountains; San Francisco, San Bruno Hills, Redwood City, Woodside, Stanford, Uvas and Llagas roads, and 8 miles east of Watsonville. April–May.

6. Leontodon L.

1. L. nudicaulis (L.) Porter. Hairy Hawkbit. Occasional in lawns, along roads, and in other disturbed habitats; San Francisco, Stanford, Boulder Creek, Graham Hill Road, and near Santa Cruz; native of Eurasia. May–September. —*L. leyseri* (Wallr.) Beck.

7. Stephanomeria Nutt.

1. S. virgata Benth. Virgate or Tall Stephanomeria. Open slopes, commonly with *Artemisia californica, Zauschneria californica, Rhus diversiloba,* and *Baccharis pilularis,* often becoming a weed along roads and highways; San Francisco, Cahill Ridge, Portola, Coal Mine Ridge, Stevens Creek, near Alviso, Loma Prieta, near Eccles, Santa Cruz, and Aptos Creek. June–November. E. L. Greene recognized several segregate species of *S. virgata.* At most they are minor varients without separate geographical ranges involving pubescence characters, internode lengths, and minor achene differences.

8. Hypochaeris L. Cat's Ear

Plants perennial; leaves usually hispid on both surfaces; marginal achenes beaked; flowering heads usually about 2 cm. in diameter.   1. *H. radicata*
Plants annual; leaves glabrous to sparingly hispid; marginal achenes usually not beaked; flowering heads about 1 cm. in diameter.   2. *H. glabra*

1. H. radicata L. Hairy or Long-rooted Cat's Ear. Very common in grasslands and disturbed areas; native of Eurasia and North Africa. February–August.

2. H. glabra L. Smooth Cat's Ear. Grasslands, pastures, and disturbed areas throughout the Santa Cruz Mountains; native of Eurasia and North Africa. March–September.

9. Picris L.

1. P. echioides L. Bristly Ox Tongue. Fairly common along the edges of salt marshes, in ditches, fallow fields, and in other disturbed habitats; native of the Mediterranean region and southwestern Asia. April–December.

10. Tragopogon L. Goat's Beard

Flowers purple; involucral bracts usually 8–9.   1. *T. porrifolius*
Flowers yellow; involucral bracts usually 10–13.   2. *T. dubius*

1. T. porrifolius L. *Fig. 220.* Oyster Root, Salsify. Weedy areas and roadsides throughout the Santa Cruz Mountains; native of the Mediterranean region. March–October.

2. T. dubius Scop. Yellow Salsify. Known locally only from near Morgan Hill; native of Europe. April–June.

11. Rafinesquia Nutt.

1. R. californica Nutt. California Chicory. Occasional on moist, brushy canyon slopes; San Francisco, Stanford, Black Mountain, Montalvo, and Swanton. May–June.

12. Taraxacum Háll. ex Zinn. Dandelion

Achenes gray, brown, or black; involucral bracts usually lacking appendages; leaves coarsely lobed.   1. *T. officinale*
Achenes reddish or rose-colored, at least a few of the involucral bracts with appendages; leaves finely lobed, usually to the midrib.   2. *T. laevigatum*

1. T. officinale Weber. Common Dandelion. Common as a lawn weed, in disturbed habitats, and occasionally on open or shaded slopes; native of Europe and Asia. To be found in flower at all seasons of the year.

2. T. laevigatum (Willd.) DC. Red-seeded Dandelion. Occasional in lawns and open areas; native of Eurasia. March–July.

13. Agoseris Raf. Mountain Dandelion

Plants annual; beak of the achene 2–3 times the length of the body, body of the achene strongly ribbed, usually conspicuously winged, the wing often crisped.   1. *A. heterophylla*
Plants perennial; body of the achene ribbed, but not winged.
  Beak of the achene less than 2 times the length of the body; leaves usually under 15 cm. long.
    Leaves lanceolate, usually conspicuously lobed, usually at least halfway to the midrib; root slender. Inland variety.
                           2a. *A. apargioides apargioides*
    Leaves oblanceolate to spatulate, usually entire to remotely toothed; root stout. Coastal sand dune variety.   2b. *A. apargioides eastwoodiae*
  Beak of the achene 2–4 times as long as the body.   3. *A. grandiflora*

1. A. heterophylla (Nutt.) Greene. Annual Agoseris. Grassy slopes, often on serpentine, more common on the eastern slopes of the Santa Cruz Mountains than on the western; San Francisco, San Mateo, Coal Mine Ridge, Black Mountain, Alma, Coyote, and near Boulder Creek. January–June. This is a variable species in which a number of varieties have been proposed.

2a. A. apargioides (Less.) Greene var. **apargioides.** Seaside Agoseris. Grasslands throughout the Santa Cruz Mountains; San Francisco, San Bruno Mountain, San Andreas Lake, Pescadero, near Portola, Wrights, Hilton Airport, Santa Cruz, and Capitola. April–November.

2b. A. apargioides (Less.) Greene var. **eastwoodiae** (Fedde) Munz. Eastwood's Agoseris. Sand dunes along the coast; San Francisco, Spanishtown, and Watsonville. March–November.

3. A. grandiflora (Nutt.) Greene. Large-flowered Agoseris, California Dandelion. Common in grasslands and in disturbed habitats along roadsides; San Francisco, San Andreas Lake, San Mateo Ravine, near Stanford, Campbell Creek, near Los Gatos, Bielawski Lookout, near Glenwood, Ben Lomond, and Watsonville. April–September.

14. Lactuca L. Lettuce

Upper cauline leaves linear, if divided the divisions also linear; margins entire; inflorescence spike-like.   1. *L. saligna*
Upper cauline leaves oblong or broader, entire or divided; margins dentate or somewhat spinose; inflorescence paniculate.
  Achenes black, not setulose; cauline leaves not held vertically.   2. *L. virosa*
  Achenes usually brown, setulose at least above on the lateral margins; cauline leaves usually held vertically.   3. *L. serriola*

1. L. saligna L. Willow Lettuce. Occasional in disturbed ground; San Francisco, Stanford, Palo Alto, Jasper Ridge, and near Saratoga Summit; native of Eurasia. August–November.
2. L. virosa L. Wild Lettuce. Occasional in disturbed areas; San Francisco and near San Andreas Lake; native of Eurasia and North Africa. August–September.
3. L. serriola L. Prickly or Wild Lettuce. Common in fields and along roadsides, San Francisco southward; native of Eurasia and North Africa. July–November. A minor varient, *Lactuca serriola* L. f. *integrifolia* Bogenh., which differs in having entire leaves, is occasionally found; San Francisco, Corte de Madera Creek, and Stanford. —*L. serriola* L. var. *integrata* Gren. & Godr.

15. Sonchus L. Sow Thistle

Auricles of the cauline leaves pointed; terminal portion of the leaves often deltoid; leaf-margins rarely spinose; achenes rugose.   1. *S. oleraceus*
Auricles of the cauline leaves rounded; terminal portion of the leaves seldom deltoid; leaf-margins usually spinose; achenes usually smooth.   2. *S. asper*

1. S. oleraceus L. *Fig. 221.* Common Sow Thistle. A very common weed along roads, coastal bluffs, edges of salt marshes, and in disturbed habitats generally; native of Europe. Flowering all year, although more commonly so from March to September.

2. S. asper (L.) Hill. Prickly Sow Thistle. Disturbed areas, especially along moist roadside drainages and cultivated fields; native of Eurasia and North Africa. March–July.

16. Malacothrix DC.

Pappus-bristles uniform, deciduous; leaves with conspicuous tufts of wool along
the margins.   1. *M. floccifera*
Pappus-bristles of two kinds, the inner deciduous, the outer persistent, 1–5 in
number; leaves lacking tufts of wool.   2. *M. clevelandii*

1. M. floccifera (DC.) Blake. Woolly Malacothrix. Rocky and sandy soils of dry hillsides; Los Gatos Canyon, Loma Prieta, and Ben Lomond Sand Hills. April–August.

2. M. clevelandii Gray. Cleveland's Malacothrix. Known locally only from the Ben Lomond Sand Hills. May–June.

17. Hieracium L. Hawkweed

1. H. albiflorum Hook. *Fig. 222.* White-flowered Hawkweed. Fairly common on shaded slopes in oak-madrone woods and less frequent in the redwood–Douglas fir forest; Sawyer Ridge, Pescadero Creek, Portola Valley, Stevens Creek, Loma Prieta, Swanton, Castle Rock, San Vicente Creek, and Santa Cruz. June–August.

18. Crepis L. Hawksbeard

Involucres 9–10 mm. long; achenes beaked.
    Plants usually over 5 dm. tall, annual or biennial; beaks about equaling the
        achenes, stout.   1. *C. vesicaria taraxacifolia*
    Plants usually less than 3 dm. tall, perennial; beaks 1.5–2 times as long as
        the achenes, slender.   2. *C. bursifolia*
Involucres 5–7 mm. long; achenes beakless.   3. *C. capillaris*

1. C. vesicaria L. ssp. **taraxacifolia** (Thuill.) Thell. Weedy Hawksbeard. Becoming very common in southern San Mateo County and northern Santa Clara County; near Woodside and Stanford; native of Europe and North Africa. April–July.

2. C. bursifolia L. Italian Hawksbeard. Occasional in disturbed areas in the vicinity of Stanford University; native of Italy. May–July. This species is probably an escape from the experimental gardens of the Carnegie Institution of Washington at Stanford, where several species of *Crepis* were grown in connection with the studies undertaken by Professor Ernest B. Babcock.

3. C. capillaris (L.) Wallr. Smooth Hawksbeard. Occasional in disturbed habitats; San Francisco, Glenwood, and Ben Lomond Sand Hills; native of Eurasia. June–September.

2. Eupatorieae. Eupatory Tribe

19. Brickellia Ell.

1. B. californica (T. & G.) Gray. California Brickellia, Brickellbush. Known locally from one specimen collected near San Jose, but to be expected along dry creek beds in the Santa Clara Valley. August–September.

20. Eupatorium L.

1. E. adenophorum Spreng. Sticky Eupatorium. Occasional in disturbed areas; San Francisco, Santa Cruz, and near Soquel; native of Mexico. April–August.

3. Astereae. Aster Tribe

21. Baccharis L.

Mature leaves cuneate to oblong, usually 1-nerved, rarely over 3 times as long as broad.
    Plants prostrate to decumbent, under 3 dm. tall. Coastal.
                       1a. B. pilularis pilularis
    Plants erect, 1–4 m. tall. Usually inland.   1b. B. pilularis consanguinea
Mature leaves ovate, lanceolate, to linear-lanceolate, 3-nerved from near the base, usually 5 or more times as long as broad.
    Stems woody; leaves ovate to lanceolate; achenes glabrous.   2. B. viminea
    Stems herbaceous, woody only near the base; leaves linear-lanceolate to linear; achenes minutely pubescent.   3. B. douglasii

1a. B. pilularis DC. var. **pilularis.** *Fig. 223.* Dwarf Chaparral Broom, Fuzzy-Wuzzy. Coastal bluffs, hills, and sand dunes; San Francisco, Pillar Point, Arroyo de los Frijoles, San Gregorio Creek, Pigeon Point, near Davenport, and Santa Cruz. March–August.

222

223

1*b*. B. pilularis DC. var. **consanguinea** (DC.) Kuntze. Coyote Brush, Chaparral Broom. A very common shrub of hills, edges of chaparral, woods, and often invading grasslands and disturbed areas; San Francisco, San Andreas Valley, near La Honda, Portola, Stanford, Black Mountain, near Mountain View, near Los Gatos, Saratoga Summit, Waddell Creek, and near Big Basin. August–October.

2. B. viminea DC. Mule Fat. Occasional along ·dry stream beds in the Santa Clara Valley; near Campbell, Alma, 6.5 miles south of San Jose, and near Coyote. April–June.

3. B. douglasii DC. Salt Marsh or Douglas' Baccharis. Moist areas along streams and in lowlands, occasionally along the edges of salt marshes; San Francisco, San Andreas Lake, Crystal Springs Lake, Menlo Park, Cooleys Landing, Alviso, near Wrights, Waddell Creek, and near Santa Cruz. April–December.

22. **Grindelia** Willd. Gum Plant

Involucral bracts pubescent.	1. *G. hirsutula*
Involucral bracts glabrous.	
Tips of the involucral bracts not sharply reflexed. Plants of salt marshes or slopes near the sea.	
Plants woody. Salt marshes along San Francisco Bay.	2. *G. humilis*
Plants herbaceous. San Francisco.	3. *G. maritima*
Tips of the involucral bracts (at least the outer) sharply reflexed. Plants of various habitats.	
Plants not succulent; stems erect. Inland plants.	4. *G. camporum*
Plants succulent; stems prostrate or decumbent, rarely erect. Coastal plants.	
Leaves pointed at the tips.	5. *G. latifolia*
Leaves rounded at the tips.	6. *G. stricta venulosa*

1. G. hirsutula H. & A. Hirsute Grindelia. Dry open slopes, commonly to the east of the crests of the Santa Cruz Mountains; San Francisco, San Bruno Hills, Cahill Ridge, Crystal Springs, near San Mateo, Pescadero–La Honda Road, Coal Mine Ridge, Stanford, Black Mountain, and near Los Gatos. April–June. Plants with densely villous stems and peduncles and very long, linear bracts beneath the heads have been called *G. hirsutula* H. & A. ssp. *rubicaulis* Keck.

2. G. humilis H. & A. Marsh Grindelia. Salt marshes along the shores of San Francisco Bay; San Francisco, Redwood City, Dumbarton Bridge, Cooleys Landing, Palo Alto, and Alviso. To be found in flower at all times of the year.

3. G. maritima (Greene) Steyerm. San Francisco Grindelia. Coastal bluffs and sandy or serpentine slopes near the ocean; endemic in San Francisco. May–September. Plants with pubescent stems have been called *G. maritima* (Greene) Steyerm. f. *anomala* Steyerm.

4. G. camporum Greene. *Fig. 224.* Great Valley Grindelia, Gum Weed. Dry slopes, road embankments, fields, and low alkaline areas, but not in salt marshes; San Francisco, San Francisco Airport, Jasper Ridge, Stanford, Mountain View, near Alviso, and near Madrone. June–December.

5. G. latifolia Kell. Coastal Gum Plant. Coastal sand dunes and salt marshes; Pebble Beach, Waddell Creek, Swanton, Santa Cruz, Seabright, Aptos, and Watsonville Slough. June–November. —*G. latifolia* Kell. ssp. *platyphylla* (Greene) Keck.

6. G. stricta DC. ssp. **venulosa** (Jeps.) Keck. Pacific Grindelia. Occasional on bluffs along the coast and in coastal salt marshes; San Francisco, Montara Mountain, Pillar Point, Half Moon Point, and Waddell Creek. March–October. —*G. arenicola* Steyerm. var. *pachyphylla* Steyerm.

23. Haplopappus Cass.

Perennial herbs; basal leaves 10–30 cm. long; heads radiate.   1. *H. racemosus*
Shrubs or subshrubs; leaves never over 6 cm. long; heads radiate or discoid.
  Leaves oblanceolate to spatulate, dentate at least at the apex; heads discoid.
                                                          2. *H. venetus vernonioides*
  Leaves linear to filiform, entire; heads radiate or discoid.
    Leaves 4–12 mm. long, fascicled; shrubs, usually less than 1 m. tall; heads radiate.
      Achenes glabrous or with a few scattered hairs; young twigs and leaves puberulent, obscurely glutinous. Coastal.
                                   3a. *H. ericoides ericoides*
      Achenes sericeous; young twigs and leaves rarely puberulent, conspicuously glutinous. Inland.   3b. *H. ericoides blakei*
    Leaves 3–6 cm. long, usually not fascicled; shrubs 0.5–3 m. tall; heads discoid.   4. *H. arborescens*

1. H. racemosus (Nutt.) Torr. Racemose Pyrrocoma. Edges of marshes, saline soils, and occasionally in disturbed areas; Cooleys Landing, near Alviso, Agnews, and San Jose. August–November. —*H. racemosus* (Nutt.) Torr. ssp. *longifolius* Hall, *Pyrrocoma elata* Greene.
2. H. venetus (HBK.) Blake ssp. **vernonioides** (Nutt.) Hall. Coastal Isocoma. Known locally only from San Francisco and from near Burlingame. September–November. —*Isocoma veneta* (HBK.) Greene var. *vernonioides* (Nutt.) Jeps.
    Haplopappus ciliatus (Nutt.) DC., a native of eastern North America, has been collected in San Francisco. It differs from the above species by its radiate heads.
3a. H. ericoides (Less.) H. & A. ssp. **ericoides.** Mock Heather. Coastal sand dunes; San Francisco, near Waddell Creek, and Pajaro River. September–November. —*Ericameria ericoides* (Less.) Jeps.
3b. H. ericoides (Less.) H. & A. ssp. **blakei** Wolf. Blake's Haplopappus. Inland marine sand deposits in Santa Cruz County; Glenwood, Eccles, and near Bonny Doon. September–November.
4. H. arborescens (Gray) Hall. Golden Fleece. Chaparral from southern San Mateo County southward; Stevens Creek, near Raymonds, Loma Prieta, Gilroy, near Bielawski Lookout, China Grade, and Big Basin. July–October. —*Ericameria arborescens* (Gray) Greene.

24. Heterotheca Cass.

1. H. grandiflora Nutt. Telegraph Weed. Occasional in sandy disturbed areas, roadsides, and coastal sand dunes; San Francisco, Stanford, San Jose, Santa Cruz Highway, near Glenwood, and Larkin Valley; perhaps introduced locally from southern California or the Central Valley. March–October.

25. Chrysopsis (Nutt.) Ell. Golden Aster

Heads radiate.
Heads solitary or few, closely subtended by leaf-like bracts; plants 1–2 dm. tall.
1*a*. *C. villosa bolanderi*
Heads several, not subtended by leafy bracts; plants 3–8 dm. tall.
Leaves and stems densely canescent, somewhat glandular.
1*b*. *C. villosa echioides*
Leaves and stems hispid, not canescent, conspicuously glandular.
1*c*. *C. villosa camphorata*
Heads discoid.   2. *C. oregona rudis*

1*a*. **C. villosa** (Pursh) Nutt. var. **bolanderi** (Gray) Gray. Bolander's Golden Aster. Occasional on open slopes, usually near the coast; San Francisco, San Bruno Hills, San Bruno Mountain, near Portola, and near Santa Cruz. April–November.

1*b*. **C. villosa** (Pursh) Nutt. var. **echioides** (Benth.) Gray. Open grassy slopes, often among rock outcroppings, usually away from the immediate coast; Menlo Park, Coal Mine Ridge, Campbell Creek, Saratoga, Saratoga Summit, Castle Rock, and Aptos Creek. September–November. Specimens intermediate between var. *echioides* and var. *bolanderi* are occasionally found.

1*c*. **C. villosa** (Pursh) Nutt. var. **camphorata** (Eastw.) Jeps. Most common on the inland marine sand deposits in Santa Cruz County, occasionally elsewhere; Stevens Creek, near Lexington, near Eccles, Glenwood, and Mount Hermon. June–September.

2. **C. oregona** (Nutt.) Gray var. **rudis** (Greene) Jeps. Oregon Golden Aster. Sandy and gravelly areas, commonly along creeks and dry washes on the eastern slopes of the Santa Cruz Mountains and in the Santa Clara Valley; near Menlo Park, Mountain View, Santa Clara, San Jose, Los Gatos Creek, near Morgan Hill, and near Gilroy. June–November.

26. Solidago L. Goldenrod

Leaves linear to linear-lanceolate, entire.   1. *S. occidentalis*
Leaves variously lanceolate to obovate, usually not entire.
Leaf-surfaces essentially glabrous.
Involucres 3–4 mm. high; leaf-margins scabrous-ciliate or glandular.
Leaves not conspicuously reduced upward, glabrous to pubescent, often toothed.   2. *S. canadensis elongata*
Leaves much reduced upward, entire, glabrous.   3. *S. confinis*
Involucres 6–7 mm. high; leaf-margins usually not ciliate.   4. *S. spathulata*
Leaf-surfaces densely pubescent.   5. *S. californica*

1. **S. occidentalis** (Nutt.) T. & G. Western Goldenrod. Edges of marshes, wet roadside ditches, and moist places along streams; San Francisco, Portola, Palo Alto, Alviso, Black Mountain, near Coyote, Waddell Creek, near Felton, and Watsonville. August–November.

2. **S. canadensis** L. ssp. **elongata** (Nutt.) Keck. Meadow Goldenrod. Occasional on ridges and brush-covered slopes; San Francisco, Kings Mountain, Los Gatos Canyon, near Los Gatos, Loma Prieta, near Felton, and Aptos. August–November. —*S. lepida* DC. var. *elongata* (Nutt.) Fern.

3. S. confinis Gray. Southern Goldenrod. Known from a Bolander collection from marshes in San Francisco made in 1863. July.

4. S. spathulata DC. Dune Goldenrod. Coastal sand hills and bluffs; San Francisco, near San Francisco, and Pebble Beach. August–October.

5. S. californica Nutt. *Fig. 225.* California Goldenrod. Dry, open or shrubby slopes and becoming a weed along roadsides; San Francisco, Sawyer Ridge, Mills Lake, near Woodside, Searsville, Coal Mine Ridge, Stevens Creek, near Saratoga Summit, Big Basin, Glenwood, Santa Cruz, and Aptos. July–November.

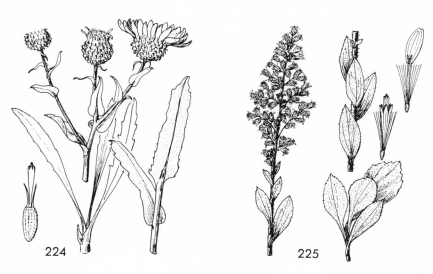

27. Bellis L. Daisy

1. B. perennis L. English Daisy. A common weed of lawns and occasionally along roads or in moist disturbed habitats; native of Europe. Plants flowering throughout the year.

28. Chaetopappa DC.

Ray-flowers present, conspicuous, white to pinkish; pappus-bristles usually 5 or
    wanting.   1. *C. bellidiflora*
Ray-flowers absent or the corollas reduced to a filiform tube; pappus-bristles
    3–5 or wanting.
    Disk-corollas dilated at the throat.   2. *C. exilis*
    Disk-corollas linear.   3. *C. alsinoides*

1. C. bellidiflora (Greene) Keck. White-rayed Chaetopappa. Occasional in the Santa Cruz Mountains, usually growing in the barer areas in grasslands; San Andreas Lake, between Scott and Waddell creeks, between Scott and Mill creeks, San Lorenzo River, near Boulder Creek, and Santa Cruz. April–May. —*Pentachaeta bellidiflora* Greene. The pappus of *C. bellidiflora* usually consists of 5 scabrous bristles, but it is often lacking. Within one population, there may be some plants with pappus and some without.

2. C. exilis (Gray) Keck. Meager Chaetopappa. Forming dense. but local stands on bare areas in grasslands; San Andreas Lake, Crystal Springs Lake, and Black Mountain. April. —*Pentachaeta exilis* Gray.

3. C. alsinoides (Greene) Keck. Tiny Chaetopappa. Occasional in grasslands on the eastern slopes of the Santa Cruz Mountains; San Francisco, Jasper Ridge, Black Mountain, Permanente Creek, near Los Gatos, and near San Jose. April–May. —*Pentachaeta alsinoides* Greene.

29. Corethrogyne DC.

Heads usually solitary; at least the outer involucral bracts somewhat to densely white-tomentose abaxially.
    Heads 2–3.5 cm. across; plants decumbent, the stems rarely over 1.5 dm. tall.
                                                   1. *C. californica*
    Heads under 2 cm. across; plants decumbent to erect, the stems usually over 2 dm. tall.   2. *C. leucophylla*
Heads several to many per stem; involucral bracts usually glandular, occasionally somewhat white-tomentose.
    Involucral bracts usually not conspicuously glandular, the outer somewhat white-tomentose abaxially. Rare.   3a. *C. filaginifolia filaginifolia*
    Involucral bracts usually conspicuously glandular, the outer not white-tomentose abaxially. Common.   3b. *C. filaginifolia rigida*

1. C. californica DC. *Fig. 226.* California Corethrogyne. Usually in bare soil on open hills, San Francisco southward; San Francisco, near Crystal Springs, Half Moon Bay Road, and Swanton. April–June.

2. C. leucophylla (Lindl.) Jeps. Branching Beach Aster. Occasional along the coast from southern San Mateo County southward; Año Nuevo Point and between Scott and Waddell creeks. April–December.

3a. C. filaginifolia (H. & A.) Nutt. var. **filaginifolia.** Common Corethrogyne. Known from southern Santa Cruz County in the vicinity of Soda Lake. May–September.

3b. C. filaginifolia (H. & A.) Nutt. var. **rigida** Gray. Rigid Corethrogyne. Coastal bluffs and sand dunes, occasionally on grassy slopes, and inland marine sand deposits; near Llagas Post Office, near Glenwood, Quail Hollow Road, Ben Lomond Sand Hills, Mount Hermon, Santa Cruz, Capitola, and near Aptos. May–October.

30. Aster L. Aster

Rays relatively inconspicuous; heads less than 1 cm. in diameter; plants annual.
                                           1. *A. exilis*
Rays conspicuous; heads usually about 1.5 cm. in diameter; plants perennial.
    Leaves linear to oblanceolate; involucral bracts glabrous abaxially, green.
        Involucre strongly gradate, the involucral bracts of several distinct lengths.
                                      2. *A. chilensis*
        Involucre not gradate, the involucral bracts all about equal in length.
                                        3. *A. subspicatus*
    Leaves obovate; involucral bracts pubescent abaxially, at least some with purple tips or margins.   4. *A. radulinus*

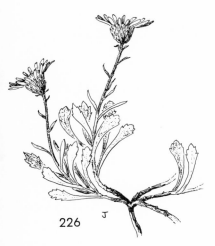

226 227

1. **A. exilis** Ell. Slim Aster. Salt marshes along San Francisco Bay and occasionally elsewhere; San Francisco, Palo Alto, Alviso, and Chittenden. October–November.

2. **A. chilensis** Nees. *Fig. 227.* Common California Aster. A common species of *Aster*, growing along the edges of salt marshes, along streams, and on wooded or brush-covered slopes; San Francisco, Crystal Springs Lake, near Pebble Beach, Menlo Park, Portola, Black Mountain, near Santa Clara, Los Gatos Canyon, near Los Gatos, Loma Prieta Road, Swanton, Big Basin, near Felton, Aptos Creek, and Watsonville. July–December. —*A. chilensis* Nees var. *sonomensis* (Greene) Jeps.

3. **A. subspicatus** Nees. Douglas' Aster. Brushy slopes, San Francisco southward, most common near the coast; San Francisco, Half Moon Bay, Stanford, and Watsonville. May–October. Occasional specimens appear to be intermediate between *A. subspicatus* and *A. chilensis* with respect to involucral characters.

4. **A. radulinus** Gray. Rough-leaved or Broad-leaved Aster. Wooded and brush-covered slopes, San Francisco southward; San Francisco, Sawyer Ridge, La Honda, Portola, Coal Mine Ridge, Black Mountain, near Los Gatos, near Loma Prieta, Glenwood, Mount Hermon, Big Basin, and Blooms Grade. July–December.

31. Conyza L.

Leaves coarsely toothed, narrowly obovate.	1. *C. coulteri*
Leaves usually entire, linear to linear-lanceolate.	
Involucral bracts essentially glabrous.	2. *C. canadensis*
Involucral bracts hirsutulous.	3. *C. bonariensis*

1. **C. coulteri** Gray. Coulter's Conyza. Occasional in rocky soils in chaparral; Loma Prieta, near Felton, and Santa Cruz. October–November.

2. **C. canadensis** (L.) Cronq. Horseweed. A widespread and common weed; San Francisco, Sawyer Ridge, Stanford, Alviso, near Coyote, Alba Grade, near Big Basin, Hilton Airport, near Santa Cruz, and Pajaro River. July–October. —*C. canadensis* (L.) Cronq. var. *glabrata* (Gray) Cronq., *C. bilbaoana* Remy, *Erigeron canadensis* L.

3. C. bonariensis (L.) Cronq. South American Conyza. A weed of disturbed areas, especially along streets and about habitations, San Francisco southward; native of South America. June–September. —*Erigeron crispus* Pourr., *E. linifolium* Willd.

<h2 style="text-align:center">32. Erigeron L. Wild Daisy, Fleabane</h2>

Plants fleshy. Coastal.   1. *E. glaucus*
Plants not fleshy. Inland.
  Leaves cordate-clasping, oblanceolate, the lower usually toothed or lobed; ray-flowers over 150.   2. *E. philadelphicus*
  Leaves not cordate-clasping, variously shaped; ray-flowers present or lacking, if present under 65.
    Ray-flowers present.
      At least the lower leaves with a few sharp-pointed teeth; plants spreading by slender rhizomes.   3. *E. karvinskianus*
      Leaves entire; plants with a short taproot.   4. *E. foliosus*
    Ray-flowers lacking.
      Leaves and stems villous.   5. *E. petrophilus*
      Leaves and stems essentially glabrous.   6. *E. inornatus angustatus*

1. E. glaucus Ker-Gawl. *Fig. 228.* Seaside Daisy. Coastal bluffs and occasionally inland on wind-swept summits; San Francisco, Montara Point, near Pescadero, Swanton, near Davenport, and Santa Cruz. March–August.
2. E. philadelphicus L. Fleabane, Philadelphia Daisy. Occasional in wet ground; San Francisco, Stanford, Almaden Ridge, near Madrone, near Gilroy, and Santa Cruz. April–June.
3. E. karvinskianus DC. Occasional as an escape from cultivation and becoming established; San Francisco, Jamison Creek, Boulder Creek, and Santa Cruz; native of Mexico and Central America. April–August.
4. E. foliosus Nutt. Leafy Daisy. Occasional in grasslands; San Francisco, Cahill Ridge, Searsville, Black Mountain, and Ben Lomond Mountain. June–July. —*E. foliosus* Nutt. var. *hartwegii* (Greene) Jeps.
5. E. petrophilus Greene. Rock Daisy. Rocky outcroppings in chaparral; Mount Umunhum and Eagle Rock. June–August.
6. E. inornatus Gray var. **angustatus** Gray. California Rayless Daisy. Known locally from serpentine outcroppings near Crystal Springs Lake. July–November.

<h2 style="text-align:center">33. Lessingia Cham.</h2>

Leaves with numerous conspicuous, spheroidal, short-stipitate glands.
    1*b*. *L. germanorum tenuipes*
Leaves rarely with such glands.
  At least the outer involucral bracts tomentose or arachnoid-tomentose abaxially; heads 8–40-flowered; involucres narrowly campanulate to turbinate.
    Heads 25–40-flowered; at least the outer involucral bracts somewhat tomentose abaxially, glabrate in age, glands present; corollas lemon yellow. Sand dunes, San Francisco and adjacent San Mateo County.
    1*a*. *L. germanorum germanorum*

Heads 8–18-flowered; involucral bracts arachnoid-tomentose abaxially, lacking glands; corollas pink to purplish. Central San Mateo County southward.

Herbage persistently white-tomentose, glabrate in age; involucres 9–12 mm. long.   *2a. L. hololeuca hololeuca*

Herbage glabrate; involucres 6–9 mm. long.

2b. L. hololeuca arachnoidea

Involucral bracts not tomentose; heads 3–5-flowered; involucres narrowly cylindro-turbinate.   *3. L. ramulosa glabrata*

1a. L. germanorum Cham. var. **germanorum.** San Francisco Lessingia. Sand dunes in San Francisco, where it is endemic, often growing with *Croton californicus* and *Baccharis pilularis.* May–November.

1b. L. germanorum Cham. var. **tenuipes** Howell. Occasional on dry slopes in foothills on the eastern side of the Santa Cruz Mountains; near Searsville, Jasper Ridge, and Saratoga. July–November. Occasional specimens, as for example one from Stevens Creek, approach *L. germanorum* Cham. var. *parvula* (Greene) Howell.

2a. L. hololeuca Greene var. **hololeuca.** *Fig. 229.* Woolly-headed Lessingia. Open grasslands in San Mateo and Santa Clara counties; near Crystal Springs, near Woodside, Portola, Los Trancos Road, Stevens Creek, near Los Gatos, and Coyote. July–November.

2b. L. hololeuca Greene var. **arachnoidea** (Greene) Howell. San Mateo Lessingia. Endemic in the vicinity of the Crystal Springs Lakes in San Mateo County, usually growing in grasslands and often in serpentine soils. August–November. —*L. micradenia* Greene var. *arachnoidea* (Greene) Ferris.

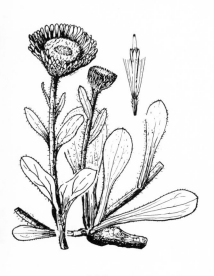

228

229

3. L. ramulosa Gray var. **glabrata** Keck. Dry open areas in oak woods, endemic on the eastern slopes of the Santa Cruz Mountains in Santa Clara County; near Los Gatos, near New Almaden, and near Madrone. July–September. —*L. micradenia* Greene var. *glabrata* (Keck) Ferris.

34. Chrysothamnus Nutt.

1. C. nauseosus (Pall.) Britt. ssp. **occidentalis** (Greene) Hall & Clements. Common Rabbit Brush. Known locally only from the serpentine slopes of Mount Umunhum. October–November.

4. Inuleae. Everlasting Tribe

35. Adenocaulon Hook.

1. A. bicolor Hook. *Fig. 230.* Trail Plant. Moist places in forests, commonly growing with *Pseudotsuga menziesii, Sequoia sempervirens, Lithocarpus densiflorus,* and *Umbellularia californica*; Cahill Ridge, Pescadero Creek, Butano Creek, Kings Mountain, Searsville, Black Mountain, Stevens Creek, Loma Prieta, Glenwood, and near Big Basin. May–September.

36. Anaphalis DC.

1. A. margaritacea (L.) Gray. Pearly Everlasting. Coastal bluffs, stabilized sand dunes, and wind-swept summits; San Francisco, Crystal Springs Lake, Pescadero Creek, near La Honda, Swanton, Whitehouse Creek, Boulder Creek, and near Santa Cruz. May–September.

37. Gnaphalium L. Everlasting, Cudweed

Heads leafy-bracted, the leaves exceeding the mature heads.
  Heads about 3 mm. high, glomerate at the tips of the stems and branches; plants annual, usually diffusely branched, usually under 2 dm. tall; pappus-bristles falling separately. Plants of moist or wet areas.
1. *G. palustre*
  Heads 4–5 mm. high, arranged in a spike-like inflorescence; plants annual or biennial, usually with a few stems, not diffusely branched, 1.5–4 dm. tall; pappus deciduous as a unit. Plants of dry areas. 2. *G. purpureum*
Heads not leafy-bracted.
  Leaves remaining white-woolly on both surfaces.
    Plants annual or rarely biennial; heads subglobose.
      Heads 3–3.5 mm. high, tips of the involucral bracts hyaline.
3. *G. luteo-album*
      Heads 4–6 mm. high; tips of the involucral bracts scarious. 4. *G. chilense*
    Plants perennial; heads longer than broad. 5. *G. beneolens*
  Leaves green or becoming so adaxially, glandular.
    Involucral bracts pink; heads longer than broad. 6. *G. ramosissimum*
    Involucral bracts white; heads subglobose.
      Leaves about equally green on both surfaces, decurrent. Common.
7. *G. californicum*
      Leaves green adaxially, white-woolly abaxially, not decurrent. Rare.
8. *G. bicolor*

1. G. palustre Nutt. Lowland Cudweed. Along the edges of lakes and ponds, along streams, and in areas in which water has stood during the early part of the year; San Francisco, San Andreas Lake, Crystal Springs Lake, Searsville Lake, Ben Lomond Mountain, near Felton, Santa Cruz, Pinto Lake, and near Pajaro River. May–October.

2. G. purpureum L. Purple Cudweed. A common plant of disturbed areas, often in sandy soil; San Francisco, San Andreas Lake, Salada Valley, Coal Mine Ridge, Stanford, Swanton, near Felton, San Lorenzo River, and near Robroy. April–June.

3. G. luteo-album L. Weedy Cudweed. A weed of gardens, roadsides, and other disturbed areas, especially common in gravelly, sandy, or hard-packed soils; San Francisco, San Andreas Lake, Burlingame, Año Nuevo Point, Stanford, Hilton Airport, and near Boulder Creek; native of Europe. To be found in flower at all times of the year.

4. G. chilense Spreng. Cotton-batting Plant. Most common near the coast on wooded slopes and on coastal bluffs and sand dunes, less frequent inland; San Francisco, Sawyer Ridge, Año Nuevo Point, Stanford, Swanton, Santa Cruz, Boulder Creek, and Watsonville Slough. Plants in flower at all seasons.

5. G. beneolens Davison. Fragrant Everlasting. Chaparral, dry brush-covered slopes, and among stands of *Baccharis*; San Francisco, Sweeney Ridge, Spanish-town, Black Mountain, Coal Mine Ridge, Stanford, Wrights, Loma Prieta, and Boulder Creek. July–September.

6. G. ramosissimum Nutt. Pink Everlasting. Coastal bluffs, wooded slopes, and in chaparral; San Francisco, Sweeney Ridge, near Pescadero, San Gregorio Beach, San Francisquito Creek, and near Felton. June–October.

7. G. californicum DC. *Fig. 231.* California Cudweed, Green Everlasting. Fairly common in poor soils, in chaparral, and on dry ridges; San Francisco, Crystal Springs, Searsville, Stanford, Black Mountain, Stevens Creek Reservoir, near Los Gatos, Swanton, Bielawski Lookout, and Santa Cruz. May–August.

8. G. bicolor Bioletti. Bioletti's Cudweed. Known from along the coast in Santa Cruz County, usually on stabilized coastal sand dunes; near Waddell Creek and Sunset Beach State Park. April–October.

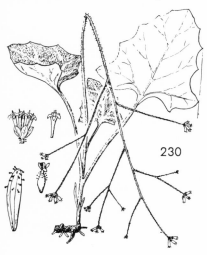

230

231

38. Psilocarphus Nutt.

Receptacular bracts about 2 mm. long. Common.   1. *P. tenellus*
Receptacular bracts about 3 mm. long. Rare.
  Pistillate flowers usually less than 80 per head; leaves broadest near the base.
                                                          2a. *P. brevissimus brevissimus*
  Pistillate flowers usually over 100 per head; leaves broadest toward the middle.
                                                          2b. *P. brevissimus multiflorus*

1. P. tenellus Nutt. Slender Woolly-heads. A common plant of dry bare ground, open grassland, and often becoming a weed in gardens and disturbed areas, often growing with *Soliva sessilis, Alchemilla occidentalis, Trifolium* spp., *Poa annua, Evax sparsiflora,* and *Micropus californicus*; San Francisco, Montara, Sawyer Ridge, Woodside, Coal Mine Ridge, Permanente Creek, Morgan Hill, near Saratoga Summit, near Big Basin, near Santa Cruz, and Soquel. March–May.
2a. P. brevissimus Nutt. var. **brevissimus.** Round or Dwarf Woolly-heads. Known locally only from the Santa Clara Valley. April–June.
2b. P. brevissimus Nutt. var. **multiflorus** Cronq. Occasional in vernal pools in the southern part of the Santa Clara Valley; near Morgan Hill and between Morgan Hill and Gilroy. April–June.

39. Filago L.

Upper leaves conspicuously exceeding the heads, linear-lanceolate.   1. *F. gallica*
Upper leaves about equaling the heads, oblong.   2. *F. californica*

1. F. gallica (L.) L. Narrow-leaved Filago. A widespread but rarely collected weed of grasslands, gardens, and disturbed areas; San Francisco, San Bruno Mountain, San Andreas Lake, Stanford, Saratoga, Morgan Hill, and near Soquel; native of Europe. April–May.
2. F. californica Nutt. California Filago. Occasional in grasslands, coastal bluffs, and brushy slopes; San Francisco, Cahill Ridge, and near Boulder Creek. April–May.

40. Evax Gaertn.

1. E. sparsiflora (Gray) Jeps. Erect Evax. Grassy areas and rocky soils, mainly on the eastern slopes of the Santa Cruz Mountains; San Francisco, Sweeney Ridge, Woodside, Coal Mine Ridge, Jasper Ridge, Stanford, near San Jose, Uvas and Llagas roads, Boulder Creek, and Jamison Creek. April–May.

41. Micropus L.

1. M. californicus F. & M. *Fig. 232.* Slender Cottonweed. Grassy slopes, rocky areas, and along the edges of chaparral; San Francisco, Cahill Ridge, near San Mateo, Kings Mountain, Coal Mine Ridge, Stanford, Loma Azule, near Los Gatos, near New Almaden, San Jose, Saratoga Summit, near Glenwood, and near Eagle Rock. April–June.

42. Stylocline Nutt.

Involucral bracts lacking; receptacular bracts subtending the central staminate
  flowers conspicuous, hooked, stellate-spreading in age; staminate flowers
  lacking pappus.   1. *S. filaginea*

Involucral bracts present; receptacular bracts subtending the staminate flowers not conspicuous, not hooked, not stellate-spreading in age; staminate flowers with a few pappus-bristles.

Receptacular bracts, enclosing the pistillate flowers, with a wide hyaline margin, the margin wider than the rest of the bract.　　2. *S. gnaphalioides*

Receptacular bracts, enclosing the pistillate flowers, with a narrow hyaline margin, the margin narrower than the rest of the bract.　　3. *S. amphibola*

1. S. filaginea Gray. Northern or Hooked Stylocline. Known locally from only one specimen collected on serpentine between Crystal Springs and Woodside. April–May.

2. S. gnaphalioides Nutt. Everlasting Stylocline, Nest-straw. Occasional in sandy or rocky soils on hilltops, ridges, and in chaparral; Stanford, Ben Lomond Sand Hills, near Zayante, and near Felton. April–May.

3. S. amphibola (Gray) Howell. Mount Diablo Cottonweed. Known locally from near Stanford, but to be expected elsewhere on hillsides with occasional oaks and rock outcroppings. April–May.

5. Heliantheae. Sunflower Tribe

43. Wyethia Nutt. Mule Ears

Outer involucral bracts ovate to ovate-lanceolate, over 4 cm. long; leaves tending to be ovate or elliptic.

Herbage shiny, not tomentose, glandular; mature disk-achenes 10–12 mm. long.
　　1. *W. glabra*

Herbage not shiny, tomentose, glabrate in age, not glandular; mature disk-achenes 12–15 mm. long.　　2. *W. helenioides*

Outer involucral bracts lanceolate, under 2.5 cm. long; leaves tending to be lanceolate.　　3. *W. angustifolia*

1. W. glabra Gray. *Fig. 233.* Mule Ears. Grassy slopes, usually in at least partial shade of evergreen trees or of *Pinus attenuata* and *P. sabiniana*; Stanford, near Los Gatos, Mount Umunhum, and Empire Grade. March–June.

2. W. helenioides (DC.) Nutt. Gray Mule Ears. Open grassy slopes from San Francisco southward; San Francisco, near La Honda, Coal Mine Ridge, Stanford, near Campbell, El Toro, near Glenwood, Zayante Valley, and San Lorenzo Valley. February–May.

3. W. angustifolia (DC.) Nutt. Narrow-leaved Mule Ears. Open grasslands; San Francisco, near Montara Mountain, Pebble Beach, near San Andreas Lake, Searsville Ridge, and Stanford. April–May. Occasional specimens of *W. angustifolia* approach *W. helenioides* in having involucral bracts that are 1.5–2 times as long as normal, slightly wider leaves, and a more tomentose pubescence.

44. Helianthella T. & G.

Involucre 1.5–2 cm. long.　　1. *H. californica*

Involucre 2.5–4 cm. long.　　2. *H. castanea*

1. H. californica Gray. California Helianthella. Rocky or somewhat grassy slopes; Jasper Ridge, near Searsville, and Black Mountain. May–June.

232   233

2. H. castanea Greene. Diablo Helianthella. Grassy slopes, known locally from San Francisco and northern San Mateo County. April–May. —*H. cannonae* Eastw.

45. Coreopsis L. Coreopsis, Tickseed

Leaves entire.   1. *C. lanceolata*
Leaves 1–2-pinnate.   2. *C. tinctoria*

1. C. lanceolata L. Garden Coreopsis. Occasional as a garden escape in Santa Cruz County; native from Michigan south to New Mexico, Texas, and Florida. May–July.
2. C. tinctoria Nutt. Calliopsis. Escaped from cultivation near Boulder Creek; native from British Columbia and Minnesota south to New Mexico and Louisiana. July–August.

46. Bidens L. Bur Marigold

Ray-flowers usually over 1.5 cm. long; plants 3–15 dm. tall.   1. *B. laevis*
Ray-flowers usually under 1.5 cm. long; plants 2–5 dm. tall.   2. *B. cernua*

1. B. laevis (L.) BSP. Bur Marigold. Occasional in marshes, roadside ditches, sloughs, and along the margins of ponds and slow-moving streams; San Francisco, Alviso, Campbell Creek, Santa Cruz, and Watsonville. August–October.
2. B. cernua L. Nodding Bur Marigold. Occasional in moist areas along standing or slowly moving water; known from Lake Merced in San Francisco. July–October.

47. Galinsoga R. & P.

1. G. parviflora Cav. Small-flowered Galinsoga. Occasional as a garden weed in San Francisco and to be expected occasionally elsewhere; native from Mexico to South America. August–February.

48. Guizotia Cass.

1. G. abyssinica (L. f.) Cass. Ramtilla. Occasional as a weed in San Francisco; native of tropical Africa.

49. Helianthus L. Sunflower

Plants perennial; involucral bracts narrowly lanceolate.   1. *H. californicus*
Plants annual; involucral bracts ovate to ovate-lanceolate.
  Involucral bracts ovate, abruptly narrowed into a long attenuate tip.
                                                     2. *H. annuus*
  Involucral bracts ovate-lanceolate, not abruptly or only slightly narrowed into
    a long attenuate tip.   3. *H. bolanderi*

1. H. californicus DC. California Sunflower. Along flood plains and rocky banks of streams; mainly on the eastern side of the crest of the Santa Cruz Mountains; Crystal Springs, Woodside, Menlo Park, Page Mill Road, near Santa Clara, Wrights, and Loma Prieta. August–September.
2. H. annuus L. Common Sunflower. Occasional in the Santa Cruz Mountains, usually in disturbed areas; San Francisco and Palo Alto. July–October. Plants with unbranched stems and disks that are over 5 cm. broad have been called *H. annuus* L. var. *macrocarpus* (DC.) Cockerell. This variety is cultivated for its seeds. Plants with branched stems and disks that are 2–3.5 cm. broad have been called *H. annuus* L. ssp. *lenticularis* (Dougl.) Cockerell.
3. H. bolanderi Gray. Bolander's Sunflower. Occasional in fields and along roadsides around the southern tip of San Francisco Bay, and to be expected further south in the Santa Clara Valley; near Mayfield, Mountain View, Moffett Field, and Alviso. July–October.

6. Madieae. Tarweed Tribe

50. Achyrachaena Schauer

1. A. mollis Schauer. Blow-Wives. Adobe or other heavy clay soils, growing in open fields and on grassy slopes; San Francisco, Baden, Stanford, near Los Gatos, Guadalupe Reservoir, and near Pajaro River. April–May.

51. Lagophylla Nutt. Hareleaf

Heads scattered along the stems, not glomerate.   1. *L. ramosissima*
Heads densely glomerate.   2. *L. congesta*

1. L. ramosissima Nutt. Common Hareleaf. Serpentine, open rocky soils, and grasslands; San Bruno Hills, San Andreas Lake, near Redwood City, Searsville, Black Mountain, Saratoga Summit, near Los Gatos, Loma Prieta, Uvas Creek, near Boulder Creek, and Santa Cruz. May–October.

2. L. congesta Greene. Rabbit's Foot. Occasional on dry chaparral slopes on the eastern side of the Santa Cruz Mountains; Black Mountain. May–October.

52. Layia H. & A.

Involucral bracts pectinate or with short, thick, hispid-ciliate hairs; each disk-flower subtended by a receptacular bract.   1. *L. chrysanthemoides*
Involucral bracts not pectinate, variously pubescent or often with tack-shaped glands but not with short, thick, hispid-ciliate hairs; only the outer disk-flowers subtended by receptacular bracts.
Ligules of the ray-flowers 2–4 mm. long, relatively inconspicuous.
Stems with dark dots, erect, branched above, 2–10 dm. tall; ligules of the ray-flowers yellow; pappus-bristles under 15. Mainly inland plants.
2. *L. hieracioides*
Stems without dark dots, freely branched from near the base, to 1.5 dm. tall; ligules of the ray-flowers white; pappus-bristles over 25. Coastal sand dune plants.   3. *L. carnosa*
Ligules of the ray-flowers 6–15 mm. long, conspicuous.
Stems with dark dots; pappus-bristles reddish brown to purplish, rarely white, plumose; ligules of the ray-flowers rarely white-tipped.
4. *L. gaillardioides*
Stems without dark dots; pappus-bristles white or tawny, scabrous; ligules of the ray-flowers usually white-tipped.
Plants semiprostrate to decumbent, succulent; stems 1–3 dm. long; tips of the involucral bracts dilated. Along the immediate coast.
5a. *L. platyglossa platyglossa*
Plants erect, not particularly succulent; stems to 5 dm. tall; tips of the involucral bracts not dilated. Usually growing away from the immediate coast.   5b. *L. platyglossa campestris*

1. L. chrysanthemoides (DC.) Gray. Smooth Layia. Open grassy summits of wind-swept hills and occasionally in low alkaline soils along San Francisco Bay; San Francisco, San Bruno Hills, Burlingame, Millbrae, Redwood City, and San Jose. April–June.
2. L. hieracioides (DC.) H. & A. Tall Layia. Chaparral, inland and coastal sand dunes, and open woods; San Francisco, San Andreas Lake, Kings Mountain, Stanford, Jasper Ridge, Ben Lomond Sand Hills, Aptos, and Sunset Beach. April–August.
3. L. carnosa Nutt. Beach Layia. Occasional on coastal sand dunes, known so far locally only from San Francisco. April–June.
4. L. gaillardioides (H. & A.) DC. Woodland Layia. Growing on grassy slopes, open fields, and brush-covered areas; San Francisco, Half Moon Bay Road, San Mateo Canyon, near La Honda, near Pescadero Creek, Black Mountain, Swanton, and near Saratoga Summit. April–July.
5a. L. platyglossa (F. & M.) Gray ssp. **platyglossa.** *Fig. 234.* Tidy Tips. Coastal bluffs and sand dunes; San Francisco, Pigeon Point, and Sunset Beach State Park. March–September.
5b. L. platyglossa (F. & M.) Gray ssp. **campestris** Keck. Field Tidy Tips. A common plant of open grasslands, serpentine, sand dunes, and occasionally on

coastal bluffs; San Francisco, San Bruno Mountain, Crystal Springs Lake, Sawyer Ridge, near La Honda, Woodside, near Stanford, Black Mountain, San Jose, Swanton, near Mill Creek, Hilton Airport, near Felton, and near Mount Hermon. March–June.

53. **Madia** Molina. Tarweed

Pappus present, of fimbriate paleae, about 1 mm. long; plants perennial.
1. *M. madioides*

Pappus lacking; plants annual.
  Disk-flowers more than 1 per head.
    Disk-achenes sterile; receptacle densely hairy; ligules of the ray-flowers conspicuous.
      Plants with a well-developed basal rosette, strongly glandular-pubescent above, flowering from August through November.
2a. *M. elegans densifolia*
      Plants lacking a well-developed basal rosette, sparingly glandular above, flowering from May through July.   2b. *M. elegans vernalis*
    Disk-achenes fertile; receptacle glabrous; ligules of the ray-flowers relatively inconspicuous.
      Plants glandular in the upper half; stems slender; involucral bracts 6–9 mm. long.   3. *M. gracilis*
      Plants glandular to the base; stems stout; involucral bracts 8–15 mm. long.
        Heads in congested glomerules, usually exceeded by the subtending leaf-like bracts; involucral bracts 8–15 mm. long.   4. *M. capitata*
        Heads scattered, rarely glomerate, not exceeded by the subtending bracts; involucral bracts 7–12 mm. long.   5. *M. sativa*
  Disk-flower 1, rarely 2 per head.   6. *M. exigua*

1. M. madioides (Nutt.) Greene. *Fig. 235.* Woodland Madia. Shade in redwood–Douglas fir forests, central San Mateo County southward; San Mateo Canyon, Pescadero Creek, Kings Mountain, near Woodside, near Los Gatos, Swanton, Mill Creek, Big Basin, Empire Grade, Santa Cruz, and Loma Prieta. April–September.

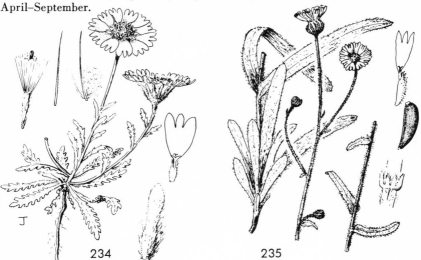

234   235

2a. M. elegans D. Don ex Lindl. ssp. **densifolia** (Greene) Keck. Common Madia. Grasslands, woodlands, and disturbed areas; San Andreas Lake, Crystal Springs Lake, Stanford, Black Mountain, near Santa Clara, Loma Prieta, Waddell Creek, Hilton Airport, Felton, Alba Road, and Aptos Creek. August–November.

2b. M. elegans D. Don ex Lindl. ssp. **vernalis** Keck. Wooded slopes and grasslands; Alpine Road, Skyline Boulevard, near Castle Rock, Uvas Creek, Felton, and Alba Grade. May–July.

3. M. gracilis (Smith) Keck. Slender Tarweed, Gumweed. Mainly east of the crests of the Santa Cruz Mountains, usually growing in grasslands and oak woodlands; San Francisco, Pilarcitos Lake, San Mateo Canyon, Redwood City, Woodside, Stanford, Jasper Ridge, Stevens Creek, near Los Gatos, near Morgan Hill, near Boulder Creek, and Mount Charlie Road. May–July.

4. M. capitata Nutt. Headland Tarweed. A widespread species of grasslands, coastal bluffs, and disturbed areas; San Francisco, near Colma, Pescadero Beach, Coal Mine Ridge, Stanford, near Boulder Creek, near Santa Cruz, and near Watsonville. May–September.

5. M. sativa Molina. Chile, Coast, or Common Tarweed. Grasslands, roadsides, and disturbed areas, becoming weedy; San Francisco, Sawyer Ridge, Menlo Park, Stanford, Los Gatos Creek, near Santa Cruz, Hilton Airport, Aptos Creek, and near Watsonville. June–November.

6. M. exigua (Smith) Gray. Small Tarweed. Grasslands, chaparral, and open woods; San Francisco, Crystal Springs Lake, Woodside, Jasper Ridge, Kings Mountain, Coal Mine Ridge, near Los Gatos, Big Basin, Blooms Grade, Ben Lomond, and Mount Hermon. May–June.

54. Hemizonia DC. Tarweed

Ligules of the ray-flowers yellow; achenes of the ray-flowers beaked; inner receptacular bracts not adnate or deliquescent.
 Leaves not spine-tipped.
  Involucral bracts not keeled; anthers yellow or black.
   Ray-flowers 5; disk-flowers 6; anthers yellow.   1. *H. kelloggii*
   Ray-flowers over 12; disk-flowers over 12; anthers black.
                                            2. *H corymbosa*
  Involucral bracts keeled; anthers black.
   Ray-flowers 3; disk-flowers 3.   3. *H. lobbii*
   Ray-flowers 5; disk-flowers 6.   4. *H. pentactis*
 Leaves spine-tipped.
  Pappus absent; anthers yellow.   5. *H. pungens maritima*
  Pappus of paleae; anthers yellow or black.
   Anthers yellow; receptacular bracts fleshy at the tips, not long-villous; paleae 3 (rarely 5); herbage with yellow glands.
    Herbage minutely glandular; receptacular bracts weakly spine-tipped.
                                            6a. *H. parryi parryi*
    Herbage not glandular; receptacular bracts not at all spine-tipped.
                                            6b. *H. parryi congdoni*
   Anthers black; receptacular bracts not fleshy at the tips, long-villous; paleae 8–12; herbage with black stipitate glands.   7. *H. fitchii*

Ligules of the ray-achenes white, purple-veined; achenes of the ray-flowers not beaked; inner receptacular bracts adnate, deliquescent.
    Inflorescence paniculate.
        Herbage conspicuously glandular at least above, villous; involucral bracts 3.5–6 mm. long; inflorescence paniculate, the heads usually scattered.
            Basal rosettes poorly developed if present; involucral bracts 5–6 mm. long.   *8a. H. luzulaefolia luzulaefolia*
            Basal rosettes well developed; involucral bracts 3.5–5 mm. long.
                                    8b. H. luzulaefolia rudis
        Herbage slightly glandular, shaggy-villous; involucral bracts 6–9 mm. long; inflorescence corymbosely paniculate, the heads usually clustered.
                                        9. H. congesta
    Inflorescence spicate.   *10. H. clevelandii*

1. H. kelloggii Greene. Kellogg's Tarweed. Occasional in fields in San Francisco and in the southern part of the Santa Clara Valley in the vicinity of Morgan Hill, usually growing with other species of *Hemizonia* and *Trichostema lanceolatum*. April–September.

2. H. corymbosa (DC.) T. & G. Coast Tarweed. The most common species of this genus locally, growing on coastal bluffs and fields, open grasslands, and in cultivated fields; San Francisco, San Bruno Hills, Crystal Springs Lake, Montara Point, Moss Beach, Belmont, Millbrae, Searsville, Black Mountain, Swanton, Big Basin, Hilton Airport, near Glenwood, Santa Cruz, Capitola, Aptos, and Hecker Pass. May–November.

3. H. lobbii Greene. Three-rayed Tarweed. Known locally from fields around Stanford, commonly growing in grainfields. September–November.

4. H. pentactis (Keck) Keck. Salinas River Tarweed. Introduced into grain fields near Stanford, perhaps no longer extant locally; native in Monterey and San Luis Obispo counties. June–October.

    Hemizonia fasciculata (DC.) T. & G., Fascicled Tarweed, has been collected as a weed in San Francisco. It is native of southern California and Baja California and may be distinguished from *H. pentactis* by the lack of pustulate hairs.

5. H. pungens (H. & A.) T. & G. ssp. **maritima** (Greene) Keck. Common Spikeweed. Lowlands bordering San Francisco Bay, and in the Santa Clara Valley, rarely elsewhere; San Francisco, San Mateo, Redwood City, Menlo Park, Palo Alto, Alviso, Santa Clara, and San Jose. May–October. —*Centromadia maritima* Greene.

6a. H. parryi Greene ssp. **parryi**. Pappose Spikeweed. Hard-packed soil of roadsides and low alkaline areas in northern San Mateo County; Salada Beach. August–November. —*Centromadia parryi* (Greene) Greene.

6b. H. parryi Greene ssp. **congdonii** (Robins. & Greenm.) Keck. Congdon's Spikeweed. Alkaline and heavy soils in the Santa Clara and Pajaro valleys; near Sunnyvale, near San Jose, and Watsonville. July–October. —*Centromadia congdonii* (Robins. & Greenm.) Smith.

7. H. fitchii Gray. Fitch's Spikeweed. Rare in the Santa Cruz Mountains; near New Almaden, near Morgan Hill, and Uvas Road. June–November. —*Centromadia fitchii* (Gray) Greene.

8a. H. luzulaefolia DC. ssp. **luzulaefolia.** Hayfield Tarweed. Grassy slopes and fields; San Francisco, near Crystal Springs Lake, Millbrae Hills, Corralitos Creek, and near Pajaro River. May–August.

8b. H. luzulaefolia DC. ssp. **rudis** (Benth.) Keck. Grain fields, pastures, and grassy slopes; near Crystal Springs Lake, Stanford, Jasper Ridge, Santa Clara, Los Gatos, Morgan Hill, and near Watsonville. July–November.

9. H. congesta DC. Hayfield Tarweed. Rare in northern San Mateo County; between Colma and San Bruno. June–July.

10. H. clevelandii Greene. Cleveland's Tarweed. Known locally from central Santa Cruz County; near Boulder Creek. July–August.

55. Holocarpha Greene.

1. H. macradenia (DC.) Greene. *Fig. 236.* Santa Cruz Tarweed. Grassy slopes and fields; Graham Hill Road, Santa Cruz, Soquel, College Lake, and Watsonville. June–November.

56. Calycadenia DC. Rosinweed

Heads usually in a terminal cluster; upper leaves exceeding the terminal cluster; involucral bracts more or less purplish, the glands black. San Mateo County.
1a. *C. multiglandulosa cephalotes*
Heads in a terminal cluster and in clusters in the axils of the upper leaves; upper leaves more or less recurved; involucral bracts greenish yellow, the glands light brown. Mainly in Santa Clara County.
1b. *C. multiglandulosa robusta*

1a. C. multiglandulosa DC. ssp. **cephalotes** (DC.) Keck. Sticky Calycadenia. Dry open hills, known locally from Emerald Lake. May–October.

1b. C. multiglandulosa DC. ssp. **robusta** Keck. Sticky Calycadenia. Southern San Mateo County and Santa Clara County, in clearings in chaparral, serpentine, and open woods; Emerald Lake, Jasper Ridge, Black Mountain, Page Mill Road, Castle Rock Ridge, and Loma Prieta. June–November.

7. Ambrosieae. Ragweed Tribe

57. Iva L.

1. I. axillaris Pursh. *Fig. 237.* Poverty Weed. Disturbed areas, usually in hard-packed soils; San Francisco, near San Jose, and southeastern Santa Cruz County. May–September.

58. Ambrosia L.

Plants perennial; leaves pinnatifid.   1. *A. psilostachya*
Plants annual; leaves bipinnatifid.   2. *A. artemisiifolia*

1. A. psilostachya DC. Common or California Ragweed. Occasional in low, moist or disturbed areas; San Francisco, Crystal Springs Lake, Aptos, and Watsonville. June–November. —*A. psilostachya* DC. var. *californica* (Rydb.) Blake.

2. A. artemisiifolia L. Annual Ragweed. Known as an occasional weed in San Francisco; native of eastern North America.

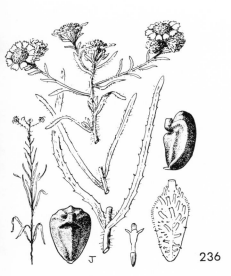

236　　　　　　　　　237

59. Xanthium L. Cocklebur

Stems with 1 or 2 trifid spines at each node; leaves ovate-lanceolate, toothed to
　　pinnately lobed, distinctly paler beneath.　　　　　　　　1. *X. spinosum*
Stems lacking spines; leaves ovate, often broadly so and sometimes cordate at
　　the base, toothed, often coarsely so, not distinctly paler beneath.
　　　　　　　　　　　　　　　　　　　　　　　　　　　2. *X. strumarium*

1. X. spinosum L. Spiny Clotbur. Fallow fields, dry lake bottoms, margins of
swamps, creek bottoms, and disturbed areas; San Francisco, Crystal Springs
Lake, Palo Alto, San Antonio Creek, San Jose, Swanton, and near Corralitos;
native of Europe. April–November.
2. X. strumarium L. Cocklebur. Low areas where water has stood or accumu-
lated during the winter and early spring, and as a weed in disturbed areas; San
Francisco, Redwood City, Stanford, near San Jose, and Pajaro River; native of
Europe. June–October. —*X. calvum* Millsp. & Sherff, *X. italicum* Moretti, *X.
canadense* Mill.

60. Franseria Cav.

Plants perennial.
　Burs 6–8 mm. long. Plants of coastal sand dunes.
　　Leaves serrate.　　　　　　　　　　1a. *F. chamissonis chamissonis*
　　Leaves 1–3 times pinnatifid.　　　　　1b. *F. chamissonis bipinnatisecta*
　Burs 2–3 mm. long. Plants of disturbed areas.　　　　2. *F. confertiflora*
Plants annual.　　　　　　　　　　　　　　　　　　　3. *F. acanthicarpa*

1a. F. chamissonis Less. ssp. **chamissonis.** *Fig. 238.* Beach-bur. Sea beaches
and sand dunes, most common in San Francisco and northern San Mateo counties.
July–November. Plants with leaves intermediate between *F. chamissonis* ssp.
chamissonis and ssp. *bipinnatisecta* are sometimes found.

1b. F. chamissonis Less. ssp. **bipinnatisecta** (Less.) Wiggins & Stockwell. Sand dunes and beaches along the coast; San Francisco, Swanton, Santa Cruz, and Palm Beach. June–October.

2. F. confertiflora (DC.) Rydb. Weak-leaved Burweed. Occasional in disturbed areas; San Francisco, San Jose, and Santa Cruz; native from southern California eastward to Colorado and Mexico. September–December.

3. F. acanthicarpa (Hook.) Cov. Annual Burweed. River bottoms and disturbed areas in the Santa Clara Valley, poorly collected; near Gilroy; perhaps introduced locally from other parts of western North America. June–August.

8. Helenieae. Sneezeweed Tribe

61. **Eriophyllum** Lag.

Involucres usually less than 8 mm. in diameter; heads in dense terminal clusters. Involucral bracts 8–12; rays 6–9. Plants of coastal sand dunes and bluffs.

1. E. staechadifolium

Involucral bracts 4–7; rays 4–6 or none. Inland plants.   *2. E. confertiflorum*

Involucres usually over 10 mm. in diameter; heads solitary or loosely corymbose. Pappus a crown of erose teeth; achenes 3–4 mm. long; adaxial leaf-surfaces becoming dark on drying. Plants of wooded areas.

3a. E. lanatum arachnoideum

Pappus of short oblong paleae; achenes 2.5–3 mm. long; adaxial leaf-surfaces not becoming dark on drying. Plants of chaparral.

3b. E. lanatum achillaeoides

1. E. staechadifolium Lag. Lizard Tail, Seaside Woolly Sunflower. Coastal bluffs and sand dunes, occasionally inland; San Francisco, Sweeney Ridge, near Pescadero, Swanton, Santa Cruz, near Seabright, and Watsonville. May–November, but individual plants may be found in flower during the remaining months of the year. —*E. staechadifolium* Lag. var. *artemisiaefolium* (Less.) Macbr.

2. E. confertiflorum (DC.) Gray. *Fig. 239.* Yellow Yarrow. Chaparral, brushy slopes, and serpentine, throughout the Santa Cruz Mountains; San Francisco, near Belmont, Redwood City, Coal Mine Ridge, Black Mountain, near Los Gatos, Loma Prieta, Glenwood, Ben Lomond Sand Hills, near Soquel, and Soda Lake. March–November.

238

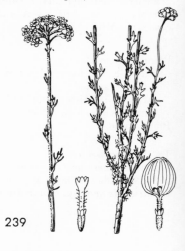

239

3*a*. E. lanatum (Pursh) Forbes var. **arachnoideum** (Fisch. & Ave-Lall.) Jeps. Common Woolly Sunflower. Wooded slopes, redwoods, and open hills, usually near the coast; near La Honda, near Woodside, and Kings Mountain. May–July. This variety apparently hybridizes with *E. confertiflorum*. The hybrids have very broad leaves that are more or less lobed like those of *E. confertiflorum*, and heads that are intermediate in size and in lax corymbs.

3*b*. E. lanatum (Pursh) Forbes var. **achillaeoides** (DC.) Jeps. Known from the vicinity of Loma Prieta, where it grows in chaparral. May–July.

62. Chaenactis DC.

1. C. tanacetifolia Gray. Inner Coast Range Chaenactis. Known locally only from near Mount Umunhum at an altitude of about 3,400 feet. April. —*C. glabriuscula* DC. var. *gracilenta* (Greene) Keck.

63. Blennosperma Less.

1. B. nanum (Hook.) Blake. Common Blennosperma. Known locally from moist open hills in San Francisco and northern San Mateo County. March–April.

64. Helenium L.

1. H. puberulum DC. *Fig. 240.* Rosilla, Sneezeweed. Creek beds and marshy meadows along streams and lakes; San Francisco, near Pescadero, Pilarcitos Canyon, San Francisquito Creek, Stevens Creek, Loma Prieta, Carnadero Creek, Swanton, and Powder Mill Gulch. April–November.

65. Monolopia DC.

Involucral bracts not united, the margins more or less imbricate.
1. *M. gracilens*
Involucral bracts united in the lower half, the margins not imbricate.
2. *M. major*

1. M. gracilens Gray. Woodland Monolopia. Shaded slopes, road cuts, serpentine, and burned chaparral from northern San Mateo County southward; Sawyer Ridge, Black Mountain, Loma Prieta, near New Almaden, Gilroy, San Lorenzo Canyon, near Big Basin, Santa Cruz, and Hecker Pass. April–June.

2. M. major DC. Cupped Monolopia. Slopes, road cuts, and serpentine outcroppings on the eastern slopes of the Santa Cruz Mountains; San Francisco, San Mateo Creek, Stanford, Black Mountain, near San Jose, and near Coyote. March–May.

66. Rigiopappus Gray.

1. R. leptocladus Gray. Rigiopappus. Serpentine and grassy slopes on the eastern face of the Santa Cruz Mountains from central San Mateo County southward; near Crystal Springs Lake, Woodside, Page Mill Road, near Coyote, and Uvas Creek. The achenes of local plants have three or five paleae. April–June.

67. Hulsea T. & G. ex Gray

1. H. heterochroma Gray. Red-rayed Hulsea. Known locally only from Loma Prieta. June–August.

68. Jaumea Pers.

1. J. carnosa (Less.) Gray. *Fig. 241.* Fleshy Jaumea. Along the ocean and the margins of San Francisco Bay in salt marshes, tidal flats, and on sea cliffs which are subjected to salt spray, commonly growing with *Frankenia grandifolia, Salicornia, Spartina foliosa,* and *Distichlis spicata*; San Francisco, Cooleys Landing, Mountain View, Laguna Salada, Swanton, Santa Cruz, and near Pajaro River. June–November.

69. Baeria F. & M. Goldfields

Some members of this genus in the Santa Cruz Mountains vary considerably, and individual specimens are hard to place in one species or another. For example, robust specimens of *B. minor* approach *B. macrantha* var. *thalassophila*; intermediates between *B. minor* and *B. chrysostoma* are occasionally encountered; and it is often difficult to assign specimens to the subspecies of *B. chrysostoma*. The characters in the following key will serve to separate most specimens found locally.

Plants perennial; involucral bracts over 9 mm. long; leaves entire. Coastal plants.   1. *B. macrantha thalassophila*
Plants annual; involucral bracts under 9 mm. long and usually under 5 mm. long; leaves entire or pinnately lobed. Coastal or inland plants.
 Receptacle conical; ray-flowers usually more than 6.
  Stems and leaves villous, villous-tomentose, or nearly glabrous.   2. *B. minor*
  Stems and leaves hirsute and/or strigose.
   Lower leaf-margins usually hirsute or strigose toward the base; achenes subclavate; paleae 2–7 or none, subulate from the base, white to brownish.
    Leaves linear, entire, acute; plants slender, when branched the branches usually ascending or erect.   *3a. B. chrysostoma chrysostoma*
    Leaves linear-oblong, entire or sometimes pinnately lobed, usually rounded at the apex; plants stout, usually branched, the branches spreading widely.   *3b. B. chrysostoma hirsutula*
   Lower leaf-margins usually lacking hirsute or strigose hairs; achenes tending to be linear; paleae 2–5 or none, usually ovate at the base, then with a subulate awn, white.   *3c. B. chrysostoma gracilis*
 Receptacle subulate; ray-flowers 1–3.   4. *B. microglossa*

1. B. macrantha (Gray) Gray var. **thalassophila** Howell. Perennial Baeria. Known locally from the vicinity of Pescadero, but to be expected elsewhere along the coast in moist areas. April–June.
2. B. minor (DC.) Ferris. Woolly Baeria, Woolly Goldfields. Mainly along the coast in moist areas; San Francisco, Pescadero Beach, Pigeon Point, Alma, near San Jose, and Sunset Beach. April–May. —*B. uliginosa* (Nutt.) Gray.
3a. B. chrysostoma F. & M. ssp. **chrysostoma.** *Fig. 242.* Goldfields. Open grassy slopes, often on rocky and wind-swept hills, and sometimes in serpentine soil or on sand dunes; San Francisco, San Bruno Hills, Sawyer Ridge, Coal Mine Ridge, Stanford, near San Jose, Edenvale, near Gilroy, and near Scott Valley. March–May.

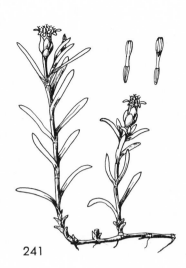

3b. B. chrysostoma F. & M. ssp. **hirsutula** (Greene) Ferris. Usually along or near the coast in moist flats; San Francisco, San Bruno Mountain, Baden, Seal Cove, Pescadero, and Santa Cruz. March–June.

3c. B. chrysostoma F. & M. ssp. **gracilis** (DC.) Ferris. Slender Goldfields. Open grassy slopes, sandy soils, and shallow soil of rocky or serpentine outcroppings; San Francisco, near Millbrae, Redwood City, Stanford, Edenvale, near Jamison Creek, Scott Valley, near Mount Hermon, and Big Trees. January–May.

4. B. microglossa (DC.) Greene. Small-rayed Baeria. Reported from San Francisco by Dr. H. H. Behr; now probably extinct locally. March–April.

70. Lasthenia Cass.

Rays inconspicuous; pappus present.	1. *L. glaberrima*
Rays conspicuous; pappus lacking.	2. *L. glabrata*

1. L. glaberrima DC. Smooth Lasthenia. A semiaquatic of small ponds or on muddy ground; San Francisquito Creek, Coal Mine Ridge, and San Jose. May–June.

2. L. glabrata Lindl. Yellow-rayed Lasthenia. Edges of salt marshes along San Francisco Bay and low moist areas of heavy clay soils in the northern part of the Santa Clara Valley; San Francisco, Millbrae, Crystal Springs, Belmont, Redwood City, Mayfield, San Jose, and 3 miles south of San Jose. March–May.

9. Calenduleae. Calendula Tribe

71. Calendula L.

1. C. officinalis L. Pot Marigold. Occasional as an escape from cultivation; San Francisco, Palo Alto, and Page Mill Road; native of Europe. March–July.

10. Anthemideae. Mayweed Tribe

72. **Achillea** L. Yarrow

Plants glabrate to arachnoid; leaves thin; involucres usually about 4 mm. high. Coastal bluffs and inland.   1*a. A. millefolium californica*
Plants tomentose; leaves thick; involucres usually about 5 mm. high. Coastal dunes.   1*b. A. millefolium arenicola*

1*a.* **A. millefolium** L. var. **californica** (Pollard) Jeps. *Fig. 243.* Common Yarrow or Milfoil. Coastal bluffs, open slopes, wooded areas, and often as weeds in disturbed places and in lawns, throughout the Santa Cruz Mountains; San Francisco, San Bruno Mountain, Coal Mine Ridge, near Los Gatos, near Big Basin, Blooms Grade, and Ben Lomond Sand Hills. April–July.
1*b.* **A. millefolium** L. var. **arenicola** (Heller) Nobs. Coastal sand dunes from San Francisco southward. May–July.

73. **Anthemis** L. Dog Fennel, Camomile

1. **A. cotula** L. Mayweed. Very common throughout the Santa Cruz Mountains in fallow and overgrazed fields, along highways and roads, and in other disturbed areas; native of Europe. Plants can be found in flower at all times of the year.

74. **Chrysanthemum** L.

Leaves bipinnatifid.
    Heads under 1.5 cm. across; rays white; ultimate divisions of the leaves
        rounded to obtuse.   1. *C. parthenium*
    Heads over 2 cm. across; rays yellow; ultimate divisions of the leaves acute.
                                                        2. *C. coronarium*
Leaves toothed or incised, but not bipinnatifid.
    Middle cauline leaves rarely over 5 cm. long; leaves incised to crenate.
    Rays white.   3. *C. leucanthemum*
    Rays yellow.   4. *C. segetum*
    Middle cauline leaves 10–15 cm. long; leaves regularly serrate.
                                                        5. *C. maximum*

1. **C. parthenium** (L.) Bernh. Feverfew. Occasional as an escape from cultivation in ditches, stream beds, marshes, and moist disturbed areas; San Francisco, Palo Alto, Stevens Creek, Laurel, and Boulder Creek; native of Europe. June–October.
2. **C. coronarium** L. Crown Daisy, Garland Chrysanthemum. Occasional as an escape from gardens; San Francisco and Half Moon Bay; native of Eurasia and North Africa. April–August.
3. **C. leucanthemum** L. Whiteweed, Ox-eye Daisy. Usually in disturbed areas along roads; near Baden, Millbrae, and Cahill Ridge; native of Eurasia. May–August.
4. **C. segetum** L. Corn Chrysanthemum, Corn Marigold. Occasional as an escape from cultivation in San Francisco; native of Eurasia. April–August.
5. **C. maximum** Ramond. Shasta Daisy. Commonly grown as a garden ornamental and occasionally becoming established; Waddell Creek; native of Europe. July–August.

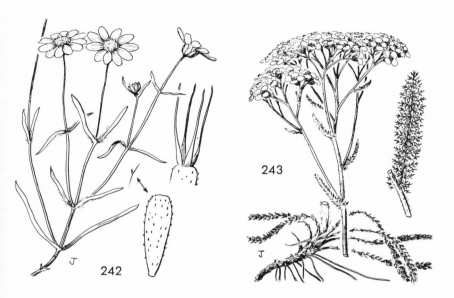

75. Santolina L.

1. S. chamaecyparissus L. Lavender Cotton. Occasional as an escape from cultivation in San Francisco and probably elsewhere; native of the Mediterranean region. June–August.

76. Soliva R. & P.

1. S. sessilis R. & P. Common Soliva. Grassy slopes, old roads, and in gardens and lawns; San Francisco, near Pescadero Creek, Pescadero Beach, Pigeon Point, Woodside, Palo Alto, Page Mill Road, Permanente Creek, Big Basin, and near Eagle Rock; native of Chile. January–June. —*S. daucifolia* Nutt.

77. Matricaria L.

Mature heads 9–11 mm. in diameter; plants 1.5–3.5 dm. tall, branched from the base or not; pappus-crown with 2 conspicuous teeth.   1. *M. occidentalis*
Mature heads 5–7 mm. in diameter; plants usually under 2 dm. tall, usually branched from the base; pappus-crown lacking 2 conspicuous teeth.
2. *M. matricarioides*

1. M. occidentalis Greene. Valley Pineapple Weed. Occasional as a weed near the waterfront in the eastern part of San Francisco; probably introduced from the Central Valley. April–June.
2. M. matricarioides (Less.) Porter. *Fig. 244.* Pineapple Weed. Very common throughout the Santa Cruz Mountains, especially in dry, hard-packed soils of disturbed areas. March–June.

78. Cotula L.

Leaves pinnate, pubescent, not succulent.   1. *C. australis*
Leaves pinnatifid, glabrous, somewhat succulent.   2. *C. coronopifolia*

1. C. australis (Sieb.) Hook. f. Australian Cotula. Common as a weed in disturbed areas, especially about habitations and along sidewalks and streets; San Francisco, Mussel Rock, Sawyer Ridge, Stanford, and Waddell Creek; native of Australia. March–October.
2. C. coronopifolia L. Brass Buttons. Common along the coast and San Francisco Bay in marshes, in moist places in sand dunes, and inland in low marshy areas, and areas of heavy soil; San Francisco, near Crystal Springs Lake, Coal Mine Ridge, Alviso, Swanton, Santa Cruz, Seabright, and near Pajaro River; native of South Africa. March–October.

79. Artemisia L. Sagewort, Sagebrush, Wormwood

Plants shrubby; leaf-divisions filiform; plants distinctly with the odor of sage.
    1. *A. californica*
Plants herbaceous or somewhat woody at base; odorless or with a slightly pungent odor.
    At least the abaxial leaf-surfaces tomentose.
    Leaves pinnate, the divisions lanceolate to ovate. Mainly inland plants.
       2. *A. douglasiana*
    Leaves once- to thrice-pinnate, the divisions linear. Plants of the sea coast
    and inland marine sand deposits.   3. *A. pycnocephala*
    Leaves glabrous to pubescent, not tomentose.
    Leaves usually entire, or if lobed, the segments entire.   4. *A. dracunculus*
    Leaves pinnate, the margins distinctly serrate.   5. *A. biennis*

1. A. californica Less. California Sage, Old Man. Coastal bluffs and exposed slopes, commonly growing with *Baccharis pilularis, Rhus diversiloba, Diplacus aurantiacus, Rubus vitifolius,* and *Anaphalis margaritacea*; San Francisco, Sweeney Ridge, Stanford, Black Mountain, near Big Basin, and Santa Cruz. July–November.
2. A. douglasiana Bess. *Fig. 245.* California or Douglas' Mugwort. Common, especially along streams, in dry stream bottoms, occasionally on shaded slopes, and often appearing as a weed in disturbed areas; San Francisco, San Andreas Lake, Portola Road, Black Mountain, near Mountain View, San Jose, near Los Gatos, Carnadero Creek, Big Basin, Rincon, and Watsonville Slough. June–November.
3. A. pycnocephala (Less.) DC. Beach Sagewort. Coastal dunes and beaches and occasionally inland; San Francisco, Mussel Rock, Pescadero Beach, Waddell Creek, Camp Evers, Sunset Beach, and Palm Beach. May–November.
4. A. dracunculus L. Tarragon, Dragon Sagewort. Occasional in disturbed areas; San Francisco, San Francisquito Creek, Palo Alto, Mountain View, Alviso, and near Santa Cruz. October–November. —*A. dracunculoides* Pursh.
5. A. biennis Willd. Biennial Sagewort. Occasional, most common in swampy or marshy areas, along streams, and where water has stood; Crystal Springs Lake,

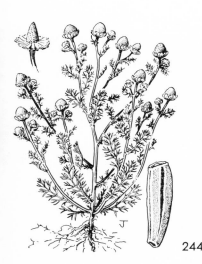

244   245

Searsville Lake, Mountain View, San Antonio Creek, near Coyote, near Santa Cruz, and Pajaro River; presumably native in the Rocky Mountains and British Columbia. September–December.

80. **Tanacetum** L. Tansy

1. **T. camphoratum** Less. Dune Tansy. Sand dunes in the western part of San Francisco and now probably restricted to the coastal dunes and the few small sand dune areas which still remain inland. June–October.

11. Senecioneae. Groundsel Tribe

81. **Arnica** L.

1. **A. discoidea** Benth. Rayless Arnica. Open slopes and edges of chaparral, often growing with *Thermopsis macrophylla, Helianthemum scoparium, Polygala californica, Pteridium aquilinum,* and various species of *Lupinus*; Pescadero Creek, near Woodside, Jasper Ridge, Alma, Sierra Azule Ridge, Eagle Rock, near Scott Creek, Big Basin, and near China Grade. May–July. Most of the specimens from the Santa Cruz Mountains are referable to *A. discoidea* Benth. var. *alata* (Rydb.) Cronq.

82. **Petasites** Mill.

1. **P. frigidus** (L.) Fries var. **palmatus** (Ait.) Cronq. *Fig. 246.* Western Coltsfoot. Along creeks and streams in the redwood–Douglas fir forest; near Pescadero Creek, Peters Creek, Stevens Creek, Alma, Swanton, Glenwood, and Hester Creek. March–May.

83. Senecio L.

Pistillate flowers if present ligulate, marginal; heads usually radiate.
Ray-flowers purple-lavender or purple-red. Introduced plants.
 Leaves pinnatifid, sparsely pubescent on both surfaces.   1. *S. elegans*
 Leaves ovate-cordate, densely tomentose abaxially, nearly glabrous and green
  above.   2. *S. cruentus*
Ray-flowers yellow or none. Introduced or native plants.
 Leaves palmately veined; plants climbing.   3. *S. mikanioides*
 Leaves pinnately veined; plants not climbing.
  Plants annual.
   Involucral bracteoles black-tipped, involucral bracts usually about 20;
    rays absent.   4. *S. vulgaris*
   Involucral bracteoles present or absent but not black-tipped, involucral
    bracts 7–13; rays present, but small.
    Involucral bracts usually 8; herbage essentially glabrous.
                                                                  5. *S. aphanactis*
    Involucral bracts usually about 12–13; herbage pubescent.
                                                                  6. *S. sylvaticus*
  Plants perennial.
   Leaves and involucral bracts densely white-tomentose, leaf-segments not
    linear. Introduced plants.   7. *S. cineraria*
   Leaves and involucral bracts not densely tomentose, or if the leaves to-
    mentose then their segments linear. Native plants.
    Leaves entire or toothed, not lobed or pinnatifid.
     Leaves entire. Plants coastal or of salt marshes. 8. *S. hydrophilus*
     Leaves dentate. Inland plants.   9. *S. aronicoides*
    Leaves lobed or pinnatifid, the lobes various.
     Segments of the leaves linear, usually tomentose. 10. *S. douglasii*
     Segments of the leaves not linear, glabrous.   11. *S. breweri*
Pistillate flowers present, marginal, lacking ligules.
 Leaves sharply but irregularly dentate.   12. *S. minimus*
 Leaves pinnatifid.   13. *S. glomeratus*

1. S. elegans L. Purple Ragwort. Ocean bluffs and sandy beach soils in San
Francisco; native of South Africa. May–June.
2. S. cruentus DC. Florist's or Garden Cineraria. Known only as an occasional
escape from cultivation in San Francisco; native of the Canary Islands. Summer
months.
3. S. mikanioides Otto. German Ivy. Occasional in moist areas along streams
or on shaded slopes, commonly climbing over other vegetation; San Francisco,
near San Bruno, Frenchman Creek, Searsville, Palo Alto, Los Altos, and near
Bonny Doon; native of South Africa. December–March.
4. S. vulgaris L. Common Groundsel, Old-Man-in-the-Spring. A very widespread
and common weed in disturbed areas, especially in orchards, where it may color
the ground with yellow, and in grasslands; native of Europe. Flowering all year.
5. S. aphanactis Greene. California Groundsel. Known locally only from hills
south of San Jose. January–March.

6. S. sylvaticus L. Wood Groundsel. Occasional in rocky or brushy areas; San Francisco, Ben Lomond Sand Hills, and Mount Hermon; native of Europe. April–June.

7. S. cineraria DC. Dusty Miller. Occasional as an escape from gardens; San Francisco and Santa Cruz; native of the Mediterranean region. June–August.

8. S. hydrophilus Nutt. Alkali Marsh Butterweed. Occasional along the margins of salt marshes of San Francisco Bay and in low wet areas along the coast; Menlo Park, Cooleys Landing, and Swanton. May–October.

9. S. aronicoides DC. *Fig. 247.* California Butterweed. Open flats in chaparral and in evergreen woods; San Francisco, Colma, San Bruno Hills, Crystal Springs, Redwood City, Kings Mountain, Searsville, Black Mountain, Castle Rock, Loma Prieta, and Ben Lomond Sand Hills. April–June.

10. S. douglasii DC. Shrubby Butterweed. Along dry stream beds, in chaparral, and on open exposed slopes from southern San Mateo County southward; Mountain View, Stevens Creek, Wrights, Soda Springs, near San Jose, and near Almaden. July–November.

11. S. breweri Davy. Brewer's Butterweed. Occasional along the edges of woodland on the eastern slopes of the Santa Cruz Mountains; near San Mateo, Kings Mountain, Stanford, and Black Mountain. May–June.

12. S. minimus Poir. Toothed Coast Fireweed. Moist soil along creeks, at the edges of chaparral, and in disturbed ground along the margins of redwood–Douglas fir forests; San Francisco, Pilarcitos Dam, Bielawski Lookout, Jamison Road, and Waddell Creek; native of Australia and New Zealand. July–January. —*Erechtites prenanthoides* (Rich.) DC.

13. S. glomeratus Desf. ex Poir. Cut-leaved Coast Fireweed. Open disturbed areas; San Francisco, Sharp Park, near Daly City, Belmont, Redwood City, near San Mateo–Santa Cruz county line, Ben Lomond, and Soda Lake; native of Australia and New Zealand. May–December. —*Erechtites arguta* DC.

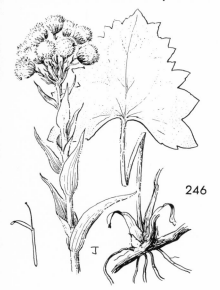

246

247

12. Cynareae. Thistle Tribe

84. Cynara L.

1. C. scolymus L. Globe or Garden Artichoke. Occasional as an escape from cultivation; Redwood City, Stanford, and near Santa Clara; native of Europe. June–November.

85. Arctium L. Burdock

Heads 2.5–4 cm. broad, long-pedunculate.	1. *A. lappa*
Heads 1.5–2.5 cm. broad, subsessile.	2. *A. minus*

1. A. lappa L. Great Burdock. Occasional in disturbed areas; San Francisco and Stanford; native of Europe. July–October.
2. A. minus (Hill) Bernh. Common Burdock. Known from near Santa Cruz; native of Europe and North Africa. July–October.

86. Centaurea L. Star Thistle

Some twenty taxa of this genus are known as introduced weeds in the three Pacific states, Washington, Oregon, and California. It is to be expected that in the due course of time additional Centaureas will be found in the area covered by this flora.

Terminal spine of the involucral bracts under 2 mm. long.
   Terminal spine 1–2 mm. long.   1. *C. diluta*
   Terminal spine lacking or shorter than the lateral divisions of the involucral bracts.
      Leaves pinnately to bipinnately divided; herbage densely white-tomentose.
         2. *C. cineraria*
      Leaves entire or sometimes laciniate; herbage not densely white-tomentose.
         Pappus-bristles about 4 mm. long; plants annual.   3. *C. cyanus*
         Pappus-paleae under 0.5 mm. long; plants perennial.   4. *C. pratensis*
Terminal spine of the involucral bracts over 5 mm. long.
   Leaf-bases not decurrent; pappus lacking; flowers blue.   5. *C. calcitrapa*
   Leaf-bases decurrent; pappus present; flowers yellow.
      Terminal spine 11–22 mm. long; pappus 3–5 mm. long, that of the marginal flowers lacking.   6. *C. solstitialis*
      Terminal spine 5–9 mm. long; pappus 1.5–3 mm. long, that of the marginal flowers present.   7. *C. melitensis*

1. C. diluta Ait. Occasional as a weed in San Francisco; native of North Africa. July–August.
2. C. cineraria L. Dusty Miller, Purple Star Thistle. Known locally as an escape from cultivation in Santa Cruz; native of southern Europe and North Africa. May–July.
3. C. cyanus L. Cornflower, Bachelor's Button. Occasional in disturbed areas, commonly cultivated in gardens; San Francisco, Stanford, Alma, and near Saratoga Summit; native of the Mediterranean region. May–August.

4. C. pratensis Thuill. Protean Knapweed. Locally introduced along Empire Grade in Santa Cruz County; native of Europe. July–August.

5. C. calcitrapa L. Purple Star Thistle. Disturbed areas; San Francisco, San Bruno Mountain, San Mateo, Crystal Springs, Belmont, Palo Alto, Stanford, near San Jose, and Boulder Creek; native of Eurasia. May–November.

6. C. solstitialis L. *Fig. 248.* Yellow Star Thistle, Barnaby's Thistle. Roadsides, fallow fields, and disturbed habitats in general; San Francisco, South San Francisco, San Andreas Lake, Stanford, Lone Hill, near Gilroy, and Corralitos; native of the Mediterranean region. June–December.

7. C. melitensis L. Tocalote, Napa Thistle. Common in disturbed areas and on brushy slopes; San Francisco, Crystal Springs Lake, Coal Mine Ridge, Stanford, near Los Gatos, Swanton, Saratoga Summit, Santa Cruz, and Sunset Beach State Park; native of Europe. May–September.

87. **Silybum** Adans.

1. S. marianum (L.) Gaertn. *Fig. 249.* Milk Thistle. Very common in disturbed habitats, along roadsides, and an indicator of overgrazing in pastures, fields, and meadows; native of the Mediterranean region. March–August.

88. **Cnicus** L.

1. C. benedictus L. Blessed Thistle. Occasional in open fields and disturbed areas; San Francisco, Stanford, Saratoga, and Ben Lomond Sand Hills; native of Europe and Asia Minor. April–November.

89. **Carthamus** L.

1. C. lanatus L. Distaff Thistle. A weed in overgrazed fields and in disturbed areas; San Francisco, San Mateo County, and near Stanford; native of Europe. April–October.

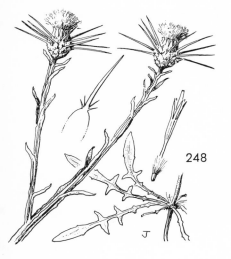

248

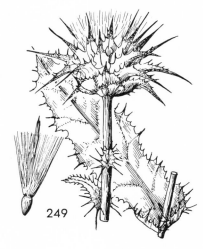

249

90. **Cirsium** Mill. Thistle

Plants dioecious; heads 2–2.5 cm. high.	1. *C. arvense*

Plants monoecious; heads usually over 2.5 cm. high.

Stems winged with spiny decurrent leaf-bases.	2. *C. vulgare*

Stems not so winged.

Involucral bracts with a conspicuous, oblong, glutinous gland on the abaxial surface.   3. *S. douglasii*

Involucral bracts lacking a gland.

Adaxial leaf-surfaces scabrous; involucral bracts 5–8 mm. wide at the widest point.   4. *C. fontinale*

Adaxial leaf-surfaces variously pubescent, but not scabrous; involucral bracts usually under 5 mm. wide.

Involucral bracts tomentose or arachnoid; corollas usually red or purple. Heads closely subtended by leaves.

Corolla-lobes exceeding the anthers, linear; flowers about equaling the involucre; involucral bracts lacking pectinate margins.   5. *C. brevistylum*

Anthers exserted beyond the corolla, its lobes oblong; flowers exceeding the involucre; involucral bracts with pectinate margins.   6. *C. andrewsii*

Heads long-pedunculate.

Outer involucral bracts reflexed; bases of the involucral bracts glabrate or naked, not obscured by the tomentum; flowers conspicuously exceeding the involucres. Mainly inland plants.   7. *C. proteanum*

Outer involucral bracts spreading but not reflexed; bases of the involucral bracts obscured by the tomentum; flowers scarcely exceeding the involucres. Most common on coastal sand dunes.   8. *C. occidentale*

Involucral bracts usually glabrous; corollas white or cream-colored.

Plants usually under 3 dm. tall; heads subsessile.   9. *C. quercetorum*

Plants 3–15 dm. tall; heads long-pedunculate.   10. *C. remotifolium*

1. C. arvense (L.) Scop. Canada Thistle. Occasional in disturbed areas; San Francisco and doubtless elsewhere; native of Eurasia. June–September.

2. C. vulgare (Savi) Tenore. Common or Bull Thistle. A weed of disturbed areas; San Francisco, Sawyer Ridge, Cooleys Landing, Stanford, near San Jose, Wrights, near Hilton Airport, Swanton, and near Watsonville; native of Eurasia. June–October.

Cirsium scabrum (Poir.) Bonnet & Barratte, a native of the Mediterranean region, has been collected near Glenwood. It is similar to *C. vulgare* but grows to a height of 4 m., while *C. vulgare* rarely is more than 2 m. tall.

3. C. douglasii DC. Douglas' or Swamp Thistle. Occasional in marshes and near sloughs, usually near the coast; Camp Evers and Watsonville. July–August. —*C. breweri* (Gray) Jeps., in part.

4. C. fontinale (Greene) Jeps. Fountain Thistle. Endemic in the Santa Cruz Mountains; Crystal Springs Lake, Spring Valley, and Alamitos Creek. May–October.

5. C. brevistylum Cronq. Indian Thistle. Wooded areas and brush-covered slopes, San Francisco southward; San Francisco, near San Bruno, Crystal Springs, San Andreas Lake, Lobitos Creek, Año Nuevo Point, Coal Mine Ridge, near Stanford, near Swanton, and Watsonville. April–July. —*C. edule* of California authors.

6. C. andrewsii (Gray) Jeps. Franciscan Thistle. Moist slopes in San Francisco and northern San Mateo County. June–October. —*C. amplifolium* (Greene) Petrak.

Cirsium praeteriens Macbr. was based on two collections made in Palo Alto in 1897 and 1901. *Cirsium praeteriens* perhaps represents a casual introduction from the Old World, and was renamed as a native. It differs from *C. andrewsii* by its long involucral spines. *Cirsium praeteriens* needs further study to establish its origin and affinities.

7. C. proteanum Howell. Red or Venus Thistle. Open areas in woods and on dry slopes; near Portola, Coal Mine Ridge, Stanford, near Los Gatos, Los Gatos Canyon, near Eagle Rock, Blooms Grade, and near Santa Cruz. May–September.

8. C. occidentale (Nutt.) Jeps. *Fig. 250.* Cobweb Thistle. Coastal sand dunes, less frequent inland; San Francisco, Sawyer Ridge, Crystal Springs Lake, near Stanford, Loma Prieta Ridge, Swanton, and Sunset Beach State Park. April–June. —*C. coulteri* Harv. & Gray.

9. C. quercetorum (Gray) Jeps. Brownie Thistle. Coastal bluffs and grassy slopes near the ocean and San Francisco Bay; San Francisco, Sawyer Ridge, Crystal Springs Lake, San Bruno Hills, near Millbrae, Pigeon Point, near Burlingame, and near Seabright. April–July. Plants with distally expanded and spiny involucral bracts may be called *C. quercetorum* (Gray) Jeps. var. *xerolepis* Petrak.

10. C. remotifolium (Hook.) DC. Remote-leaved Thistle. Known locally on wooded slopes in San Francisco. June–August.

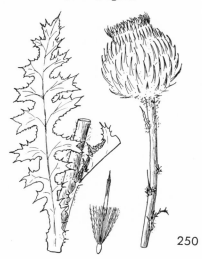

250

91. Carduus L. Plumeless Thistle

Involucral bracts scarious-margined, the tips not scabrous abaxially.
                                                                    1. *C. tenuiflorus*
Involucral bracts not scarious-margined, the tips scabrous abaxially.
                                                              2. *C. pycnocephalus*

1. C. tenuiflorus Curtis. Slender-flowered Thistle. Occasional in disturbed areas; San Francisco, San Andreas Valley, San Mateo–Santa Cruz county line, Jasper Ridge, Guadalupe Creek, San Jose, near Davenport, and near Boulder Creek; native of Europe. May–July.

2. C. pycnocephalus L. Italian Thistle. Occasional in disturbed areas; San Francisco and Moffett Field; native of the Mediterranean region. May–June.

ADDITIONS

p. 75.   **Arundo donax** L. Giant Reed. Known from a brackish marsh near San Pedro Point and being spread by its thick rhizomes along highway construction areas; native of North Africa and Eurasia.
Lemmas naked; rachilla hairy.   *Phragmites*
Lemmas hairy; rachilla glabrous.   *Arundo*

p. 108.   **Cyperus rotundus** L. Purple Nut Grass. Known only as a weed at Stanford; native of Eurasia. July–November.
Scales dull, yellowish-brown.   *C. esculentus*
Scales shiny, reddish.   *C. rotundus*

p. 303.   **Solanum nigrum** L. Black Nightshade. Rare locally, known only from near Boulder Creek; native of Europe. April–October.
Anthers 1.8–2.4 mm. long. Rare.   *S. nigrum*
Anthers 1–1.2 mm. long. Common.   *S. nodiflorum*

p. 308.   **Kickxia elatine** (L.) Dumort. Sharp-leaved Fluellin. Occasional in disturbed areas; Moffett Field and near Freedom; native of Europe. June–September.
Corollas 6–8 mm. long; leaves rounded to cordate at the base.
                                                            K. spuria
Corollas about 5 mm. long; leaves hastate-lobed at the base. *K. elatine*

p. 320.   **Proboscidea louisianica** (Mill.) Thell. Common Unicorn Plant. Known from one specimen collected in Menlo Park, but to be expected occasionally elsewhere; native from California to Mexico and eastward. August–September. *Proboscidea louisianica* is a member of the Martyniaceae and may be distinguished from families 120–122 by its capsules, which are 10 cm. or more long.

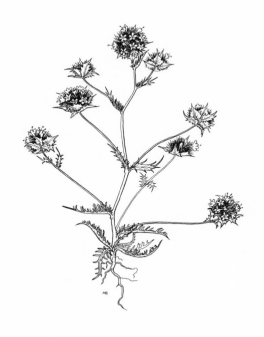

PART III

References and Glossary

General References

Abrams, L., and R. S. Ferris. 1923–60. *Illustrated Flora of the Pacific States.* 4 vols. Stanford University Press, Stanford, California.

Anderson, C. L. 1892. Catalogue of Flowering Plants and Ferns of Santa Cruz County, California. Pp. 118–35 in E. S. Harrison, *History of Santa Cruz County, California.* Pacific Press Publishing Co., San Francisco, California.

Barry, M. A. 1940. A Floristic and Ecological Study of Coal Mine Ridge. *Contributions from the Dudley Herbarium,* **3**:1–40.

Behr, H. H. 1891. Botanical Reminiscences. *Zoe,* **2**:2–6.

Bowerman, M. L. 1944. *The Flowering Plants and Ferns of Mount Diablo, California.* The Gillick Press, Berkeley, California.

Brandegee, K. 1892. Catalogue of the Flowering Plants and Ferns Growing Spontaneously in the City of San Francisco. *Zoe,* **2**:334–86.

Campbell, D. H., and I. L. Wiggins. 1947. Origins of the Flora of California. Stanford University Series, Biological Sciences, **10**:1–20.

Cooper, W. S. 1922. The Broad-Sclerophyll Vegetation of California. *Carnegie Institution of Washington Publications,* **319**:1–124.

Daly, J. 1935. A Botanical Survey of San Gregorio Creek. M.A. Thesis, University of California, Berkeley, California.

Eastwood, A. 1939. Early Botanical Explorers of the Pacific Coast and the Trees They Found There. *California Historical Society Quarterly,* **18**:335–45.

Felton, E. L. 1954. United States Department of Commerce, *Climatological Data, California,* **57**:311–35.

Greene, E. L. 1894. *Manual of the Botany of the Region of San Francisco Bay.* Cubery & Co., San Francisco, California.

Hall, H. M., and J. Grinnell. 1919. Life-Zone Indicators in California. *Proceedings of the California Academy of Sciences,* **9**:37–67.

Hayward, I. R. 1931. The Marsh and Aquatic Plants of the Pajaro Valley. M.A. Thesis, Stanford University, Stanford, California.

Howell, J. T. 1949. *Marin Flora.* University of California Press, Berkeley, California.

Howell, J. T., P. H. Raven, and P. Rubtzoff. 1958. A Flora of San Francisco, California. *Wasmann Journal of Biology,* **16**:1–157.

Jenkins, O. P., editor. 1951. *Geologic Guidebook of the San Francisco Bay Counties.* State of California, Dept. of Mines & Natural Resources, San Francisco, California.

Jepson, W. L. 1909–43. *A Flora of California.* 3 vols., incomplete. Associated Students Store, University of California, Berkeley, California.

Jepson, W. L. 1925. *A Manual of the Flowering Plants of California.* Associated Students Store, University of California, Berkeley, California.

Mason, H. L. 1957. *A Flora of the Marshes of California.* University of California Press, Berkeley, California.

McKelvey, S. D. 1955. *Botanical Exploration of the Trans-Mississippi West, 1790–1850.* Arnold Arboretum, Jamaica Plain, Massachusetts.

McMinn, H. E. 1939. *An Illustrated Manual of California Shrubs.* J. W. Stacey, Inc., San Francisco, California.

Mock, B. A. 1941. Some Ecological Relationships of the Vegetation of Lagunita Lake Bed. M.A. Thesis, Stanford University, Stanford, California.

Moeur, J. C. 1947. An Ecological and Taxonomic Survey of the Spermatophytes of Jasper Ridge. M.A. Thesis, Stanford University, Stanford, California.

Munz, P. A., and D. D. Keck. 1949. California Plant Communities. *El Aliso*, **2**:87–105.
Munz, P. A., and D. D. Keck. 1950. California Plant Communities.—Supplement. *El Aliso*, **2**:199–202.
Munz, P. A., in collaboration with D. D. Keck. 1959. *A California Flora*. University of California Press, Berkeley, California.
Oberlander, G. T. 1953. The Taxonomy and Ecology of the Flora of the San Francisco Watershed Reserve. Ph.D. Dissertation, Stanford University, Stanford, California.
Russell, R. J. 1926. Climates of California. *University of California Publications in Geography*, **2**:73–84.
Sharsmith, H. K. 1945. Flora of the Mount Hamilton Range of California. *American Midland Naturalist*, **34**:289–367.
Springer, M. E. A Floristic and Ecologic Study of Jasper Ridge. M.A. Thesis, Stanford University, Stanford, California.
Thomas, J. H. 1958. The Vascular Plants of the Santa Cruz Mountains of Central California. 2 vols. Ph.D. Dissertation, Stanford University, Stanford, California. University Microfilm, L.C. Card No. Mic 59-1457.
Thomas, J. H. 1961. The History of Botanical Collecting in the Santa Cruz Mountains of Central California. *Contributions from the Dudley Herbarium*, **5**: 147–68.
Voss, Sister Anna. 1950. An Ecologic and Taxonomic Study of the Spermatophytes on the Campus of the College of Notre Dame, Belmont. M.A. Thesis, Stanford University, Stanford, California.

Glossary of Technical Terms

Abaxial. The surface away from the axis; in a leaf the lower or dorsal surface.

Acaulescent. Without a well-developed aerial stem.

Achene. A dry, 1-seeded, 1-celled, indehiscent fruit derived from either an inferior or a superior ovary and with the ovary-wall closely enveloping the seed.

Acicular. Needle-shaped.

Actinomorphic. Radially symmetrical; usually used with respect to flowers, actinomorphic flowers are the same as regular flowers.

Acumen. A gradually tapered point.

Acuminate. Gradually tapering to a point.

Acute. Sharp-pointed.

Adaxial. The surface toward the axis; in a leaf the upper or ventral surface.

Adnate. Fusion of unlike parts.

Adventitious roots. Roots arising from vegetative parts of the plant other than roots.

Alternate. Placed singly at different levels on an axis; usually used with respect to leaves which are neither opposite nor whorled.

Androecium. A collective term for all the stamens in a flower.

Androgynous. With the staminate flowers distal to the pistillate flowers, commonly used in *Carex*.

Annulus. A ring or ring-like structure; in ferns a partial ring of cells in the sporangium which is responsible for its dehiscence.

Anther. The portion of the stamen which contains pollen.

Anthesis. The time at which pollination of a flower takes place.

Apiculate. With a short, sharp point.

Approximate. Near or close together.

Arachnoid. Having soft entangled hairs; cobwebby.

Arcuate. Curved or arched.

Aril. A fleshy covering or a fleshy projection of a seed.

Articulation. The joint between two structures or between two parts of the same structure.

Attenuate. Gradually drawn out or tapering.

Auricle. An ear-shaped structure.

Awn. A needle-like projection; a stiff, hair-like structure.

Axil. The angle between the stem and the leaf or between an axis and a structure arising from it.

Axile placentation. Arrangement of the ovules on a central longitudinal axis in an ovary.

Axillary. Occurring in an axil.

Banner. The uppermost petal in papilionaceous flowers.

Barbed. With sharp, often hook-like projections.

Barbellate. Minutely barbed.

Basal. Toward the base or bottom.

Beak. A long projection, usually used in relation to fruits.

Berry. A fleshy, 1- to several-seeded, indehiscent fruit derived from one pistil.

Bi-. A prefix meaning two or twice.

Bifid. Forked or deeply divided into two parts.

Blade. The expanded portion of a leaf.

Bloom. A white waxy coating.

Bract. A small leaf-like structure, usually subtending a flower or a portion of an inflorescence.

Bracteole. A small bract.

Bractlet. A small bract on a secondary axis.

Bristle. A stiff hair-like structure.

Bud. An embryonic axis with its appendages; an unopened flower.

Bulb. A fleshy modified vegetative bud, often underground.

Bulblet. A small bulb-like structure.

Bulbous. Having bulbs or bulb-like in character.

Bur. A spiny fruit.

Burl. A knotty growth on a tree trunk.

Caducous. Falling away early.

Caespitose. Matted; growing in tufts.

Callosity. A hard cartilagenous structure or thickening.

Callous grains. The hard swellings on the abaxial surface of the perianth-segments of some species of *Rumex.*

Callus. The base of the individual floret in a grass spikelet; a thickened structure.

Calyptra. A hood or lid.

Calyx. A collective term for the outermost whorl of the perianth, composed of sepals which may be distinct or fused to form a synsepalous calyx, usually green but in some flowers colored or completely absent.

Campanulate. Bell-shaped.

Canescent. Gray-pubescent or hoary.

Capillary. Very slender.

Capitate. Arranged in a dense or compact head-like structure.

Capsule. A dry dehiscent fruit containing two or more carpels.

Carnose. Fleshy or pulpy.

Carpel. The structural unit of the gynoecium—*see* pistil.

Caruncle. A hard warty appendage near the hilum of a seed.

Caryopsis. The kind of fruit or achene found in the grass family.

Catkin. A small scaly spike or spike-like inflorescence, usually flexuous and pendulous.

Caudate. With a tail.

Caudex. The basal woody part of herbaceous perennials.

Cauline. Of or pertaining to the stem.

Cell. The basic unit of structure and function of plants and animals; the cavity or locule of an ovary or of an anther.

Chaffy. Dry and membranous, usually used in relation to scale-like bracts.

Chartaceous. With the texture of paper, usually not green.

Ciliate. Fringed with hairs.

Ciliolate. Fringed with short hairs.

Circinate. Coiled from the top down so that the apex is at the center of the coil.

Circumscissile. Dehiscent in a ring around a fruit with the upper part coming off as a cap.

Clavate. Club-shaped.

Claw. The long narrowed basal part of a petal.

Cleft. Divided to about the middle.

Cleistogamous. Said of flowers that are self-pollinated and that do not open.

Collar. The region of the abaxial surface of a grass leaf where the blade and sheath meet.

Conical. Cone-shaped.

Compound leaf. A leaf composed of two or more distinct laminae or leaflets.

Compressed. Flattened.

Connate. Like parts grown together.

Connective. The tissue between the two cells of an anther.

Connivent. Coming together, but not grown together.

Contracted. Shortened or condensed.

Convolute. With overlapping margins.

Cordate. Heart-shaped.

Coriaceous. Leathery.

Corm. A solid bulb-like portion of a stem, usually underground.

Corniculate. Having a terminal horn-like process.

Corolla. A collective term for the second whorl of the perianth, composed of petals that may be distinct or fused to form a sympetalous corolla, usually colored but in some flowers green or lacking.

Corymb. A more or less flat-topped inflorescence with the outer flowers opening first.

Corymbose. Arranged in a corymb.

Costa. A rib.

Cotyledon. The seed leaves; the first leaves of the embryo.

Crenate. With shallow rounded teeth; scalloped.

Crenulate. Minutely crenate.

Crest. An elevated ridge.

Crisped. Curled.

Cruciferous. The type of flower in the mustard family, usually characterized by 4 sepals, 4 petals arranged in the form of a cross or of an X, 6 stamens, and 1 pistil.

Cruciform. Cross-shaped.

Culm. The jointed stem of grasses and sedges.

Cuneate. Wedge-shaped.

Cuspidate. With a sharp point at the tip.

Cyme. A more or less flat-topped inflorescence in which the central flower opens first.

Deciduous. Not persistent; woody plants that lose their leaves in the fall are said to be deciduous.

Decumbent. Prostrate but bent upwards toward the tip.

Decurrent. Extending downward on the stem and fused to it.

Decussate. Leaves that are opposite but successive pairs on a stem are at right angles to each other.

Deflexed. Bent downward abruptly.

Dehiscent. Splitting open.

Deliquescent. Becoming semiliquid; repeated branching of the main axis in a more or less dichotomous fashion.

Deltoid. Triangular.

Dentate. Toothed.

Denticulate. Minutely toothed.

Depauperate. Dwarfed or reduced.

Diadelphous. Said of the stamens of one flower when they are united in two groups by their filaments.

Dichotomous. Branched or forked in pairs.

Digitate. Finger-like, with the members arising from one point.

Dimorphic. Of two kinds.

Dioecious. Plants with the pistillate and the staminate flowers on separate plants —*see* monoecious.

Disarticulating. Coming apart, usually at a joint.

Disk-flower. The tubular actinomorphic flower in the center of the head of members of the sunflower family—*see* rayflower.

Discoid heads. Heads or inflorescences of the sunflower family with only diskflowers.

Dissected. Deeply divided or cut into segments.

Distal. Remote, at the far end of.

Divaricate. Widely divergent.

Divided. Segmentation which extends nearly to the base or center.

Dorsal. The surface away from the axis; in a leaf the lower or abaxial surface.

Drupe. A single-seeded, fleshy, indehiscent fruit with the seed enclosed within the woody endocarp, the pit.

Drupelet. A single drupe in an aggregate fruit.

Ellipsoid. An elliptic solid.

Elliptic. A shape in the form of a flattened circle that is more than twice as long as broad.

Emarginate. With a notch at the apex.

Entire. A margin that is smooth, lacking teeth or incisions.

Epidermis. The outermost layer of a plant; the epidermis is one cell layer thick and is often shed as secondary growth occurs.

Epigynous. Calyx, corolla, and stamens arising from the top of the ovary, the ovary thus inferior.

Epipetalous. Borne on a petal or on the corolla.

Erose. Irregularly toothed or gnawed.

Exserted. Sticking out beyond; the opposite of included.

Exfoliate. To come off in scales or flakes.

Fascicle. A bundle or cluster.

Fasciculate. Arranged in a bundle or cluster.

Filament. The stalk of the stamen which supports the anther.

Filamentose. Formed of fibers or filaments.

Filiform. Thread-like.

Fimbriate. Coarsely fringed.

Fistulose. Hollow and usually somewhat swollen.

Floret. The individual flower in the grass family; a small flower.

Floccose. With tufts of woolly hairs.

Flower. The structure in angiosperms composed of an androecium and a gynoecium and often including a perianth (calyx and corolla), concerned with reproduction and the formation of seeds; unisexual or imperfect flowers lack either an androecium or a gynoecium; perfect flowers contain both an androecium and a gynoecium.

Foliaceous. Leaf-like.

Foliage. A collective term for the leaves of a plant.

Foliolate. With leaflets.

Follicle. A 1-carpellate dry fruit, dehiscing along the adaxial suture.

Frond. The leaves of ferns and palms.

Fruit. The mature ovary, or ovaries containing mature seeds.

Funnelform. Gradually widening upward; funnel-shaped.

Furfuraceous. Covered with small scales; scurfy.

Fusiform. Shaped like a spindle, thickened at the center and tapering toward the ends.

Galea. A hood-like petal or a hood-like structure formed by the fusion of 2 petals.

Galeate. Having a galea.

Geniculate. Abruptly bent; resembling a knee-joint.

Glabrate. Almost or nearly glabrous.

Glabrous. Lacking hairs or pubescence.

Gland. A structure that secretes; glands may be on the surface, imbedded, at the ends of hairs, or may consist of swellings.

Glandular. Having glands.

Glaucous. Covered with a waxy coating that usually appears white or gray.

Globose. Spherical.

Glomerate. Aggregated into dense clusters.

Glomerule. A compact head-like cyme.

Glume. The basal bract of a grass spikelet; each spikelet usually has two glumes.

Glutinous. Sticky.

Gynaecandrous. With the pistillate flowers distal to the staminate flowers, commonly used in *Carex*.

Gynoecium. A collective term for all the pistils in a flower—*see* pistil.

Hastate. Halbred-shaped; arrow-shaped but with the lower points spreading.

Head. A dense cluster of sessile or subsessile flowers; commonly the inflorescence of the sunflower family.

Herbaceous. Not woody.

Herbage. A collective term for the green parts of a plant.

Hirsute. With stiff coarse hairs.

Hirsutulous. Slightly hirsute.

Hispid. With rough or stiff hairs or bristles.

Hyaline. Thin and translucent.

Hypanthium. An enlarged cup-like structure below the calyx.

Hypogynous. Calyx, corolla, and stamens arising below the ovary, the ovary thus superior.

Imbricate. Overlapping, as the shingles on a roof.

Included. Not prolonged beyond the surrounding structure; the opposite of exserted.

Indehiscent. Not splitting open.

Indurate. Hard.

Indusium. The covering of the sorus in the ferns.

Inferior ovary. An ovary that is below the attachment of the calyx, corolla, and stamens.

Inflated. Swollen or bladdery.

Inflorescence. The arrangement of flowers on an axis or a series of axes.

Inserted. Attached to.

Internode. The portion of the stem between adjacent nodes.

Involucel. A secondary involucre.

Involucrate. Having an involucre.

Involucre. A series of bracts subtending a flower or an inflorescence.

Involucral bract. An individual bract of an involucre.

Involute. With the edges rolled toward the upper side.

Irregular. Usually said of flowers that are bilaterally symmetrical—*see* zygomorphic.

Keel. The two lower, partly fused petals in a papilionaceous flower.

Laciniate. Cut into narrow, usually sharp-pointed lobes or segments.

Lamelliform. Composed of thin plates.

Leaf. The green photosynthetic organ of a plant, usually composed of a flat expanded blade or lamina, a stalk or petiole, and often 2 stipules, and attached to the stem or branch at a node.

Leaflet. A single division of a compound leaf.

Lanceolate. Lance-shaped with the broader end toward the base.

Legume. The fruit of a member of the pea family, a 1-celled, 1-carpellate fruit derived from a superior ovary.

Lemma. The lower sterile bract of the floret of the grass flower.

Lenticular. Shaped like a biconvex lens.

Ligule. The scarious adaxial tissue at the junction of the blade and sheath of a grass leaf; the strap-shaped corolla in some sunflowers.

Limb. The expanded portion of a sympetalous corolla.

Linear. Long and narrow with parallel margins.

Lip. One of the two principal divisions in the calyx or corolla of zygomorphic flowers.

Lobe. A segment or part of an organ.

Locule. The cavity or one of the cavities within the gynoecium or the anther—*see also* cell.

Loculicidal. Splitting along the abaxial surface of a carpel between adjacent partitions.

Loment. A flat indehiscent legume which is constricted between the seeds and which breaks into 1-seeded segments.

Lunate. Crescent-shaped.

Mammillate. Having nipple-like structures.

Membranous or membranaceous. With the texture of parchment.

-merous. A suffix meaning having parts.

Midrib, midvein. The main rib or vein of a leaf, sepal, or petal.

Monadelphous. Said of stamens of one flower when they are united in one group by their filaments.

Monoecious. Plants with the pistillate and staminate flowers on the same plant —*see* dioecious.

Moniliform. Like beads on a string.

Mucro. A sharp terminal tip.

Mucronate. Having a mucro.

Muricate. Roughened with hard sharp points.

Nascent. Developing or in the process of developing.

Nerve. A vein.

Node. The part of the stem which bears a leaf.

Nut. A hard, 1-seeded, indehiscent fruit.

Nutlet. A small nut, a nut-like fruit.

Ob-. A prefix meaning the inverse or opposite.

Obconic. With the shape of an inverted cone.

Oblong. Longer than broad with more or less parallel sides.

Obtuse. Blunt to rounded at the end.

Opposite. Placed at the same level on an axis and opposite each other; one part in front of another part.

Orbicular. With a circular outline.

Orifice. An opening.

Oval. Broadly elliptic.

Ovary. The portion of the pistil containing the ovules.

Ovate. With the outline of a hen's egg and with the wider end down.

Ovoid. With the shape of a hen's egg.

Ovule. The structure that contains the egg and that, after fertilization of the egg, becomes the seed.

Palate. A projection that closes the orifice of a corolla.

Palea. The second sterile bract of the grass flower; the chaffy bract-like scales of the receptacle of some sunflowers.

Paleaceous. Having paleae or palea-like bracts.

Palmate. Palm-like, with the members arising from one point.

Palustrine. Of or growing in marshes.

Panicle. A compound racemose inflorescence.

Paniculate. Arranged in panicles.

Papery. With the texture of paper.

Papilionaceous. The kind of flower with a keel, two wings, and a banner, found in the pea family and several related families.

Papillate. With minute conical projections.

Pappus. The modified calyx in the sunflower family, consisting variously of scales, hairs, bristles, or absent.

Pappus-crown. A pappus reduced to a low, often irregular rim.

Parietal placentation. Arrangement of the ovules along the outer walls of a 1-celled ovary.

Parted. Divided about one-half the way to the midrib, and with broad sinuses between the divisions.

Pectinate. Comb-like.

Pedicel. The stalk of an individual flower.

Pedicellate. Having pedicels.

Peduncle. The stalk of an inflorescence.

Peltate. With the attachment of a stalk at the center rather than at the margin.

Perfoliate. With a leaf completely surrounding the stem.

Perforate. With holes.

Perianth. The floral envelope, composed of the calyx and the corolla.

Perianth-segment. A division or lobe of the perianth.

Perigynium. The sac-like bract surrounding the pistillate flower in *Carex*.

Perigynous. Calyx, corolla, and stamens arising from the top of a cup-like structure that surrounds the ovary, which may be superior or partly inferior.

Petal. One unit of the corolla, usually colored.

Petaloid. Resembling a petal.

Petiole. The stalk-like part of a leaf.

Petiolule. The stalk of a leaflet.

Phyllode. A flattened expanded petiole that carries out the functions of a leaf.

Pilose. Bearing soft hairs.

Pinna. The primary division of a pinnately compound leaf.

Pinnate. With leaflets arranged on either side of a rachis.

Pinnatifid. Divided in a pinnate manner.

Pinnule. The primary division of a pinna.

Pistil. A unit of the gynoecium; a simple pistil consists of 1 carpel; a compound pistil consists of 2 or more carpels; a single flower may contain 1 to several simple pistils or 1 compound pistil.

Pistillate. Having 1 or more pistils.

Pistillate flower. A flower having pistils but not stamens.

Placenta. That portion of the ovary to which the ovules are attached.

Placentation. The arrangement of the ovules in an ovary.

Plano-convex. A structure that is convex on one side and flat on the other.

Plicate. Folded.

Plumose. Feathery.

Pod. Any dry dehiscent fruit.

Pollen. The microgametophyte; the plant of the gametophytic generation that produces male gametes; produced in the anther.

Pollen-sac. The structure in the anther that produces the pollen.

Polygamodioecious. A plant that is functionally dioecious, but also has a few flowers of the opposite sex and a few perfect flowers.

Polygamous. A plant with both perfect and imperfect flowers.

Pome. A fleshy fruit developed from a several-carpellate inferior ovary, as for example an apple.

Prismatic. Prism-shaped.

Procumbent. Lying on the ground but not rooting.

Prostrate. Lying on the ground.

Proximal. Near, at the near end of.

Puberulent. Minutely hairy.

Pubescence. A general term for all kinds of hairs.

Pubescent. A general term for hairiness.

Punctate. Marked with translucent glands, most clearly seen when held against a bright light.

Pungent. Sharp-pointed or -tipped.

Quadrangular. Nearly square.

Quadrifoliate. With four leaflets.

Quadrate. Nearly square.

Raceme. An inflorescence containing pedicellate flowers arranged along an axis.

Racemose. Arranged in a raceme.

Rachilla. A secondary axis; in grasses and sedges the axis that bears the flowers or florets.

Rachis. An axis of an inflorescence or of a compound leaf.

Radiate heads. Heads of the sunflower family that contain ray-flowers.

Ray. One of the primary branches of an umbel.

Ray-flower. The ligulate, zygomorphic flower of the sunflower family—*see* diskflower.

Receptacle. The portion of an axis that bears the flower parts; the expanded portion of the axis that bears the flowers in the sunflower family.

Reflexed. Abruptly bent backward.

Regular. Usually said of flowers which are radially symmetrical—*see* actinomorphic.

Remote. Far apart.

Reniform. Kidney-shaped.

Resinous. Coated with a sticky resin.

Reticulate. With a network-like pattern.
Retrorse. Bent backward or downward.
Retuse. Notched at an obtuse apex.
Revolute. With the edges rolled toward the lower side.
Rhizome. An underground stem.
Rhomboidal. Diamond-shaped.
Rootstock. Rhizome.
Rosette. Leaves arranged in a radiating cluster, usually at the ground.
Rosulate. Arranged in a rosette.
Rotate. Wheel-shaped.
Rugose. Wrinkled.
Rugulose. Minutely wrinkled.

Saccate. Bag-shaped.
Sagittate. Arrow-shaped.
Salverform. A corolla with a slender tube and an abruptly expanded limb.
Samara. An indehiscent winged fruit.
Saprophyte. A plant living on dead organic matter.
Scabrous. Rough to the touch.
Scale. A thin, membranous, reduced, scale-like structure.
Scape. A long leafless peduncle, often arising from near the ground.
Scapose. Having a scape.
Scarious. Thin, dry, and membranous, not green, often translucent.
Scorpioid. A coiled inflorescence in which the flowers are 2-ranked and are borne alternately on either side of the rachis.
Scurfy. With small, bran-like scales.
Secund. With the flowers turned toward one side of the axis.
Seed. The mature fertilized ovule.
Segment. One of the parts of a simple leaf, calyx, corolla, or perianth; a section.
Sepal. One unit of the calyx, usually green.
Sepaloid. Resembling a sepal.
Septate. Having partitions or septa.
Septicidal. Splitting along the abaxial surface of a carpel along the septum.
Septum. A partition.
Sericeous. Silky.
Serrate. Toothed like the edge of a saw.
Sessile. Not stalked, lacking a petiole or pedicel.
Setaceous. With bristles.
Setose. With bristles.

Sheath. A tubular structure, in grasses the lower part of the leaf which surrounds the stem.
Simple. Of one part.
Sinuate. With a deep wavy margin.
Sinus. The space between adjacent lobes, segments, or divisions.
Sorus. An aggregation of sporangia in ferns.
Spadix. The fleshy spike of the Araceae surrounded by a spathe.
Spathe. The bract or bracts surrounding a spadix.
Spatulate. Shaped like a spatula.
Spiculate. With fine sharp points.
Spike. An inflorescence containing sessile flowers arranged along an axis.
Spicate. Having spikes.
Spikelet. A secondary spike.
Spinescent. With spines.
Spinose. Spiny.
Sporangiophore. The structure bearing sporangia in *Equisetum*.
Sporangium. The structure that produces spores.
Spore. A haploid cell that gives rise to the gametophytic generation.
Sporocarp. A structure containing sporangia; in the Marsileaceae it is hard and seed-like.
Sporophyll. A spore-bearing leaf.
Spur. A slender tubular projection on the perianth.
Stamen. The male part of the flower consisting of filament, connective, and anther.
Staminate. Having stamens.
Staminate flower. A flower that contains stamens, but not pistils.
Staminodium. A sterile stamen.
Standard. An erect petal.
Stellate. Star-shaped.
Sterile. Not producing functional ovules or pollen.
Stigma. The part of the pistil that receives pollen.
Stipe. The stalk of an ovary; the petiole of a fern.
Stipitate. With a stipe.
Stipule. One of a pair of appendages borne at the base of some leaves.
Stolon. A horizontal creeping stem that tends to root at the nodes.
Stoloniferous. Having stolons.

Stoma. The pore in a leaf or in other parts of a plant that allows for gas exchange with the atmosphere.

Stone. A pit or hard endocarp.

Striate. With fine parallel lines or ridges.

Strigose. With appressed hairs.

Strobilus. An aggregation of sporangia into a cone-like structure.

Stump sprouting. Producing vegetative growth from stumps.

Style. The portion of the pistil between the stigma and the ovary.

Stylopodium. An enlargement at the base of the style.

Sub-. A prefix meaning nearly or almost.

Subtend. To be situated closely below.

Subulate. Tapering to a point.

Succulent. Fleshy.

Suffrutescent. Somewhat shrubby.

Superior ovary. An ovary that is situated above the point of origin of the calyx, corolla, and stamens.

Suture. A junction or seam.

Sympetalous. The petals fused laterally at least at the base.

Synsepalous. The sepals fused laterally at least at the base.

Taxon, pl. taxa. A taxonomic category of any rank.

Taxonomy. The systematic classification of organisms.

Tepal. A perianth-segment.

Tendril. A thread-like structure derived from stem or leaves by which certain plants may attach themselves.

Terete. Cylindrical and tapering.

Ternate. Arranged or grouped in threes.

Thallose. Consisting of a thallus.

Thallus. A more or less flat, leaf-like structure, not differentiated into stem and leaves.

Throat. The opening of a sympetalous corolla.

Tomentose. With matted woolly hairs.

Tomentum. A dense, matted, woolly pubescence.

Toothed. With symmetrical triangular projections along a margin.

Torus. The receptacle.

Translucent. Allowing the passage of some light.

Trichome. A hair.

Trifid. Divided into three parts.

Trifoliolate. With three leaflets.

Trigonous. Three-angled.

Triquetrous. Three-angled in cross section.

Truncate. Abruptly or squarely cut off at the end.

Tube. The united part of a synsepalous calyx or a sympetalous corolla.

Tuber. A short thickened underground stem.

Tubercle. A small protruding structure.

Tuberculate. With tubercles.

Turbinate. Top-shaped.

Turion. A fleshy scaly shoot.

Umbel. An inflorescence in which the branches or pedicels arise from the same point.

Umbellate. With umbels.

Umbellet. A secondary umbel.

Uncinate. Hooked.

Undulate. Wavy.

Urceolate. Urn-shaped.

Utricle. A small, 1-seeded, bladdery fruit, usually indehiscent.

Valvate. Having valves.

Valve. One of the pieces into which a dehiscent fruit separates.

Vascular. Pertaining to woody tissue.

Vein. A fine strand of vascular tissue evident in leaves and floral parts.

Velum. A thin veil-like tissue.

Ventral. The surface toward the axis; in a leaf the upper or adaxial surface.

Vernation. The arrangement of leaves in a bud.

Verticil. A whorl.

Villous. With long, soft, rather weak hairs.

Virgate. Willowy or wand-like.

Viscid. Sticky.

Vitiform. Shaped like a grape leaf.

Whorl. Several structures arranged in a circle about an axis.

Whorled. Being in a whorl.

Wing. A thin lateral expansion; the lateral petals in a papilionaceous flower.

Zygomorphic. Bilaterally symmetrical; usually used with reference to flowers, zygomorphic flowers are the same as irregular flowers.

PART IV

———

Indexes

Index of Place Names

The names in parentheses indicate the counties. Place names in boldface type are shown on the map of the Santa Cruz Mountains area (p. 4).

Adobe Creek (Santa Clara). *See* San Antonio Creek.

Alamitos Creek (Santa Clara). A creek draining into Guadalupe Creek.

Alba Grade or Road (Santa Cruz). A road between Ben Lomond and Empire Grade.

Alba School (Santa Cruz). A school along Alba Grade.

Alma (Santa Clara). A small community about 3 miles south of Los Gatos.

Almaden (Santa Clara). A quicksilver mining town in the west-central part of the county.

Almaden Canyon (Santa Clara). The canyon of Almaden Creek.

Almaden Creek (Santa Clara). A creek passing through Almaden.

Alma–Soda Springs (Santa Clara). Springs about 3 miles south of Los Gatos.

Alpine Creek (San Mateo). A small creek about 3 miles southeast of La Honda.

Alpine Creek Road (San Mateo). A road near Alpine Creek.

Alviso (Santa Clara). A town near the southern end of San Francisco Bay.

Alviso Slough (Santa Clara). The slough between San Francisco Bay and Alviso.

Año Nuevo Point (San Mateo). A point along the coast in the southern part of the county.

Aptos (Santa Cruz). A town along the coast in the southern part of the county.

Aptos Creek (Santa Cruz). The creek flowing through Aptos.

Aptos Hills (Santa Cruz). Hills near Aptos.

Arroyo de Los Frijoles (San Mateo). An arroyo about 3 miles southward from Pescadero.

Baden (San Mateo). A town about 3.5 miles south of San Francisco.

Barrocal, Berrocal, or Berrogal Canyon (Santa Clara). A canyon between Loma Prieta and Almaden.

Basin Way (Santa Cruz). A road near Boulder Creek.

Bayview Hills (San Francisco). Hills in the southeastern part of the city.

Bean Creek. The creek flowing through Glenwood.

Bean Hollow (San Mateo). The same as Arroyo de Los Frijoles.

Bear Creek (Santa Cruz). A creek to the northeast of Boulder Creek.

Bear Creek Canyon (Santa Cruz). The canyon of Bear Creek.

Bear Creek Road (Santa Cruz). A road along Bear Creek.

Belmont (San Mateo). A town between San Mateo and Redwood City.

Ben Lomond (Santa Cruz). A town along the San Lorenzo River.

Ben Lomond Mountain, Ben Lomond Ridge (Santa Cruz). A mountain extending from south of Big Basin to near Santa Cruz.

Ben Lomond Sand Hills (Santa Cruz). A series of sand hills to the east of Ben Lomond.

Bielawski Lookout (Santa Cruz). A fire lookout tower about 3 miles southeast of Saratoga Summit.

Big Basin, Big Basin Redwoods State Park (Santa Cruz). A state park in the northern part of the county.

Big Basin Highway (Santa Cruz). The highway between Boulder Creek and Big Basin.

Big Trees (Santa Cruz). A grove of redwood trees about 1 mile southeast of Felton, now part of Henry Cowell Redwoods State Park.

Black Mountain (Santa Clara). The highest point on Monte Bello Ridge, about 4 miles southward from Los Altos.

Blooms Grade (Santa Cruz). An old lumber road from Hilton Airport to near the southern entrance of Big Basin.

Bonny Doon (Santa Cruz). A town about 9 miles northwest of Santa Cruz on Ben Lomond Mountain.

Bonny Doon Road (Santa Cruz). The road from Bonny Doon to near Davenport.

Boulder Creek (Santa Cruz). A town along the San Lorenzo River; a creek to the northwest of Boulder Creek.

Brackenbrae (Santa Cruz). A residential community about 1.5 miles northwest of Boulder Creek.

Brisbane (San Mateo). A town just south of San Francisco.

Brookdale (Santa Cruz). A town about 1 mile south of Boulder Creek.

Burlingame (San Mateo). A town to the northwest of San Mateo.

Buri or Buriburi Ridge (San Mateo). The ridge immediately to the east of Lower Crystal Springs Lake and extending to the southern end of San Andreas Lake.

Burrell (Santa Clara). A small community about 1.5 miles southeast of Wrights.

Butano Canyon (San Mateo). The canyon of Butano Creek.

Butano Creek (San Mateo). A creek in the southern part of the county.

Butano Ridge (San Mateo). The ridge to the north of Butano Creek.

Camp Evers (Santa Cruz). A marsh and community about 5 miles north of Santa Cruz.

Campbell (Santa Clara). A town about 4 miles southwest of San Jose.

Campbell Creek (Santa Clara). The creek flowing from near Saratoga to Alviso.

Capitola (Santa Cruz). A town along the coast about 3.5 miles east of Santa Cruz.

Carnadero Creek (Santa Clara). A creek south of Gilroy.

Casserly Creek (Santa Cruz). A small creek running from near Mount Madonna to near Watsonville.

Castle Rock (Santa Cruz). A high point about 2.5 miles southeast of Saratoga Summit.

Castle Rock Ridge (Santa Clara, Santa Cruz). A ridge about 10 miles long extending southeast from Saratoga Summit along the crests of the Santa Cruz Mountains.

Chalks (Santa Cruz). A region of Monterey shale between Big Basin and Año Nuevo Point.

Chalks Lookout (Santa Cruz). A fire lookout about 4.5 miles northeast of Año Nuevo Point.

China Grade (Santa Cruz). A road extending from about 3.5 miles northwest of Boulder Creek to near the northern end of Big Basin.

Chittenden (Santa Cruz). A small community near Soda Lake.

Chittenden Station (Santa Cruz). A railroad stop near Chittenden.

Chittenden Pass (Santa Cruz). A low pass near Chittenden.

Cliff House (San Francisco). A resort area along the coast at the northwest corner of the city.

Coal Mine Canyon (San Mateo). A canyon near Coal Mine Ridge.

Coal Mine Ridge (San Mateo). A ridge between Corte Madera Creek and Los Trancos Creek about 6 miles southward from Stanford.

College Lake (Santa Cruz). A small lake about 3 miles north of Watsonville.

Colma (San Mateo). A town west of Brisbane.

Congress Springs (Santa Clara). Springs near Saratoga.

Cooleys Landing (San Mateo). An old landing on San Francisco Bay north of Palo Alto.

Corralitos (Santa Cruz). A town in the southern part of the county.

Corralitos Creek (Santa Cruz). The creek flowing through Corralitos.

Corte Madero Creek, Corte de Madera Creek, Corte Madera Creek (San Mateo). A creek to the south of and flowing into Searsville Lake; a tributary of San Gregorio Creek near San Gregorio.

Coyote (Santa Clara). A town in the Santa Clara Valley.

Coyote Creek (Santa Clara). One of the main creeks in the Santa Clara Valley.

Crystal Springs (San Mateo). Originally some springs near what is now Crystal Springs Lakes.

Crystal Springs Dam (San Mateo). A dam

on San Mateo Creek at Crystal Springs Lakes.

Crystal Springs Lakes (San Mateo). Two partly artificial lakes or reservoirs in the northern part of the county.

Crystal Springs Reservoir (San Mateo). *See* Crystal Springs Lakes.

Cupertino (Santa Clara). A town about 4 miles north of Saratoga.

Daly City (San Mateo). A city just south of San Francisco.

Davenport (Santa Cruz). A small town along the coast.

Davenport Landing (Santa Cruz). A small landing near Davenport.

Deer Ridge Farm (Santa Cruz). An old farm on a ridge near Deer Creek, about 2 miles south of Bielawski Lookout.

Dumbarton Bridge (San Mateo). A bridge across the southern end of San Francisco Bay about 3 miles north of the San Mateo – Santa Clara county line.

Eagle Rock (Santa Cruz). A peak about 4.5 miles northwest of Boulder Creek.

Eccles (Santa Cruz). A community along Zayante Creek about 2.5 miles from Felton.

Edenvale (Santa Clara). A community in the Santa Clara Valley about 6 miles southeast of San Jose.

Edgemar (San Mateo). A town along the coast about 4 miles south of San Francisco.

El Toro (Santa Clara). A hill near Morgan Hill.

Emerald Lake (San Mateo). A small lake back of Redwood City.

Empire Grade (Santa Cruz). The main road running the length of Ben Lomond Mountain from near Eagle Rock to Santa Cruz.

Evergreen (Santa Clara). A small town about 6 miles southeast of San Jose.

Felt Lake (Santa Clara). A small lake on the Stanford University Campus.

Felton (Santa Cruz). A town along the San Lorenzo River.

Freedom (Santa Cruz). A town about 1.5 miles north of Watsonville.

Frenchman Creek (San Mateo). A creek to the north of Pilarcitos Creek.

Gazos Creek (San Mateo). The first main creek south of Butano Creek.

Gilroy (Santa Clara). A town at the southern end of the Santa Clara Valley.

Glenarbor (Santa Cruz). A community between Ben Lomond and Felton.

Glencarry Road (Santa Cruz). A road about 2 miles south of Felton.

Glenwood (Santa Cruz). An old community northeast of Felton.

Goertz or Gaertz Ravine (Santa Clara). A ravine on the southwestern slope of Loma Prieta.

Golden Gate Park (San Francisco). A park in the western part of the city.

Granada (San Mateo). A community along the coast at Half Moon Bay.

Graham Hill (Santa Cruz). The ridge parallel to and east of the San Lorenzo River and extending from near Felton to Santa Cruz.

Graham Hill Road (Santa Cruz). The road along Graham Hill.

Guadalupe (Santa Clara). An old mining town about 9 miles south of San Jose.

Guadalupe Canyon (Santa Clara). The canyon of Guadalupe Creek.

Guadalupe Creek (Santa Clara). The creek flowing from Guadalupe to San Jose and Alviso.

Guadalupe Hills (Santa Clara). Hills near Guadalupe.

Guadalupe Mine or Mines (Santa Clara). The mines near Guadalupe.

Guadalupe Reservoir (Santa Clara). A reservoir along Guadalupe Creek about one-half the way between Guadalupe and Almaden.

Half Moon Bay (San Mateo). A town near the coast; the bay in this vicinity.

Half Moon Bay Road (San Mateo). The road from Crystal Springs Lakes to Half Moon Bay.

Half Moon Point (San Mateo). Probably the same as Pillar Point.

Harkins Slough (Santa Cruz). A slough about 2.5 miles west of Watsonville.

Hecker Pass (Santa Clara, Santa Cruz). The pass on the highway between Gilroy and Watsonville, about 1 mile south of Mount Madonna.

Hester Creek (Santa Cruz). One of the tributaries of Soquel Creek.

Hidden Villa (Santa Clara). An old estate on Moody Road near Los Altos.

Highland Way (Santa Cruz). A road along the crests of the Santa Cruz Mountains from near the headwaters of Soquel Creek to near the headwaters of Corralitos Creek.

Hillsborough (San Mateo). A town between Lower Crystal Springs Lake and San Mateo.

Hilton Airport (Santa Cruz). A small private landing field along the Big Basin Highway about 3 miles northwest of Boulder Creek.

Holy City (Santa Clara). A religious community about 5 miles south of Los Gatos.

Jamison Creek (Santa Cruz). A small creek flowing eastward from near Eagle Rock to Boulder Creek.

Jamison Creek Road (Santa Cruz). The road near Jamison Creek.

Jasper Ridge (San Mateo). A ridge on the Stanford University Campus near Searsville Lake; utilized as a biological reserve and experimental area by the Department of Biological Sciences of the University.

Kelly Lake (Santa Cruz). A small lake about 2 miles northeast of Watsonville.

Kings Mountain (San Mateo). A mountain in the central part of the county.

Laguna Creek (Santa Cruz). A creek to the northwest of Santa Cruz on the western slope of Ben Lomond Mountain.

La Honda (San Mateo). A town in the southern part of the county.

La Honda Creek (San Mateo). The creek flowing through La Honda.

La Honda Grade (San Mateo). The road northward from La Honda.

Lake Lagunita (Santa Clara). A small seasonal lake on the Stanford University Campus.

Lake Merced (San Francisco). A series of dune lakes in the southwestern part of the city.

Langley Hill (San Mateo). A hill about 2 miles northeast of La Honda.

Larkin Valley (Santa Cruz). A valley parallel to the coast about 5 miles northwest of Watsonville.

La Selva Beach (Santa Cruz). Another name for Robroy.

Laurel (Santa Cruz). A small community about 2 miles southwest of Wrights.

Laurel Hill Cemetery (San Francisco). The site of an old cemetery in the northern part of the city.

Lexington (Santa Clara). An old community about 2 miles south of Los Gatos.

Liddell Creek (Santa Cruz). A creek between San Vicente Creek and Laguna Creek on the western slope of Ben Lomond Mountain.

Little Arthur Creek (Santa Clara). A tributary of Carnadero Creek in the southern part of the county.

Little Basin (Santa Cruz). A deep canyon along the upper part of Scott Creek.

Little Los Gatos Creek (Santa Clara). Presumably a creek near Los Gatos.

Llagas Creek (Santa Clara). A large creek in the southern part of the county.

Llagas Post Office (Santa Clara). A locality along Llagas Creek about 4 miles west of Madrone.

Lobitos Creek (San Mateo). A creek north of Tunitas Creek.

Locatelli Ranch (Santa Cruz). A ranch and vineyard near Eagle Rock.

Loma Azule (Santa Clara, Santa Cruz). See Sierra Azule.

Loma Prieta (Santa Clara). The highest peak in the Santa Cruz Mountains.

Loma Prieta Ridge (Santa Clara). The ridge extending northwest from Loma Prieta and ending in Mount Umunhum.

Los Altos (Santa Clara). A town in the northern part of the county.

Los Gatos (Santa Clara). A town in the west-central part of the county.

Los Gatos Canyon (Santa Clara). The canyon of Los Gatos Creek.

Los Gatos Creek (Santa Clara). The large creek flowing through Los Gatos.

Los Gatos Hills (Santa Clara). Hills near Los Gatos.

Los Gatos Reservoir (Santa Clara). A large reservoir south of Los Gatos, now known as Lexington Reservoir.

Los Trancos Creek (San Mateo, Santa Clara). A creek along the San Mateo–Santa Clara county line south of Stanford.

Lower Crystal Springs Lake (San Mateo). The northernmost of the two Crystal Springs Lakes.

Lower Zayante (Santa Cruz). A locality near Zayante.

Madrone (Santa Clara). A town in the southern part of the Santa Clara Valley.

Madrone Springs (Santa Clara). Springs to the east of Madrone.

Madrone Station (Santa Clara). The railway station at Madrone.

Matadero Creek (Santa Clara). A small creek to the east of Palo Alto.

Mayfield (Santa Clara). Once a separate town, but now part of Palo Alto, located in the vicinity of the South Palo Alto railway station.

Menlo Park (San Mateo). A town just across the county line to the northwest of Palo Alto.

Menlo Golf Club (San Mateo). A golf club near Menlo Park.

Menlo Heights (San Mateo). An area near Menlo Park.

Millbrae (San Mateo). A town along San Francisco Bay about halfway between Brisbane and San Mateo.

Mill Creek (Santa Cruz). A creek in the western part of the county.

Mill Creek Dam (Santa Cruz). A small dam on Mill Creek.

Mindego Creek (San Mateo). A creek between La Honda and Mindego Hill.

Mindego Hill (San Mateo). A hill about 2.5 miles east of La Honda.

Miramar (San Mateo). A small town along the coast about 2.5 miles north from Half Moon Bay.

Mission Hills (San Francisco). Hills near the center of the city.

Montalvo (Santa Clara). The former home of the late U.S. Senator James D. Phelan near Saratoga and now a center for creative arts.

Montara Mountain or **Mountains** (San Mateo). A subdivision of the Santa Cruz Mountains near the coast.

Montara Point (San Mateo). A point along the coast about 4.5 miles south of San Pedro Point.

Moffett Field (Santa Clara). A naval air station between Palo Alto and Alviso near the bay.

Monte Bello Ridge (Santa Clara). The ridge extending from Black Mountain to the southwest.

Morgan Hill (Santa Clara). A town in the southern part of the Santa Clara Valley.

Moss Beach (San Mateo). A town near the coast in the northern part of the county.

Mountain View (Santa Clara). A town between Palo Alto and Sunnyvale.

Mount Charlie Road (Santa Cruz). A road from near Glenwood running northward to the summit of the Santa Cruz Mountains.

Mount Hermon (Santa Cruz). A religious community about 1 mile east of Felton.

Mount Madonna (Santa Clara). A high peak about 7.5 miles west of Gilroy.

Mount Umunhum (Santa Clara). A peak at the western end of Loma Prieta Ridge, about 5 miles northwest of Loma Prieta.

Mussel Rock (San Mateo). Rocks along the coast about 3 miles south of San Francisco.

Neary Lagoon (Santa Cruz). A lagoon in Santa Cruz.

New Almaden (Santa Clara). An old community about 2 miles north of Almaden.

New Brighton (Santa Cruz). A beach town about 5 miles east of Santa Cruz.

Oakhill Cemetery (Santa Clara). A cemetery about 3 miles south of San Jose.

Ocean View (San Francisco). A section of the city just east of Lake Merced.

Oil Creek (San Mateo). A small creek with its headwaters near Saratoga Summit flowing southwest into Pescadero Creek.

Old Page Mill (Santa Clara). An old mill along Page Mill Road.

Olympia (Santa Cruz). A community along Zayante Creek about 2 miles from Felton.

Page Mill Road (Santa Clara). A road running southward from Palo Alto to near the northern end of Monte Bello Ridge.

Pajaro River (Santa Clara, Santa Cruz). The river that forms the southern boundary of the Santa Cruz Mountains.

Palm Beach (Santa Cruz). A small beach resort near the mouth of the Pajaro River.

Palo Alto (Santa Clara). A city in the northern part of the county.

Peavine Creek Canyon (Santa Cruz). A canyon near Boulder Creek.

Pebble Beach (San Mateo). A beach near Pescadero.

Pedro Point (San Mateo). *See* San Pedro Point.

Pedro Valley (San Mateo). A small town and valley near San Pedro Point.

Permanente Creek (Santa Clara). A creek flowing northward from Black Mountain.

Permanente Ravine (Santa Clara). A ravine along the upper part of Permanente Creek.

Pescadero (San Mateo). A town near the coast.

Pescadero Creek (San Mateo). The creek flowing through Pescadero.

Peters Creek (San Mateo). A small tributary of Pescadero Creek.

Pigeon Point (San Mateo). A point along the coast in the southern part of the county.

Pilarcitos Canyon (San Mateo). The canyon of Pilarcitos Creek.

Pilarcitos Creek (San Mateo). A creek draining into Half Moon Bay.

Pilarcitos Dam (San Mateo). A dam on Pilarcitos Lake.

Pilarcitos Lake (San Mateo). A lake near the head of Pilarcitos Creek.

Pilarcitos Stone Dam (San Mateo). *See* Pilarcitos Dam.

Pillar Point (San Mateo). A point along the coast.

Pine Mountain (Santa Cruz). A small mountain about 2 miles south of the center of Big Basin.

Pinto Lake (Santa Cruz). A lake about 3 miles north of Watsonville.

Portola (San Mateo). A community southwest of Palo Alto.

Portola Valley (San Mateo). The valley in which Portola is situated.

Portola Road (San Mateo). A road in Portola Valley.

Powder Mill Canyon or Gulch (Santa Cruz). A canyon near Santa Cruz.

Presidio (San Francisco). A military reservation in the northwestern part of the city.

Princeton (San Mateo). A town on the north shore of Half Moon Bay.

Purisima Creek (San Mateo). The first major creek north of Tunitas Creek.

Quail Hollow (Santa Cruz). A valley about 1.5 miles east of Ben Lomond.

Quail Hollow Road (Santa Cruz). The road running through Quail Hollow.

Rancho del Oso (Santa Cruz). The ranch along the lower part of Waddell Creek.

Raymonds, Raymonds Ranch (Santa Clara). A collecting locality of W. R. Dudley, probably on the west side of Los Gatos Creek about 2 miles south of Los Gatos at an altitude of 1,400 feet.

Redwood City (San Mateo). A city in the east-central part of the county.

Rincon Canyon (Santa Cruz). A canyon near the San Lorenzo River about 3 miles north of Santa Cruz.

Robroy (Santa Cruz). A town along the coast in the southern part of the county.

Salada (San Mateo). A community along the coast about 5 miles south of San Francisco, now known as Sharp Park.

San Andreas Dam (San Mateo). A dam at San Andreas Lake.

San Andreas Lake (San Mateo). A partly artificial lake in the northern part of the county.

San Andreas Valley (San Mateo). The valley in which San Andreas Lake lies.

San Antonio Creek (Santa Clara). The first main creek to the west of Permanente Creek, now known as Adobe Creek.

San Bruno Hills or San Bruno Mountain (San Mateo). A low range of hills in the northern part of the county.

San Carlos (San Mateo). A town between Redwood City and Belmont.

San Francisquito Creek (San Mateo, Santa Clara). A creek from Searsville Lake to the Bay, forming a part of the boundary between San Mateo and Santa Clara counties.

San Francisco (San Francisco). The city and county at the northern end of the Santa Cruz Mountains.

San Francisco Watershed Reserve (San Mateo). The area including San Andreas, Crystal Springs, and Pilarcitos lakes, serving as a part of the San Francisco water supply system.

San Gregorio (San Mateo). A small town near the coast.

San Gregorio Creek (San Mateo). The creek flowing through San Gregorio.

San Jose (Santa Clara). A city in the Santa Clara Valley.

San Juan Hills or San Juan Bautista Hills (Santa Clara). A series of low hills in the Santa Clara Valley between San Jose and Edenvale.

San Lorenzo River (Santa Cruz). The largest river in the county.

San Lorenzo Road (Santa Cruz). An old road along the San Lorenzo River.

San Martin (Santa Clara). A town between Morgan Hill and Gilroy.

San Mateo (San Mateo). A city in the northern part of the county.

San Mateo Creek (San Mateo). The creek flowing through San Mateo.

San Mateo Ravine (San Mateo). The ravine of San Mateo Creek.

San Miguel Hills (San Francisco). Hills in the central part of the city.

San Pedro (San Mateo). *See* Pedro Valley.

San Pedro Point (San Mateo). A point along the coast.

Santa Clara (Santa Clara). A city in the northern part of the Santa Clara Valley.

Santa Cruz (Santa Cruz). The largest city in the county.

San Vicente Creek (Santa Cruz). A creek on the western slope of Ben Lomond Mountain emptying into the ocean at Davenport.

San Vicente Road (Santa Cruz). A road along San Vicente Creek.

Saratoga (Santa Clara). A town in the northern part of the county.

Saratoga Springs (Santa Clara). Springs near Saratoga.

Saratoga Summit (Santa Clara, Santa Cruz). A low gap along the crest of the Santa Cruz Mountains to the west of Saratoga.

Sargents (Santa Clara). A small community about 6 miles south of Gilroy.

Sawyer Ridge (San Mateo). A ridge to the west of Lower Crystal Springs Lake and extending to the southern end of San Andreas Lake.

Scott Creek (Santa Cruz). A creek in the northwestern part of the county.

Scott Valley (Santa Cruz). A community and valley about 5 miles north of Santa Cruz.

Seabright (Santa Cruz). The eastern part of Santa Cruz.

Sea Cliff (Santa Cruz). A beach near Aptos, now part of Seacliff Beach State Park.

Seal Cove (San Mateo). A small community along the coast about one-half mile south of Moss Beach.

Searsville (San Mateo). An old community near Searsville Lake.

Searsville Lake (San Mateo). An artificial lake on the Stanford University Campus.

Searsville Ridge (San Mateo). The same as Jasper Ridge.

Sierra Azule, Sierra Azule Ridge (Santa Clara, Santa Cruz). The southern part of the Santa Cruz Mountains—*see* the section in the Introduction on the description of the Santa Cruz Mountains.

Sierra Morena (San Mateo, Santa Clara, Santa Cruz). The high ridge about 3 miles southwest of Woodside; a name for the northern part of the Santa Cruz Mountains—*see* the section in the Introduction on the description of the Santa Cruz Mountains.

Sharp Park (San Mateo). *See* Salada.

Skyline Boulevard (San Mateo, Santa Clara, Santa Cruz). The highway along the summit of the Santa Cruz Mountains from near San Francisco to Saratoga Summit.

Slate Creek (San Mateo). A tributary of Pescadero Creek between Peters Creek and Oil Creek.

Smith Grade (Santa Cruz). A road on Ben Lomond Mountain running southeast from Bonny Doon.

Soda Lake (Santa Cruz). A small alkaline lake in the southeastern part of the county.

Soda Springs (Santa Clara). Springs about 3 miles south of Los Gatos.

Soquel (Santa Cruz). A town near the coast west of Santa Cruz.

Soquel Creek (Santa Cruz). The creek flowing through Soquel.

Soquel Road (Santa Cruz). The road along Soquel Creek.

Soquel Valley (Santa Cruz). The valley of Soquel Creek.

South San Francisco (San Mateo). A city along the San Francisco Bay about 3.5 miles south of San Francisco.

Spanishtown (San Mateo). An old community near the cemeteries in the northern part of the county.

Spring Valley (San Mateo). The valley in which Pilarcitos Lake lies.

Squealer Gulch (San Mateo). A gulch between Kings Mountain and Woodside.

Stanford, Stanford University (Santa Clara, San Mateo). A private university and community.

Stevens Creek (Santa Clara). A large creek in the northern part of the county.

Stevens Creek Reservoir (Santa Clara). A reservoir along the upper part of Stevens Creek.

Sunset Beach (Santa Cruz). A beach near the mouth of the Pajaro River, now a part of Sunset Beach State Park.

Sunnyvale (Santa Clara). A town in the northern part of the county.

Swanton (Santa Cruz). An old community near the coast in the western part of the county.

Sweeney Ridge (San Mateo). The ridge west of San Andreas Lake.

Tunitas Creek (San Mateo). A creek emptying into the ocean north of San Gregorio.

Tuxedo (Santa Cruz). A locality in the Ben Lomond Sand Hills near Felton.

Upper Crystal Springs Lake (San Mateo). The southernmost of the two Crystal Springs Lakes.

Uvas Canyon (Santa Clara). The canyon of Uvas Creek.

Uvas Creek (Santa Clara). A large creek in the southern part of the county.

Uvas Creek Road (Santa Clara). A road along Uvas Creek.

Waddell Creek (Santa Cruz). A creek in the northwestern part of the county.

Warrenella (Santa Cruz). An old community on the western slope of Ben Lomond Mountain about 6 miles north of Davenport.

Watsonville (Santa Cruz). A city in the southern part of the county.

Watsonville Junction (Monterey). A railroad junction about 1 mile south of Watsonville.

Watsonville Slough (Santa Cruz). A slough extending from near the mouth of the Pajaro River to Watsonville.

Whitehouse Creek (San Mateo, Santa Cruz). A creek between Gazos Creek and Waddell Creek.

Woodside (San Mateo). A town in the central part of the county.

Woodside serpentine (San Mateo). An area of serpentine outcroppings north of Woodside.

Wrights (Santa Clara). An old community near the summit of the Santa Cruz Mountains.

Wrights Station (Santa Clara). The railway station at Wrights.

Zayante (Santa Cruz). An old community along Zayante Creek about 2 miles west of Glenwood.

Zayante Creek (Santa Cruz). A creek to the northeast of Felton.

Zayante Valley (Santa Cruz). The valley of Zayante Creek.

Index of Common Plant Names

Abronia, 155
Acacia
  Bailey, 202
  Black, 202
  Blackwood, 202
  Star, 202
  Whorl-leaf, 202
Acaena, 196
  California, 197
Adder's Tongue Family, 56
Adder's Tongue
  California Fetid, 121
Agoseris
  Annual, 342
  Eastwood's, 342
  Large-flowered, 343
  Seaside, 342
Alder, 134
  Oregon, 134
  Red, 134
  White, 134
Alfalfa, 211
Alkali Heath, 240
Alkali Weed, 276
Alkanet
  Everlasting, 288
Allocarya
  Artist's, 292
  Bracted, 292
  California, 292
  Coast, 292
  Diffuse, 292
  Glabrous, 291
  Hickman's, 292
Alum Root, 190
Alyssum
  Sweet, 187
  Yellow, 187
Amaranth Family, 154
Amaranth, 154
  California, 155
  Green, 154
  Low, 154
  Powell's, 154
  Prostrate, 155
  Spleen, 154
Amaryllis Family, 123
American Brooklime, 314
Ammannia
  Long-leaved, 244
Ammi
  Toothpick, 260
Amole, 117
Amsinckia
  Seaside, 289
Androsace
  California, 269
Anemone, 169
  Western Rue, 167

Western Wood, 169
Angelica
  California, 262
  Coast, 262
  Henderson's, 262
  Wood, 262
Apple Family, 201
Aralia Family, 252
Arnica
  Rayless, 373
Arrow Grass Family, 68
Arrow Grass, 68
  Seaside, 68
  Slender, 68
  Three-ribbed, 68
Arrow Weed Family, 68
Arrowhead
  Broad-leaved, 69
Artichoke
  Garden, 376
  Globe, 376
Arum Family, 111
Ash, 271
  Flowering, 271
  Foothill, 271
  Oregon, 271
Asparagus
  Garden, 121
Aster Tribe, 345
Aster, 350
  Bolander's Golden, 348
  Branching Beach, 350
  Broad-leaved, 351
  Common California, 351
  Douglas', 351
  Golden, 348
  Oregon Golden, 348
  Rough-leaved, 351
  Slim, 351
Athysanus
  Dwarf, 184
Australian Tea Tree, 244
Azalea
  Western, 264
Azolla
  Fern-like, 55

Baby Blue Eyes, 284
Baby's Tears, 139
Baccharis
  Douglas', 346
  Salt Marsh, 346
Bachelor's Button, 376
Baeria
  Perennial, 368
  Small-rayed, 369
  Woolly, 368
Balm-of-Gilead, 299
Bamboo Tribe, 75

Baneberry, 167
  Western Red, 167
Barberry Family, 171
Barberry, 171
  Coast, 172
  Darwin's, 172
Barley Tribe, 85
Barley, 85
  California, 86
  Common, 86
  Low, 86
  Meadow, 86
  Mediterranean, 86
  Tufted, 86
  Wall, 86
Bayberry Family, 133
Beach-bur, 365
Bear Brush, 263
Beard-Tongue, 310
Bedstraw, 323, 325
  California, 324
  Climbing, 324
  Fragrant, 325
  Phlox-leaved, 324
  Rough-fruited, 324
  Sweet-scented, 325
  Trifid, 324
  Wall, 324, 325
  Yellow, 324
Beech Family, 135
Bee-Plant
  California, 310
Beet
  Garden, 148
Belladonna, 304
Bellardia, 316
Bellflower Family, 330
Bent
  Colonial, 97
  Common, 97
  Creeping, 97
Berberis
  California, 172
  Holly-leaved, 172
Bidi-bidi, 197
Billberry
  Red, 268
Bindweed, 275
  Black, 146
  Field, 276
Birch Family, 134
Birds Eyes, 282
  Hairy, 316
  Salt Marsh, 316
  Soft, 316
  Stiffly-branched, 316
Birds' Foot, 216
Birds'-beak, 316
Birthwort Family, 140

Bishop's Weed, 260
Bitter Sweet, 303
Bitterroot, 160
Blackberry, 198
　California, 199
　Pacific, 199
Bladderwort Family, 320
Bladderwort
　Common, 320
Blazing Star, 242
　Lindley's, 242
Bleeding Heart, 175
　Pacific, 175
Blennosperma
　Common, 367
Blow-Wives, 359
Blue Blossom, 235
Blue Brush, 235
Blue Curls, 297
Blue Dicks, 125
Blue Gem, 315
Blue Gum, 244
Blue Witch, 303
Bluecup
　Common, 331
Blue-eyed Grass, 127
　California, 127
Blue-eyed Mary
　Franciscan, 310
Bluegrass, 80
　Annual, 80
　Canada, 80
　Douglas', 80
　Dune, 80
　Howell's, 80
　Kellogg's, 80
　Kentucky, 80
　Malpais, 81
　Pine, 81
　San Francisco, 81
Boisduvalia
　Dense-flowered, 247
　Narrow-leaved, 248
　Smooth, 248
Borage Family, 287
Borage, 288
Bouncing Bet, 166
Box Elder
　California, 232
Boykinia, 190
　Coast, 190
Bracken, 60
Bract-scale, 150
Brass Buttons, 372
Brickellia
　California, 345
Brickellbush, 345
Broad Bean, 219
Brodiaea
　Common, 125
　Dwarf, 126
　Golden, 125

Harvest, 126
　Many-flowered, 125
　White, 126
Brome
　California, 78
　Chilean, 78
Brook Foam, 190
Broom, 208
　California, 210
　Chaparral, 346
　Dwarf Chaparral, 345
　French, 208
　Prickly, 207
　Scotch, 208
　Spanish, 208
Broomrape Family, 320
Broomrape, 320
　California, 321
　Chaparral, 321
　Clustered, 321
　Franciscan, 321
　Gray's, 321
　Jepson's, 321
　Naked, 321
　Pine, 321
Brownies, 121
Bryony, 330
Buck Brush
　Blue, 236
　Common, 235
　Santa Cruz, 235
Buckbean Family, 272
Buckbean, 272
Buckeye Family, 232
Buckeye, 232
　California, 232
Buckhorn, 322
Buckthorn Family, 233
Buckthorn, 233
　Holly-leaved, 234
Buckwheat Family, 140
Buckwheat
　Brush, California, 143
　Wild, 142, 143
Buffalo Bur, 303
Bull-Nettle, 303
Bulrush, 109
　American Great, 110
　California, 110
　Olney's, 110
　Panicled, 110
　Prairie, 110
　River, 110
　Small-fruited, 110
　Viscid, 110
Bunch Flower Family, 116
Bur Chervil, 257
Bur Marigold, 358 ·
　Nodding, 358
Burdock, 376
　Common, 376
　Great, 376

Burgrass, 100
Burhead
　Upright, 69
Burnet, 197
Burning Bush Family,
Burning Bush, 231
　Western, 231
Burreed Family, 65
Burreed
　Broad-fruited, 65
Burweed
　Annual, 366
　Weak-leaved, 366
Butter-and-Eggs, 308, 318
Buttercup Family, 167
Buttercup, 170
　Bermuda, 224
　Bloomer's, 171
　California, 171
　Creeping, 171
　Low, 170
　Lobb's Water, 170
　Prickle-fruited, 171
　Pubescent-fruited, 171
　Water, 170
Butterweed
　Alkali Marsh, 375
　Brewer's, 375
　California, 375
　Shrubby, 375

Cabbage, 179
Cacao Family, 238
Calandrinia
　Brewer's, 158
Calendula Tribe, 369
California Bay, 173
California Bottle Brush, 8
California Coffee Berry, 2
California Fuchsia, 245
California Golden-eyed G
　127
California Mountain Balm
California Nutmeg, 62
California Privet, 271
California Rose Bay, 264
California Tea, 211
Calla Family, 111
Calla
　Common, 111
Calla Lily, 111
Calliopsis, 358
Calliprora
　Common, 125
Calochortus, 119
Caltrop Family, 226
Caltrop, 226
　Land, 226
Calycadenia
　Sticky, 364
Calyptridium
　Hesse's, 160

Camomile, 370
Campanula
  Eastwood's, 331
Camphor Weed, 297
Campion, 164
  Bladder, 165
  Dolores, 166
  Lemmon's, 166
  Scouler's Large, 166
  Rose, 164
Canary Balm, 299
Canary Grass Tribe, 98
Candy Flower, 159
Candytuft, 185
Carex, 103
Carpet Weed Family, 156
Carpet Weed, 156
Carrot Family, 252
Carrot, 257
Cascara, 233
Castor Bean, 229
Catchfly, 164
  Common, 165
  Sleepy, 165
  Snapdragon, 165
Cat's Ear, 341
  Hairy, 341
  Long-rooted, 341
  Smooth, 341
Cat-tail Family, 64
Cat-tail, 64
  Broad-leaved, 65
  Narrow-leaved, 65
Catmint, 299
Catnip, 299
Ceanothus, 234
  Coast, 236
  Coyote, 236
  Cropleaf, 235
  Dwarf, 235
  Ferris', 236
  Varnishleaf, 235
  Warty-leaved, 235
  Wavyleaf, 235
Celery, 259
  Wild, 256
Centaury, 272
  Alkali, 272
  Davy's, 272
  June, 272
  Monterey, 272
Chaenactis
  Inner Coast Range, 367
Chaetopappa
  Meager, 350
  Tiny, 350
  White-rayed, 349
Chaffweed, 269
Chamise, 198
Charlock, 179
  Jointed, 178
  White, 179

Cheat
  Downy, 80
  Downy-sheathed, 79
  Smooth-flowered Soft, 79
Checker, 236
Checker Bloom, 236
Cheeses, 237
Cheese-weed, 237
Cherry, 200
  Bitter, 200
  Holly-leaved, 200
  Western Choke, 200
Chess
  Australian, 79
  Downy, 80
  Foxtail, 79
  Hairy, 79
  Soft, 79
Chia, 296
Chicory Tribe, 339
Chicory, 339
  California, 342
Chickweed, 162
  Common, 162
  Field, 161
  Indian, 156
  Large Mouse-ear, 161
  Meadow, 161
  Mouse-ear, 161
  Shiny, 162
Chinese Caps, 229
Chinese Houses, 309
  Purple-and-White, 309
  Round-headed, 309
  White, 309
Chinese Pusley, 290
Chinquapin, 135
  Golden, 135
Chorizanthe, 141
  Knotweed, 141
  Pink, 141
  San Francisco, 142
Christmas Berry, 201
Chrysanthemum
  Corn, 370
  Garland, 370
Chu-Chu-Pate, 261
Chufa, 108
Cineraria
  Florist's, 374
  Garden, 374
Cinquefoil, 196
  Diffuse, 196
  Hickman's, 196
  River, 196
Clarkia, 248
  Brewer's, 249
  Elegant, 249
  Lovely, 249
  Modest, 249
  Rhomboid, 249
  San Francisco, 250

Cleavers, 325
Clematis, 169
  Chaparral, 169
Cliffbrake, 61
Clintonia
  Red, 122
Clocks, 223
Clotbur
  Spiny, 365
Clover, 212
  Alsatian, 215
  Alsike, 215
  Bearded, 214
  Bowl, 215
  Branched Indian, 216
  Bur, 211
  Carolina, 215
  Clammy, 215
  Coast, 215
  Common Indian, 216
  Common Owl's, 318
  Cow, 215
  Creek, 215
  Crimson, 216
  Deceptive, 216
  Dwarf Sack, 214
  Elk, 252
  Few-flowered, 215
  French, 216
  Gray's, 214
  Italian, 216
  Low Hop, 215
  MacRae's, 216
  Narrow-leaved, 215
  Notch-leaved, 215
  Owl's, 316, 318
  Pale Sack, 214
  Pinole, 215
  Pin-point, 216
  Rancheria, 216
  Red, 216
  Small-headed, 215
  Sour, 214
  Spanish, 209
  Spineless Bur, 211
  Spotted Bur, 211
  Subterranean, 216
  Sweet, 212
  Tomcat, 215
  Tree, 216
  Valparaiso, 215
  White, 215
  White Sweet, 212
  White-tipped, 215
  Wormskjold's, 215
  Yellow Sweet, 212
Club Moss
  Bigelow's, 53
Club Rush, 109
  Dwarf, 110
  Keeled, 110
  Low, 110

Cocklebur, 365
Collinsia
  Iodine, 309
Collomia
  Large-flowered, 281
  Varied-leaved, 281
Coltsfoot
  Western, 373
Columbine, 167
  Northwest Crimson, 167
  Van Houtte's, 167
Compass Plant, 228
Conyza
  Coulter's, 351
  South American, 352
Coral Root, 131
  Spotted, 131
  Striped, 131
Cordgrass
  California, 98
  Pacific, 98
Coreopsis, 358
  Garden, 358
Corethrogyne
  California, 350
  Common, 350
  Rigid, 350
Coriander, 260
Corn Gromwell, 294
Corn Salad, 329
Cornflower, 376
Cotoneaster, 201
Cotton-batting Plant, 355
Cottonweed
  Mount Diablo, 357
  Slender, 356
Cottonwood, 132
  Black, 132
  Fremont's, 132
Cotula
  Australian, 372
Coyote Brush, 346
Cranesbill, 222
Cream Bush, 197
Cream Cups, 174
  California, 174
Cream Sacs, 318
Cress
  American Winter, 183
  Bitter, 183
  Brewer's Rock, 182
  Coast Rock, 182
  Early Winter, 183
  Few-seeded Bitter, 183
  Hoary, 186
  Rock, 181
  Water, 182
  Western Yellow, 182
  Winter, 183
  Yellow, 182
Cressa, 276
Croton, 227

California, 227
Crowfoot Family, 167
Crowfoot, 171
Crypsis
  Sharp-leaved, 93
Cryptantha
  Cleveland's, 291
  Coast, 291
  Flaccid, 291
  Minute-flowered, 290
  Prickly, 290
  Tejon, 291
  Torrey's, 291
Cudweed, 354
  Bioletti's, 355
  California, 355
  Lowland, 355
  Purple, 355
  Weedy, 355
Currant, 192
  Bugle, 192
  California Black, 192
  Chaparral, 192
  Flowering, 192
  Winter, 192
Cyperus
  Brown, 107
  Red-rooted, 108
  Smooth, 107
  Straw-colored, 108
  Tall, 108
Cypress Family, 63
Cypress, 63
  Abrams', 64
  Monterey, 64
  Santa Cruz, 64
  Summer, 151
Cypselea, 156

Daisy, 349
  California Rayless, 352
  Crown, 370
  English, 349
  Leafy, 352
  Ox-eye, 370
  Philadelphia, 352
  Rock, 352
  Seaside, 352
  Shasta, 370
  Wild, 352
Dandelion, 342
  California, 343
  Common, 342
  Mountain, 342
  Red-seeded, 342
Daphne Family, 243
Darnel, 88, 89 ·
Datisca Family, 243
Death Camas, 116
Deer Brush, 235
Deerweed, 210
Dew Cup, 196

Dichondra, 276
Dobie Pod, 183
Dock, 144
  Bitter, 145
  California, 145
  Curly, 145
  Fiddle, 145
  Golden, 145
  Green, 145
  Willow-leaved, 145
  Yellow, 145
Dodder, 274
  Alkali, 275
  California, 274
  Canyon, 275
  Chaparral, 274
  Common, 274
  Fringe, 275
  Salt Marsh, 275
  Western, 274
  Western Field, 275
Dogbane Family, 272
Dogbane, 273
  Common, 273
  Western, 273
Dogwood Family, 262
Dogwood, 262
  Brown, 262
  Creek, 262
  Mountain, 262
  Smooth, 262
  Western Red, 262
Douglas Fir, 63
Dove Weed, 227
Downingia
  Flat-faced, 331
  Maroon-spotted, 331
Dropseed, 98
Duckweed Family, 112
Duckweed, 112
  Gibbous, 112
  Greater, 112
  Ivy-leaved, 112
  Least, 112
  Smaller, 112
  Valdivia, 112
Dudleya, 188
  Lax, 189
  Powdery, 189
  Setchell's, 189
  Spreading, 189
Durango Root, 243
Dusty Miller, 375, 376
Dutchman's Pipe, 140

Eardrops
  Golden, 175
Echidiocarya
  California, 293
Eelgrass Family, 67
Eelgrass, 67
Elderberry, 325

Blue, 325
Coast Red, 326
Elm Family, 139
  Elm, 139
  English, 139
Elodea, 70
Encina, 137
Endive, 339
Eragrostis
  Creeping, 76
  Orcutt's, 76
  Pilose, 76
Eriastrum
  Abrams', 280
Eriogonum
  Clay-loving, 143
  Coast, 142
  Hairy-flowered, 143
  Naked-stemmed, 142
  Rock, 143
  Virgate, 143
  Wicker, 143
Eryngo
  Coast, 254
  Ground, 254
  Jepson's, 254
  Prickly, 254
Escallonia Family, 192
Escallonia, 192
Erysimum
  Terete-leaved, 181
Escobita, 318
Eucrypta
  Common, 284
Eupatory Tribe, 345
Eupatorium
  Sticky, 345
Evax
  Erect, 356
Evening Primrose Family,
  245
Evening Primrose, 250
  Hooker's, 251
  Jones', 251
  Lamarck's, 251
  Monterey, 251
Evening Snow, 277
Everlasting Tribe, 354
Everlasting, 354
  Fragrant, 355
  Green, 355
  Pearly, 354
  Pink, 355

Fairy Bells, 122
  Hooker's, 122
  Large-flowered, 122
Fairy Lantern, 122
  White, 119
False Mermaid Family, 230
Farewell-to-Spring, 249
Fat Hen, 150

Feathertop, 100
Fennel
  Dog, 370
  Hog, 260
  Sweet, 258
Fern Family, 56
Fern
  American Water, 55
  Birds Foot, 61
  Brittle, 59
  California Grape, 56
  California Lace, 61
  California Shield, 58
  Clover, 55
  Coastal Lip, 61
  Cooper's Lip, 61
  Coastal Wood, 59
  Coffee, 61
  Deer, 59
  Dudley's Shield, 58
  Five-finger, 60
  Giant Chain, 58
  Goldenback, 57
  Grape, 56
  Lace, 61
  Leather, 57
  Licorice, 57
  Lip, 61
  Maidenhair, 60
  Spreading Wood, 59
  Western Chain, 58
  Western Lady, 59
  Western Sword, 58
Fescue Tribe, 75
Fescue, 83
  Alta, 85
  California, 85
  Coast Range, 85
  Eastwood's, 84
  Elmer's, 85
  Few-flowered, 85
  Foxtail Six-weeks, 84
  Gray's, 84
  Hairy-leaved, 84
  Meadow, 85
  Nuttall's, 85
  Pacific, 84
  Rattail, 84
  Red, 85
  Reed, 85
  Six-weeks, 84
  Slender, 84
  Western, 85
  Western Six-weeks, 84
Feverfew, 370
Fiddleneck, 289
  Common, 289
  Macbride's, 290
  Menzies', 289
  Rigid, 289
Fiesta Flower
  Common, 283

White, 283
Fig
  Hottentot, 157
  Sea, 157
Figwort Family, 306
Figwort, 310
  Coast, 310
Filago
  California, 356
  Narrow-leaved, 356
Filaree, 223
  Broad-leaved, 223
  Large-leaved, 223
  Long-beaked, 223
  Musk, 223
  Red-stemmed, 223
  White-stemmed, 223
Fimbristylis
  Grass-like, 109
Fireweed
  Cut-leaved Coast, 375
  Toothed Coast, 375
Five-finger, 196
Flag, 128
Flannel Bush, 238
Flat-top, 143
Flax Family, 224
Flax, 225
  Common, 225
  Congested Dwarf, 226
  Dwarf, 225
  Marin Dwarf, 226
  Narrow-leaved, 225
  Slender Dwarf, 225
  Small-flowered, 225
  Small-flowered Dwarf, 225
Fleabane, 352
Flixweed, 184
**Flowering Quillwort Fam-
  ily,** 68
Flowering Quillwort, 68
Fluellin, 308
  Round-leaved, 308
  Sharp-leaved, 380
Fog-Weed
  Mohave, 150
Footsteps-of-Spring, 256
Forget-Me-Not
  White, 290
  Wood, 288
Four-O'clock Family, 155
Four-O'clock, 155
Foxglove
  Purple, 311
Foxtail, 94
  Farmer's, 86
  Howell's Meadow, 94
  Perennial, 100
Frankenia Family, 240
Freesia
  Common, 129
Fremontia, 238

California, 238
Fringe Cups, 191
Fringe Pod, 184
  Crenate, 185
  Elegant, 185
  Hairy, 185
  Ribbed, 185
Fritillary, 118
  Black, 119
  Fragrant, 119
  Ill-scented, 119
  Purple Rice-bulbed, 118
  White, 119
Frogbit Family, 70
Fumewort Family, 174
Fumitory Family, 174
Fumitory, 175
  Common, 175
  Small-flowered, 175
Furse, 207
Fuzzy-Wuzzy, 345

Galinsoga
  Small-flowered, 359
Gamble Weed, 256
Garden Balm, 296
Garlic
  False, 126
  Grace, 126
Gentian Family, 271
Geranium Family, 221
Geranium, 221
  Anemone-leaved, 223
  Carolina, 222
  Cut-leaved, 222
  Dove's-foot, 222
  Feather, 152
  Garden, 221
  New Zealand, 222
  Small-flowered, 222
Giant Reed, 380
Gilia Family, 276
Gilia, 281
  Blue Field, 281
  California, 282
  Dune, 281
  Globe, 281
  Range, 282
  Slender-flowered, 282
  Tricolor, 282
Ginger
  Long-tailed Wild, 140
  Wild, 140
Ginseng Family, 252
Giraffe Head, 299
Glasswort, 148
  Woody, 148
Gnome Plant, 262
Goat's Beard, 341
Godetia, 248
  Blasdale's, 249
  Davy's, 249

Four-spotted, 250
  Large, 249
  Willow Herb, 249
Golden Dicentra, 175
Golden Eggs, 251
Golden Fleece, 347
Goldenrod, 348
  California, 349
  Dune, 349
  Meadow, 348
  Southern, 349
  Western, 348
Goldentop, 76
Goldfields, 368
  Slender, 369
  Woolly, 368
Goose Grass, 325
Gooseberry Family, 192
Gooseberry, 193
  Bay, 193
  California, 193
  Canyon, 193
  Hill, 193
  Hillside, 193
  Santa Cruz, 193
  Straggly, 193
Goosefoot Family, 148
Goosefoot, 151
  Berlandier's, 153
  California, 153
  Coast, 153
  Cut-leaved, 153
  Nettle-leaved, 153
  Tasmanian, 152
  Wall, 153
  White, 153
Gopher Plant, 228
Gorse, 207
Gourd Family, 329
Grama Tribe, 98
Grass Family, 70
Grass
  Alaska Onion, 81
  Alkali, 83
  Alkali Rye, 88
  American Dune, 88
  Annual Beard, 94
  Annual Hair, 92
  Australian Brome, 79
  Barnyard, 102
  Beach, 97
  Bear, 116
  Beard, 94
  Bent, 96
  Bermuda, 98
  Big Quaking, 77
  Blue Bunch, 85
  Bristle, 100
  Brome, 77
  California Bent, 97
  California Canary, 99
  California Hair, 92

California Vanilla, 98
California Wild Oat, 91
Canary, 99, 100
Candy, 76
Cheat, 80
Chilean Brome, 78
Crab, 101
Crested Dog's Tail, 76
Dallis, 102
Davy's Manna, 82
Ditch, 67
Elk, 116
Elongate Wheat, 89
Few-flowered Manna, 8
Geyer's Onion, 81
Giant Rye, 88
Gould's Rye, 88
Green Bristle, 100
Hair, 92
Hairy-flowered Bent, 97
Hall's Bent, 97
Harding, 99
Hare's Tail, 95
Indian Basket, 116
Johnson, 102
June, 92
Kikuyu, 100
Knot, 102
Knotroot Bristle, 100
Large Mountain Brome
Leafy Bent, 97
Lemmon's Canary, 100
Little Alkali, 83
Little Quaking, 77
Love, 76
Manna, 82
Mediterranean Canary,
Melic, 81
Narrow Canary, 100
Narrow-flowered Brome
Needle, 94
Nit, 95
Nuttall's Alkali, 83
Oat, 91
Old Witch, 101
Orchard, 77
Pacific Alkali, 83
Pacific Bent, 97
Pacific Reed, 96
Pacific Rye, 88
Pampas, 75
Paradox Canary, 99
Pine, 95
Prickle, 93
Purple-awned Wallaby,
Quack, 89
Quaking, 76
Rabbit's Foot, 94
Rattlesnake, 77
Reed, 95
Rescue, 78
Rice, 94

Ripgut, 80
Rough Hair, 97
Rough-stalked Meadow, 81
Rush-like Wheat, 89
Rye, 87
Saint Augustine, 101
Salt, 75
Small-leaved Bent, 97
Scribner's, 90
Seashore Bent, 97
Seaside Brome, 78
Semaphore, 82
Shield, 186
Sickle, 90
Silvery Hair, 93
Slender Cotton, 109
Slender Hair, 92
Slender Wheat, 89
Slough, 98
Smilo, 94
Spanish Brome, 79
Spear, 94
Sprangletop, 98
Sudan, 102
Sweet Velvet, 98
Sweet Vernal, 98
Tall Brome, 79
Tall Oat, 91
Tufted Hair, 92
Tufted Pine, 96
Vancouver's Rye, 88
Velvet, 92
Water Bent, 97
Water Manna, 82
Western Bent, 97
Western Rye, 87
Western Vanilla, 98
Wheat, 89
Wire, 109
Woodland Brome, 79
Yellow Bristle, 100
Grass Nut, 125
Grass Poly, 244
Grass-of-Parnassus Family, 192
Grass-of-Parnassus, 192
California, 192
Greasewood, 198
Grindelia
Great Valley, 346
Hirsute, 346
Marsh, 346
Pacific, 347
San Francisco, 346
Ground Cone
California, 321
Ground Cherry, 303
Low Hairy, 304
Groundsel Tribe, 373
Groundsel
California, 374
Common, 374

Wood, 375
Gum Plant, 346
Coastal, 346
Gum Weed, 302, 346

Haplopappus
Blake's, 347
Harebell
California, 330
Swamp, 331
Hareleaf, 359
Common, 359
Hawkbit
Hairy, 341
Hawksbeard, 344
Italian, 344
Smooth, 344
Weedy, 344
Hawkweed, 344
White-flowered, 344
Hazel Family, 135
Hazel, 135
California, 135
Heal-all, 297
Heart's Ease
Western, 241
Heath Family, 264
Hedge Nettle, 299
Bugle, 300
California, 300
Coast, 300
Rigid, 300
Short-spiked, 300
Hedge-hyssop
Bractless, 313
Hedypnois
Crete, 339
Helianthella
California, 357
Diablo, 358
Heliotrope
Seaside, 290
Helleborine
Giant, 129
Hemitomes, 263
Hemp
California, 211
Indian, 273
Mountain, 273
Henbit, 299
Common, 299
Red, 299
Herb Robert, 223
Herb-of-Gilead, 299
Herb-Mercury, 228
Herniaria
Gray, 156
Heterocodon, 331
Heuchera, 190
Seaside, 190
Small-flowered, 190

Hill Star, 191
Himalaya Berry, 199
Hoarhound
Common, 297
Cut-leaved Water, 298
White, 297
Hollyhock, 237
Wild, 236
Honesty, 187
Honeysuckle Family, 325
Honeysuckle, 326
Chaparral, 326
Garden, 326
Hairy, 326
Japanese, 326
Hop Tree, 226
Western, 226
Horkelia
Bolander's, 195
California, 195
Leafy, 195
Wedge-leaved, 195
Hornwort Family, 166
Hornwort, 166
Horse Bean, 219
Horse Chestnut Family, 232
Horse Chestnut, 232
Horsetail Family, 54
Horsetail, 54
California, 55
Common, 54
Giant, 54
Horseweed, 351
Hound's Tongue
Grand, 288
Western, 288
Huckleberry Family, 268
Huckleberry
Evergreen, 268
Red, 268
Short, 268
Hulsea
Red-rayed, 367
Hutchinsia
Prostrate, 185
Hyacinth
Common, 117
Grape, 117
Water, 113
Wild, 125, 126
Hydrangea Family, 192

Ice Plant, 157
Common, 157
Slender-leaved, 157
Indian Pipe Family, 263
Indian Warrior, 315
Indigo Brush, 235
Innocence, 309
Inside-out Flower, 172
Iris Family, 127
Iris, 128

Coast, 128
Douglas', 128
Field, 128
Ground, 128
Long-petaled, 128
Mountain, 128
Slender-tubed, 128
Yellow, 129
Islay, 200
Isocoma
    Coastal, 347
Ithuriel's Spear, 125
Ivy, 252
    English, 252
    German, 374
    Ground, 299
    Kenilworth, 308
Ixia
    African, 129

Jaumea
    Fleshy, 368
Jerusalem Oak, 152
Jewel Flower
    Common, 180
Jim Bush, 235
Jimson Weed, 204
Johnny-Jump-Up, 241
Johnny-Tuck, 318
June Berry, 201
Jupiter's Beard, 328

Karo, 193
Klamath Weed, 239
Knapweed
    Protean, 377
Knotweed, 145
    Common, 146
    Dooryard, 146
    Dune, 146
    Silversheath, 146
    Swamp, 147
Knotwort Family, 155
Koeleria
    Bristly, 92

Labrador Tea
    Coastal, 264
Lace Pod, 184
Lady's Comb, 255
Lady's Mantle
    Western, 196
Lady's Slipper, 129
    Clustered, 129
    Mountain, 129
Lady's Thumb, 147
Lady's Tresses, 131
    Hooded, 131
    Western, 131
Lamb's Lettuce, 329
Lamb's Quarter, 153
Larkspur, 168

Blue, 169
Coast, 169
Parry's 169
Red, 168
Rocket, 168
Royal, 169
Western, 169
Lasthenia
    Smooth, 369
    Yellow-rayed, 369
Laurel Family, 172
Laurel, 173
    California, 173
    Sticky, 235
Lavatera
    Cretean, 237
Lavender Cotton, 371
Layia
    Beach, 360
    Smooth, 360
    Tall, 360
    Woodland, 360
Leadwort Family, 270
Leather Root, 211
Leatherwood, 243
    Western, 243
Legenere, 331
Lemon Balm, 296
Lentil, 221
Lessingia
    San Francisco, 353
    San Mateo, 353
    Woolly-headed, 353
Lettuce, 343
    Bluff, 189
    Prickly, 343
    Sea, 189
    Wild, 343
    Willow, 343
Libertia
    Showy, 127
Licorice
    American, 216
    Wild, 216
Lilac
    California, 234
    Summer, 271
Lily Family, 117
Lily
    California Tiger, 118
    Checker, 118
    Chocolate, 119
    Cow, 166
    Fire, 116
    Fremont's Star, 116
    Leopard, 118
    Mariposa, 119
    Panther, 118
    Tiger, 118
    Triplet, 125
    White Globe, 119

Lily-of-the-Valley Family,
    120
Lily-of-the-Valley Vine, 306
Linanthus
    Bicolored, 279
    Bristly, 279
    Bristly-leaved, 279
    Common, 278
    Flax-flowered, 278
    Large-flowered, 278
    Pigmy, 277
    Serpentine, 278
    Shower, 278
    Small-flowered, 279
Lindernia
    Long-stalked, 313
Linseed, 225
Lippia
    Garden, 294
Live-forever, 188
Lizard Tail Family, 132
Lizard Tail, 366
Loasa Family, 242
Lobelia Family, 331
Locoweed, 217
    Gambell's Dwarf, 217
    Marsh, 217
Locust, 207
    Black, 207
Loeflingia
    California, 163
Logania Family, 271
Lomatium
    California, 261
    Caraway-leaved, 261
    Common, 261
    Large-fruited, 261
    Woolly-fruited, 261
London Rocket, 184
Loosestrife Family, 243
Loosestrife, 244
    California, 244
    Hyssop, 244
Lousewort, 315
    Dudley's, 316
Lovage, 260
    Celery-leaved, 260
    Pacific, 260
    Wood, 260
Lucerne, 211
Lupine, 204
    Bentham's Bush, 205
    Blue Beach, 205
    Bridges', 205
    Broad-leaved, 205
    Bush Beach, 205
    Chamisso's Bush, 205
    Davy's Bush, 205
    Douglas' Annual, 206
    Dudley's, 205
    False, 211
    Hilly Bush, 206

Large-leaved, 205
Late, 205
Lindley's Annual, 206
Lindley's Varied, 205
Mount Diablo Annual, 206
Nettle Annual, 206
Nuttall's Annual, 206
Silver Bush, 205
Sky, 206
Small-flowered Annual, 207
Succulent Annual, 206
Summer, 205
Tree Beach, 205
White-foliaged Bush, 205
Yellow Beach, 205
Yellow-flowered Beach, 205

Madder Family, 323
Madder
Blue Field, 323
Wild, 324
Madeira Vine, 160
Madia
Common, 362
Woodland, 361
Madrone, 265
Madroño, 265
Maidenhair
California, 60
Western, 60
Malacothamnus
Hall's, 238
Northern, 238
Malacothrix
Cleveland's, 344
Woolly, 344
Mallow Family, 236
Mallow, 237
Alkali, 238
Bull, 238
High, 238
Tree, 237
Wheel, 236
Manroot
Coast, 330
Valley, 330
Manzanita, 265
Big-berried, 267
Brittle-leaved, 268
Eastwood's, 267
Glutinous, 267
Heart-leaved, 267
Hoary, 267
Hooker's, 267
Monterey, 267
Rose's, 268
San Francisco, 267
Santa Cruz Mountains, 267
Sensitive, 267
Silver-leaved, 267
Maple Family, 232
Maple, 232

Big-leaved, 232
Canyon, 232
Mare's Tail, 252
Marigold
Corn, 370
Pot, 369
Mariposa
Butterfly, 120
White, 120
Yellow, 120
Marjoram
Wild, 301
Marsh Pennywort, 257
Floating, 258
Spike, 258
Whorled, 258
Marsh Rosemary
California, 271
Marsilea Family, 55
Marvel-of-Peru, 155
Mat Grass, 294
Matrimony Vine, 302
Mattress Vine, 147
Mayweed Tribe, 370
Mayweed, 370
Meadow Foam Family, 230
Meadow Foam, 230
Common, 230
Snow White, 230
Meconella
California, 174
Narrow-leaved, 174
Medick, 211
Black, 211
Spotted, 211
Medusa Head, 87
Melica
Harford's, 82
Small-flowered, 82
Torrey's, 81
Western, 81
Melilot, 212
Indian, 212
White, 212
Yellow, 212
Mercury, 228
Metake, 75
Mexican Tea, 152
Mezereum Family, 243
Microcala
American, 272
Microseris, 239
Bigelow's, 341
Coast, 341
Derived, 340
Douglas', 341
Elegant, 340
Thomas', 340
Mignonette Family, 187
Mignonette, 187
White, 187
Milfoil, 370

American, 252
Whorl-leaved, 252
Milkmaids, 182
Milkvetch, 217
Milkweed Family, 274
Milkweed, 274
Narrow-leaved, 274
Milkwort Family, 226
Milkwort, 226
California, 226
Sea, 269
Millet Tribe, 100
Millet
Broomcorn, 101
Mimetanthe
Downy, 312
Mimosa Family, 202
Mind-your-own-Business, 139
Miner's Lettuce, 159
Mint Family, 295
Mint, 298
Apple, 299
Bergamot, 299
Coyote, 297
Field, 299
Franciscan Coyote, 298
Marsh, 299
Mirrorbush, 323
Mission Bells, 118, 119
Mistletoe Family, 139
Mistletoe, 139
Hairy, 139
Oak, 139
Pine, 140
Western Dwarf, 140
Mock Heather, 347
Modesty, 192
Monardella, 297
Curly-leaved, 297
Fenestra, 297
Monkey Flower, 312
Androsace, 313
Bush, 311
Common Large, 312
Congdon's, 313
Floriferous, 313
Rattan's, 313
Scarlet, 313
Sticky, 311
Monolepis
Nuttall's, 151
Monolopia
Cupped, 367
Woodland, 367
Montia
Coast Range, 159
Common, 159
Hall's, 159
Siberian, 159
Small-leaved, 159
Vernal, 159
Moonwort, 187

Morning-Glory Family, 274
Morning-Glory, 275, 276
  Beach, 275
  Bush, 275
  Chaparral, 275
  Common, 276
  Hill, 275
  Orchard, 276
  Western, 275
  Woolly, 275
Mother-of-Thousands, 139
Mountain Mahogany, 197
  California, 197
Mountain Rice
  Millet, 94
Mourning Bride, 327
Mouse Tail, 170
  Common, 170
Mouse-Ears
  Purple, 313
Mugwort
  California, 372
  Douglas', 372
  Northern, 309
Muilla, 124
  Common, 124
Mule Ears, 357
  Gray, 357
  Narrow-leaved, 357
Mule Fat, 346
Mullein, 307
  Common, 307
  Moth, 307
  Showy, 307
  Turkey, 227
  Wand, 307
  Woolly, 307
Musk Flower, 313
Mustard Family, 176
Mustard, 178
  Black, 180
  California, 184
  Chinese, 180
  Common, 179
  Desert, 184
  Field, 179
  Hedge, 184
  Indian, 180
  Mediterranean, 179
  Oriental, 180
  Summer, 179
  Tansy, 184
  Tower, 182
  Tumble, 184
  Western Tansy, 184
  White, 179
Myrtle Family, 244

Naias Family, 67
Nail Rod, 65
Nasturtium, 224
  Garden, 224

Navarretia
  Calistoga, 280
  Holly-leaved, 280
  Honey-scented, 280
  Sticky, 280
Nemophila
  Canyon, 284
  Meadow, 284
  Small-flowered, 284
  Variable-leaved, 284
  Woodland, 284
Nest-straw, 257
Nettle Family, 138
Nettle, 138
  California, 138
  Coast, 138
  Dead, 299
  Dwarf, 138
  Hoary, 138
  Silver-leaved, 303
  Western, 138
Nievitas, 290, 291
Nigger Babies, 127
Nightshade Family, 301
Nightshade, 302
  Black, 380
  Climbing, 303
  Deadly, 304
  Douglas', 303
  Forked, 303
  Hairy, 303
  Purple, 303
  Small-flowered, 303
  White-margined, 302
Ninebark, 197
  Pacific, 197
Nipplewort, 339
Nut Grass
  Purple, 380
Nutgrass
  Yellow, 108

Oak Family, 135
Oak, 136
  Blue, 137
  California Black, 136
  California Live, 137
  California Scrub, 137
  California White, 137
  Canyon, 137
  Douglas', 137
  Garry's, 137
  Gold Cup, 137
  Interior Live, 137
  Kellogg's Black, 136
  Leather, 137
  Maul, 137
  Oracle, 137
  Oregon, 137
  Post, 137
  Sierra Live, 137
  Valley, 137

Oat Tribe, 90
Oat, 90
  Cultivated, 90
  Slender Wild, 90
  Wild, 90
Ocean Spray, 197
Oenanthe
  American, 259
  Pacific, 259
Old Man, 372
Old-Man-in-the-Spring, 374
Olive Family, 271
Onion
  Brewer's, 124
  Coastal, 124
  Narrow-leaved, 124
  Neapolitan, 123
  One-leaved, 124
  Paper-flowered, 124
  Pitted, 124
  Serrated, 124
  Wild, 123
Ookow, 125
Orache
  Garden, 150
  Halberd-leaved, 150
  Red, 150
  Spear, 150
Orchid Family, 129
Orchid
  Phantom, 131
  Rein, 130
  Sparsely-flowered Bog, 130
  Stream, 129
  White-flowered Bog, 130
  White Rein, 130
Oregon Grape, 172
Oregon Myrtle, 173
Oregon Pepperwood, 173
Orpin, 188
Orthocarpus, 316
  Dwarf, 318
  Field, 318
  Narrow-leaved, 318
  Paint Brush, 318
  San Francisco, 318
  Smooth, 318
Oso Berry, 200
Ox Tongue
  Bristly, 341
Oxalis Family, 223
Oxalis
  Cape, 224
  Flesh-colored, 224
  Windowbox, 224
Oyster Root, 342

Paint Brush
  Franciscan, 319
  Great Red, 319
  Indian, 319, 320
  Large-flowered, 319
  Monterey, 319

Seaside, 319
Wight's, 320
Woolly, 319
Palm Family, 111
Palm
  Canary Island Date, 111
  Guadalupe Island, 111
Panicum
  Pacific, 101
Pansy
  Wild, 241
  Yellow, 241
Parrot Feather, 252
Parsley Family, 252
Parsley, 258
  California Hedge, 257
  Hemlock, 262
  Knotted Hedge, 255
  Mock, 256
Parsnip, 260
  Alkali, 261
  Bladder, 261
  Coast, 261
  Cow, 260
  Cut-leaved Water, 259
  Lace, 261
  Sheep, 261
Pawnbroker, 221
Pea Family, 203
Pea, 220, 221
  Bolander's, 221
  Buff, 220
  Chaparral, 207
  Common Pacific, 220
  Common Sweet, 220
  Everlasting, 220
  Garden, 221
  Redwood, 220
  Silky Beach, 220
  Tangier, 220
  Torrey's, 220
  Turkey, 257
Peach Family, 200
Pearlwort, 161
  Beach, 162
  Procumbent, 162
  Sticky, 161
  Western, 162
Pectocarya
  Little, 290
  Winged, 290
Pelargonium
  Grape-leaf, 221
Pelican Flower, 318
Pellitory, 138
  Florida, 138
Penstemon
  Foothill, 311
  Rattan's, 311
  Redwood, 311
Pennyroyal, 297, 299
Pepper Grass, 186

Common, 186
Dwarf, 187
Round-leaved, 186
Sharp-podded, 186
Shining, 186
Wayside, 186
Wild, 186
Pepper Tree, 231
  Peruvian, 231
Peppermint, 299
Pepperwort Family, 55
Pepperwort
  Hairy, 55
Periwinkle, 273
Petunia
  Common Garden, 305
  Wild, 305
Phacelia Family, 283
Phacelia
  Branching, 285
  California, 285
  Common, 286
  Divaricate, 286
  Douglas', 286
  Great Valley, 286
  Imbricate, 286
  Rattan's, 286
  Shade, 285
  Stinging, 285, 286
  Sweet-scented, 286
Phlox Family, 276
Phlox
  Slender, 277
Pickerel Weed Family, 113
Pickleweed, 148
Pigmyweed, 188
  Sand, 188
  Water, 188
Pigweed, 151, 154
  Rough, 154
  Slender, 154
Pillwort
  American, 56
Pimpernel, 269
  False, 269
  Scarlet, 269
Pincushion, 327
Pine Family, 62
Pine, 62
  Digger, 62
  Knobcone, 63
  Monterey, 63
  Ponderosa, 62
  Western Yellow, 62
Pineapple Weed, 371
  Valley, 371
Pine-sap
  Fringed, 263
Pink Family, 160
Pink
  California Indian, 165
  Mullein, 164

Windmill, 165
Pinkweed, 147
Pipe Vine
  California, 140
Piperia
  Alaska, 130
  Coast, 130
  Elegant, 131
Pittosporum Family, 193
Pittosporum
  Thick-leaved, 193
Plane Tree Family, 193
Plane Tree
  California, 197
Plantago, 321
  Annual Coast, 322
Plantain Family, 321
Plantain
  California, 322
  California Seaside, 322
  Common, 322
  Cut-leaved, 322
  English, 322
  Menzies Rattlesnake, 131
  Mexican, 322
  Pacific Seaside, 322
  Rattlesnake, 131
  Sand, 322
Platycarpos
  Dense-flowered, 206
  Intermediate, 206
Plectritis
  Long-spurred, 329
  Pink, 328
  White, 329
Pleuropogon
  California, 82
Plum, 200
  Pacific, 200
  Sierra, 200
Plumbago Family, 270
Pogogyne
  Thyme-leaved, 301
Poison Hemlock, 260
Poison Oak, 230
  Pacific, 230
Pokeberry, 155
  Mexican, 155
Pokeweed Family, 155
Pokeweed, 155
Polemonium
  Great, 280
Polycarp
  Four-leaved, 163
Polygala, 226
  California, 226
Polypody, 57
  California, 57
  Coast, 57
Pond Lily
  Indian, 166
Pondweed Family, 65

Pondweed, 66
Broad-leaved, 66
Common American, 66
Common Floating, 66
Curl-leaved, 66
Fennel-leaved, 66
Horned, 67
Illinois, 66
Leafy, 66
Long-leaved, 66
Sego, 66
Small, 66
Various-leaved, 66
Popcorn Flower, 293, 318
Fulvous, 293
Rusty, 293
Slender, 293
Valley, 293
Poplar, 132
Lombardy, 132
Poporo, 303
Poppy Family, 173
Poppy
Bush, 173
California, 174
Sea, 174
Tree, 173
Wind, 174
Yellow-horned, 174
Potato, 303
Squaw, 258
Poverty Weed, 364
Pretty Face, 125
Primrose Family, 268
Primrose
Beach, 251
Contorted, 251
Slender-flowered, 251
Small, 251
Psoralea
Loma Prieta, 211
Round-leaved, 211
Pterostegia, 140
Puccoon, 294
Pumpkin
Field, 330
Puncture Vine, 226
Puncture Weed, 226
Purselane Family, 157
Purselane, 160
Marsh, 251
Pacific Marsh, 251
Sea, 150
Pussy Ears, 119
Pussy Paws, 160
Pyrola
Leafless, 263
Pyrrocoma
Racemose, 347

Quassia Family, 226
Queen Anne's Lace, 257

Quillwort Family, 53
Quillwort
Howell's, 53
Nuttall's, 53
Quinine Bush, 263

Rabbit Brush
Common, 354
Rabbit's Foot, 360
Radish, 178
Wild, 178
Rainbells, 182
Ragweed Tribe, 364
Ragweed
Annual, 364
California, 364
Common, 364
Purple, 374
Ramtilla, 359
Raspberry, 198
Western, 199
White-stemmed, 199
Rattlesnake Weed, 257
Rattleweed, 217
San Francisco, 217
Slender, 217
Red Berry, 234
Red Canker, 263
Red Maids, 158
Red Robin, 223
Red-Hot-Poker Plant, 118
Redscale, 150
Redtop, 97
Redwood
Coast, 64
Reed
Common, 75
Rhododendron
California, 264
Ribwort, 322
Rigiopappus, 367
Roble, 137
Rock-Rose Family, 240
Romanzoffia
Suksdorf's, 286
Rose Family, 194
Rose, 199
Broom, 240
California, 199
California Wild, 199
Common Rush, 240
Ground, 200
Malva, 237
Wood, 200
Rose Sundrops, 251
Rosilla, 367
Rosinweed, 364
Rue Family, 226
Rue, 226
Meadow, 171
Rush Family, 113
Rush, 113

Baltic, 114
Bog, 114
Brown-headed, 115
Bolander's, 115
Common, 114
Common Wood, 115
Iris-leaved, 115
Kellogg's, 115
Mariposa, 115
Mexican, 114
Salt, 114
Sickle-leaved, 115
Sharp-fruited, 115
Slender, 115
Spreading, 114
Toad, 115
Western, 115
Wood, 115
Rye
Blue Wild, 87
Cereal, 90
Ryegrass, 88
Australian, 89
English, 89
Italian, 89
Perennial, 89

Sage, 296
California, 372
California Black, 296
Canon, 296
Crimson, 296
Pitcher, 296, 300
Sagebrush, 372
Sagewort, 372
Beach, 372
Biennial, 372
Dragon, 372
St. John's Wort Family, 238
St. John's Wort, 238
Aaronsbeard, 239
Common, 239
Creeping, 239
Scouler's, 239
Saitas
Common, 125
Many-flowered, 125
Narrow, 125
Salal, 265
Salmon Berry, 199
Salsify, 342
Yellow, 342
Saltbush Family, 148
Saltbush, 149
Australian, 149
Beach, 149
Bracted, 150
Brewer's, 149
California, 150
Red, 150
San Joaquin, 150
Spear, 150

Salvinia Family, 55
Saltwort
  Black, 269
Sand Mat, 156
Sand Verbena, 155
  Beach, 155
  Pink, 155
  Yellow, 155
Sandbur
  Small-flowered, 100
Sandwort, 163
  California, 163
  Douglas', 163
  Dwarf, 163
  Large-leaved, 163
  Swamp, 163
Sanicle, 256
  Adobe, 256
  Bear's Foot, 256
  Coast, 256
  Dobie, 256
  Pacific, 256
  Poison, 257
  Purple, 257
  Saltmarsh, 256
  Tuberous, 257
Saxifrage Family, 189
Saxifrage, 191
  California, 191
Scorzonella
  Marsh, 340
Scouring Rush, 54
  Western, 54
Sea Lavender, 271
Sea Lyme, 88
Sea Pink, 270
Sea Rocket, 178
  California, 178
  Pacific, 178
Sea-Blite, 151
  California, 151
Sea-scale, 149
Sedge Family, 103
Sedge, 103
  Ample-leaved, 106
  Beaked, 107
  Bifid, 107
  Bolander's, 105
  Bristly, 107
  Clustered Field, 105
  Coastal Stellate, 105
  Cotton, 109
  Cusick's, 105
  Deceiving, 106
  Dense, 105
  Dudley's, 105
  Foothill, 105
  Harford's, 106
  Hasse's, 106
  Monterey, 106
  Olney's Hairy, 106
  Rough, 106

Round-fruited, 106
Rusty, 106
Sand Dune, 105
Santa Barbara, 106
Schott's, 106
Short-beaked, 105
Short-scaled, 106
Short-stemmed, 106
Sierra Slender, 106
Slender, 106
Slough, 106
Small-bracted, 106
Torrent, 106
Walking, 109
Western Inflated, 107
Woolly, 106
Sedum
  White, 188
Seep-Weed, 151
Selaginella Family, 53
Self-heal, 297
Senna Family, 202
Senna, 202
Service Berry, 201
  Shamrock, 215
Shepherd's Needle, 255
Shepherd's Purse, 185
Shooting Star, 270
  Henderson's, 270
  Lowland, 270
  Padres', 270
Sidalcea
  Fringed, 236
  Maple-leaved, 236
Sierra Sap, 263
Silk Tassel Family, 262
Silk Tassel
  Coast, 263
  Fremont's, 263
Silk Tassel Bush, 263
Silverweed
  Pacific, 196
Simarouba Family, 226
Sisymbrium
  Oriental, 184
Skull Cap, 296
  Bolander's, 296
  Dannie's, 296
  Sierra, 296
Skunk Cabbage
  Yellow, 111
Skunkweed, 280
Slink Pod, 121
Smallage, 259
Smartweed, 145
  Water, 147
Smutgrass, 98
Snake Root, 256
  Button, 254
Snapdragon, 308
  Corn, 309
  Garden, 308

Lax, 308
Wiry, 309
Withered, 308
Sneezeweed Tribe, 366
Sneezeweed, 367
Snowberry, 326
  Common, 326
  Creeping, 326
  Trailing, 326
Snowdrops
  Foothill, 293
Soap Plant, 117, 153
Soapwort, 166
Soft Flag, 65
Soliva
  Common, 371
Solomon
  Fat, 122
  Slim, 122
Solomon's Seal
  False, 122
  Western, 122
Sorghum Tribe, 102
Sorghum, 102
Sorrel, 144
  Redwood, 224
  Sheep, 145
Spearmint, 299
Spearwort, 170
Speedwell, 314
  Broad-fruited Water, 314
  Corn, 314
  Germander, 314
  Persian, 314
  Purselane, 315
  Thyme-leaved, 314
Spiderwort Family, 113
Spiderwort, 113
Spikenard, 252
  California, 252
Spike-rush, 108
  Beaked, 109
  Blunt, 109
  Common, 109
  Creeping, 109
  Dombey's, 109
  Needle, 109
Spikeweed
  Common, 363
  Congdon's, 363
  Fitch's, 363
  Pappose, 363
Spinach, 150
  New Zealand, 157
  Sea, 157
Spindleroot
  Dobie, 258
Spine-flower, 141
  Diffuse, 142
  Hartweg's, 142
  Robust, 141
Spurge Family, 227

Spurge, 228
  Caper, 228
  Large, 229
  Petty, 228
  Reticulate-seeded, 228
  Spotted, 229
  Sun, 228
  Thyme-leaved, 229
  Wart, 228
Spurry, 163
  Sand, 163
  Boccane's Sand, 164
  Large-flowered Sand, 164
  Middle-sized Sand, 164
  Purple Sand, 164
  Salt-marsh Sand, 164
  Villous Sand, 164
Squirreltail, 86
  Big, 87
  Hansen's, 86
Stachys
  Coast, 300
  Swamp, 300
Staff-Tree Family, 231
Star Tulip
  Hairy, 119
  Large-flowered, 119
  Oakland, 119
  Tolmie's, 119
Starflower, 269
  Pacific, 269
Starwort, 162
  Beach, 162
  Lesser, 162
Stephanomeria
  Tall, 341
  Virgate, 341
Sterculia Family, 238
Stick-leaf, 242
  Nada, 243
  Small-flowered, 242
Stingaree Bush, 207
Stink Bean, 202
Stink Bells, 119
Stipa
  Nodding, 94
  Small-flowered, 94
Stock, 180
Stone Crop Family, 188
Stone Crop, 188
  Pacific, 188
  Star-fruited, 188
Stone-fruits, 200
Storksbill, 223
Straggle Bush, 193
Strawberry, 195
  Beach, 195
  California, 195
  Chilean, 195
  Indian, 195
  Mock, 195
Stylocline

Everlasting, 357
  Hooked, 357
  Northern, 357
Subclover, 216
Sugar-scoop, 190
Sumac Family, 230
Sumac, 230
Suncup, 251
Sunflower Family, 332
Sunflower Tribe, 357
Sunflower, 359
  Bolander's, 359
  California, 359
  Common, 359
  Common Woolly, 367
  Seaside Woolly, 366
Surfgrass
  Scouler's, 67
  Torrey's, 67
Sweet Bay, 173
Sweet Cicely, 255
  California, 255
  Mountain, 255
  Wood, 255
Sweet Gale Family, 133
Sycamore Family, 193
Sycamore, 193

Tan Bark Oak, 136
Tan Oak, 136
Tansy, 373
  Dune, 373
Tapegrass Family, 70
Tare, 217, 218
Tarragon, 372
Tarweed Tribe, 359
Tarweed, 361, 362
  Chile, 362
  Cleveland's, 364
  Coast, 362, 363
  Common, 362
  Fascicled, 363
  Hayfield, 364
  Headland, 362
  Kellogg's, 363
  Salinas River, 363
  Santa Cruz, 364
  Slender, 362
  Small, 362
  Three-rayed, 363
Tauschia
  Hartweg's, 259
  Kellogg's, 259
Taxodium Family, 64
Teasel Family, 327
Teasel, 327
  Fuller's, 327
  Wild, 327
Telegraph Weed, 347
Thelypodium, 184
Thermopsis
  California, 211

Thimble Berry, 199
Thin-tail, 90
Thistle Tribe, 376
Thistle, 378
  Barnaby's, 377
  Blessed, 377
  Brownie, 379
  Bull, 378
  Canada, 378
  Cobweb, 379
  Common, 378
  Common Sow, 344
  Coyote, 254
  Distaff, 377
  Douglas', 378
  Fountain, 378
  Franciscan, 379
  Golden, 339
  Indian, 379
  Italian, 380
  Milk, 377
  Napa, 377
  Plumeless, 380
  Prickly Sow, 344
  Purple Star, 376, 377
  Red, 379
  Remote-leaved, 379
  Russian, 151
  Sow, 343
  Slender-flowered, 380
  Star, 376
  Swamp, 378
  Venus, 379
  Yellow Star, 377
Thorn Apple
  Purple, 304
Thornmint
  Dutton's, 296
Thousand Mothers, 190
Three Square, 110
Thrift Family, 270
Thrift
  California, 270
Tibinagua, 142
Tickseed, 358
Tidy Tips, 360
  Field, 360
Tillaea, 188
Timothy Tribe, 93
Timothy, 95
  Cultivated, 95
  Mountain, 95
Timwort, 272
Tinker's Penny, 239
Toadflax, 307
  Blue, 308
Tobacco, 304
  Indian, 305
  Many-flowered, 304
  Mexican, 304
  Tree, 304
Tobacco Bush, 235

Tocalote, 377
Tolguacha, 304
Tomatillo, 304
Tomato, 303
Tonella
  Small-flowered, 310
Toothwort, 182
  California, 182
Torch Flower, 118
Toyon Berry, 201
Trail Plant, 354
Travelers' Joy, 169
Tree-of-Heaven, 226
Trefoil
  Bentham's, 210
  Bird's Foot, 208, 210
  Broad-leaved, 209
  Chile, 210
  Coast, 209
  Coastal, 209
  Pursh's, 209
  Rush, 210
  Short-podded, 210
  Slender, 209
  Small-flowered, 209
  Stipulate, 209
  Strigose, 209
  Torrey's 209
  Woolly, 210
Trillium
  Coast, 121
  Common, 121
  Western, 121
Trisetum
  Nodding, 93
  Tall, 93
Triteleia
  Common, 125
Tropaeolum Family, 224
Tropidocarpum
  Slender, 183
Tule, 109, 110
  California, 110
  Common, 110
Tumbleweed, 154
Turpentine Weed, 297
Twinberry, 326
Twisted-Stalk, 122
  Clasping-leaved, 122

Umbrella Plant, 108
Unicorn Plant
  Common, 380
Uropappus, 340

Valerian Family, 328
Valerian
  Red, 328
Vancouveria, 172
  Small-flowered, 172
Venus' Comb, 255
Venus' Looking-glass

Small, 331
Verbena Family, 294
Verbena
  Western, 294
Vervain Family, 294
Vervain, 294
  California, 294
  Robust, 294
Vetch, 217
  American, 219
  Common, 219
  Giant, 218
  Hairy, 218
  Linear-leaved, 219
  Slender, 218
  Smaller Common, 219
  Spring, 219
  Thick-fruited, 218
  Truncate-leaved, 219
  Winter, 218
  Woolly, 218
  Yellow, 219
Vinegar Weed, 297
Violet Family, 240
Violet, 240
  Blue, 241
  California Golden, 241
  English, 241
  Evergreen, 242
  Mountain, 242
  Redwood, 242
  Smooth Yellow, 242
  Stream, 242
  Sweet, 241
  Two-eyed, 241
  Western Dog, 241
  Yellow, 241
Viper's Bugloss, 288
Virgin's Bower, 169
  Western, 169

Wake Robin, 121
  Giant, 121
  Western, 121
Wallflower, 180
  Coarse-leaved, 181
  Coast, 181
  Douglas', 181
  Franciscan, 181
Wandering Jew, 113
Wapato, 69
Wartcress
  Lesser, 185
Wartweed, 228
Water Fern Family, 55
Water Hawthorn Family, 67
Water Hawthorn
  Cape, 67
Water Hemlock, 259
  Douglas, 259
  Western, 259
Water Lily Family, 166

Water Lily
  Yellow, 166
Water Milfoil Family, 251
Water Milfoil, 252
Water Nymph Family, 67
Water Nymph
  Common, 67
  Guadalupe, 67
Water Plantain Family, 69
Water Plantain
  Common, 69
Water Starwort Family, 229
Water Starwort, 229
  Autumnal, 229
  Bolander's, 229
  California, 229
  Hermaphroditic, 229
  Northern, 229
  Vernal, 229
Water Weed
  Yellow, 245
Waterleaf Family, 283
Watermelon, 330
Waterweed, 70
  Brazilian, 70
  Canadian, 70
Waterwort Family, 239
Waterwort, 239
  Short-seeded, 239
Wattle
  Cootamundra, 202
  Golden, 202
  Green, 202
  Water, 202
Wax Myrtle Family, 133
Wax Myrtle
  California, 133
Western Turkey Beard, 116
Western Wahoo, 231
Wheat, 89
Whisker Brush, 279
Whispering Bells, 286
White Man's Foot, 322
Whitethorn
  Coast, 234
Whiteweed, 370
Whitlowwort
  California, 155
Wild Cucumber, 330
  Western, 330
Willow Family, 132
Willow, 133
  Arroyo, 133
  Coulter's, 133
  Hinds', 133
  Red, 133
  Scouler's, 133
  Valley, 133
  Yellow, 133
Willow Herb, 246
  California, 247
  Hall's, 247

Minute, 246
Northern, 247
Panicled, 246
San Francisco, 247
Watson's, 247
Willow Weed, 147
Wind Flower, 169
Wintergreen Family, 263
Wintergreen, 263
Wiregrass, 113
Wolffiella
    Tongue-shaped, 112
Wood Sorrel Family, 223
Wood Sorrel, 223
    Creeping, 224

Hairy, 224
Oregon, 224
Woodland Star, 191
Woolly-heads
    Dwarf, 356
    Round, 356
    Slender, 356
Wormseed, 153
Wormwood, 372
Wrack, 67

Yampah
    Gairdner's, 258
    Kellogg's, 258
Yarrow, 370

Common, 370
    Yellow, 366
Yellow Mats, 256
Yellow-eyed Grass, 127
Yerba Buena, 301
Yerba de Selva, 192
Yerba Mansa, 132
Yerba Reuma, 240
Yerba Santa, 287
Yew Family, 62
Youth-on-Age, 190

Zygadene, 116
    Deadly, 116
    Fremont's, 116

Index of Scientific Plant Names

Abronia, 155
  latifolia, 155
  umbellata, 155
Acacia, 202
  baileyana, 202
  decurrens, 202
  longifolia, 202
  melanoxylon, 202
  retinodes, 202
  verticillata, 202
Acaena, 196
  californica, 197
  sanguisorbae, 197
Acanthomintha obovata dut-
    tonii, 296
Acer, 232
  macrophyllum, 232
  negundo californicum, 232
Aceraceae, 232
Achillea, 370
  millefolium arenicola, 370
  millefolium californica, 370
Achyrachaena mollis, 359
Actaea, 167
  arguta, 167
  rubra arguta, 167
Adenocaulon bicolor, 354
Adenostoma fasciculatum, 198
Adiantum, 60
  jordani, 60
  pedatum aleuticum, 60
Aesculus californica, 232
Agoseris, 342
  apargioides, 342
  apargioides eastwoodiae, 342
  grandiflora, 343
  heterophylla, 342
Agropyron, 89
  arenicola, 88
  elongatum, 89
  junceum, 89
  pauciflorum, 89
  repens, 89
  tenerum pauciflorum, 89
  trachycaulum, 89
Agrostideae, 93
Agrostis, 96
  alba, 97
  avenacea, 97
  californica, 97
  diegoensis, 97
  exarata, 97
  exarata pacifica, 97
  hallii, 97
  hiemalis, 97
  microphylla, 97
  pallens, 97
  palustris, 97

retrofracta, 97
  scabra, 97
  semiverticillata, 97
  stolonifera, 97
  tenuis, 97
  verticillata, 97
Ailanthus altissima, 226
Aira, 93
  caespitosa, 92
  caryophyllea, 93
  danthonioides, 92
  elongata, 92
  holciformis, 92
Aizoaceae, 156
Albizia, 202
  distachya, 202
  lophantha, 202
Alchemilla occidentalis, 196
Alisma, 69
  plantago-aquatica, 69
Alismataceae, 69
Allium, 123
  amplectens, 124
  breweri, 124
  dichlamydeum, 124
  hyalinum, 124
  lacunosum, 124
  neapolitanum, 123
  serratum, 124
  triquetrum, 123
  unifolium, 124
Allocarya, 291
  bracteata, 292
  californica, 292
  chorisiana, 292
  chorisiana hickmanii, 292
  chorisiana myriantha, 292
  cusickii vallicola, 292
  diffusa, 292
  glabra, 291
  undulata, 292
Allophyllum divaricatum, 283
Alnus, 134
  oregona, 134
  rhombifolia, 134
  rubra, 134
Alopecurus, 94
  howellii, 94
  pratensis, 94
Althaea rosea, 237
Alyssum, 187
  alyssoides, 187
  maritimum, 187
Amaranthaceae, 154
Amaranthus, 154
  albus, 154
  blitoides, 155
  californicus, 155

deflexus, 154
  gracilis, 154
  graecizans, 154, 155
  hybridus, 154
  powellii, 154
  retroflexus, 154
Amaryllidaceae, 123
Ambrosia, 364
  artemisiifolia, 364
  psilostachya, 364
  psilostachya californica, 364
Ambrosieae, 364
Amelanchier pallida, 201
Ammannia coccinea, 244
Ammi visnaga, 260
Ammophila, 97
  arenaria, 97
  breviligulata, 98
Amsinckia, 289
  intermedia, 289
  lycopsoides, 289
  lunaris, 290
  menziesii, 289
  retrorsa, 289
  spectabilis, 289
Amygdalaceae, 200
Anacardiaceae, 230
Anacharis canadensis, 70
Anagallis, 269
  arvensis, 269
  arvensis azurea, 269
  arvensis caerulea, 269
  minima, 269
Anaphalis margaritacea, 354
Anchusa sempervirens, 288
Andropogoneae, 102
Androsace, 269
  acuta, 269
  elongata acuta, 269
Anemone quinquefolia grayi,
    169
Anemopsis californica, 132
Angelica, 261
  hendersonii, 262
  tomentosa, 262
Angiospermae, 64
Anthemideae, 370
Anthemis cotula, 370
Anthoxanthum odoratum, 98
Anthriscus, 257
  neglecta scandix, 257
  scandicina, 257
  vulgaris, 257
Antirrhinum, 308
  elmeri, 309
  kelloggii, 308
  majus, 308
  multiflorum, 308

orontium, 309
vexillo-calyculatum, 309
Apiastrum angustifolium, 256
Apium graveolens, 259
Apocynaceae, 272
Apocynum, 273
　cannabinum, 273
　cannabinum floribundum,
　　273
　cannabinum pubescens, 273
　medium floribundum, 273
　pumilum, 273
　pumilum rhomboideum, 273
Aponogeton distachyus, 67
Aponogetonaceae, 67
Aquilegia, 167
　eximia, 167
　formosa truncata, 167
Arabis, 181
　blepharophylla, 182
　breweri, 182
　glabra, 182
Araceae, 111
Aralia californica, 252
Araliaceae, 252
Arbutus menziesii, 265
Arceuthobium campylopodum,
　140
Arctium, 376
　lappa, 376
　minus, 376
Arctostaphylos, 265
　andersonii, 267
　andersonii imbricata, 267
　canescens, 267
　crustacea, 268
　crustacae rosei, 268
　crustacea tomentosiformis,
　　268
　franciscana, 267
　glandulosa, 267
　glauca, 267
　glutinosa, 267
　hookeri, 267
　hookeri franciscana
　nummularia sensitiva, 267
　sensitiva, 267
　silvicola, 267
Arenaria, 163
　californica, 163
　douglasii, 163
　macrophylla, 163
　paludicola, 163
　pusilla diffusa, 163
Aristolochia californica, 140
Aristolochiaceae, 140
Armeria, 270
　arctica californica, 270
　maritima californica, 270
Arnica, 373
　discoidea, 373
　discoidea alata, 373

Arrhenatherum elatius, 91
Artemisia, 372
　biennis, 372
　californica, 372
　douglasiana, 372
　dracunculoides, 372
　dracunculus, 372
　pycnocephala, 372
Arundo donax, 380
Asarina stricta, 308
Asarum caudatum, 140
Asclepiadaceae, 274
Asclepias, 274
　fascicularis, 274
　mexicana, 274
Asparagus officinalis, 121
Aspidotis californica, 61
Aspris caryophyllea, 93
Aster, 350
　chilensis, 351
　chilensis sonomensis, 351
　exilis, 351
　radulinus, 351
　subspicatus, 351
Astereae, 345
Astragalus, 217
　franciscanus, 217
　gambellianus, 217
　menziesii virgatus, 217
　nigrescens, 217
　nuttallii virgatus, 217
　pycnostachyus, 217
　tener, 217
Asyneuma prenanthoides, 330
Athyrium, 59
　filix-femina, 59
　filix-femina californicum, 59
　filix-femina cyclosorum, 59
Athysanus pusillus, 184
Atriplex, 149
　californica, 150
　expansa mohavensis, 150
　hastata, 150
　hortensis, 150
　joaquiniana, 150
　lentiformis breweri, 149
　leucophvlla, 149
　patula, 150
　patula hastata, 150
　patula spicata, 150
　rosea, 150
　semibaccata, 149
　serenana, 150
Atropa belladonna, 304
Avena, 90
　barbata, 90
　brevis, 90
　fatua, 90
　fatua glabrata, 90
　sativa, 90
Aveneae, 90
Azalea californica, 264

Azolla filiculoides, 55

Baccharis, 345
　douglasii, 346
　pilularis, 345
　pilularis consanguinea, 346
　viminea, 346
Baeria, 368
　chrysostoma, 368
　chrysostoma gracilis, 369
　chrysostoma hirsutula, 369
　macrantha thalassophila, 368
　microglossa, 369
　minor, 368
　uliginosa, 368
Bambuseae, 75
Barbarea, 183
　americana, 183
　orthoceras, 183
　verna, 183
Basellaceae, 160
Bassia hyssopifolia, 151
Beckmannia, 98
　erucaeformis, 98
　syzigachne, 98
Bellardia trixago, 316
Bellis perennis, 349
Berberidaceae, 171
Berberis, 171
　aquifolium, 172
　darwinii, 172
　nervosa, 172
　pinnata, 172
Berula erecta, 259
Beta vulgaris, 148
Betulaceae, 134
Bidens, 358
　cernua, 358
　laevis, 358
Blechnum spicant, 59
Blennosperma nanum, 367
Boisduvalia, 247
　densiflora, 247
　densiflora imbricata, 247
　glabella, 248
　stricta, 248
Boraginaceae, 287
Borago officinalis, 288
Boschniakia strobilacea, 321
Botrychium, 56
　multifidum intermedium, 56
　multifidum silaifolium, 56
　silaifolium californicum. 56
Boussingaultia gracilis, 160
Bowlesia, 258
　incana, 258
　lobata, 258
Boykinia elata, 190
Brassica, 178
　adpressa, 179
　campestris, 179
　fruticulosa, 180

geniculata, 179
hirta, 179
juncea, 180
kaber, 179
nigra, 180
oleracea, 179
Brickellia californica, 345
Briza, 76
  maxima, 77
  minor, 77
Brodiaea, 125
  appendiculata, 126
  capitata, 125
  congesta, 125
  coronaria, 126
  coronaria macropoda, 126
  elegans, 126
  hyacinthina, 126
  laxa, 125
  lutea, 125
  multiflora, 125
  pulchella, 125
  terrestris, 126
Bromus, 77
  arenarius, 79
  breviaristatus, 78
  carinatus, 78
  catharticus, 78
  commutatus, 79
  grandis, 79
  hordeaceus, 79
  laevipes, 79
  madritensis, 79
  marginatus, 79
  maritimus, 78
  mollis, 79
  pseudolaevipes, 79
  racemosus, 79
  rigidus, 80
  rubens, 79
  stamineus, 78
  tectorum, 80
  tectorum glabratus, 80
  trinii, 78
  unioloides, 78
  vulgaris, 79
  willdenowii, 78
Bryonia dioica, 330
Buddleia davidii, 271

Caesalpiniaceae, 202
Cakile, 178
  edentula californica, 178
  maritima, 178
Calais lindleyi, 340
Calamagrostis, 95
  koelerioides, 96
  nutkaensis, 96
  rubescens, 95
Calamophyta, 54
Calandrinia, 158
  breweri, 158

ciliata menziesii, 158
Calendula officinalis, 369
Calenduleae, 369
Calliprora ixioides, 126
Callitrichaceae, 229
Callitriche, 229
  autumnalis, 229
  bolanderi, 229
  hermaphroditica, 229
  heterophylla bolanderi, 229
  marginata, 229
  palustris, 229
  verna, 229
Calochortus, 119
  albus, 119
  albus rubellus, 119
  luteus, 120
  tolmiei, 119
  umbellatus, 119
  uniflorus, 119
  venustus, 120
Calycadenia, 364
  multiglandulosa cephalotes, 364
  multiglandulosa robusta, 364
Calyptridium parryi hesseae, 160
Campanula, 331
  angustiflora, 331
  californica, 331
  linnaeifolia, 331
  prenanthoides, 330
Campanulaceae, 330
Caprifoliaceae, 325
Capsella bursa-pastoris, 185
Cardamine oligosperma, 183
Cardaria draba, 186
Cardionema ramosissimum, 156
Carduus, 380
  pycnocephalus, 380
  tenuiflorus, 380
Carex, 103
  amplifolia, 106
  barbarae, 106
  bifida, 107
  bolanderi, 105
  brevicaulis, 106
  comosa, 107
  cusickii, 105
  densa, 105
  dudleyi, 105
  exsiccata, 107
  globosa, 106
  gracilior, 106
  gynodynama, 106
  harfordii, 106
  hassei, 106
  lanuginosa, 106
  leptopoda, 106
  monterevensis, 106
  nudata, 106
  obnupta, 106

pansa, 105
phyllomanica, 105
praegracilis, 105
rostrata, 107
salinaeformis, 106
schottii, 106
senta, 106
serratodens, 107
simulata, 105
subbracteata, 106
subfusca, 106
teneraeformis, 106
tumulicola, 105
vicaria, 105
Carthamus lanatus, 377
Carum kelloggii, 258
Caryophyllacea, 160
Cassia tomentosa, 202
Castanopsis chrysophylla minor, 135
Castilleja, 318
  affinis, 320
  foliolosa, 319
  franciscana, 319
  latifolia, 319
  latifolia wightii, 320
  miniata, 319
  stenantha, 319
  wightii rubra, 320
Caucalis microcarpa, 257
Ceanothus, 234
  andersonii, 235
  cuneatus dubius, 235
  dentatus, 235
  ferrisae, 236
  foliosus, 235
  incanus, 234
  integerrimus, 235
  papillosus, 235
  ramulosus, 236
  × regius, 235
  sorediatus, 235
  thyrsiflorus, 235
  velutinus laevigatus, 235
Cedronella canariensis, 299
Celastraceae, 231
Cenchrus pauciflorus, 100
Centaurea, 376
  calcitrapa, 377
  cineraria, 376
  cyanus, 376
  diluta, 376
  melitensis, 377
  pratensis, 377
  solstitialis, 377
Centaurium, 272
  davyi, 272
  floribundum, 272
  muhlenbergii, 272
  trichanthum, 272
Centranthus ruber, 328
Centromadia

congdonii, 363
fitchii, 363
maritima, 363
parryi, 363
Centunculus minimus, 269
Cephalanthera austinae, 131
Cerastium, 161
 arvense, 161
 arvense maximum, 161
 glomeratum, 161
 holosteoides, 161
 viscosum, 161
 vulgatum, 161
Ceratophyllaceae, 166
Ceratophyllum demersum, 166
Cercocarpus betuloides, 197
Chaenactis, 367
 glabriuscula gracilenta, 367
 tanacetifolia, 367
Chaetochloa
 geniculata, 100
 lutescens, 100
 viridis, 100
Chaetopappa, 349
 alsinoides, 350
 bellidiflora, 349
 exilis, 350
Chasmanthe aethiopica, 127
Cheilanthes, 61
 californica, 61
 cooperae, 61
 intertexta, 61
Chenopodiaceae, 148
Chenopodium, 151
 album, 153
 ambrosioides, 152
 ambrosioides anthelminticum, 153
 ambrosioides chilense, 152
 ambrosioides vagans, 152
 berlandieri, 153
 botrys, 152
 californicum, 153
 carinatum, 152
 farinosum, 153
 humile, 153
 macrospermum farinosum,
  153
 multifidum, 153
 murale, 153
 pumilio, 152
 rubrum humile, 153
 strictum glaucophyllum, 153
Chlorideae, 98
Chlorogalum, 117
 pomeridianum, 117
 pomeridianum divaricatum,
  117
Chlorophytum capense, 117
Chorizanthe, 141
 cuspidata, 142
 cuspidata marginata, 142

diffusa, 142
 membranacea, 141
 polygonoides, 141
 pungens hartwegii, 142
 robusta, 141
Chrysanthemum, 370
 coronarium, 370
 leucanthemum, 370
 maximum, 370
 parthenium, 370
 segetum, 370
Chrysopsis, 348
 oregona rudis, 348
 villosa bolanderi, 348
 villosa camphorata, 348
 villosa echioides, 348
Chrysothamnus nauseosus
  occidentalis, 354
Cichorieae, 339
Cichorium, 339
 endivia, 339
 intybus, 339
Cicuta, 259
 bolanderi, 259
 douglasii, 259
Cirsium, 378
 amplifolium, 379
 andrewsii, 379
 arvense, 378
 brevistylum, 379
 breweri, 378
 coulteri, 379
 douglasii, 378
 edule, 379
 fontinale, 378
 occidentale, 379
 praeteriens, 379
 proteanum, 379
 quercetorum, 379
 quercetorum xerolepis, 379
 remotifolium, 379
 scabrum, 378
 vulgare, 378
Cistaceae, 240
Citrullus vulgaris, 330
Clarkia, 248
 affinis, 250
 breweri, 249
 concinna, 249
 davyi, 249
 elegans, 249
 epilobioides, 249
 franciscana, 250
 modesta, 249
 purpurea quadrivulnera, 250
 purpurea viminea, 249
 rhomboidea, 249
 rubicunda, 249
 rubicunda blasdalei, 249
 unguiculata, 249
Clematis, 169
 lasiantha, 169

ligusticifolia, 169
 vitalba, 169
Clintonia andrewsiana, 122
Cnicus benedictus, 377
Collinsia, 309
 bartsiaefolia hirsuta, 309
 bicolor, 309
 corymbosa, 309
 franciscana, 310
 heterophylla, 309
 multicolor, 309
 solitaria, 309
 sparsiflora collina, 310
 sparsiflora franciscana, 310
 tinctoria, 309
Collomia, 281
 grandiflora, 281
 heterophylla, 281
Commelinaceae, 113
Compositae, 332
Conioselinum chinense, 262
Conium maculatum, 260
Convallariaceae, 120
Convolvulaceae, 274
Convolvulus, 275
 arvensis, 276
 malacophyllus collinus, 275
 occidentalis, 275
 occidentalis purpuratus, 275
 occidentalis solanensis, 275
 soldanella, 275
 subacaulis, 275
Conyza, 351
 bilbaoana, 351
 bonariensis, 352
 canadensis, 351
 canadensis glabrata, 351
 coulteri, 351
Coprosma repens, 323
Corallorhiza, 131
 maculata, 131
 striata, 131
Cordylanthus, 316
 maritimus, 316
 mollis, 316
 pilosus, 316
 rigidus, 316
Coreopsis, 358
 lanceolata, 358
 tinctoria, 358
Corethrogyne, 350
 californica, 350
 filaginifolia, 350
 filaginifolia rigida, 350
 leucophylla, 350
Coriandrum sativum, 260
Cornaceae, 262
Cornus, 262
 californica, 262
 glabrata, 262
 nuttallii, 262
Coronopus, 185

didymus, 185
procumbens, 185
squamatus, 185
Cortaderia selloana, 75
Corylaceae, 135
Corylus, 135
californica, 135
cornuta californica, 135
rostrata californica, 135
Cotoneaster, 201
franchetii, 201
pannosa, 201
Cotula, 372
australis, 372
coronopifolia, 372
Crassulaceae, 188
Crepis, 344
bursifolia, 344
capillaris, 344
vesicaria taraxacifolia, 344
Cressa, 276
cretica truxillensis, 276
truxillensis vallicola, 276
Croton californicus, 227
Cruciferae, 176
Crypsis nilaca, 93
Cryptantha, 290
flaccida, 291
hispidissima, 291
leiocarpa, 291
micromeres, 290
microstachys, 291
muricata jonesii, 290
torreyana pumila, 291
Cucurbita pepo, 330
Cucurbitaceae, 329
Cupressaceae, 63
Cupressus, 63
abramsiana, 64
macrocarpa, 64
Cuscuta, 274
californica, 274
campestris, 275
ceanothi, 275
occidentalis, 274
salina major, 275
suaveolens, 275
subinclusa, 275
Cymbalaria muralis, 308
Cynara scolymus, 376
Cynareae, 376
Cynodon dactylon, 98
Cynoglossum grande, 288
Cynosurus, 75
cristatus, 76
echinatus, 76
Cyperaceae, 103
Cyperus, 107
alternifolius, 108
brevifolius, 107
difformis, 108
eragrostis, 108

erythrorhizos, 108
esculentus, 108
laevigatus, 107
melanostachyus, 108
niger capitatus, 107
rotundus, 380
strigosus, 108
vegetus, 108
Cypripedium, 129
fasciculatum, 129
montanum, 129
Cypsela humifusa, 156
Cystopteris fragilis, 59
Cytisus, 208
maderensis, 208
monspessulanus, 208
scoparius, 208

Dactylis glomerata, 77
Danthonia, 91
californica, 91
californica americana, 91
pilosa, 91
Datisca glomerata, 243
Datiscaceae, 243
Datura, 304
inoxia, 304
meteloides, 304
stramonium, 304
stramonium tatula, 304
Daucus, 257
carota, 257
pusillus, 257
Delphinium, 168
ajacis, 168
californicum, 169
decorum, 169
decorum patens, 169
hesperium, 169
hesperium seditosum, 169
nudicaule, 168
parryi seditosum, 169
patens, 169
variegatum, 169
variegatum superbum, 169
Dendromecon rigida, 173
Dentaria, 182
californica, 182
california integrifolia, 182
Deschampsia, 92
caespitosa, 92
caespitosa holciformis, 92
danthonioides, 92
danthonioides gracilis, 92
elongata, 92
Descurainia, 184
pinnata menziesii, 184
sophia, 184
Dicentra, 175
chrysantha, 175
formosa, 175
Dichelostemma

congestum, 125
multiflorum, 125
pulchellum, 125
Dichondra repens, 276
Dicotyledoneae, 132
Digitalis purpurea, 311
Digitaria, 101
ischaemum, 101
sanguinalis, 101
Diplacus aurantiacus, 311
Dipsacaceae, 327
Dipsacus, 327
fullonum, 327
sativus, 327
sylvestris, 327
Dirca occidentalis, 243
Disporum, 122
hookeri, 122
smithii, 122
Distichlis, 75
spicata nana, 75
spicata stolonifera, 75
Dodecatheon, 270
clevelandii patulum, 270
clevelandii sanctarum, 270
hendersonii, 270
hendersonii cruciatum, 270
patulum bernalinum, 270
Downingia, 331
concolor tricolor, 331
pulchella, 331
Dryopteris, 59
arguta, 59
dilatata, 59
Duchesnea indica, 195
Dudleya, 188
cymosa, 189
cymosa setchellii, 189
farinosa, 189
laxa, 189
setchellii, 189

Eburophyton austinae, 131
Echidiocarya californica, 293
Echinochloa, 102
crusgalli, 102
crusgalli frumentacea, 102
crusgalli mitis, 102
crusgalli zelayensis, 102
Echinocystis
fabaceus, 330
oreganus, 330
Echinodorus, 69
berteroi, 69
cordifolius, 69
Echinopsilon hyssopifolium, 151
Echium fatuosum, 288
Eichornia crassipes, 113
Elatinaceae, 239
Elatine, 239
brachysperma, 239

triandra brachysperma, 239
Eleocharis, 108
  acicularis, 109
  acicularis radicans, 109
  macrostachya, 109
  montana, 109
  montevidensis, 109
  obtusa, 109
  palustris, 109
  radicans, 109
  rostellata, 109
  rostellata congdonii, 109
Ellisia
  aurita, 283
  membranacea, 283
Elodea, 70
  canadensis, 70
  densa, 70
Elymus, 87
  caput-medusae, 87
  condensatus, 88
  glaucus, 87
  glaucus jepsonii, 87
  glaucus virescens, 88
  mollis, 88
  pacificus, 88
  triticoides, 88
  triticoides multiflorus, 88
  × vancouverensis, 88
  virescens, 88
Emex australis, 143
Emmenanthe penduliflora, 286
Epilobium, 246
  adenocaulon occidentale, 247
  adenocaulon parishii, 247
  californicum, 247
  californicum occidentale,
    247
  franciscanum, 247
  halleanum, 247
  minutum, 246
  paniculatum, 246
  paniculatum adenocladon,
    247
  paniculatum jucundum, 247
  paniculatum laevicaule, 247
  watsonii, 247
  watsonii franciscanum, 247
Epipactis gigantea, 129
Equisetaceae, 54
Equisetum, 54
  arvense, 54
  × ferrissii, 55
  funstoni, 55
  hyemale affine, 55
  hyemale californicum, 54
  hyemale elatum, 55
  hyemale robustum, 55
  kansanum, 55
  laevigatum, 55
  praeltum, 55
  telmateia braunii, 54

Eragrostis, 76
  caroliniana, 76
  cilianensis, 76
  diffusa, 76
  hypnoides, 76
  megastachya, 76
  orcuttiana, 76
  pilosa, 76
Erechtites
  arguta, 375
  prenanthoides, 375
Eremocarpus setigerus, 227
Eremochloa ciliaris, 102
Eriastrum abramsii, 280
Ericaceae, 264
Ericameria
  arborescens, 347
  ericoides, 347
Erigeron, 352
  canadensis, 351
  crispus, 352
  foliosus, 352
  foliosus hartwegii, 352
  glaucus, 352
  inornatus angustatus, 352
  karvinskianus, 352
  linifolium, 352
  petrophilus, 352
  philadelphicus, 352
Eriodictyon californicum, 287
Eriogonum, 142
  argillosum, 143
  fasciculatum foliolosum, 143
  hirtiflorum, 143
  latifolium, 142
  latifolium decurrens, 143
  latifolium nudum, 143
  nudum, 142
  saxatile, 143
  vimineum, 143
  vimineum caninum, 143
  virgatum, 143
Eriophorum
  gracile, 109
Eriophyllum, 366
  confertiflorum, 366
  lanatum achillaeoides, 367
  lanatum arachnoideum, 367
  staechadifolium, 366
  *staechadifolium artemisiae-
    folium*, 366
Erodium, 223
  botrys, 223
  cicutarium, 223
  macrophyllum, 223
  moschatum, 223
  obtusiplicatum, 223
Erucastrum gallicum, 180
Eryngium, 254
  aristulatum, 254
  armatum, 254
Erysimum, 180

ammophilum, 181
  capitatum, 181
  filifolium, 181
  franciscanum, 181
  franciscanum crassifolium,
    181
  teretifolium, 181
Erythea edulis, 111
Escallonia macrantha, 192
Escalloniaceae, 192
Eschscholzia californica, 174
Eucalyptus, 244
  globulus, 244
  × mortoniana, 244
  viminalis, 244
Eucrypta chrysanthemifolia,
    284
Eugenia apiculata, 244
Euonymus occidentalis, 231
Eupatorieae, 345
Eupatorium adenophorum, 345
Euphorbia, 228
  crenulata, 229
  exigua, 228
  helioscopia, 228
  lathyris, 228
  maculata, 229
  peplus, 228
  serpens, 229
  serpyllifolia, 229
  spathulata, 228
  supina, 229
Euphorbiaceae, 227
Evax sparsiflora, 356

Fabaceae, 203
Fagaceae, 135
Festuca, 83
  arundinacea, 85
  bromoides, 84
  californica, 85
  confusa, 84
  dertonensis, 84
  eastwoodae, 84
  elatior, 85
  elmeri, 85
  elmeri luxurians, 85
  grayi, 84
  idahoensis, 85
  megalura, 85
  microstachys, 85
  myuros, 84
  occidentalis, 85
  octoflora, 84
  octoflora hirtella, 84
  pacifica, 84
  reflexa, 85
  rubra, 85
  subuliflora, 85
Festuceae, 75
Filago, 356
  californica, 356

gallica, 356
Filicinae, 55
Fimbristylis miliacea, 109
Foeniculum vulgare, 258
Fragaria, 195
  californica, 195
  chiloensis, 195
  indica, 195
Frankenia grandifolia, 240
Frankeniaceae, 240
Franseria, 365
  acanthicarpa, 366
  chamissonis, 365
  chamissonis bipinnatisecta, 366
  confertiflora, 366
Fraxinus, 271
  dipetala, 271
  latifolia, 271
Freesia refracta, 129
Fremontia californica crassifolia, 238
Fremontodendron californicum crassifolium, 238
Fritillaria, 118
  agrestis, 119
  biflora, 119
  biflora inflexa, 119
  lanceolata, 118
  lanceolata floribunda, 119
  liliacea, 119
  mutica, 119
Fumaria, 175
  officinalis, 175
  parviflora, 175
Fumariaceae, 174

Galinsoga parviflora, 359
Galium, 323
  andrewsii, 324
  aparine, 325
  californicum, 324
  mollugo, 324
  murale, 324
  nuttallii, 324
  parisiense, 325
  spurium echinospermum, 325
  tinctorium subbiflorum, 324
  tricorne, 324
  tricornutum, 324
  trifidum pacificum, 324
  trifidum subbiflorum, 324
  triflorum, 325
  vaillantii, 325
  verum, 324
Garrya, 262
  elliptica, 263
  fremontii, 263
Garryaceae, 262
Gastridium ventricosum, 95
Gaultheria shallon, 265

Gaura, 245
  odorata, 245
  sinuata, 245
Gentianaceae, 272
Geraniaceae, 221
Geranium, 222
  anemonifolium, 223
  carolinianum, 222
  dissectum, 222
  molle, 222
  pusillum, 222
  retrorsum, 222
  robertianum, 223
Gilia, 281
  achilleaefolia, 282
  achilleaefolia chamissonis, 281
  achilleaefolia staminea, 282
  capitata, 281
  capitata chamissonis, 281
  capitata staminea, 282
  clivorum, 282
  divaricata, 283
  gilioides volcanica, 283
  gracilis, 277
  multicaulis clivorum, 282
  peduncularis, 282
  tenuiflora, 282
  tricolor, 282
Githopsis specularioides, 331
Glaucium flavum, 174
Glaux maritima, 269
Glecoma hederacea, 299
Glyceria, 82
  fluitans, 82
  leptostachya, 82
  occidentalis, 82
  pauciflora, 82
Glycyrrhiza lepidota glutinosa, 216
Gnaphalium, 354
  beneolens, 355
  bicolor, 355
  californicum, 355
  chilense, 355
  luteo-album, 355
  palustre, 355
  purpureum, 355
  ramosissimum, 355
Godetia
  blasdalei, 249
  epilobioides, 249
  quadrivulnera, 250
  quadrivulnera davyi, 249
  viminea, 249
Goodyera, 131
  decipiens, 131
  oblongifolia, 131
Gramineae, 70
Gratiola ebracteata, 313
Grindelia, 346
  arenicola pachyphylla, 347

camporum, 346
  hirsutula, 346
  hirsutula rubicaulis, 346
  humilis, 346
  latifolia, 346
  latifolia platyphylla, 346
  maritima, 346
  maritima anomala, 346
  stricta venulosa, 347
Grossularia, 193
  californica, 193
  divaricata, 193
  leptosma, 193
  menziesii, 193
  senilis,·193
Grossulariaceae, 192
Guizotia abyssinica, 359
Gymnogramme triangularis, 57
Gymnospermae, 62

Habenaria, 130
  dilatata leucostachys, 130
  elegans, 131
  elegans maritima, 131
  greenei, 131
  michaelii, 131
  sparsiflora, 130
  unalascensis, 130
  unalascensis elata, 131
  unalascensis maritima, 130
Haloragidaceae, 251
Haplopappus, 347
  arborescens, 347
  ciliatus, 347
  ericoides, 347
  ericoides blakei, 347
  racemosus, 347
  racemosus longifolius, 347
  venetus vernonioides, 347
Hebe
  × *franciscana,* 315
  speciosa, 315
Hedera helix, 252
Hedypnois cretica, 339
Helenieae, 366
Helenium puberulum, 367
Heliantheae, 357
Helianthella, 357
  californica, 357
  cannonae, 358
  castanea, 358
Helianthemum, 240
  scoparium, 240
  scoparium vulgare, 240
Helianthus, 359
  annuus, 359
  annuus lenticularis, 359
  annuus macrocarpus, 359
  bolanderi, 359
  californicus, 359
Heliotropium curassavicum oculatum, 290

Helxine soleirollii, 139
Hemitomes congestum, 263
Hemizonia, 362
    clevelandii, 364
    congesta, 364
    corymbosa, 363
    fasciculata, 363
    fitchii, 363
    kelloggii, 363
    lobbii, 363
    luzulaefolia, 364
    luzulaefolia rudis, 364
    parryi, 363
    parryi congdonii, 363
    pentactis, 363
    pungens maritima, 363
Heracleum, 260
    lanatum, 260
    maximum, 260
Herniaria cinerea, 156
Hesperocnide tenella, 138
Hesperolinon, 225
    congestum, 226
    micranthum, 225
    spergulinum, 225
Hesperomecon
    linearis, 174
Hesperoscordum hyacinthi-
    num, 126
Heterocodon rariflorum, 331
Heteromeles arbutifolia, 201
Heterosporium, 287
Heterotheca grandiflora, 347
Heuchera, 190
    hartwegii, 190
    micrantha, 190
    micrantha pacifica, 190
    pilosissima, 190
Hieracium albiflorum, 344
Hierochloe, 98
    macrophylla, 98
    occidentalis, 98
Hippocastanaceae, 232
Hippuris vulgaris, 252
Holcus, 92
    halepensis, 102
    lanatus, 92
Holocarpha macrandenia, 364
Holodiscus, 197
    discolor, 197
    discolor franciscanus, 197
Hordeae, 85
Hordeum, 85
    brachyantherum, 86
    californicum, 86
    depressum, 86
    glaucum, 86
    gussoneanum, 86
    hystrix, 86
    jubatum caespitosum, 86
    leporinum, 86
    murinum, 86

    stebbinsii, 86
    vulgare, 86
Horkelia, 195
    bolanderi parryi, 195
    californica, 195
    cuneata, 195
    cuneata sericea, 195
    frondosa, 195
    marinensis, 196
Hosackia
    americana, 209
    brachycarpa, 210
    crassifolia, 209
    cytisoides, 210
    glabra, 210
    gracilis, 209
    juncea, 210
    maritima, 209
    parviflora, 209
    stipularis, 209
    strigosa, 209
    subpinnata, 210
    tomentosa, 210
    torreyi, 209
Howellia limosa, 331
Hugelia abramsii, 280
Hulsea heterochroma, 367
Hutchinsia procumbens, 185
Hyacinthus orientalis, 117
Hydastylus californicus, 127
Hydrangeaceae, 192
Hydrocharitaceae, 70
Hydrocotyle, 257
    prolifera, 258
    ranunculoides, 258
    sibthorpioides, 258
    verticillata, 258
    verticillata triradiata, 258
Hydrophyllaceae, 283
Hypericaceae, 238
Hypericum, 238
    anagalloides, 239
    calycinum, 239
    formosum scouleri, 239
    perforatum, 239
    scouleri, 239
Hypochaeris, 341
    glabra, 341
    radicata, 341
Hystrix, 87
    californica, 87

Iberis umbellata, 185
Ibidium
    porrifolium, 131
    romanzoffianum, 131
Illecebraceae, 155
Inuleae, 354
Ipomoea, 276
    mutabilis, 276
    purpurea, 276
Iridaceae, 127

Iris, 128
    californica, 128
    douglasiana, 128
    fernaldii, 128
    longipetala, 128
    macrosiphon, 128
    pseudacorus, 129
Isocoma veneta vernonioides,
    347
Isoetaceae, 53
Isoetes, 53
    howellii, 53
    nuttallii, 53
Isopyrum occidentale, 167
Iva axillaris, 364
Ixia maculata, 129

Jaumea carnosa, 368
Juncaceae, 113
Juncaginaceae, 68
Juncus, 113
    acuminatus, 115
    balticus, 114
    balticus montanus, 115
    bolanderi, 115
    bolanderi riparius, 115
    bufonius, 115
    bufonius congestus, 115
    bufonius fasciculatus, 115
    capitatus, 115
    dubius, 115
    effusus brunneus, 114
    effusus pacificus, 114
    falcatus, 115
    kelloggii, 115
    leseurii, 114
    mexicanus, 114
    occidentalis, 115
    patens, 114
    phaeocephalus, 115
    phaeocephalus glomeratus,
        115
    phaeocephalus paniculatus,
        115
    tenuis, 115
    tenuis congesta, 115
    xiphioides, 115
    xiphioides auratus, 115
Jussiaea, 245
    californica, 245
    repens peploides, 245

Kickxia, 308
    elatine, 380
    spuria, 308
Kniphofia uvaria, 118
Kochia, 151
    scoparia, 151
    scoparia subvillosa, 151
Koeleria, 91
    cristata, 92
    gerardi, 92

gracilis, 92
  macrantha, 92
Koniga maritima, 187

Labiatae, 295
Lactuca, 343
  saligna, 343
  serriola, 343
  serriola integrata, 343
  serriola integrifolia, 343
  virosa, 343
Lagophylla, 359
  congesta, 360
  ramosissima, 359
Lagurus ovatus, 95
Lamarckia, 76
  aurea, 76
Lamium, 299
  amplexicaule, 299
  purpureum, 299
Lapsana communis, 339
Lasthenia, 369
  glaberrima, 369
  glabrata, 369
Lathyrus, 220
  cicera, 220
  jepsonii californicus, 220
  latifolius, 220
  littoralis, 220
  odoratus, 220
  tingitanus, 220
  torreyi, 220
  vestitus, 220
  vestitus bolanderi, 221
  vestitus puberulus, 221
  violaceus, 221
  watsonii, 220
Lauraceae, 172
Laurus nobilis, 173
Lavatera, 237
  arborea, 237
  assurgentiflora, 237
  cretica, 237
Layia, 360
  carnosa, 360
  chrysanthemoides, 360
  gaillardioides, 360
  hieracioides, 360
  platyglossa, 360
  platyglossa campestris, 360
Ledum glandulosum colum-
    bianum, 264
Legenere limosa, 331
Lemna, 112
  cyclostasa, 112
  gibba, 112
  minima, 112
  minor, 112
  trisulca, 112
  valdiviana, 112
Lemnaceae, 112
Lens culinaris, 221

Lentibulariaceae, 320
Leontodon, 341
  leyseri, 341
  nudicaulis, 341
Lepechinia calycina, 300
Lepidium, 186
  bipinnatifidum, 186
  densiflorum, 186
  draba, 186
  latipes, 187
  nitidum, 186
  oxycarpum, 186
  perfoliatum, 186
  pinnatifidum, 186
  pubescens, 186
  strictum, 186
  virginicum pubescens, 186
Lepidophyta, 53
Leptochloa fascicularis, 98
Leptospermum laevigatum, 244
Leptotaenia californica, 261
Lepturus cylindricus, 90
Lessingia, 352
  germanorum, 353
  germanorum parvula, 353
  germanorum tenuipes, 353
  hololeuca, 353
  hololeuca arachnoidea, 353
  micradenia arachnoidea, 353
  micradenia glabrata, 354
  ramulosa glabrata, 354
Lewisia rediviva, 160
Leycesteria formosa, 327
Libertia formosa, 127
Ligusticum apiifolium, 260
Ligustrum ovalifolium, 271
Lilaea, 68
  scilloides, 68
  subulata, 68
Lilaeaceae, 68
Liliaceae, 117
Lilium pardalinum, 118
Limnanthaceae, 230
Limnanthes, 230
  douglasii, 230
  douglasii nivea, 230
Limnorchis
  leucostachys, 130
  sparsiflora, 130
Limonium, 270
  californicum, 271
  perfoliatum, 271
Limosella aquatica, 309
Linaceae, 224
Linanthus, 277
  acicularis, 279
  ambiguus, 278
  androsaceus, 278
  androsaceus croceus, 279
  androsaceus luteus, 279
  bicolor, 279
  ciliatus, 279

  dichotomus, 277
  grandiflorus, 278
  liniflorus, 278
  liniflorus pharnaceoides, 278
  longituba, 279
  parviflorus, 279
  pygmaeus, 277
  rattanii ambiguus, 278
  rosaceus, 279
Linaria, 307
  bipartita, 308
  canadensis texana, 308
  texana, 308
  vulgaris, 308
Lindernia anagallidea, 313
Linum, 225
  angustifolium, 225
  bienne, 225
  congestum, 226
  micranthum, 225
  spergulinum, 225
  usitatissimum, 225
Lippia nodiflora rosea, 294
Lithocarpus densiflora, 136
Lithophragma, 191
  affinis, 191
  heterophylla, 191
Lithospermum arvense, 294
Loasaceae, 242
Lobeliaceae, 331
Lobularia maritima, 187
Loeflingia squarrosa, 163
Loganiaceae, 271
Lolium, 88
  multiflorum, 89
  multiflorum muticum, 89
  multiflorum ramosum, 89
  perenne, 89
  perenne cristatum, 89
  temulentum, 89
  temulentum leptochaeton, 89
Lomatium, 260
  californicum, 261
  caruifolium, 261
  dasycarpum, 261
  macrocarpum, 261
  parvifolium, 261
  utriculatum, 261
Lonicera, 326
  hispidula, 326
  hispidula vacillans, 326
  interrupta, 326
  involucrata, 326
  japonica, 326
  ledebourii, 326
Loranthaceae, 139
Lotus, 208
  americanus, 209
  benthamii, 210
  corniculatus, 210
  crassifolius, 209
  eriophorus, 210

formosissimus, 209
heermannii eriophorus, 210
humistratus, 210
junceus, 210
micranthus, 209
oblongifolius nevadensis, 209
oblongifolius torreyi, 209
purshianus, 209
rubellus, 209
salsuginosus, 209
scoparius, 210
stipularis, 209
strigosus, 209
subpinnatus, 210
Ludwigia palustris pacifica, 251
Lunaria annua, 187
Lupinus, 204
affinis, 206
albifrons, 205
albifrons collinus, 206
apricus, 206
arboreus, 205
arboreus eximius, 205
bicolor, 206
bicolor microphyllus, 207
bicolor pipersmithii, 206
bicolor tridentatus, 207
bicolor trifidus, 206
bicolor umbellatus, 207
chamissonis, 205
densiflorus, 206
formosus, 205
formosus bridgesii, 205
franciscanus, 205
hirsutissimus, 206
latifolius, 205
latifolius dudleyi, 205
micranthus, 207
nanus, 206
nanus carnosulus, 206
nanus latifolius, 206
pachylobus, 206
polyphyllus grandifolius, 205
subvexus, 206
succulentus, 206
truncatus, 206
variicolor, 205
Luzula, 115
campestris congesta, 115
multiflora, 115
subsessilis, 115
Lychnis coronaria, 164
Lycium halimifolium, 302
Lycopersicon esculentum, 303
Lycopus americanus, 298
Lysichiton, 111
americanum, 111
kamtschatcensis, 111
Lythraceae, 243
Lythrum, 244

californicum, 244
hyssopifolia, 244

Madia, 361
capitata, 362
elegans densifolia, 362
elegans vernalis, 362
exigua, 362
gracilis, 362
madioides, 361
sativa, 362
Madieae, 359
Mahonia
aquifolium, 172
nervosa, 172
pinnata, 172
Maianthemum dilatum, 122
Malaceae, 201
Malacothamnus, 238
arcuatus, 238
hallii, 238
Malacothrix, 344
clevelandii, 344
floccifera, 344
Malvaceae, 236
Malva, 237
neglecta, 238
nicaeensis, 238
parviflora, 237
sylvestris, 238
Malvastrum
arcuatum, 238
hallii, 238
Marah, 330
fabaceus, 330
oreganus, 330
Marrubium vulgare, 297
Marsilea vestita, 55
Marsileaceae, 55
Martyniaceae, 380
Matricaria, 371
matricarioides, 371
occidentalis, 371
Matthiola incana, 180
Meconella, 174
californica, 174
linearis, 174
Meconopsis heterophylla, 174
Medicago, 211
apiculata, 211
arabica, 211
hispida, 211
hispida confinis, 211
lupulina, 211
lupulina cupaniana, 211
polymorpha vulgaris, 211
polymorpha vulgaris tuberculata, 211
sativa, 211
Melanthaceae, 116
Melica, 81
bulbosa, 81

californica, 81
geyeri, 81
harfordii, 82
imperfecta, 82
imperfecta refracta, 82
subulata, 81
torreyana, 81
Melilotus, 212
albus, 212
indicus, 212
officinalis, 212
Melissa officinalis, 296
Mentha, 298
arvensis lanata, 299
citrata, 299
piperita, 299
pulegium, 299
rotundifolia, 299
spicata, 299
Mentzelia, 242
dispersa, 243
lindleyi, 242
micrantha, 242
Menyanthaceae, 272
Menyanthes trifoliata, 272
Mercurialis annua, 228
Mesembryanthemum, 157
chilense, 157
cordifolium, 157
crassifolium, 157
edule, 157
elongatum, 157
floribundum, 157
nodiflorum, 157
Microcala quadrangularis, 272
Micromeria douglasii, 301
Micropus californicus, 356
Microseris, 339
bigelovii, 341
decipiens, 340
douglasii, 341
douglasii tenella, 341
elegans, 340
heterocarpa, 340
paludosa, 340
linearifolia, 340
Microsteris gracilis, 277
Mimetanthe pilosa, 312
Mimosaceae, 202
Mimulus, 312
androsaceus, 313
aurantiacus, 311
cardinalis, 313
cleistogamus, 313
congdonii, 313
douglasii, 313
floribundus, 313
glareosus, 313
guttatus, 312, 313
guttatus arenicola, 313
guttatus arvensis, 312
guttatus depauperatus, 313

guttatus gracilis, 313
guttatus grandis, 313
guttatus litoralis, 313
guttatus micranthus, 312
langsdorfii, 313
moschatus, 313
moschatus longiflorus, 313
moschatus sessifolius, 313
nasutus, 313
pilosus, 312
platycalyx, 313
rattanii, 313
rattanii decurtatus, 313
scouleri, 313
Mirabilis jalapa, 155
Modiola caroliniana, 236
Mollugo verticillata, 156
Monardella, 297
douglasii, 297
undulata, 297
villosa, 297
villosa franciscana, 298
villosa subglabra, 297
Monerma cylindrica, 90
Monocotyledoneae, 64
Monolepis nuttalliana, 151
Monolopia, 367
gracilens, 367
major, 367
Monotropaceae, 263
Montia, 158
exigua, 160
fontana, 159
gypsophiloides, 159
hallii, 159
parvifolia, 159
perfoliata, 159
sibirica, 159
spathulata, 159
tenuifolia, 160
verna, 159
Muhlenbeckia complexa, 147
Muilla maritima, 124
Muscari botryoides, 117
Myosotis, 288
latifolia, 288
sylvatica, 288
Myosurus, 170
minimus, 170
minimus filiformis, 170
Myrica californica, 133
Myricaceae, 133
Myriophyllum, 252
brasiliense, 252
exalbescens, 252
spicatum exalbescens, 252
verticillatum, 252
Myrtaceae, 244

Naiadaceae, 67
Naias guadalupensis, 67
Nasturtium officinale, 182

Navarretia, 280
abramsii, 280
atractyloides, 280
heterodoxa, 280
mellita, 280
squarrosa, 280
viscidula, 280
Nemophila, 283
aurita, 283
heterophylla, 284
membranacea, 283
menziesii, 284
menziesii atomaria, 284
parviflora, 284
pedunculata, 284
Nepeta, 299
cataria, 299
hederacea, 299
Newberrya congesta, 263
Nicandra physalodes, 305
Nicotiana, 304
acuminata multiflora, 304
alata, 305
bigelovii, 305
forgetiana, 305
glauca, 304
× sanderae, 305
sylvestris, 305
Nothoscordum, 126
fragrans, 126
inodorum, 126
Nuphar polysepalum, 166
Nyctaginaceae, 155
Nymphaea polysepala, 166
Nymphaeaceae, 166

Oenanthe sarmentosa, 259
Oenothera, 250
cheiranthifolia, 251
contorta epilobioides, 251
contorta strigulosa, 251
× erythrosepala, 251
graciliflora, 251
hookeri, 251
hookeri montereyensis, 251
micrantha, 251
micrantha jonesii, 251
ovata, 251
rosea, 251
Oleaceae, 271
Onagraceae, 245
Ophioglossaceae, 56
Orchidaceae, 129
Origanum vulgare, 301
Ornithopus, 216
pinnatus, 216
roseus, 216
Orobanchaceae, 320
Orobanche, 320
bulbosa, 321
californica, 321
fasciculata, 321

fasciculata franciscana, 321
grayana jepsonii, 321
grayana violacea, 321
pinorum, 321
uniflora, 321
uniflora minuta, 321
uniflora occidentalis, 321
uniflora purpurea, 321
uniflora sedi, 321
Orthocarpus, 316
attenuatus, 318
campestris, 318
castillejoides, 318
densiflorus, 318
erianthus, 318
erianthus roseus, 318
faucibarbatus albidus, 318
floribundus, 318
lithospermoides, 318
purpurascens, 318
purpurascens latifolius, 318
pusillus, 318
Oryzopsis miliacea, 94
Osmaronia cerasiformis, 200
Osmorhiza, 255
brachypoda, 255
chilensis, 255
nuda, 255
Oxalidaceae, 223
Oxalis, 223
cernua, 224
corniculata, 224
corniculata atropurpurea, 224
corniculata purpurea, 224
incarnata, 224
oregana, 224
pes-caprae, 224
pilosa, 224
rubra, 224

Palmae, 111
Paniceae, 100
Panicum, 101
capillare, 101
capillare occidentale, 101
miliaceum, 101
pacificum, 101
Papaver heterophyllum, 174
Papaveraceae, 173
Parapholis incurva, 90
Parietaria, 138
floridana, 138
judaiea, 138
Parnassia, 192
californica, 192
palustris californica, 192
Parnassiaceae, 192
Paronychia franciscana, 155
Pasania densiflora, 136
Paspalum, 102
dilatatum, 102

distichum, 102
Pastinaca sativa, 260
Pectocarya, 290
  penicillata, 290
  pusilla, 290
Pedicularis, 315
  densiflora, 315
  dudleyi, 316
Pelargonium, 221
  grossularioides, 221
  vitifolium, 221
  zonale, 221
Pellaea, 61
  andromedaefolia, 61
  mucronata, 61
Pennisetum, 100
  clandestinum, 100
  villosum, 100
Penstemon, 310
  corymbosus, 311
  heterophyllus, 311
  rattanii kleei, 311
Pentachaeta
  alsinoides, 350
  bellidiflora, 349
  exilis, 350
Peramium decipiens, 131
Perideridia, 258
  gairdneri, 258
  kelloggii, 258
Petasites frigidus palmatus,
  373
Petroselinum crispum, 258
Petunia, 305
  hybrida, 305
  parviflora, 305
Phacelia, 284
  californica, 285
  ciliata, 286
  distans, 286
  divaricata, 286
  douglasii, 286
  imbricata, 286
  magellanica calycosa, 286
  malvaefolia, 286
  nemoralis, 285
  ramosissima, 285
  rattanii, 286
  suaveolens, 286
Phalarideae, 98
Phalaris, 99
  angusta, 100
  californica, 99
  canariensis, 100
  lemmoni, 100
  minor, 100
  paradoxa, 99
  paradoxa praemorsa, 99
  tuberosa stenoptera, 99
Phleum, 95
  alpinum, 95
  pratense, 95

Phlox gracilis, 277
Phoenix canariensis, 111
Pholistoma, 283
  auritum, 283
  membranaceum, 283
Pholiurus incurvus, 90
Phoradendron, 139
  flavescens villosum, 139
  villosum, 139
Photinia arbutifolia, 201
Phragmites, 75
  communis, 75
  communis berlandieri, 75
  phragmites, 75
Phyla nodiflora rosea, 294
Phyllospadix, 67
  scouleri, 67
  torreyi, 67
Physalis, 303
  ixocarpa, 304
  pubescens, 304
Physocarpus capitatus, 197
Phytolacca, 155
  heterotepala, 155
  icosandra, 155
Phytolaccaceae, 155
Piaropus crassipes, 113
Pickeringia montana, 207
Picris echioides, 341
Pilularia americana, 56
Pinaceae, 62
Pinus, 62
  attenuata, 63
  × attenuradiata, 63
  ponderosa, 62
  radiata, 63
  sabiniana, 62
Piperia
  elegans, 131
  maritima, 131
  unalascensis, 130
Pisum sativum, 221
Pittosporaceae, 193
Pittosporum crassifolium, 193
Pityrogramma triangularis, 57
Plagiobothrys, 293
  bracteatus, 292
  californicus, 293
  campestris, 293
  canescens, 293
  chorisianus, 292
  chorisianus hickmanii, 292
  diffusus, 292
  fulvus campestris, 293
  glaber, 291
  nothofulvus, 293
  reticulatus rossianorum, 292
  tenellus, 293
  undulatus, 292
Plantaginaceae, 321
Plantago, 321
  bigelovii, 322

  coronopus, 322
  erecta, 322
  hirtella galeottiana, 322
  indica, 322
  hookeriana californica, 322
  juncoides, 322
  juncoides californica, 322
  lanceolata, 322
  major, 322
  maritima californica, 322
  maritima juncoides, 322
  subnuda, 322
Platanaceae, 193
Platanus racemosa, 193
Platystemon californicus, 174
Plectritis, 328
  anomala, 329
  aphanoptera, 329
  californica, 329
  ciliosa, 329
  ciliosa insignis, 329
  congesta, 328
  congesta brachystemon, 328
  congesta nitida, 329
  gibbosa, 329
  macrocera, 329
  macroptera, 329
  magna, 329
  samolifolia, 329
Pleuricospora fimbriolata, 263
Pleuropogon californicus, 82
Plumbaginaceae, 270
Poa, 80
  annua, 80
  bolanderi howellii, 80
  compressa, 80
  douglasii, 80
  howellii, 80
  kelloggii, 80
  nemoralis, 81
  pratensis, 80
  scabrella, 81
  trivialis, 81
  unilateralis, 81
Pogogyne serpylloides, 301
Polemoniaceae, 276
Polemonium carneum, 280
Polycarpon tetraphyllum, 163
Polygala californica, 226
Polygalaceae, 226
Polygonaceae, 140
Polygonum, 145
  amphibium stipulaceum, 147
  argyrocoleon, 146
  aviculare, 146
  aviculare erectum, 146
  aviculare littorale, 146
  coccineum, 147
  coccineum natans, 147
  coccineum terrestris, 147
  convolvulus, 146

hydropiperoides, 147
hydropiperoides asperi-
folium, 147
lapathifolium, 147
lapathifolium salicifolium,
147
nutans, 147
paronychia, 146
patulum, 146
pennsylvanicum, 147
persicaria, 147
punctatum, 147
Polypodiaceae, 56
Polypodium, 57
californicum, 57
californicum kaulfussii, 57
glycyrrhiza, 57
scouleri, 57
Polypogon, 94
interruptus, 94
littoralis, 94
lutosus, 94
monspeliensis, 94
Polystichum, 58
aculeatum dudleyi, 58
californicum, 58
dudleyi, 58
munitum, 58
munitum imbricans, 58
Pontederiaceae, 113
Populus, 132
fremontii, 132
nigra italica, 132
trichocarpa, 132
Portulaca oleracea, 160
Portulacaceae, 157
Potamogeton, 66
americanus, 66
crispus, 66
foliosus macellus, 66
gramineus, 66
heterophyllus, 66
illinoensis, 66
lutescens, 66
natans, 66
nodosus, 66
pectinatus, 66
pusillus, 66
pusillus minor, 66
Potamogetonaceae, 65
Potentilla, 196
egedii grandis, 196
frondosa, 195
glandulosa, 196
hickmanii, 196
lindleyi, 195
lindleyi sericea, 195
millegrana, 196
pacifica, 196
recta, 196
rivalis, 196
rivalis millegrana, 196

Primulaceae, 268
Proboscidea louisianica, 380
Prunella, 296
vulgaris, 297
vulgaris atropurpurea, 297
vulgaris lanceolata, 297
vulgaris parviflora, 297
Prunus, 200
demissa, 200
emarginata, 200
ilicifolia, 200
subcordata, 200
virginiana demissa, 200
Pseudosassa japonica, 75
Pseudotsuga, 63
menziesii, 63
taxifolia, 63
Psilocarphus, 356
brevissimus, 356
brevissimus multiflorus, 356
tenellus, 356
Psoralea, 210
douglasii, 211
macrostachya, 211
orbicularis, 211
physodes, 211
strobilina, 211
Ptelea crenulata, 226
Pteridium, 60
aquilinum lanuginosum, 60
aquilinum pubescens, 60
Pterophyta, 55
Pterostegia drymarioides, 140
Puccinellia, 83
airoides, 83
distans, 83
grandis, 83
nutkaensis, 83
nuttalliana, 83
pauciflora, 82
Pyrola picta aphylla, 263
Pyrolaceae, 263
Pyrrocoma elata, 347

Quercus, 136
agrifolia, 137
chrysolepis, 137
chrysolepis nana, 137
douglasii, 137
dumosa, 137
durata, 137
garryana, 137
kelloggii, 136
lobata, 137
× morehus, 137
wislizeni, 137
wislizeni frutescens, 137

Rafinesquia californica, 342
Ranunculaceae, 167
Ranunculus, 170
aquatilis capillaceus, 170

aquatilis hispidulus, 170
bloomeri, 171
californicus, 171
californicus cuneatus, 171
californicus gratus, 171
hebecarpus, 171
lobbii, 170
muricatus, 171
pusillus, 170
repens, 171
repens erectus, 171
Raphanus, 178
raphanistrum, 178
sativus, 178
Rapistrum rugosum, 180
Reseda alba, 187
Resedaceae, 187
Rhamnaceae, 233
Rhamnus, 233
californica, 234
californica tomentella, 234
crocea, 234
crocea ilicifolia, 234
Rhododendron, 264
californicum, 264
macrophyllum, 264
occidentale, 264
Rhus diversiloba, 230
Ribes, 192
aureum gracillimum, 192
californicum, 193
divaricatum, 193
glutinosum, 192
gracillimum, 192
malvaceum, 192
menziesii leptosmum, 193
menziesii senile, 193
sanguineum glutinosum, 192
Ricinus communis, 229
Rigiopappus leptocladus, 367
Robinia pseudo-acacia, 207
Romanzoffia, 286
californica, 286
suksdorfii, 286
Rorippa, 182
curvisiliqua, 182
islandica occidentalis, 182
nasturtium-aquaticum, 182
palustris occidentalis, 182
Rosa, 199
aldersonii, 199
californica, 199
gymnocarpa, 200
sonomensis, 200
spithamea, 200
Rosaceae, 194
Roubieva multifida, 153
Rubiaceae, 323
Rubus, 198
leucodermis, 199
parviflorus velutinus, 199
procerus, 199

spectabilis franciscanus, 199
spectabilis menziesii, 199
ulmifolius inermis, 199
ursinus, 199
ursinus glabratus, 199
ursinus sirbenus, 199
vitifolius, 199
Rumex, 144
acetosella, 145
californicus, 145
conglomeratus, 145
crassus, 145
crispus, 145
fenestratus, 145
fueginus, 145
obtusifolius, 145
obtusifolius agrestis, 145
occidentalis procerus, 145
persicarioides, 145
pulcher, 145
salicifolius, 145
salicifolius crassus, 145
salicifolius ecallosus, 145
salicifolius transitorius, 145
transitorius, 145
Ruppia maritima, 67
Ruta chalepensis, 226
Rutaceae, 226

Sagina, 161
apetala barbata, 161
crassicaulis, 162
occidentalis, 162
procumbens, 162
Sagittaria latifolia, 69
Salicaceae, 132
Salicornia, 148
depressa, 148
pacifica, 148
rubra, 148
virginica, 148
Salipchroa rhomboidea, 306
Salix, 133
coulteri, 133
hindsiana, 133
laevigata, 133
lasiandra, 133
lasiolepis, 133
lasiolepis bigelovii, 133
scouleriana, 133
Salsola kali tenuifolia, 151
Salvia, 296
columbariae, 296
mellifera, 296
spathacea, 296
Salviniaceae, 55
Sambucus, 325
callicarpa, 326
coerulea, 325
glauca, 325
mexicana, 325
pubens arborescens, 326

Samolus, 270
floribundus, 270
parviflorus, 270
Sanguisorba minor, 197
Sanicula, 256
arctopoides, 256
bipinnata, 257
bipinnatifida, 257
crassicaulis, 256
laciniata, 256
laciniata serpentina, 256
maritima, 256
menziesii, 256
tuberosa, 257
Santolina chamaecyparissus, 371
Saponaria officinalis, 166
Satureja douglasii, 301
Saururaceae, 132
Saxifraga californica, 191
Saxifragaceae, 189
Scabiosa atropurpurea, 327
Scandix pectin-veneris, 255
Schinus molle, 231
Scirpus, 109
acutus, 110
americanus, 110
americanus polyphyllus, 110
californicus, 110
carinatus, 110
cernuus californicus, 110
fluviatilis, 110
koilolepis, 110
microcarpus, 110
olneyi, 110
paludosus, 110
robustus, 110
rubiginosus, 110
validus, 110
Scleropoa rigida, 85
Scoliopus bigelovii, 121
Scolymus hispanicus, 339
Scorzonella paludosa, 340
Scribneria bolanderi, 90
Scrophularia, 310
californica, 310
californica floribunda, 310
multiflora, 310
Scrophulariaceae, 306
Scutellaria, 296
bolanderi, 296
tuberosa, 296
Secale cereale, 90
Sedum, 188
album, 188
dendroideum, 188
radiatum, 188
spathulifolium, 188
stenopetalum radiatum, 188
Selaginella bigelovii, 53
Selaginellaceae, 53
Senecio, 374

aphanactis, 374
aronicoides, 375
breweri, 375
cineraria, 375
cruentus, 374
douglasii, 375
elegans, 374
glomeratus, 375
hydrophilus, 375
mikanioides, 374
minimus, 375
sylvaticus, 375
vulgaris, 374
Senecioneae, 373
Sequoia sempervirens, 64
Setaria, 100
geniculata, 100
glauca, 100
viridis, 100
Sherardia arvensis, 323
Sida, 238
hederacea, 238
leprosa hederacea, 238
Sidalcea, 236
diploscypha, 236
malachroides, 236
malvaeflora, 236
malvaeflora laciniata, 237
Silene, 164
antirrhina, 165
californica, 165
cucubalus, 165
gallica, 165
laciniata major, 166
latifolia, 165
lemmonii, 166
pacifica, 166
scouleri grandis, 166
verecunda, 166
verecunda platyota, 166
Silybum marianum, 377
Simaroubaceae, 226
Sinapis
alba, 179
arvensis, 179
incana, 179
Sisymbrium, 183
altissimum, 184
irio, 184
officinale, 184
orientale, 184
sophia, 184
Sisyrinchium, 127
bellum, 127
californicum, 127
Sitanion, 86
hanseni, 86
jubatum, 87
Smilacina, 122
racemosa amplexicaulis, 122
stellata sessilifolia, 122
Solanaceae, 301

Solanum, 302
  aviculare, 303
  douglasii, 303
  dulcamara, 303
  elaeagnifolium, 303
  furcatum, 303
  gayanum, 303
  marginatum, 302
  nigrum, 303, 380
  nodiflorum, 303
  rostratum, 303
  sarrachoides, 303
  tuberosum, 303
  umbelliferum, 303
  umbelliferum incanum, 303
  xanti intermedium, 303
Solidago, 348
  californica, 349
  canadensis elongata, 348
  confinis, 349
  lepida elongata, 348
  occidentalis, 348
  spathulata, 349
Soliva, 371
  daucifolia, 371
  sessilis, 371
Sonchus, 343
  asper, 344
  oleraceus, 344
Sorghum, 102
  bicolor, 102
  halepense, 102
  sudanense, 102
  vulgare, 102
Sparganiaceae, 65
Sparganium, 65
  eurycarpum, 65
  greenei, 65
Spartina foliosa, 98
Spartium junceum, 208
Specularia biflora, 331
Spergula arvensis, 163
Spergularia, 163
  bocconii, 164
  macrotheca, 164
  marina, 164
  marina tenuis, 164
  media, 164
  rubra, 164
  villosa, 164
Sphacele calycina, 300
Spinacia oleracea, 150
Spiranthes, 131
  romanzoffiana, 131
  romanzoffiana porrifolia, 131
Spirodela polyrhiza, 112
Sporobolus poiretii, 98
Spraguea umbellata, 160
Stachys, 299
  ajugoides, 300
  ajugoides quercetorum, 300
  bullata, 300

chamissonis, 300
  pycnantha, 300
  rigida quercetorum, 300
Statice arctica californica, 270
Stellaria, 162
  graminea, 162
  littoralis, 162
  media, 162
  nitens, 162
Stenotaphrum secundatum, 101
Stephanomeria virgata, 341
Sterculiaceae, 238
Stipa, 94
  cernua, 94
  lepida, 94
  lepida andersonii, 94
  pulchra, 94
Streptanthus, 180
  glandulosus, 180
  glandulosus albidus, 180
Streptopus amplexifolius
  denticulatus, 122
Struthiopteris spicant, 59
Stylocline, 356
  amphibola, 357
  filaginea, 357
  gnaphaloides, 357
Stylomecon heterophylla, 174
Suaeda, 151
  californica, 151
  depressa erecta, 151
Symphoricarpos, 326
  albus laevigatus, 326
  mollis, 326
  rivularis, 326
Syntherisma
  ischaemum, 101
  sanguinalis, 101

Tanacetum camphoratum, 373
Taraxacum, 342
  laevigatum, 342
  officinale, 342
Tauschia, 258
  hartwegii, 259
  kelloggii, 259
Taxaceae, 62
Taxodiaceae, 64
Taxus brevifolia, 62
Tellima grandiflora, 191
Tetragonia, 157
  expansa, 157
  tetragonioides, 157
Thalictrum polycarpum, 171
Thelypodium, 184
  lasiophyllum, 184
  lasiophyllum inalienum, 184
Thermopsis, 211
  macrophylla, 211
  velutina, 211
Thymelaeaceae, 243
Thysanocarpus, 184

curvipes, 185
  curvipes elegans, 185
  emarginatus, 185
  lacinatus crenatus, 185
  radians, 185
Tiarella unifoliata, 190
Tillaea, 188
  aquatica, 188
  erecta, 188
Tillaeastrum aquaticum, 188
Tolmiea menziesii, 190
Tonella tenella, 310
Torilis, 255
  arvensis, 256
  nodosa, 255
Torreya californica, 62
Torreyochloa pauciflora, 82
Tradescantia fluminensis, 113
Tragopogon, 341
  dubius, 342
  porrifolius, 342
Tribulus terrestris, 226
Trichostema lanceolatum, 297
Trientalis, 269
  europaea latifolia, 269
  latifolia, 269
Trifolium, 212
  albopurpureum, 216
  albopurpureum neolagopus,
    216
  amplectens, 214
  angustifolium, 215
  barbigerum, 214
  barbigerum andrewsii, 214
  bifidum, 215
  bifidum decipiens, 216
  carolinianum, 215
  ciliolatum, 216
  cyathiferum, 215
  depauperatum, 214
  dichotomum, 216
  dichotomum turbinatum,
    216
  dubium, 215
  fimbriatum, 215
  flavulum, 214
  fucatum, 214
  gracilentum, 216
  gracilentum inconspicuum,
    216
  grayi, 214
  hybridum, 215
  hydrophilum, 214
  incarnatum, 216
  involucratum, 215
  lilacinum, 214
  macraei, 216
  microcephalum, 215
  microdon, 215
  obtusiflorum, 215
  oliganthum, 215
  pratense, 216

procumbens, 215
repens, 215
stenophyllum, 214
subterraneum, 216
tridentatum, 215
truncatum, 214
variegatum, 215
variegatum major, 215
variegatum melananthum,
 215
variegatum pauciflorum, 215
wormskjoldii, 215
wormskjoldii kennedianum,
 215
Triglochin, 68
concinna, 68
maritima, 68
striata, 68
Trillium, 121
chloropetalum, 121
ovatum, 121
sessile giganteum, 121
Triodanus biflora, 331
Trisetum, 93
canescens, 93
cernuum, 93
cernuum canescens, 93
Triteleia
hyacinthina, 126
ixioides. 126
laxa, 125
Triticum aestivum, 89
Tropaeolaceae, 224
Tropaeolum majus, 224
Tropidocarpum gracile, 183
Turritis glabra, 182
Typha, 64
angustifolia, 65
domingensis, 65
latifolia, 65
Typhaceae, 64

Ulex europaeus, 207
Ulmaceae, 139
Ulmus procera, 139
Umbelliferae, 252
Umbellularia californica, 173
Uropappus
heterocarpus, 340
linearifolius, 340
Urtica, 138
californica, 138
holosericea, 138

urens, 138
Urticaceae, 138
Utricularia vulgaris macro-
rhiza, 320

Vaccaria, 166
segetalis, 166
vulgaris, 166
Vacciniaceae, 268
Vaccinium, 268
ovatum, 268
ovatum saporosum, 268
parvifolium, 268
Valerianaceae, 328
Valerianella, 329
locusta, 329
olitoria, 329
Vancouveria planipetala, 172
Velaea
hartwegii, 259
kelloggii, 259
Verbascum, 307
blattaria, 307
speciosum, 307
thapsus, 307
virgatum, 307
Verbena, 294
lasiostachys, 294
robusta, 294
Verbenaceae, 294
Veronica, 314
americana, 314
arvensis, 314
buxbaumii, 314
chamaedrys, 314
comosa, 314
connata glaberrima, 314
decussata, 315
filiformis, 314
× franciscana, 315
peregrina xalapensis, 315
persica, 314
serpyllifolia, 314
speciosa, 315
Viburnum tinus, 327
Vicia, 217
americana minor, 219
americana oregana, 219
americana truncata, 219
angustifolia, 219
angustifolia segetalis, 219
atropurpurea, 218
benghalensis, 218

dasycarpa, 218
exigua, 218
faba, 219
gigantea, 218
hassei, 218
hirsuta, 218
linearis, 219
lutea, 219
sativa, 219
tetrasperma, 218
villosa, 218
Vinca major, 273
Viola, 240
adunca, 241
glabella, 242
ocellata, 241
odorata, 241
pedunculata, 241
purpurea, 242
quercetorum, 242
sarmentosa, 242
sempervirens, 242
tricolor, 241
Violaceae, 240

Whipplea modesta, 192
Wolffiella lingulata, 112
Woodwardia, 58
chamissoi, 58
fimbriata, 58
Wyethia, 357
angustifolia, 357
glabra, 357
helenioides, 357

Xanthium, 365
calvum, 365
canadense, 365
italicum, 365
spinosum, 365
strumarium, 365
Xerophyllum tenax, 116

Zannichellia palustris, 67
Zantedeschia aethiopica, 111
Zauschneria californica, 245
Zostera marina, 67
Zosteraceae, 67
Zygadenus, 116
fremontii, 116
fremontii minor, 116
venenosus, 116
Zygophyllaceae, 226

Chicory tribe - all ray flowers, milky juice (dandelions)

PLANT FAMILY CHARACTERISTICS

APIACEAE (Umbelliferae)

Flowers 5 merous = 5 sepals, 5 petals, 5 stamen

Flowers arranged in umbels

Fruit called a schizocarp - carpels separating from one another into 1 seeded segments.

Leaves usually compound.

Asteraceae (Composite) Sun Flower Family

Flowers either or both ray or disk flowers arranged in a head

Ray flowers are strap shaped called ligulate

Disk flowers are are tubular shaped

pappus surrounds the flowers and is a modified calyx (sepal) - looks like bristles or hairs

Phyllaries are bracts that surround the flowers

Receptacle is the structure that the flower sits on

Fruit is called an achene

Brassicaceae (Cruciferae) Mustard Family

Flowers consist of 4 sepals and 4 petals in the shape of a cross

Flower is Actinomorphic - regurally symetrical

Stamen 6

Fabaceae

Flowers 5 sepals and 5 petals - Zygomorphic (Bilateral)

Petals composed of wings, banner and keel the wings are 2 fused petals and the banner is 2 fused petals

Stamen 10 - either monadelphous (united on one structure) or diadelphous (united in 2 structures

Fruit a legume (pod)

Laminaceae (Labiatae) Mint Family

Flowers consist of 5 united petals - 2 fused upper and 3 fused lower petals

Flowers arranged in whorls

Flowers zygomorphic - Bilaterally symetrical

Stamen 2 or 4, style 1

Fruit a nutlet many 4 nutlets

Leaves opposit or whorled

Ranunculus - Buttercup Family

Flower either actinomorphic (regular) or zygomorphic (bilateral)

Sepals and Petals united or not

Stamen numerous

Fruit - follicle or achene or berry